BOTANICAL PESTICIDES *in* AGRICULTURE

Anand Prakash
Jagadiswari Rao

Division of Entomology
Central Rice Research Institute
Cuttack, India

Acquiring Editor: Ken McCombs
Project Editor: Suzanne Lassandro
Marketing Manager: Greg Daurelle
Direct Marketing Manager: Arline Massey
Cover design: Denise Craig
Manufacturing: Sheri Schwartz

Library of Congress Cataloging-in-Publication Data

Prakash, A. (Anand), 1952–
Botanical pesticides in agriculture / Anand Prakash and Jagadiswari Rao
p. cm.
Includes bibliographical references and index.
ISBN 1-87371-825-9 (alk. paper)
1. Botanical pesticides. I. Rao, Jagadiswari, 1948 - . II. Title.
SB951.145.B68P73 1996
632′.95—dc20 96-3071
CIP

International Standard Book Number 1-87371-825-9
Library of Congress Card Number 96-3071
Printed in the United States of America 1 2 3 4 5 6 7 8 9 0
Printed on acid-free paper

CONTENTS

FOREWORD

Living organisms have existed on earth for at least 3.5 billion years. They have evolved complex relationships including a bewildering variety of predator-prey, parasite-host and competitive interactions. Anyone examining these relationships can not help but be intrigued by how predators, parasites and competitors have evolved mechanisms to capture prey, attack hosts and exploit their neighbors.

Even more intriguing are the various defenses against these attacks. Prey, hosts and competitors are not without their own strategies of survival. Some of these are anatomical defenses such as the tough bark on a Mediterranean oak or the spines on a cactus. Others are behavioural, such as the rolling up of the armor plated armadillo during an attack, the escape flight of a jackrabbit or the tail flicking of a horse to prevent flies from nibbling at its flesh. Still others are biochemical, such as the production of toxic or distasteful chemicals by plants to dissuade predators and parasites and defeat competitors.

Chemical defenses against pests are little known to the public. They are subtle in nature, and are poorly understood because the science of biochemistry is young. Yet some of these effects have been seen for thousands of years. Early humans discovered the medicinal qualities of certain plant fungi and animal tissues. They did not appreciate why these substances had therapeutic value but they worked. They did not realize that the plants, fungi and animals evolved and used these chemicals for their own defense or for attack. They simply worked by magic, by ritual, by God.

Today the search for medicines is big business. Pharmaceutical companies screen thousands of biochemical products yearly looking for cures to Acquired Immunity Deficiency Syndrome (AIDS), cancer and other debilitating disease. Similarly, pesticide companies search the biological world for natural chemicals, which organisms have evolved for their own use on the off chance that some of these agents may be used in pest management. This search is modest compared to the search for medicines, yet it is no less important for billions of tons of potential food are lost each year because of the ravages of insects, nematodes and rodents. Both searches are threatened by the destruction of large regions of the world's jungles. For as we cut down our forests of India, Africa and South America, we are destroying the diversity of the pool of organisms which have evolved biochemical 'solutions' to pest attack. The race is on to identify as many of these 'life saving' chemicals before the organisms which produce them are gone forever.

The working scientist and student are always in need of reference books to keep them abreast of the development in the discipline. This is especially true for the applied agriculture scientist who must deal with a puzzling array of choices of pest management products which are routinely tested across the world. It is vital to have an up to date dictionary or encyclopedia which expertly catalogues the options available. The reference book which you hold in your hands is written by two experts, Drs. Anand Prakash and Jagadiswari Rao of the Central Rice Research Institute, and is a fine compendium for the specialist in pesticide management. It is arranged alphabetically according to the species which produce chemicals useful against insects (Section A), nematodes (Section B), mites, rodents and molluscs (Section C) and biological active chemicals components (Section D). Of particular importance will be the references which Prakash and Rao have provided for each pesticide product so that the curious may pursue the trail further.

The reference work on **Botanical Pesticides in Agriculture** is an important milestone for agricultural scientists who daily struggle with decisions which affect the lives of billions of people.

Clyde Freeman Herreid
Distinguished Teaching Professor
State University of New York at Buffalo

ABOUT THE AUTHORS

Dr. Anand Prakash: Born on July 5, 1952 in the Village of Barkali in district Muzaffarnagar, Uttar Pradesh (India), completed matriculation in 1968 from Amrit Inter College, Rohana Mills; I.Sc. in 1970 from S.D. Inter College, Muzaffarnagar affiliated to U.P. Board of Education, Allahabad; graduated in 1972 from S.D. College and post-graduated in 1974 from D.A.V. College at Muzaffarnagar under Meerut University. He was awarded Ph.D. degree in 1986 in zoology from Utkal University for his research work on 'Botanical Pesticides in Insect Pest Management of Stored Rice'. He joined as Research Fellow in Institute of Cecidology, Allahabad in 1975 and as Scientist-1 at Central Rice Research Institute, Cuttack in 1977. Since 1983 Prakash is working as Senior Scientist in the field of rice grain entomology at the same Institute. He has published over a hundred research papers in journals of national/international repute, attended several national and international scientific conferences and has guided six Ph.D. students. He has written a reference book *Rice Storage and Insect Pest Management*, PP 337 (1987) published by B.R. Publishing Corporation, New Delhi and is associated with several professional societies. He founded the Applied Zoologists Research Association in 1989 and has edited *Journal of Applied Zoological Researches* since 1990. However, most of his research is shared by his wife and co-author Dr. (Mrs.) Jagadiswari Rao.

Dr. (Mrs.) Jagadiswari Rao: Born on September 16, 1948 in the Cuttack town of Orissa (India), completed matriculation in 1965, I.Sc. in 1967 from Secondary Board of Education of Orissa; graduated in 1970; completed M.Sc. in 1972 and was awarded Ph.D. degree in 1990 in zoology from Utkal University. She was trained in insects and disease control in rice at Kobe University, Japan during May-Dec. 1985. She joined the Central Rice Research Institute, Cuttack as Research Assistant in 1972, as Scientist-1 in 1976 and as Senior Scientist in 1983. She has been actively engaged in research on the 'Bio-deterioration of paddy seed quality and its control'. She has published over eighty research papers and co-authored the book *Rice Storage and Insect Pest Management* and attended many scientific national/international conferences. She is associated with several professional societies and also founded jointly the Applied Zoologists Research Association.

PREFACE

Pest management is one of the vital factors for stepping up agricultural production. During the past few decades application of synthetic pesticides to control of agricultural pests has been a standard practice. However with the growing evidences regarding the detrimental effects of many of the conventional pesticides on health and environment, the need for safer means of pest management has become very crucial. Thus there is an urgent need and demand to develop strategies of pest management.

During the course of evolution plants have acquired effective defense mechanisms against arthropods that feed on them and some of these are based on their chemical components. An understanding of these natural processes which govern the insect-plant interactions is expected to provide the clues for designing ecologically acceptable means of pest management. Only a few reviews/books have been written in the early nineteen seventies on exploitation and evaluation of natural plant products for pest management in agriculture. Books dealing with up-to-date research informations on exploitation of pesticidal agents in plants and their utilization as alternative means of agricultural pest management are not available.

The text of the present reference book is written for entomology students at post graduate and higher levels, for students in the fields of agricultural pest management, economic botany, pesticidal chemistry and also for workers and agencies interested in protecting the environment. Efforts have been made to compile information on research work carried out on botanical pesticides to date with special reference to the use of higher plants for controlling pests of agricultural importance, including research work carried out by the authors at Central Rice Research Institute, Cuttack-753006, India.

The text is divided into four sections, viz., botanical pesticides against insects (Section A), nematodes (Section B), mites, rodents and molluscan pests (Section C) and biologically active chemical components (Section D). Of 1079 plants described in the text, biological activities were reported in 866 plants against insects, 150 plants against nematodes, 30 plants against mites, 20 plants against rodents and 13 plants against snails. Plants have been arranged in alphabetical order of their botanical names in English including the names of the authors, families, common names, habit and habitat and their geographical distributions. The plant products exploited for their bio-pesticidal properties in terms of toxicity, repellency, antifeedancy, chemosterility and growth inhibition have been dealt with in the text. The active components of the effective plant materials isolated, identified and evaluated for their bio-efficacy and reported to possess pest control properties have also been mentioned with special reference to their chemical structures.

We are grateful to Dr. Martin Jacobson, Biologically Active Natural Plant Products Laboratory, Agricultural & Environmental Institute, ARS-USDA Beltsville, MD (USA); Dr. Helen C.F. Su, Stored Products Insects Research and Development Laboratory, ARS-USDA, Savannah, GA (USA); Dr. R.C. Saxena, International Rice Research Institute, Manila, Philippines; Dr. Heinz Rembold, Max-Planck Institute for Biochemistry, Martinsried, Germany; Prof. Arthur A. Muka, Cornell University, Ithaca, NY (USA); Mrs. Sabitri Rao Tripathy and Mrs. Sarojini R. Misra, 160 Redwood Terrace, Williamsville, NY 14221 (USA); Dr. T.D. Yadav, Indian Agricultural Research Institute, New Delhi; Prof. G. Srimmannarayan, Osmania University, Hyderabad, for providing research literature, which helped us to prepare the manuscript of the book. We remember with much gratitude the encouragement and help we received from our colleague scientists and friends Drs. K.C. Mathur, S.N. Tiwari, M.S. Panwar, S. Rajamani and K.S. Behera of Central Rice Research Institute, Cuttack (India); Dr. S.P. Gupta, Central Integrated Pest Management Station, Bhubaneswar.

We are grateful to Dr. Clyde Freeman Herreid, Distinguished Teaching Professor, State University of New York at Buffalo (USA) for writing the foreword. We are also thankful to Prof. B. Senapati, Head, Division of Entomology, Orissa University of Agriculture and Technology, Bhubaneswar, India and Dr. B.V. David, Director, Jai Research Foundation, Valvad, Gujarant, India for their useful suggestions in preparing this book. We would greatly appreciate suggestions for further improvements so that they may be incorporated in the next edition.

Anand Prakash
Jagadiswari Rao
Division of Entomology
Central Rice Research Institute
Cuttack-753006, India

INTRODUCTION

The plant kingdom has supported mankind in more than one way. Perhaps the first attempt to utilise this resource by ancient man was for food. Civilised man developed agriculture to use plant resources primarily as food and then with further development for fibre and fodder. Soon human diseases took heavy tolls of populations and again it was the plant kingdom which came to the rescue resulting in development of ayurvedic, homeopathic and unani systems of medicine.

The development of agriculture brought the phenomenon of increases in the populations of pests and diseases and competition for food. Encouraged by the results of treatment of human disease it was quite natural that the civilised man again turned towards the plant kingdom to find solutions for pests and diseases of agricultural crops.

Plants have evolved over some 400 million years and acquired effective defence mechanisms during their evolution, which secure their survival in the presence of hostile environment and enemies. Some of the morphological protective measures such as thick cuticular waxes, thorns and prickles and sticky hairs are easily identifiable, although less obvious, more important and subtle defence mechanisms are based on chemicals which protect roots, stems, foliage and flowers from insects and other bioagents. As well as the production of toxins which protect the plants from attacks by insects and other herbivorous animals, a major sophisticated type of defence is provided by the micro-molecular constituents which may not cause instantaneous mortality but their effects are manifested by an effect on normal biochemical and physiological functions. An understanding of these natural defence processes might provide clues for the development of new biotechnical procedures for pest control, which are based on ecologically acceptable biocides. Thus a large number of different plant species contain natural pesticidal materials, and some of these have been used by man as pesticides since very early times, although many of them could not be extracted profitably. However, several of these extracts have provided valuable contact insecticides.

Synthetic pesticides were produced because of the ease with which these could be used. Their effectiveness and the possibility of storage and use as and when needed became very powerful agents in the control of pests and diseases in agriculture, horticulture, forestry and even human health programs.

In recent years the use of synthetic pesticides in crop protection programs around the world has resulted in disturbances of the environment, pest resurgence, pest resistance to pesticides and lethal effects on non-target organisms agroecosystems in addition to direct toxicity to users. The problem has been compounded by the energy crises which brought about a sharp increase in the prices of commercial petroleum-based pesticides. Thus the pesticides, apart from producing undesirable effects, became unavailable to small farmers, particularly in developing and underdeveloped countries. Therefore, now it has become necessary to search for a alternative means of pest control which can minimise the use of these synthetic chemicals.

Use of botanical pesticides/natural plant products in an agroecosystem is now emerging as one of the prime means to protect crop produce and the environment from pesticidal pollution, which is a global problem. Botanical pesticides possess an array of properties including insecticidal activity, repellency to pests, antifeedancy, insect growth regulation, toxicity to nematodes, mites and other agricultural pests and also antifungal, antiviral and antibacterial properties against pathogens (Prakash and Rao 1986, 1987; Prakash et al., 1987, 1989, 1990b). Some of these indigenous resources have been in use for over a century to minimise losses due to pests and disease in agricultural production (Prakash et al., 1990b; Parmar and Devkumar, 1993). However, plant products have many advantages over synthetic chemicals, which are as follows:

1. Botanical pesticides in general possess low mammalian toxicity and thus constitute least/no health hazards and environmental pollution.
2. There is practically no risk of developing pest resistance to these products, when used in natural forms.
3. These cause less hazards to non-target organisms and pest resurgence has not been reported except synthetic pyrethrins.
4. No adverse effect on plant growth, seed viability and cooking quality of grains.
5. Botanical pesticides are less expensive and easily available because of their natural occurrence especially in oriental countries.

Due to the prohibitive cost of synthetic pesticides and the problems of environmental pollution caused by continuous use of these chemicals, there is a renewed interest in the use of botanicals in crop protection. A number of agricultural entomologists, nematologists and pathologists all over the world are now actively engaged in research into usage of plant products against agricultural pests and diseases to minimise losses caused by them. We have attempted in this book to review the research work on exploitation of botanical pesticides to minimise losses caused by pests of agricultural importance with special reference to higher plants. It is expected that this will serve as baseline information for future research and some of the botanicals like neem, bael, begunia, pyrethrum, tobacco, karanj, mahuwa etc. may become an integral part of pest control programs being developed. It is believed that usage of pesticides of botanical origin will go a long way in minimising the undesirable side effects of synthetic pesticides and help to preserve the environment for the future generations.

The book has been divided into four sections. Section A deals with evaluation of higher plant products against agricultural insect pests for their insecticidal, antifeedant, repellent and insect growth regulator properties. Section B deals with plant products found toxic to the nematodes of agricultural importance and Section C deals with plant products possessing toxic properties against the mites, molluscs and rodents. In Section D 250 biologically active chemical components are presented in tabular form mentioning their chemical grouping, biological activity and plant from which these chemicals were isolated. The plants are arranged in alphabetical order mentioning in brief their botanical names, common name, families, distribution, habits and habitats. Plant products possessing antiviral, antifungal and antibacterial properties were excluded from the scope of this book. We expect that the information presented in this book will serve as a single source for all available information on the subject and will stimulate further research on the subject. The book should also prove useful to the students of agricultural entomology, economic botany, pesticidal chemistry and those interested in protecting the environment for the future generations.

A BOTANICAL PESTICIDES AGAINST INSECTS

1. *Abies balsamea* (Linn.) Mill. (Pinaceae)

Common name: Balsam fir

This is a beautiful perennial small tree found in tropical and sub-tropical regions and commonly available in the Himalayan tract in India. Balsam fir tree is also distributed in Sri Lanka, Malaysia, China and the Philippines. Extracts from its leaves were reported to show strong juvenile hormone activity against the European bug, *Pyrrocoris apterus* in laboratory tests (Slama and Williams, 1966). Banerji (1988) found this tree to be a source of 'juvabione', an active component, which also showed strong juvenile hormone mimic activity against insects. He also emphasized its commercial use in the control of white flies.

2. *Abrus precatorius* Linn. (Fabaceae)

Common name: Red bean vine or Indian liquorice or Jecquirity seeds

This plant is a perennial vine or woody climber found in tropical regions distributed throughout India and neighboring countries. Jacobson and Crosby (1971) reported a contact insecticidal activity in its aqueous extract prepared of crushed fruits, when sprayed on the host plant against a grasshopper, *Poecilocera picta*.

3. *Acacia arabica* Willd. (Mimosaceae)

Common name: Babool

This is a perennial shrub or small tree found in tropical regions and commonly available on the Indian plains. Its gum extract sprayed/admixed with stored wheat grains was reported to reduce the adult emergence and male-to-female ratio of the rice moth, *Corcyra cephalonica* and prolonged developmental period of this moth (Pandey et al., 1985a). It was also found to act as ovipositional deterrent and reduced its fecundity and fertility (Pandy et al., 1985b).

4. *Acacia catechu* (l.f.) Willd. (Mimosaceae)

(Syn.-*A. catechuoides* Benth. *Mimosa catechuoides* Roxb.)

Common name: Black cutch or Catechu

This is a perennial shrub found in tropical and sub-tropical regions and commonly available in coastal States of India. Natural proanthocyanidin monomer called epicatechin isolated from black cutch exhibited moderate antifeedant activity against lepidopteran larvae when tested under laboratory conditions (Srimmannaryan and Rao, 1985).

5. *Acacia concinna* D.C. (Mimosaceae)

Common name: Shekakai

This is a perennial shrub found in tropical and sub-tropical regions and commonly available in the sub-Himalayan tract from Oudh eastward, Assam, Bihar, Western Peninsula in India and in Burma and Malaysia, mainly in evergreen forests. Quadri (1973) reported antifeedant activity in its seed extracts when tested against the rice weevil, *Sitophilus oryzae* and the red flour beetle, *Tribolium castaneum*. However, Chellappa and Chelliah (1976) found its seed powder ineffective to protect stored paddy against infestation of lesser grain borer, *Rhyzopertha dominica* and Angoumois grain moth, *Sitotroga cerealella*. Insecticidal and antifeedant properties in its leaves and pods were reported against *R. dominica*, when used as a dried powder with stored grains (Balasubramanian, 1982). Prakash et al. (1978–1987) found its seed extract and dried leaves powder ineffective to protect paddy grains in storage against *S. cerealella*, *R. dominica* and *S. oryzae*. Seed extract of this shrub in petroleum ether was also reported to reduce oviposition of the brown planthopper, *Nilaparvata lugens* at 2% and 5% concentrations of this solution when sprayed on 40-day-old rice seedlings (Reddy and Urs, 1988).

6. *Acacia pennata* (Linn.) Willd. (Mimosaceae)

Common name: Aila

This is a perennial woody climber/shrub found in tropical, sub-tropical and arid zones. In India, it is available throughout the country on riversides and streams and also found in Sri Lanka and Malaysia. Water extracts of its bark and fruit were reported to show insecticidal activity against the maggots of dipteran pests of agricultural importance (Jacobson, 1975).

7. *Acalypha indica* Linn. (Euphorbiaceae)

Common name: Indian nettle or Indian acalypha

This is an annual/perennial herb found in tropical and sub-tropical regions and distributed throughout India. Its stem and bark extracts showed toxicity to the grasshopper, *Epacromia* sp. (Puttarudriah and Bhatia, 1955). McIndoo (1983) reported insecticidal activity in its bark and leaf powders and their ethanol extracts against the hairy caterpillar, *Euproctis fraterna* and the diamond-back moth, *Plutella xylostella*.

8. *Acer carpinifolium* Linn. (Aceraceae)

Common name: Hornbean maple

This is a perennial tree found in temperate regions. Its leaves extract in water was reported to show insecticidal activity against the vinegar fruit fly, *Drosophila hydei* (Jacobson, 1958).

9. *Acer rabrum* Linn. (Aceraceae)

Common name: Red maple

This is a perrennial tree found in temperate regions. Red maple is commonly found in North America. Its leaf extract in water was reported to show repellent activity against the forest tent caterpillar, *Malocosoma disstria* (Jacobson, 1975).

10. *Achillea sibrica var. ptarmicoides* Linn. (Asteraceae)

(Syn.-*A. ptarmica* Linn.)

This is a perennial hedge plant found in temperate regions and commonly available in Europe, Siberia and North America. Its aqueous leaf, stem and flower extracts were toxic to the vinegar fly, *Drosophila melanogaster* (Jacobson, 1958).

11. *Acokanthera spectabilis* Hook. (Apocynaceae)

This is a perennial woody herb found in tropical and sub-tropical regions. This plant was reported to possess phago-stimulant properties in its leaf extract, which contains alkaloids, triterpenes, cardinolids, glycosoids and sterols against the 3rd instar larvae of *Spodoptera littoralis*, when sprayed on the castor leaves. The larvae fed on the treated leaves pupated earlier than those fed on untreated leaves (Abassay *et al.*, 1977). Abassay et al. (1978) isolated triterpene ‘friedelin’ as an active component possessing antifeedant property against 3rd instar larvae of *S. littoralis*.

12. *Aconitum chinese* Monks. (Ranunculaceae)

Common name: China aconite

This is a perennial creeping herb found in temperate regions and commonly available in North China and USSR. Jacobson (1958) reported toxicity in its root and rhyzome aqueous extracts to Mexican bean beetle, *Epilachna varivestis*.

13. *Aconitum ferox* Wall. ex. Ser. (Ranunculaceae)

(Syn.-*A. virosum* Don., *A. napellus* var. *rigidium* Hook. & F. Thomas)

Common name: Indian aconite

This is an annual/perennial herb found in temperate regions. In India, it is commonly available in Kumaun hills and Alpine Himalayas. Its crude aqueous leaf extract showed toxicity to the red pumpkin beetle, *Aulacophora forveicollis*; wheat aphid, *Aphis maidis*; mustard fly, *Athalia proxima*; kharif grasshopper, *Ilieroglyphus nigrorepletus*; radish aphid, *Rhopalosiphum psedobrassicae*; and mustard aphid, *Siphocoryne indobrassicae* (Jacobson, 1975).

A number of toxic alkaloids, i.e., pseudaconitine, chasmaconitine, indaconitine, bikhaconitine and diacetylpseudaconitine, were isolated to be active components from this herb (Anonymous, 1995).

14. *Aconitum japonicum* Monks. (Ranunculaceae)

Common name: Japanese aconite

This is an annual/perennial herb found in temperate regions. Jacobson (1958) registered toxicity in its root and leaf aqueous extracts to the vinegar fruit fly, *Drosophila hydei*.

15. *Aconitum napellus* Linn. (Ranunculaceae)

Common name: Helmet flower

This is a perennial herb found in temperate regions. Its root extract in water was toxic to the rice brown plant hopper, *Nilaparvata lugens* (Waespe and Hans, 1979) and also against the stage beetle, *Leucanus cervus* (McIndoo, 1983).

16. *Aconitum villosum* Monks. (Ranunculaceae)

This is a perennial herb found in temperate regions. Water extract of its tuberous roots was toxic to Mexican bean beetle, *Epilachna varivestis* (Jacobson, 1958).

17. *Acorus calamus* Linn. (Araceae)

Common name: Sweet flag

This is a perennial herb/shrub found in tropical, temperate and Mediterranean regions. Sweet flag shrub is distributed throughout South-East Asia and Australia. In India, this plant is available mainly in Jammu and Kashmir, Manipur, Mysore and Northern Himalayas. Sweet flag has been intensively studied for its biological activity against insect pests of agricultural importance and found to possess insecticidal, insect repellent, antifeedant, attractant and chemosterilant properties against the pests in storage as well as in field ecosystems as mentioned below.

Biological Activity Against Stored Grain Pests

Root/rhizome powder of sweet flag and also its extracts were found to be the effective grain protectants when tested with paddy grains stored in metallic bins against the lesser grain borer, *R. dominica*; the rice weevil, *S. oryzae*; and the red flour beetle, *T. castaneum* (Israel and Vedamoorty, 1953; Israel, 1965). Similarly, its root powder and extracts in water and other organic solvents were reported to show insecticidal activity against *T. castaneum, S. oryzae* and the flat grain beetle, *Latheticus oryzae* (Paul et al., 1965). However, its dried root powder admixed with stored paddy minimized the infestation of Angoumois grain moth, *Sitotroga cerealella* and showed promising grain protection (Mammen et al., 1968). Yadav (1971) found its crude rhizome extract in the absolute alcohol obtained from steam distillation to show insecticidal activity against the pulse beetle, *Callosobruchus chinensis*. Abraham et al. (1972) reported its root powder to reduce the multiplication of *S. cerealella* when admixed with stored paddy in the gunny bags. Similarly, against the paddy insect pests in general its rhizome extract was reported to protect paddy grain from their infestation in bagged storage (Savitri and Subba Rao, 1976).

Sweet flag oil was reported to be toxic to *C. chinensis* (Pandey and Singh, 1976) and also against *S. cerealella* and *S. oryzae* (Teotia and Tewari, 1977; Teotia and Pandey, 1979) in tests under controlled conditions. Saxena et al. 1977 reported its oil vapours to show antigonadal (sterilising) activity against stored grain beetles. Chemosterilant activity of its oil was also found in khapra beetle, *Trogoderma granarium* (Koul et al., 1977; Kalpana et al., 1978a) and in *C. chinensis* (Kalpana et al., 1977b). However, rhizome powder admixed with stored sorghum grains in the metallic bins and bags protected them against infestation of the storage insects (Anonymous, 1979). Its rhizome extract and oil both registered toxicity to the pulse beetles, *Callosobruchus* spp. (Anonymous, 1980–1981; Ghosh et al., 1981).

Prakash et al. (1978–1987) and Prakash and Rao (1986) reported its rhizome extract in methanol to reduce the larval and pupal development of the saw-toothed grain beetle, *Oryzaephilus surinamensis* and *T. castaneum* in a laboratory test. Chander and Ahmed (1986) reported rhizome powder to show toxicity to 1st instar larvae of the rice moth, *Corcyra cephalonica*, when admixed at the rate of 1, 2 and 5% w/w with wheat grains. Similarly, Khan (1986) reported its rhizome powder to show residual toxicity to the pulse beetle, *C. chinensis* in stored chickpea when applied using 0.2% formulation in water and reduced its multiplication up to 120 days in a laboratory test. Jilani and Saxena (1987) found its oil as an effective repellent to *R. dominica* on treated milled rice. Similarly, its oil was also reported as a repellent to *T. castaneum* and *R. dominica* on treated grains in a test under controlled conditions (Jilani et al., 1987). Chauhan et al. (1987) found its ethanolic plant extract to show toxicity to the adults and the larvae of the rice moth, *Corcyra cephalonica*. Its rhizome powder extract in benzene exhibited an effective grain protection against the attack of *C. chinensis* in greengram and pigeonpea in storage and reduced the oviposition and the development of this beetle (Reddy and Reddy, 1987; Rao et al., 1990).

Biological Activity Against Field Insect Pests

Biological activity of this plant as an insecticide against field insect pests was reported as early as 1939 (Mirono, 1940) and later Subramanian (1949) found sweet flag as a potential source of valuable insecticide and suggested its systematic exploitation.

Ether and alcohol extracts of its rhizome extract were found to be insecticidal to the termite, *Heterotermes indicola* (Paul et al., 1965). Saxena and Srivastava (1972) found insecticidal activity in the vapours of its oil extracted from the rhizome against eggs and nymphs of the red cotton bug, *Dysdercus koenigii*. Hatching of the eggs was greatly reduced when 0–12 hrs old eggs were placed on the filter paper treated with its oil in the closed petri-plates and its 1st instar larvae were found to be more susceptible than older instars. Both the male and the female of Mediterranean fruit fly, *Ceratitis capitata* and the Oriental fruit fly, *Dacus dorsalis* showed attracting behaviour towards ethyl ether extract of its rhizome (Keiser et al., 1975).

Pandey et al. (1977) found its rhizome extract to show antifeedant, repellent and insecticidal activities when tested against the mustard saw fly, *Athalia proxima* larvae. Rajendran and Gopalan (1979) reported insecticidal property in acetone extract of its rhizome and observed to show mortality to 5th instar nymphs of the red cotton bug, *Dysdercus cingulatus*, 3rd instar larvae of *Spodoptera litura* and *Pericallia ricini*. Pandey et al. (1979) reported its rhizome extract in ether to show larvicidal activity against *Athalia proxima*. Pandey et al. (1982) reported insecticidal activity in its rhizome extract to 5th instar larvae of the potato tuber moth, *Gnorimoschema operculella* and recorded 42% mortality under laboratory tests. Custard apple oil (10%) emulsified with 0.1% liquid detergent greatly reduced the life span of the rice leafhopper, *Nephotettix veriscens* and also transmission of the rice tungro virus (Mariappan et al., 1982b). In a laboratory test, its rhizome extract was also reported to show antifeedant activity to the larvae and the adults of *Leptinotarsa decemlineata* on potato leaves (Britski, 1982). Insecticidal activity in its rhizome extracts was also reported against *S. litura* (Balasubramaniam, 1982); against the adults and the nymphs of the turnip aphid, *Hydaphis erysimi* (Yingchol, 1983); against the adults of the red cotton bug, *Dysdercus cingulatus* (Kareem, 1984); and against 3rd instar larvae of the sugarcane leafhopper, *Pyrilla perpusilla* (Pandey et al., 1984). Chandel et al. (1987) reported acetone extract of its rhizome to show significant toxicity to *Henosepilachna vigintioctopunctata* in a laboratory test. Similarly, Koul (1987) reported distillates and essential oil of its rhizome to show gustatory repellency at 0.5 to 1.0% concentrations to *S. litura*. Bandra et al. (1987) found dichloromethane and methanol extracts of its rhizome to show toxicity to the aphids. Gupta and Dogra (1993) reported that sweet flag oil showed fumigation activity @ 10–200 μl oil/100 cc and caused death to *H. vigintioctopunctata* due to anorexia within 6 days of ecdysis but showed antifeedancy at sub-lethal dose.

Sweet flag oil being a mixture of several acidic and neutral compounds like Beta-asarone, acoragermacrone and agarylaldehyde etc. it was difficult to define the components responsible for specific biological activity (Jacobson et al., 1976).

Active Components

From the distillates of its root/rhizome oil four components were isolated and reported to show biological activity against insect pests.

1. *Methyl eugenol* (3,4-dimethoxyallylbenzene) (Figure 1) as a powerful attractant to the females of the Oriental fruit fly, *Dacus dorsalis* was isolated by Fujita (1971).

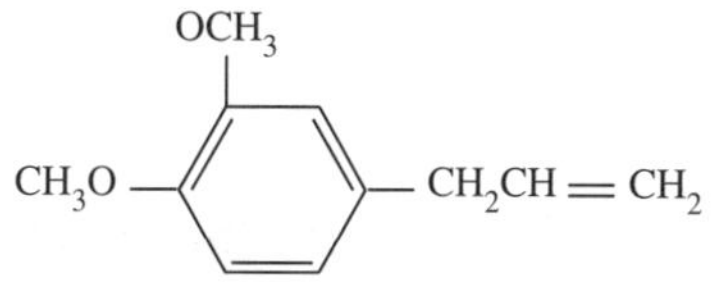

OCH_3

CH_3O — [benzene ring] — $CH_2CH=CH_2$

Figure 1. Methyl eugenol.

2. β-asarone [(Z)-2,4,5-trimethoxypropenyl benzene] (Figure 2) was found to be an attractant to females of the Oriental fruit fly, *D. dorsalis* (Jacobson, 1976), as fumigant and insecticidal to the aphids (Bandara et al., 1987) and also as sterilent in *H. vigintioctopunctata* (Gupta and Dogra, 1993).

3. *Acoragermacrone* [(E,E)-10-isopropyl-3,7-dimethyl-2, 6-cyclodec-adien-1-one] (Figure 3) as an attractant to the female melon fly, *D. cucurbitae* (Jacobson, 1976).

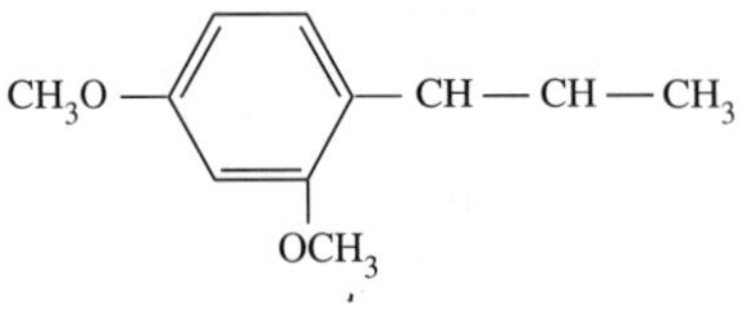

Figure 2. β-asarone.

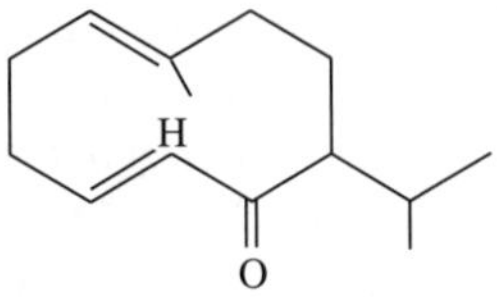

Figure 3. Acoragermacrone.

4. *Asarylaldehyde*, [2,4,5-trimethoxybenzaldehyde] (Figure 4) as an attractant to both male and female of *D. dorsalis* and *C. capitata* (Jacobson, 1976).

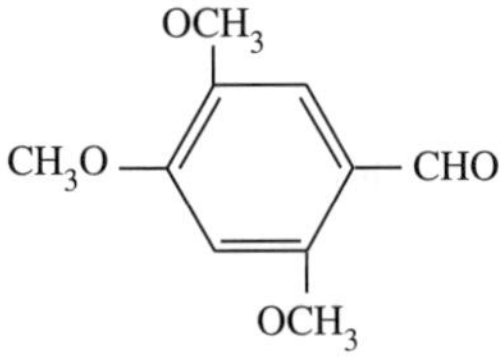

Figure 4. Asaryladehyde.

Muckenstrum et al. (1981) isolated phenyl propanoids from ether extracts of its rhizome and found them to show antifeedant activity against *Leptinotarsa decemlineata* and *Mythimna unipunctata*.

18. *Acorus gramineus* Linn. (Araceae)

This is a perennial shrub found in tropical regions and commonly available in India, Sri Lanka, Malaysia, China and the Philippines. Its aqueous rhizome extract was toxic to the sugarcane top shoot borer, *Tryporyza nivella* (Pandey, 1988).

19. *Adhatoda vasica* Nees. (Acanthaceae)

(Syn.-*Justica adhatoda* Linn.)

Common name: Malabar nut tree or Ajuba

This is a perennial shrub in tropical and sub-tropical regions. Malabar nut tree is commonly available in South India, Sri Lanka, South China, Malaysia, the Philippines and West Africa. Ajuba is known for possessing insecticidal, anti-feedant and insect repellent properties in its roots and leaf powder and water extracts against the stored grain pests of rice and paddy. Chellappa and Chelliah

(1976) reported its root powder as a paddy grain protectant against *S. oryzae*, *R. dominica* and *S. cerealella*. Pandey et al. (1981) reported its root oil to protect greengram against the attack of the pulse beetle, *C. maculatus* when admixed with the grams. Its root powder was found to be insect repellent to *R. dominica* and *S. cerealella* when dusted or sprayed on the surface of stored grains (Balasubramaniam, 1982). Mathur et al. (1985) also found its root powder and extract as pulse grain protectant against pulse beetle, *C. chinensis*. Its leaf extract was found to be toxic and reported to protect stored grains against infestation of the pulse beetle, *C. chinensis* (Pandey and Singh, 1976; Chander and Ahmed, 1982). Prakash et al. (1978–1987) reported its leaf extract in water to show promising grain protection when rice/paddy grains were treated or its root powder was admixed with the grains and tested against *S. cerealella, S. oryzae* and *T. castaneum*. Similarly, Bhaduri et al. (1985) reported its leaf extracts in petroleum ether, benzene and alcohol to show toxicity to *C. maculatus* when tested on cowpea seeds in a laboratory test. Methanol and dichloromethane (1:1) extract of its leaf sprayed on the host plant leaves showed antifeedant and insecticidal activities when tested against 3rd and 4th instars of tobacco caterpillar, *Spodoptera litura* under no choice test (Mathew and Chauhan, 1994).

20. *Adina cardifolia* (Roxb.) Hook f. ex Brandis (Rubiaceae)

(Syn.-*Nauclea cardifolia* Roxb.)

Common name: Karam or Yellow teak or Saffron teak

This is a perennial shrub or tree found in tropical and sub-tropical zones and available in forests all over India. Insecticidal activity was found in its root and leaf aqueous extracts and powders when tested against the beetles of agricultural importance (Jacobson, 1975) and also against the dipteran maggots (Usher, 1973).

21. *Aegle gultinosa* Corr. (Rutaceae)

Common name: Tabon

This is a perennial tree in tropical and sub-tropical zones. This plant is known to show repellent activity in its branch and leaf extracts against the rice field insects such as rice brown planthoppers and green leafhoppers (Litsinger et al., 1978).

22. *Aegle marmelos* (Linn.) Corr. (Rutaceae)

(Syn.-Crataeva marmelos Linn.)

Common name: Bael

This is a perennial tree in tropical and sub-tropical zones. Bael tree is available throughout India, Bangladesh, Nepal, Malaysia, China, the Philippines and

East Africa. This tree is known to show insect repellent activity against the storage pests of rice, viz., *S. cerealella*, *S. oryzae* and *R. dominica*, under controlled conditions and also significantly protected grains from the infestation of these pests under natural conditions of storage (Prakash et al., 1982, 1983, 1978–1987). Its leaf extract in water reduced survival of the green leafhopper, *Nephotettix virescens* when sprayed on the rice plant during vegetative growth stage (Satpathy, 1983).

23. *Aeschynomene sensitiva* Linn. (Leguminosae)

Common name: Swamp grass

This is an annual grass found in tropical and sub-tropical zones and available in India, China, Bangladesh and Australia. Swamp grass is reported to show insecticidal activity in its branch and seed aqueous and ethanolic extracts and also in its latex or juice against the melon worm, *Diaphania hyalinata* (Jacobson, 1958).

24. *Aesculus californica* Linn. (Hippocastanaceae)

Common name: California buckeye

This is a perennial tree found in temperate climates and commonly available in North India, Canada and Northern USA. This plant is known to show toxicity in its seed extract against Mexican bean beetle, *Epilachna varivestis* (Jacobson, 1958).

25. *A. hippocastanum* Linn. (Hippocastanaceae)

Common name: Horse chestnut

This is a perennial tree found in temperate climates and commonly available in North India. Its fruit extract possessed insecticidal property against the eastern sub-terranean termite, *Reticulitermes flavipes* (Jacobson, 1975).

26. *A. pavia* Linn. (Hippocastanaceae)

Common name: Dwarf buckeye

This is a perennial tree or shrub found in temperate climates. Jacobson (1958) reported attractancy in its fruit extract to the Japanese beetle, *Popillia japonica*.

27. *Afrormosia laxiflora* Harms. (Leguminosae)

Common name: Legumin

This is a perennial tree found in tropical climates and commonly available in coastal Tamil Nadu and Andhra Pradesh in India. Its dried leaf powder reduced the infestation of the insects in stored paddy (Ahmed, 1984).

28. *Agava americana* Linn. (Amaryllidaceae)

Common name: Century plant
This is a perennial shrub found in tropical regions and a native of South America. Century plant was reported to show repellent activity in its leaf extract when tested against the rice weevil, *S. oryzae* and other stored grain pests (Hameed, 1982, 1983).

29. *Ageratum conyzoides* Linn. (Asteraceae)

Common name: Goat weed
This is an annual or perennial plant found in tropical and temperate regions. Goat weed is distributed in all the South Coast Asia countries, Australia, Europe and West Africa. In India, it is available in Uttar Pradesh, Orissa, Bihar and other coastal states. Its different plant parts are known to possess insecticidal, ovicidal and antifeedant properties against agricultural insect pests.

Leaf, flower and root extracts of goat weed were reported to show toxicity to the cotton stainer, *Dysdercus cingulatus*; corn weevil, *Sitophilus zeamais*; and red flour beetle, *Tribolium castaneum* (Carino, 1981). Its leaf and branch extracts in water also showed insecticidal activity against cotton stainer, *Dysdercus flavidus* (Fagoonee, 1982), 5th instar larvae of the potato tuber moth, *Gnorimoschema operculella* (Pandey et al., 1982) and the adults of the vinegar fly, *Drosophila melanogaster* (Padolina, 1983). Its whole plant powder admixed with stored wheat seed significantly reduced the grain damage and population of the rice weevil, *S. oryzae* (Rout, 1986). Its leaf, flower and bud extracts in benzene were found to be repellent to pulse beetle, *Callosobruchus chinensis* and also reduced its oviposition in stored greengram (Pandey et al., 1986).

Fagoonee and Umrit (1981) found a crude lipid extract from this plant to show ovicidal activity and also reduced fertility of the cotton stainer, *Dysdercus flavidus* when applied topically on its 5th instar larvae and the adult females. This crude extract contained two biologically active components, viz., precocene I [7-methoxy-2,2-dimethyl-2H-1-benzopyran] and precocene II [6,7-dimethoxy-2,2-dimethyl-2H-1-benzopyran]. Precocenes (ageratochromenes) were toxic to stored beetle, *Oryzaephilus surinamensis* (Saleem and Wilkins, 1984), and showed inhibition of juvenile hormone dependent reproduction in *Epilachna varivestis*, induction of adult-diapause in *L. decemlineata* (Bowers et al., 1976), flight activity inhibition in *Hippodania convergens* (Rankin and Rankin, 1980), juvenilization of 5th instar of *Locusta migratoria* (Miall and Mordue, 1980) and prolongation of larval-pupal period of *S. mauritia* (Mathai and Nair, 1983).

Lu (1982) reported precocene I and precocene II highly toxic to the rice weevil, *Sitophilus oryzae* and rice earhead bug, *Leptocorisa chinensis*. Precocene II was also found to cause morphological abnormalities in the treated pupae of *Epilachna vigintioctopunctata* when applied topically (Gupta and Dogra, 1990).

30. *Ageratum houstonianum* Mill. (Asteraceae)

(Syn.-*A. mexicanum* Sims.)

Common name: Floss flower or Mist flower

This is an annual/perennial ornamental shrub found in sub-tropical zones and commonly available at hill stations. This is native to Mexico but also available in South India. Bowers et al. (1976) reported anti-allatic property of ageratochromene (6,7-dimethoxy-2,2-dimethyl,-3-chromene) isolated from the whole plant extract of this shrub and coined the term 'precocene' for such a biologically active compound, which induced precocious development in treated insects. Precocene I and precocene II isolated from the leaf extract of this plant were reported to show anti-juvenile hormone (AJH) activity when tested against the cotton stainer, *Dysdercus koenigii*. These components effectively malformed the ovaries of the adult female bugs when their eggs were treated and allowed to develop up to the adult emergence (Gawade et al., 1990).

31. *Aglaia odorata* Kostern (Meliaceae)

This is a perennial tree found in temperate and sub-tropical zones. Its leaf extract was reported to inhibit the larval development of the variegated cutworm, *Peridroma saucca* (Champagne et al., 1989).

32. *Ailanthus altissima* Desf. (Simgroubaceae)

(Syn.-*Ailanthus galandulosa* Desf.)

Common name: Tree of heaven

This is a perennial woody tree found in Australia and in all Asian countries. A toxic alkaloid called quassin ($C_{22}H_{28}O_6$) isolated from this tree was found to show toxicity to the insect pests of agricultural importance (Crosby, 1971).

33. *Ajuga remota* Benth. (Labiatae)

Common name: Indian ajuga

This is a perennial creeping herb found in tropical and temperate regions and commonly available in North and Coastal Indian States. Its leaf extract was reported to show antifeedant activity against *Spodoptera exempta*, *Schistocerca gregaria* and insecticidal activity against *Spodoptera littoralis* (Kubo and Nakanishi, 1978). Leaf extract and powder of whole plant and its leaf were found to be antifeedant against the pink cotton bollworm, *Pectinophora gossypiella* and the fall armyworm, *Spodoptera frugiperda* (Kubo and Klocke, 1981). Leaf extract in water of this herb was also reported to show toxicity to the African armyworm, *Spodoptera exempta* and the cotton leaf armyworm, *Spodoptera litura* (Michael et al., 1985).

Active Components

From the leaf extract of this plant, 3 terpenoid components, viz., ajugarins I, II and III (Figures 5, 6, 7), were isolated, identified and evaluated. Ajugarin I showed toxicity to *S. littoralis*, whereas all the three components exhibited antifeedant property to *S. exempta* and *S. gregaria* (Kubo and Nakanishi, 1978).

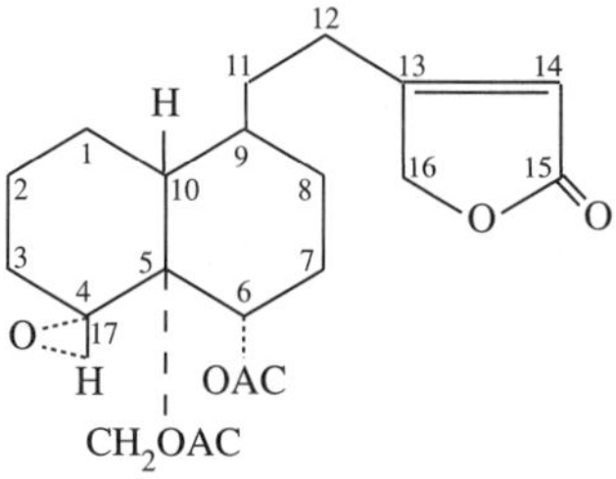

Figure 5. Ajugarin I.

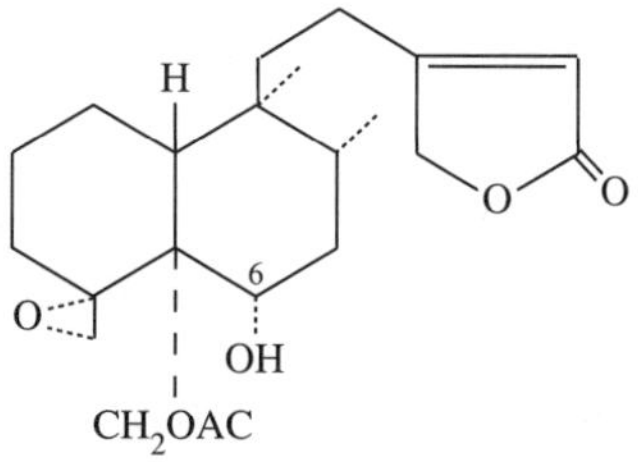

Figure 6. Ajugarin II.

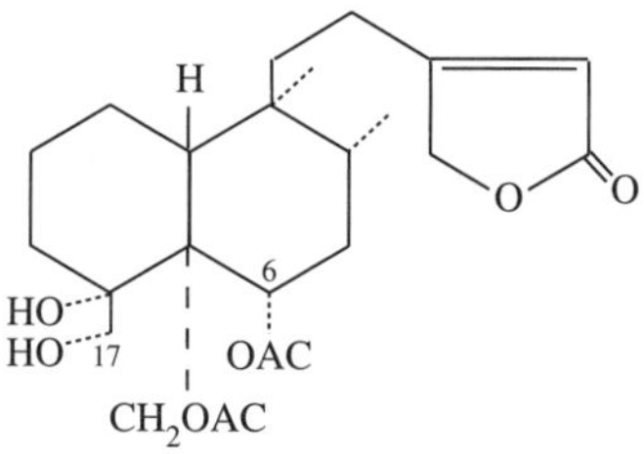

Figure 7. Ajugarin III.

Further, Jacobson (1989) referred to a number of isomers of ajugarins of clerodane diterpenes group from this plant, which showed juvenile hormone analogue activity in Mexican bean beetle, *E. varivestis*. These components were also found in its other species, i.e., *A. reptans*, *A. chamalpitys*, *A. nipponenes*, *A. iva* and *A. pseudoria*.

34. *Alangium salviifolium* Lif. Wanger (Allangiaceae)

Common name: Alangi

This is a perennial herb found in tropical and temperate zones and available in South Indian States, Sri Lanka, Indonesia and Malaysia. Ethyl ether extracts of its stem and bark showed attractant properties to the male and female of

Mediterranean fruit fly, *Ceratitis capitata* and the melon fly, *Dacus cucurbitae* (Keiser et al., 1975).

35. *Albizia lebbek* Linn. (Mimosaceae)

(Syn.-*Mimosa lebbeck* Linn., *Acacia lebbek* Willd.)

Common name: Woman tongue tree or Kokko or Lebbek tree

This is a perennial tree found in tropical and subtropical zones. In India, it is available in West Bengal, Assam, Tamil Nadu, Punjab, Uttar Pradesh and Madhya Pradesh. This plant is known for insecticidal activity in its leaf extract against the leaf eating caterpillars of agricultural importance (Jacobson, 1958).

36. *Aleurites fordii* Hemsl. (Euphoribaceae)

Common name: Tung oil tree or China wood oil tree

This is a perennial tree in tropical and sub-tropical zones and a native of China. In India, this tree is grown in Assam, Himachal Pradesh and West Bengal for the shade to tea bushes. Its different parts are known to show antifeedant and growth inhibiting activities to a wide range of the insects of agricultural importance. Tung seed oil and aqueous fruit extract were reported to show antifeedant activity to the sugarcane woolly aphid, *Ceratovacuna lanigera* and the boll weevil, *Anthonomus grandis* when sprayed on the host plants (Jacobson, 1958). McIndoo (1983) also reported growth inhibiting activity in its seed oil against the stink bugs.

From its seed oil, two components, viz., α-eleosteric acid and erythro-9,10-dihydroxy octadecyl acetate, were isolated and identified as responsible for feeding deterrency to the adults of the boll weevil, *Anthonomus grandis* (McIndoo, 1983).

37. *Allium cepa* Linn. (Amaryllidaceae)

Common name: Onion

This is a seasonally cultivated herb in tropical and sub-tropical climates and available all over the world as a vegetable. In India, it is grown mainly in North and Central States and used as a vegetable throughout the country.

Bulb extracts in water and its purified fractions are known to possess insecticidal and repellent properties against stored grain insect pests. Crude extract of its bulb was found to show insecticidal property against pulse beetle, *Callosobruchus chinensis* (Pandey and Singh, 1976) and the larvae of the rice moth, *Corcyra cephalonica* (Prakash et al., 1978–1987), whereas its aqueous extract showed repellent activity to *S. cerealella* and *R. dominica* but failed to provide significant grain protection to the paddy grains against these insects under natural conditions of storage (Prakash et al., 1980).

From the aqueous extract of onion bulb, an active component, i.e., 1-3 diphenyl thiourea, was isolated as an insecticide against the rice weevil, *S. oryzae* (Chatterjee et al., 1980a).

38. *Allium sativum* Linn. (Amaryllidaceae)

Common name: Garlic

This is also a seasonally cultivated herb in tropical and sub-tropical zones but available all over the world as a spice. Garlic is a native of Central Asia and commonly grown in Northern and Central States of India. Bulb extract and oil of garlic are known to possess insecticidal, insect repellent, nematicidal and fungicidal properties against a wide range of agricultural pests. This plant is well explored for its biological properties against the insects infesting grains in storage as well as insects infesting crops under field conditions.

Biological Activity Against Stored Grain Insect Pests

Garlic clove extract was reported to be toxic to the pulse beetle, *C. chinensis* (Quadri, 1970; Quadri and Rao, 1980). Garlic oil showed insecticidal activity against the khapra beetle, *Trogoderma granarium* (Bhatnagar-Thomas and Pal, 1974). Garlic powder and extract, however, protected wheat grains from infestation of the lesser grain borer, *R. dominica* when admixed/treated with its powder and extract and did not adversely affect the viability of the grains (Zag and Bhardwaj, 1976). Chatterjee et al. (1980b) reported garlic oil and its fractions to show toxicity to rice weevil, *S. oryzae*.

When tested under laboratory conditions, Prakash et al. (1980) found garlic extract as an effective paddy grain protectant against Angoumois grain moth, *S. cerealella* and the lesser grain borer, *R. dominica*. Under natural conditions of insect infestation in farm godowns, garlic extract also effectively minimized the losses in stored paddy caused by *R. dominica*, *S. oryzae* and *S. cerealella* (Prakash et al., 1982). Yadav and Bhatnagar (1987) reported garlic powder to show significant mortality to pulse beetle, *C. chinensis* when admixed with cowpea seeds. Garlic also showed repellency to *T. castaneum* (Mohiuddin et al., 1987) Prusty et al. (1989) reported garlic extract to inhibit larval population of the rice moth, *C. cephalonica* when milled rice was treated with the extract and stored.

Biological Activity Against Field Insect Pests

Bulb extract of garlic in acetone was found to show repellent and insecticidal activities against the red cotton bug, *Dysdercus cingulatus* (Sundramurthy, 1979; Rajendran and Gopalan, 1979); cotton leaf armyworm, *Spodoptera litura*; and castor pest, *Pericallia ricini* (Rajendran and Gopalan, 1979). An oil fraction obtained from steam distillation of crushed cloves and crude extract in methanol reported to show toxicity to 3rd instar larvae of *S. litura* (Debkirtaniya et al., 1980a). Crude extract of garlic showed strong antifeedancy and also affected the metamorphosis during development of the larvae and the pupae of *Epilachna varivestis*. A pupal-larval intermediary was formed and showed deformities of varying extents when this crude extract was eaten up through its larval food (Nasseh, 1981). Garlic extract in water and its powder showed repellent activity against *S. litura* (Balasubramanian, 1982), whereas Kareem (1984) reported insecticidal activity in its aqueous extract against *D. clingulatus*. Garlic oil 2% protected the okra fruits from infestation of *Earias vittela* when sprayed on the crop (Sardana and Kumar, 1989). Ram Prasad et al. (1990) reported methanolic fractions of its bulb extract highly toxic to the tobacco caterpillar, *Spodoptera litura* when tested individually against its 3rd instar larvae by leaf disc method and also enhanced the activity of nuclear polyhedrosis virus (NPV) when tested in combination. Similarly, its 5% bulb extract was found to be an effective toxicant against 2nd instar larvae of Gujrat hairy caterpillar, *Amsacta moorei* (Patel et al., 1990a).

Active Components

Chatterjee et al. (1980b) isolated an active compound, i.e., 1-3,diphenyl thiourea, as an insecticide against the rice weevil, *S. oryzae*. The active component from garlic extract was isolated and identified as 'allitin,' a mixture of diallyl di- and trisulphides (Amonkar and Banerjee, 1971), which was found to inhibit cholinesterase activity in insects (Bhatnagar-Thomas and Pal, 1974) and also showed promising grain protection against insects in stored paddy when admixed with the grains (Prakash et al., 1984). But its unpleasant odour and low persistence restricted its use as a general insecticide (Chadha, 1986). Prabhakar and Rao (1993) reported allitin to be 2.8 times more toxic to the 2nd instar larvae of *S. litura* than annmet.

39. ***Alnus ferma*** **Mill. (Betulaceae)**

Common name: Alder tree

This is a perennial tree or shrub available in sub-tropical and temperate zones. The root and rhizome aqueous extracts of this plant were reported to show insecticidal activity against the vinegar fly, *Drosophila hydei* (Jacobson, 1958).

40. ***Alnus hirsuta*** **Mill. (Betulaceae)**

Common name: Manchurian alder

This is a perennial tree found in temperate climates. Its stem, branch and leaf extracts in water were reported to be insecticidal against dipteran pests of agricultural importance (Jacobson, 1958).

41. *Alocasia macrorrhiza* (Linn.) Schoott. (Araceae)

Common name: Taro or Giant taro

This is an annual herb found in tropical and sub-tropical regions and commonly available in the Philippines. This plant was reported to protect field pests of paddy when its whole plant powder was dusted on the crop (Litsinger et al., 1978).

42. *Aloe barbadensis* Mill. (Liliaceae)

(Syn.-*A. vera* [Linn.] Web. & Berth. [non Mill.], *A. perfoliate* var. *vera* Linn.)

Common name: Medical aloe or Barbados aloe

This is a perennial herb found in tropical and sub-tropical regions and commonly available in North India. Rhizome extract of this aloe in ether was reported to show larvicidal activity against the mustard sawfly, *Athalia proxima* (Pandey et al., 1977, 1979).

43. *Aloe succotrina* Linn. (Liliaceae)

This is a perennial shrub or herb found in tropical and sub-tropical regions. Latex and juice or extract of its whole plant were found to be insecticidal when tested against the scale insects (McIndoo, 1983).

44. *Aloysia tripjhylla* Ort. & Palem. (Verbenaceae)

Common name: Lemon verbena

This is a perennial shrub found in tropical and sub-tropical regions. This plant is known to possess insecticidal property in its leaf extract against the cotton aphid, *Aphis gossypii* (Jacobson, 1975).

45. *Alpina galanga* Willd. (Zingiberaceae)

Common name: Greater galangal

This is a robust perennial herb native to Indonesia and Malaysia but distributed now in Eastern Himalayas and South Western India. Its rhizome powder admixed with stored grains @ 1%, w/w was found to protect the grains against infestations of rice weevil, *Sitophilus oryzae* and rice moth, *Corcyra cephalonica* and showed 100% mortality to both the insects up to 45 days (Ahamad and Ahmad, 1991).

46. *Alternathera polygonoides* Forsk. (Amaranthaceae)

This is an annual/seasonal herb found in tropical and sub-tropical regions and commonly available in South India. Ether extract obtained from its leaves showed significant juvenile hormone analogue activity to the red cotton bug, *Dysdercus koenigii* (Banerji, 1988).

47. *Amanita muscaria* Linn. (Agaricaceae)

Common name: Fly agarie
This is a poisonous mushroom found in tropical zones. The insecticidal property was reported in water extract of its fruiting bodies against the peach aphid, *Aphis persicae* (McIndoo, 1983).

48. *Amaranthus spinosus* Linn. (Amaranthaceae)

Common name: Spiny or prickly amaranthus
This plant is an annual herb found in tropical and sub-tropical zones and available in East Indian States. Its leaf extract in methanol showed larvicidal activity to newly hatched larvae of Angoumois grain moth, *S. cerealella* (Prakash et al., 1978–1987).

49. *Ambrosia psilostachya* Linn. (Asteraceae)

Common name: Western ragwort
This is an annual or perennial herb found in sub-tropical and temperate zones. This plant is reported to show insecticidal activities in its dried root powder against the pea aphid, *Acyrthosiphum pisum* and the armyworm, *Mythimna (Pseudoletia) unipunctata* (Jacobson, 1958).

50. *Amianthium muscitoxicum* A. Gray (Liliaceae)

Common name: Fly poison
This is a perennial herb found in temperate zones. This plant is reported to possess insecticidal and antifeedant properties in its bulb, leaf and flower extracts in water, alcohol and petroleum ether and their powders against the aphids and the grasshoppers of agricultural importance (McIndoo, 1983); European corn borer, *Ostrinia nubilalis* (Jacobson, 1958); and the tent caterpillars (Jacobson, 1958; McIndoo, 1983).

51. *Amomum aculeatum* Roxb. (Zingiberaceae)

This is an annual and biannual herb found in tropical and sub-tropical zones. This plant is found in deep forests in shady and moist places in South Andaman, Malaysia and Java. Crushed/squeezed stems of this plant or sap of its flowers, when rubbed on the body of the honey collectors, was reported to act as tranquilizer for the giant honey bee, *Apis dorsata* (Dutta et al., 1985).

52. *Amorpha fuiticosa* Linn. (Leguminosae)

Common name: False indigo

This is a perennial shrub available in arid zones. This plant has been reported to show insecticidal and growth inhibiting activities when its root, flower and fruit extracts in water and acetone were tested against the cotton aphid, *Aphis gossypii*; bean leaf beetle, *Cerotoma trifurcata*; chinch bug, *Blissus leucopterus*; potato leafhopper, *Empoasca fabae* (Jacobson, 1958; Khan, 1982); kharif grasshopper, *Hieroglyphus nigrorepletus*; and tarnished plant beetle, *Lygus lineolaris* (Jacobson, 1958). Aqueous extract of its root was reported to show antifeedancy to the larvae and the adults of Colorado beetle, *Leptinotarsa decemlineata* on potato leaves (Britski, 1982).

Gombos (1985) reported '6a,12a-dihydro-α-toxicanol', an aromatic active component isolated from its roots extract, to show insecticidal and repellent properties to *L. decemlineata*, *Locusta migratoria migratorioides* and *Plutella maculipennis* when tested under laboratory conditions using its 1% formulation.

53. *Amorphophallus campanulatus* (Roxb.) Blume ex Dene (Araceae)

(Syn.-*Arum campanulatum* Roxb.)

Common name: Elephant foot yam or White spot giant yam

This is a perennial herb found in tropical and temperate zones. It is grown as vegetable throughout India. Its corn extract in water was found to control the insect pests of paddy fields (Litsinger et al., 1978).

54. *Amsonia elleptica* Walt. (Apocynaceae)

Common name: Bluster genus

This is a perennial herb found in temperate zones. Leaf aqueous extract of bluster genus showed insecticidal activity against the vinegar fly, *Drosophila hydei* (Jacobson, 1958).

55. *Anabasis aphylla* Linn. (Chenopodiaceae)

This is a perennial shrub found in temperate zones and commonly available in North America and USSR. The aqueous extract of its branch, stem and its whole plant powder showed insecticidal activity against the aphids and jassids of agricultural importance (Roark, 1947; Rusbey, 1950; Poe, 1980).

Anabasine ($C_{10}H_{14}N_2$), chemically called 3-(2-piperidyl)-pyridine (Figure 8), an alkaloid was isolated from its stem extract. Anabasine was also known as 'neonicotine' having the same empirical formula as nicotine and was highly toxic to soft bodied insects like the bean aphid, *Aphis fabae* (De-Ong, 1971). Commercially it was used as 40% anabasine sulfate and was as toxic as nicotine.

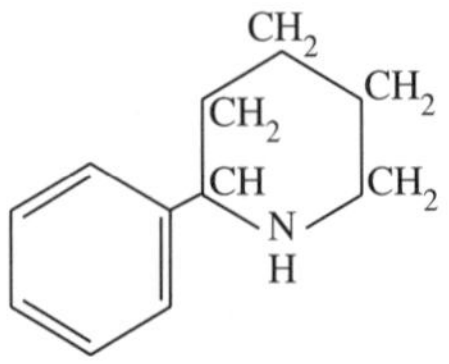

CH_2
CH_2 CH_2
CH CH_2
N
H

Figure 8. Anabasine.

56. *Anacardium occidentale* Linn. (Anacardiaceae)

Common name: Cashew

This is a perennial evergreen tree found in tropical and semi-arid zones. In India, it is available in Malabar, Kerala, Karnataka, Tamil Nadu, Andhra Pradesh, Orissa, West Bengal, Goa and Maharashtra. This plant is reported to show insecticidal and repellent properties when its bark, stem, branch, fruit and seed extracts were tested against termites (Jacobson, 1975) and the white coffee borer, *Anthores leuconotus* (McIndoo, 1983). Oil of its seeds showed toxicity to the red cotton bug, *D. cingulatus* and the cotton leaf armyworm, *S. litura* (Kareem, 1984). Mathew and Chauhan (1994) also found its leaf extract in methanol and dichloromethane (1:1) to show antifeedant and insecticidal activities against 3rd and 4th instar larvae of *S. litura* when tested under no choice test on the treated host plant leaves.

57. *Anacyclus pyrethrum* D.C. (Asteraceae)

Common name: Bertram root or Pellitory root

This is an annual herb found in semi-arid and tropical zones. It is a commonly available medicinal plant in North Africa, Algeria, India, Saudi Arabia, Syria and Spain. From ethanolic extract of its roots an unsaturated isobutylamide (N-isobutyl-(2E 4E)-2, 4-4 decadienamide) called 'pellitorine' (Figure 9) was isolated and evaluated to show insecticidal activity. Along with this component another chemical called 'anacyclin' ($C_{18}H_{25}$ NO) (Figure 10), an unsaturated isobutylamide (N-isbutyl-trans-2, trans-4-tetradecadien-8, 10-dyamide), was also reported occurring in its roots and it showed toxicity to the aphids (Jacobson and Crosby, 1971). Both these components were reported to show toxicity to the yellow mealworm, *Tenebrio molitor* (Crombie, 1955) and the aphids of agricultural importance (Jacobson and Crosby, 1971) and also to the mustard beetle, *Phaedon cochleariae* (Miyakado et al., 1989).

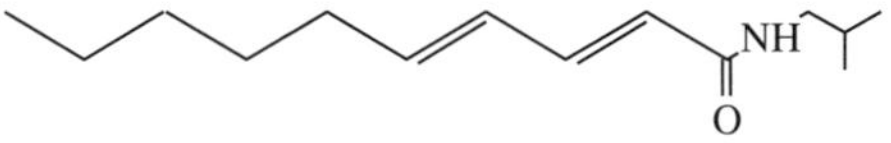

NH
O

Figure 9. Pellitorine.

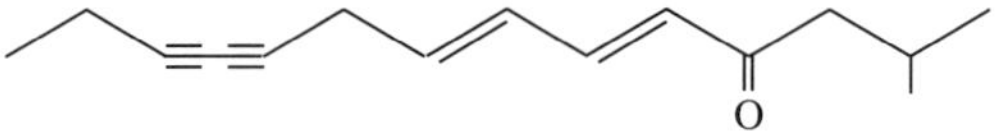

O

Figure 10. Anacyclin.

58. *Andrographis paniculata* (Burm f.) Wall. ex Ness (Acanthaceae)

(Syn-*Justica paniculata* Burm. f.)

Common name: Kariyat or King of bitters

This is an annual and perennial branched herb found in tropical zones and commonly available in Orissa and other coastal States of India. Nematicidal, insecticidal and antifeedant properties have been reported in its leaf and seed extracts. Leaf extract in water was found to be insecticidal to the grasshoppers, whorl maggots and other pests in paddy fields (Satpathy, 1983). Gunasekran and Chelliah (1985b) and Rajasekaran and Kumaraswami (1985b) reported antifeedant property in its leaf extract in acetone against the cotton leaf armyworm, *Spodoptera litura*. Seed extract of kariyat showed strong antifeedancy to 3rd instar larvae of *S. litura* (Rajasekaran and Kumarswami, 1985b). Its leaf extract also reduced the weight loss of treated and stored rice/paddy and minimized the multiplication of Angoumois grain moth, *S. cerealella* (Prakash et al., 1989) and the rice weevil, *S. oryzae* (Prakash et al., 1990c). Gupta et al. (1990) reported its leaf extract to significantly protect paddy grains of the caged plants against *Leptocorisa acuta* when sprayed on the plants.

59. *Anemone radeleana* Linn. (Ranunculaceae)

This is a perennial herb found in tropical, temperate and Mediterranean zones. Insecticidal activity in its aqueous extracts of branch and leaf was reported against the vinegar fly, *Drosophila hydei* (Jacobson, 1958).

60. *Anethum graveolus* Linn. (Apiaceae or Umbelliferae)

Common name: Dill plant

This is an annual or perennial herb found in the corn fields of South Europe, Egypt and the Cape of Good Hope. Lichtenstein et al. (1974) reported apiol, dill-apiol, miristicin and d-carvone as active components in its plant extract, which were toxic to the fruit flies, *Dacus spp.* Su and Horvat (1988) also reported acetone extract of its seed to show repellency to the red flour beetle, *T. castaneum* and two active components, i.e., carvone and dill-apiol, were found to show repellency to this beetle.

61. *Angelica sylvestris* Linn. (Umbelliferae)

This is an annual/perennial herb found in sub-tropical and temperate regions. Bisabolangelone, a sesquiterpenoid isolated from the seed of this plant, was reported to show a powerful antifeedancy against the armyworm, *Mythimna unipunctata* (Robert et al., 1987).

62. *Annona glabra* Linn. (Annonaceae)

Common name: Pond apple

This is a perennial tree found in tropical and sub-tropical zones. Aqueous extracts of its leaf and seed were reported to show insecticidal and antifeedant activities against the milkweed bug, *Oncopeltus fasciatus* (Heal et al., 1950) and the chrysanthemum aphid, *Macrosiphoniella sanborni* (McIndoo, 1983). Jacobson (1989) reported liriodenine, an alkaloid isolated from its stem and bark extracts, to show toxicity to the insects.

63. *Annona montana* Linn. (Annonaceae)

Common name: Mountain sour sop

This is a perennial tree found in tropical zones. Water extract and powder of its seed were reported to be toxic when tested against the milkweed bug beetle, *Oncopeltus faciatus* (Heal et al., 1950).

64. *Annona muricata* Linn. (Annonaceae)

Common name: Sour sop

This is a perennial small evergreen tree found in tropical and sub-tropical zones but a native of the West Indies. In South India, it is popularly grown for its fruits. Insecticidal property in its aqueous seed extract or seed powder was reported against the pea aphid, *Acyrthosiphum pisum* and the southern armyworm, *Spodoptera eridania* (Jacobson, 1958). Crosby (1971) reported a toxic alkaloid 'anonaine' ($C_{17}H_{15}NO_2$) (Figure 11) as an active component isolated from this plant and found it to show insecticidal property against the general insects of agricultural importance when tested under controlled as well as natural conditions of their infestation in the fields.

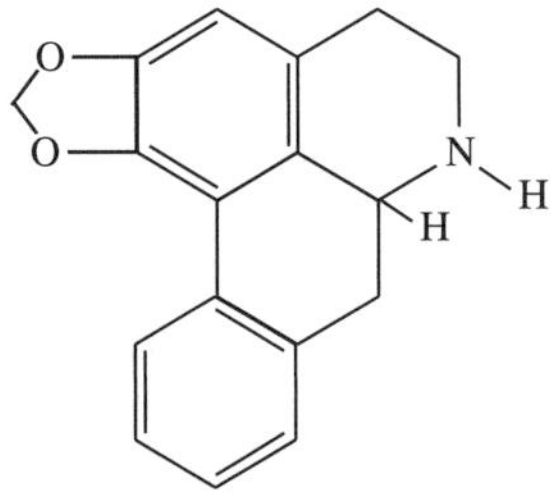

Figure 11. Anonaine.

65. *Annona palustris* Linn. (Annonaceae)

Common name: Alligator apple

This is a perennial tree found in tropical zones. Aqueous extract of its seed was reported to show insecticidal activity against the milkweed bug, *Oncopeltus fasciatus* (Hael et al., 1950).

66. *Annona reticulata* Linn. (Annonaceae)

Common name: Bullocks heart or Custard apple

This plant is a perennial tree of tropical and sub-tropical zones. This is a native of the West Indies but also available in West Bengal, Assam, Khasi hills and South India as edible fruit. This plant is also commonly available in Sri Lanka, Bangladesh, Malaysia, Indonesia, the Philippines and Pakistan. Its bark, leaf, fruit and seed have been reported to show insecticidal, antifeedant and repellent activities against a number of agricultural insect pests infesting grains in storage as well as crops in field conditions.

Biological Activity Against Storage Insects

Powder of custard apple seed protected mungbean seeds from infestation of the pulse beetle, *Callosobruchus maculatus* (Pandey and Verma, 1977). Its seed extract showed promising results in controlling the pulse beetle, *C. chinensis* and the lesser grain borer, *R. dominica* (McIndoo, 1983). Rout et al. (1986) reported significant wheat seed protection against the rice weevil, *Sitophilus oryzae* when seed powder of custard apple was admixed with the grains and stored for a period of six months.

Biological Activity Against Field Insects

Seed extract and powder of this plant was reported to show insecticidal activity against the insect pests infesting rice fields (Litsinger et al., 1978).

However, its seed extract was found to be an effective toxicant against the chrysanthemum aphid, *Macrosiphoniella sanborni* (Morallo-Rajessus, 1982). Jacobson (1975) reported repellent activity in its root extract against the green bug, *Leanium* sp. Aqueous extract of its seed was also reported to possess insecticidal property when tested against the croton caterpillar, *Achoaea janata*; hair caterpillar, *Euproctis fraterna*; diamond-back moth, *Plutella xylostella*; and the cotton leaf armyworm, *Spodoptera litura* (Jacobson, 1975).

Active Components

Harper et al. (1947) found insecticidal components such as glycerides of hydroxylated unsaturated acids in its seed extract. Crosby (1971) reported 'anonaine' ($C_{17}H_{15}NO_2$), a toxic alkaloid, which was found to show insecticidal activity. Anonaine was also reported in other species of *Annona*, i.e., *A. squamosa*, *A. musicata*, *A. palistrus*, and other plant species such as *Nelumbo mufura*, *Michelia compressa*, *Roemesia refraeta*, *Neolitsea sericea* along with other relative alkaloids, i.e., roemerine, nuciferine (Figure 12), recticuline (Figure 13), and 6-methoxy, 7-hydroxy aporphine etc., which were toxic to the insects.

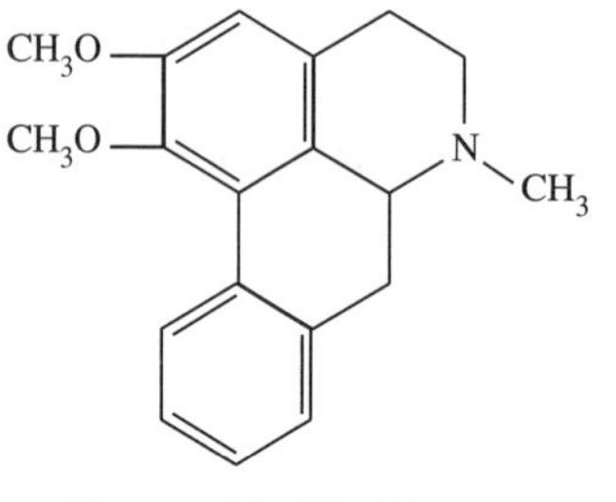

Figure 12. Nuciferine.

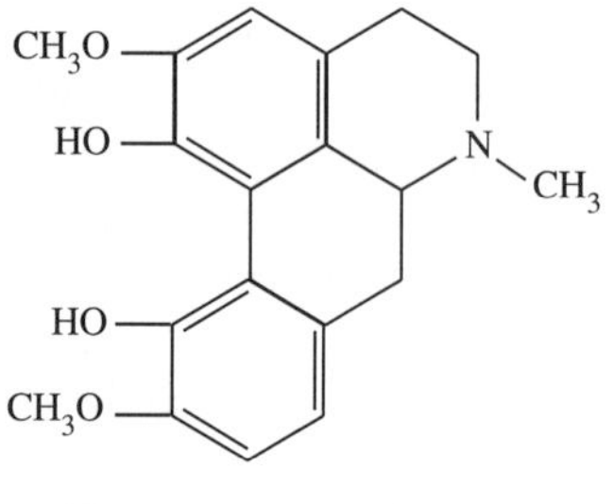

Figure 13. Recticuline.

67. *Annona senegalensis* (Linn) Gn. (Annonaceae)

This is a perennial tree found in tropical and sub-tropical regions. Its dried leaf powder was reported to show promising grain protection to stored grains against insect pests (Giles, 1964).

68. *Annona squamosa* Linn. (Annonaceae)

Common name: Sweet sop or Supper apple or Custard apple

This plant is a perennial shrub or small tree found in tropical and sub-tropical zones. This is a native of South America and the West Indies but available throughout India. Its root, leaf, fruit and seed extracts in water, ether and alcohol and also their powders have been reported to show insecticidal, antifeedant and repellent activities against a number of insect pests of agricultural importance in the field and storage ecosystems.

Biological Activity Against Storage Insects

Its root powder and extract were reported to be toxic to the pulse beetle, *Callosobruchus* sp. when grains were treated with these products and stored (Puttarudirah and Bhatia, 1955; Quadri and Rao, 1977). Its seed powder protected stored wheat from infestation of *S. oryzae* and *T. castaneum* when admixed with the grains and showed repellent activity (Quadri, 1973). Prakash et al. (1978–1987) found its aqueous leaf extract and root powder to inhibit the development of *T. castaneum* and *O. surinamensis* when tested under controlled conditions. Fruit and seed extracts in alcohol, water and ether showed insecticidal activity against *O. surinamensis* (Khan, 1982) and also against *T. castaneum* (Jacobson, 1975). Supper apple seed powder was reported to be an insecticidal to the pulse beetle, *C. chinensis* and significantly protected Bengal gram grains in storage (Ali et al., 1981; Satpathy, 1983). Chauhan et al. (1987) also found its seed powder extract to be toxic to the rice moth, *Corcyra cephalonica.*

Biological Activity Against Field Insect Pests

Leaf and seed extracts in water and alcohol showed toxicity to the milkweed bug, *Oncopeltus fasciatus* (Heal et al., 1950); mustard sawfly, *Athalia proxima*; cabbage aphid, *Brevicoryne brassicae*; and diamond-back moth, *Plutella xylostella* (Jacobson, 1958, 1975). Its seed powder was toxic to *Spodoptera litura, Crocidolomia binotalis, Plutella maculipennis* and also to the mango hopper, *Idiocerus* sp. (Puttarudirah and Bhatia, 1955). Borle (1982) reported insecticidal activity in its root extract to the cotton stainer, *Dysdercus koenigii* and *Lipaphis erysimi.* Against paddy field pests, the rice green leafhopper, *Nephotettix virescens* and the brown planthopper, *Nilaparvata lugens*, insecticidal activity in its leaf and fruit extracts was reported by Mariappan et al. (1982a,b) and Epino and Saxena (1982). Mariappan and Saxena (1983) found its seed oil to effectively reduce the survival of *N. virescens* and transmission to rice tungro virus when sprayed on the rice plants. Khanvilkar (1983) reported its leaf and root extracts in water and alcohol to possess insecticidal and antifeedant properties against the green bug, *Coccus viridis* and *Spodoptera littoralis.* Neem oil and custard apple oil (1:4) either in combination or alone reduced population of the green leafhopper, *N. verescens* (Kareem et al., 1987a). Custard apple seed oil sprayed on rice also reduced the survival of the leaffolder, *Cnaphalocrosis medinalis* and the green leafhopper, *N. virescens* and checked transmission of tungro virus (Narsimhan and Mariappan, 1988; Mariappan et al., 1988). Its seed extract in petroleum ether reduced oviposition of the rice brown plant hopper when sprayed on the rice plants at 2 and 5% concentrations (Reddy and Urs, 1988). Further, Reddy et al. (1990) and Chitra et al. (1991a) reported petroleum ether extracts of its leaf to effectively reduce the population of the grubs of *Henosepilachna vigintio-ctopunctata* when tested under greenhouse conditions on potted eggplants. Purohit (1989) reported its leaf extract in petroleum ether to show antifeedant activity against 3rd instar larvae of the castor semi-looper, *Achoea janata*. Seed and leaf suspension (5%) sprays of custard apple significantly reduced the population of 2nd instar larvae of Gujrat hairy caterpillar, *Amsacta moorei* and caused 33.74% mortality in the test under net-house (Patel et al., 1990a). Similarly, Vir (1990) reported its leaf extract to reduce the survival of the whitefly, *Bemisia tabaci* and also transmission of yellow mosaic virus (YMV) in mungbean crop using foliar spray applications under field conditions. Rao et al. (1990) found 15% aqueous extract of its leaf to give 100 per cent protection to the treated host plant leaves against 2nd instar larvae of *H. vigintioctopunctata* and showed antifeedant activity. Aqueous crude leaf extract of this plant at 0.5% concentration also significantly reduced the body weight of 1st and 3rd instar grubs of *H. viginctioctopunctata* when tested for its antifeedant activity (Chitra et al., 1991a,b). Chitra et al. (1993) also found its 0.1% petroleum ether leaf extract to control *H. vigintioctopunctata* on the brinjal crop in the field condition.

Anonaine ($C_{37}H_{15}NO_2$), an alkaloid isolated from this plant, showed insecticidal activity (Crosby, 1971). Prabhakar and Rao (1993) found annmet, an active formulation of this plant, toxic to 2nd instar larvae of *S. litura*.

69. *Anodendron offine* D.C. (Apocynaceae)

This plant is a woody climber found in tropical and sub-tropical zones and commonly available in Sri Lanka, Japan, Taiwan, Malaysia and the Solomon Islands. Insecticidal property in its branch and leaf aqueous extracts was reported against the soft bodied insect pests of agricultural importance (Jacobson, 1958).

70. *Anthriscus sylvestris* Pers. (Umbelliferae)

This is an annual/perennial herb found in tropical, sub-tropical and temperate zones. Root extract of this herb was reported to show larvicidal activity against *Epilachna sprasa* and 'anthricin' (deoxypodophyllotoxin), anthriscinol methyl ether and (Z)-2-angeloyloxy-methyl-2-betenoic acid were reported to be the active components in its root extract and were found to show insecticidal activity against *E. sprasa* when evaluated under natural and control conditions of its infestation (Kozawa et al., 1982).

71. *Aphanamixis polystachya* (Wall.) Parker (Meliaceae)

(Syn.-*Amoora rohifuka* Wight & Arn.)

Common name: Pitharaj

This is a perennial tree in the tropics and distributed in sub-Himalyan tracts, Sikkim, Assam, West Bengal, Andaman and Western Ghats in India. Insecticidal properties in dried leaves, pressed fruits and oil of pitharaj were explored against Angoumois grain moth, *Sitotroga cerealella* (Michael et al., 1985). Prakash et al. (1978–1987) reported its leaf powder protecting paddy grains from infestation of the grain boring insects like *S. cerealella*, *R. dominica* and *S. oryzae* in bulk containers for a period of 3–6 months. Lidert et al. (1985), however, reported antifeedant activity in its seed extract and oil against the tobacco budworm, *Heliothis virescens*. Active component 'rohituka-7', a limonoid isolated from its seed extract, was found to show antifeedant activity against *H. virescens* (Lidert et al., 1985).

72. *Apium graveolens* Linn. (Umbelliferae)

Common name: Ajmodh or Celery

This is an annual herb found in tropical regions and commonly available in South India. Its aqueous leaf extract was reported to be toxic to the pulse beetles, *C. chinensis* and *C. maculatus* (Anonymous, 1980–1981; Chander and Ahmed, 1982).

73. *Arachis hypogaea* Linn. (Fabaceae)

Common name: Peanut or Groundnut

This is an annual herb found in tropical and sub-tropical regions. Groundnut is a native of Brazil but now grown almost in all the plain and coastal areas in India. Its seed oil has been reported to show insecticidal, repellent, attractant and nematicidal activities against a wide range of pests of agricultural importance.

Groundnut oil admixed with cowpea was reported to protect the grains safely from attack of the pulse beetles, *C. chinensis* and *C. maculatus* during storage (Mittal, 1971). Edible groundnut oil was found to inhibit the population build up of stored grain insect pests (Anonymous, 1975) and protected maize grains from damage by the grain weevil, *Sitophilus zeamais* and showed toxicity to this weevil resulting to 100% mortality within 2 days when admixed with grains using 10 ml oil/1 kg grains (Ivbijaro, 1984). Groundnut oil admixed with greengram @ 0.25 to 0.5% w/w was reported to be toxic to pulse beetles, *C. chinensis* and *C. maculatus* and promisingly protected the grains in storage (Majumder and Amla, 1977; Singh et al., 1983; Golob and Webley, 1980; Ali et al., 1983; Doharey et al., 1983, 1984a,b, 1987). Sujata and Punnaiah (1984) reported blackgram treatment with its oil using 0.25% w/w to check multiplication of *C. chinensis* in a laboratory test.

Groundnut oil treatment to the greengram grains @ 8ml/kg reduced egg laying capacity of the pulse beetle, *C. chinensis* by 90% and the rest of the eggs attached to the treated grains died within 48 hrs (Sharma and Srivastava, 1984). Coating of stored greengram seeds with groundnut oil at the rate of 0.3 to 1.0% w/w was also reported to protect them from *C. chinensis* and *C. maculatus* and checked multiplication of these beetles (Sujatha and Punnaiah, 1985; Babu et al., 1989; Rao, 1990; Doharey et al., 1990; Kumari et al., 1990; Singh et al., 1993). Groundnut oil was also reported as a carrier for pyrethrin dust when tested against *R. dominica* and *T. castaneum* in a laboratory test and it enhanced pyrethrin toxicity to the test insects (Trivedi, 1987). Singal and Singh (1990) found groundnut oil treatment to the chickpea seeds @ 3 and 5 ml/kg to significantly reduce the hatching of the eggs and development of *C. chinensis* and to protect the grains.

74. *Archangelica officinalis* Hoffm. (Apiaceae)

(Syn.-*Angelica archangelica* Linn.)
This is a perennial/biennial aromatic herb found in temperate zones in North Europe. It is a native of Syria but now available in Jammu and Kashmir in India. Seed oil of this herb showed insecticidal activity against the fruit fly, *Dacus* sp. (Jacobson, 1975).

75. *Arctostaphylos uva-ursi* (Linn.) Spreng. (Ericaceae)

Common name: Bearberry
This is a perennial shrub found in temperate zones and commonly available in barren gravelly hills and dry sandy woods in the North of Europe and America. Its leaf, fruit and seed extracts were found to be repellent against the Japanese beetle, *Popillia japonica* (McIndoo, 1983).

76. *Ardisia crispa* var. *dielsii* Sw. (Myrsinaceae)

Common name: Meta-ajan
This plant is a perennial shrub in sub-tropical and temperate zones. Insecticidal property has been reported in its aqueous leaf extract against Mexican bean beetle, *Epilachna varivestis* (Michael et al., 1985).

77. *Arenaria leptoclados* Linn. (Caryophyllaceae)

Common name: Sandwort
This is an annual or biannual herb found in temperate zones. Leaf extract in

water of this herb was reported to be insecticidal against the vinegar fly, *Drosophila hydei* (Jacobson, 1958).

78. *Argemone mexicana* Linn. (Papaveraceae)

Common name: Mexican prickly poppy
This is an annual prickly herb or undershrub found in tropical and subtropical zones. This herb is known to possess insecticidal, repellent, nematicidal and bactericidal properties in its whole plant, root, branch, stem, leaf, flower and seed extracts and oil against a wide range of pests of agricultural importance.

Biological Activity Against Storage Insects

Its leaf and whole plant extracts showed repellent activity against the rice weevil, *S. oryzae* (Jacobson, 1975). However, Prakash et al. (1978–1987) reported 78% rice grain protection against *S. oryzae* when grains were treated with its leaf extract and stored under controlled conditions.

Biological Activity Against Field Insects

Whole plant aqueous extract and its seed oil showed insecticidal activity against the cotton stainer, *Dysdercus koenigii* (Roark, 1947); mustard aphid, *Lipaphis erysimi* (Borle, 1982); cotton leaf armyworm, *Spodoptera litura* (Balasubramaniam, 1982); and termites (McIndoo, 1983). Leaf and stem extracts effectively controlled the population of *S. litura* and showed antifeedant activity (Kareem, 1984). Patel et al. (1990b) reported its leaf extract to show high toxicity to Gujrat hair caterpillar, *Amsacta moorei* within 24 h of exposure. Vir (1990) also reported its leaf extract in acetone to show toxicity and repellency to *Apis craccivora* on cowpea crop and also to *Cyrtozemia cognata* on mung bean crop when tested using foliar spray under field conditions. Chitra et al. (1991a,b, 1993) also reported its leaf extracts in petroleum ether and water to protect leaves of eggplant from the attack of brinjal spotted leaf beetle, *Henosepilachna vigintioctopunctata* when tested under net-house and field conditions.

79. *Arisaema consangiuneum* Mart. (Araceae)

This is an annual herb found in temperate zones. Its leaf and seed extracts showed insecticidal activity against Mexican beetle, *Epilachna varivestis* (Michael et al., 1985).

80. *Arisaema erabescens* Mart. (Araceae)

This is an annual herb found in temperate zones and commonly available in the United States and Canada. Its leaf powder was reported to be an effective insecticide against Mexican bean beetle, *Epilachna varievestis* when tested under controlled laboratory conditions (Jacobson, 1958).

81. *Arisaema purpureoglaeatum* Mart. (Araceae)

This is an annual herb found in temperate zones. Its whole plant extract in water showed insecticidal and repellent activities against *Aphis fabae* (Jacobson, 1958) and *Epilachna varivestis* (Michael et al., 1985).

82. *Aristolochia elegans* Motch. (Aristolochiaceae)

Common name: Timbangan

This is a perennial woody climber found in tropical climates. Its leaf extract in water was reported to be antifeedant against Asian corn borer, *Ostrinia furnacalis* and the common cutworm, *Spodoptera litura* and also regulated the growth during their development when incorporated with the diets and fed to the test insects under controlled conditions (Caasi, 1983).

83. *Aristolochia bracteata* Retz. (Aristolochiaceae)

This is a perennial woody climber found in tropical zones and commonly available in Coromondal coasts in India. Its root powder showed toxicity to the pulse beetle, *Callosobruchus chinensis* and protected stored pulse grains (Puttarudirah and Bhatia, 1955).

84. *Aristolochia grandifolia* Swartz. (Aristolochiaceae)

Common name: Poison hog meat or Pelicon flower

This is an ornamental woody climber found in tropical climates and commonly available in mountain thickets in Jamaica. Bark and seed aqueous extracts and their powders showed insecticidal activity against the cabbage worm, *Pieris rapae* (McIndoo, 1983).

85. *Aristolochia indica* Linn. (Aristolochiaceae)

Common name: Indian birthwort

This is a twining shrub found in tropical and sub-tropical regions. In India, it is available throughout the country in plains and hilly regions. Its branch, leaf and fruit aqueous extracts showed insecticidal activity against the soft green scale, *Coccus viridis* (Puttarudirah and Bhatia, 1955); the hairy caterpillar,

Euproctis fraterna; cotton leaf armyworm, *Spodoptera litura*; and mango hopper, *Idiocerus* sp. (McIndoo, 1983).

86. *Aristolochia mexima* Linn. (Aristolochiaceae)

Common name: Contra capetano

This plant is a shrub found in tropical and sub-tropical regions and is commonly available in West Africa and South America. Its bark and seed aqueous extracts showed insecticidal activity against the cabbage worm, *Pieris rapae* (McIndoo, 1983).

87. *Aristolochia tagala* Cham. (Aristolochiaceae)

Common name: Malanbi

This is a perennial tropical shrub found in South India, Sri Lanka, Malaysia and South China. Its leaf extract in water showed antifeedant activity against the corn borer, *Ostrinia furnacalis* and the common cutworm, *Spodoptera litura* when sprayed on the host plant (Caasi, 1983).

88. *Artabotyrs* sp. (Annonaceae)

Common name: Hari-champa

This is a perennial shrub or small tree found in tropical zones and commonly available in Karnataka in India. Its seed powder and aqueous extract were reported to be toxic when tested against the pulse beetles, *Callosobruchus chinensis* and *C. maculatus* when tested under the control conditions of insect infestations (Anonymous, 1980–1981).

89. *Artemisia absinthium* Linn. (Asteraceae)

Common name: Wormwood or Vilayati afrantin

This plant is a perennial herb found in temperate, arid and Mediterranean regions. In India, it is available in Jammu and Kashmir. Its leaf extract in water showed insecticidal, antifeedant and repellent activities to the rice weevil, *Sitophilus oryzae* and the paddy moth, *Spoctoptera cerealella* (Jacobson, 1985) and also against the cotton leaf armyworm, *Spotoptera litura* (Waespe and Hans, 1983).

90. *Artemisia capillaris* (Linn.) Boiss. (Asteraceae)

This is an annual/perennial herb found in tropical and sub-tropical regions. The active components, viz., capillin, capillarin, methyl-eugenol, ar-curcumine and bornyl acetate, isolated from its leaf extract were reported to show antifeedant activity to the cabbage butterfly, *Pieris rapae* (Yano, 1987).

91. *Artemisia kurramensis* Linn. (Asteraceae)

This is an annual/perennial herb found in tropical and sub-tropical zones and commonly available in North India and Pakistan. Its seed oil showed insecticidal activity against rice pests, viz., *Sogatella longifurcifera*, *Sogata striatus*, *Perkinsiella insigns* and *Toya attenuata*, when tested alone as well as in combinations with organic synthetic insecticides (Khan and Khan, 1985).

92. *Artemisia saissanica* Linn. (Asteraceae)

Common name: Zaisan wormwood

This is an annual/perennial herb found in temperate zones. Chloroform extracts of its leaves, buds and budding twigs were reported to show repellency to the rice weevil, *Sitophilus oryzae* and the lesser grain borer, *Rhyzopertha dominica.* Active component isolated from these extracts was identified as alpha-santonin (Adekonov et al., 1990).

93. *Artemisia vulgaris* Auct. non Linn. (Asteraceae)

(Syn.-*A. nilagirica* [Clarke] Pamp, *A. valgaris nilagirica* O.B. Clarke)

Common name: Mugwort or Indian wormwood fleabane

This plant is a perennial aromatic herb found in arid zones. In India, it is commonly available in Himalayas, Sikkim, Khasi Hills, Darjeeling, Manipur, Western Ghats from Konkan southward. Leaf extract of this herb was reported to show repellent activity against stored grain pests, i.e., *S. cerealella* and *R. dominica* (Hameed, 1982). Morallo-Rajessus (1984) also found insecticidal property with leaf extract against the flour beetle, *T. castaneum* and Asiatic corn borer, *Ostrinia furnacalis* under controlled conditions testing.

94. *Arundo donax* Linn. (Poaceae)

(Syn.-*A. sativa* Lamk., *Donax aundinacedus* P. Beauv., *Cynodon donax* [Linn.] Raspail)

This plant is a perennial woody weed found in tropical and Mediterranean zones. This is a native of the Mediterranean region and commonly available in Jammu and Kashmir, Assam and Nilgiris. Insecticidal property in its aqueous leaf extract was reported against the desert locust, *Schistocerca gregaria* (Waespe and Hans, 1983).

95. *Asclepias curassavica* Linn. (Asclepiadaceae)

Common name: Milkweed

This is a perennial herb found in tropical and sub-tropical zones that originated in South Africa. It is commonly available in the West Indies and tropical

American continent. Its leaf extract in ethyl ether showed attractancy to the male and the female adults of melon fly, *Dacus cucurbitae* (Keiser et al., 1975).

96. *Asclepias incarnata* Linn. (Asclepiadaceae)

Common name: Swamp milkweed

This is a perennial herb found in temperate zones. Insecticidal activity in water extracts of its root, stem, branch, leaf, flower and also in their powders was reported against the vinegar fly, *Drosophila melanogaster* (Michael et al., 1985).

97. *Asclepias labiuformis* Linn. (Asclepiadaceae)

Common name: Milkweed

This is a perennial herb found in tropical and temperate zones. Its dried powders and extracts in alcohol or petroleum ether of the whole plant, branch and stem showed insecticidal and antifeedant activities against European corn borer, *Ostrinia nubilalis* (Jacobson, 1958).

98. *Asclepias speciosa* Linn. (Asclepiadaceae)

Common name: Showy milkweed

This is a perennial herb found in temperate zones. Insecticidal activity in its leaf, flower and dried seed powder and also its aqueous, alcohol and petroleum ether extracts were reported to show insecticidal activity against the vinegar fly, *Drosophila melanogaster* when tested under controlled laboratory conditions (Patterson et al., 1975).

99. *Asclepias syriaca* Linn. (Asclepiadaceae)

Common name: Common milkweed

This is a perennial herb found in temperate and semi-arid zones. Its root and branch extracts in water, ether, alcohol and petroleum ether were reported to show insecticidal activity against the aphids and the mealy bugs (Khan, 1987) and the vinegar fly, *Drosophila melanogaster* (Michael et al., 1985).

100. *Asimina tribola* (Linn.) Duin. (Anonaceae)

Common name: Pawpaw

This is a perennial tree found in tropical and sub-tropical zones and commonly available in South, East and North America. From its seed extract an alkaloid, 'anonanine', was isolated and found to show insecticidal activity against Colorado potato beetle, *Leptinotarsa decemlineata* (Crosby, 1971). Further, from its seed and bark ethanolic extracts an acetogenin compound, 'asimicin', was isolated and found to show toxicity to the melon aphid, *Aphis gossypii* when tested under controlled laboratory conditions (Alkofahi et al., 1989).

101. *Astragalus canadensis* Linn. (Fabaceae)

Common name: Canada milkvetch

This is a perennial herb found in sub-tropical and temperate zones. Its leaf and fruit aqueous extracts were reported to show insecticidal activity and also to inhibit insect growth when tested against the vinegar fly, *Drosophila melanogaster* (Michael et al., 1985).

102. *Astragalus pectinatus* Linn. (Fabaceae)

Common name: Milkvetch

This is a perennial herb found in temperate zones. Its root and fruit dried powders and aqueous extracts were reported to show insecticidal and growth inhibiting activities against the vinegar fly, *Drosophila melanogaster* (Michael et al., 1985).

103. *Atalantia monophylla* (Roxb.) D.C. (Rutaceae)

(Syn.-*Limonia monohylla* Roxb. non Linn.)

This plant is a perennial shrub or small tree found in tropical and temperate zones. It is a native of South East Asia and commonly available in South India. Its branch and leaf powders admixed with stored grains showed insecticidal and repellent activities against the paddy moth, *Sitotroga cerealella* and rice weevil, *Sitophilus oryzae* (Nagarajah, 1983).

104. *Atalantia recemosa* Weitht. & Arnott. (Rutaceae)

This is a perennial shrub or small tree found in tropical regions and commonly available in Mahabaleshwar and Bhimashankar areas in Maharastra (India). Antifeedant property in this plant was discovered by Banerji et al. (1987) against *Spodoptera litura*. Banerji (1988) and Luthria et al. (1989) also reported coumarins, viz., luvangetin, xanthyletin and recemosin, from the petroleum ether extracts of its leaf, bark and flower which showed antifeedancy to lepidopteran larvae under laboratory tests.

105. *Athrium pterorachis* Roth. (Polypodiaceae)

This is a fern found in tropical zones. Insecticidal activity in its branch and leaf aqueous extracts was reported against the vinegar fly, *Drosophila hydei* (Jacobson, 1958).

106. *Atractylis gummifera* Linn. (Asteraceae)

Common name: Add-add

This is an annual/perennial herb found in semi-arid and Mediterranean zones. Root extract of this plant in water showed toxicity to the tomato moth, *Hadena oleracea* and repellent activity to the mustard beetle, *Phaedon cochleariae* (Jacobson, 1958; Khan, 1982).

107. *Atractylis ovata* Linn. (Asteraceae)

This is annual/perennial herb found in temperate and Mediterranean zones. Its leaf and root extracts and also dried powders showed repellent activity to stored grain insect pests, viz., *Sitotroga cerealella*, *Rhyzopertha dominica*, and *Sitophilus oryzae*, when tested under controlled conditions (McIndoo, 1983).

108. *Atriplex lentiformis* Linn. (Chenopodiaceae)

This is an annual bushy herb found in sub-tropical and temperate zones and commonly available in desert tracts near salt marshes and on seacoasts. Petroleum ether extract of its leaf was reported to show grain protectant activity on cowpea seeds against the pulse beetle, *Callosobruchus chinensis* when tested under laboratory conditions (El-Ghar and El-Sheikh, 1987).

109. *Atropa belladonna* Linn. (Solanaceae)

Common name: Belladonna or Deadly night shade

This plant is a biennial herb found in tropical and temperate zones. It is a native of Europe but also available in Jammu and Kashmir in India. Its whole plant, root, leaf and seed extracts in water were reported to show insecticidal, antifeedant and repellent activities against Colorato potato beetle, *Leptinotarsa decemlineata* when tested under laboratory conditions (Jacobson, 1975).

110. *Aureolaria virginica* Rafib. (Crophylariaceae)

Common name: Downy false foxglove

This is an annual herb found in temperate zones. Its whole plant and leaf extracts in water were reported to show repellent activity against the Japanese beetle, *Popillia japonica* in a test under laboratory conditions (McIndoo, 1983).

111. *Azardirachta indica* A. Juss. (Meliaceae)

(Syn.-*Antelaea azardirachta.*)

Common name: Neem or Margosa tree

This is a perennial tree found in tropical, sub-tropical, semi-arid and arid zones. Neem tree is available throughout India and also distributed in Pakistan, the Philippines, Sri Lanka, Bangladesh and East Africa. Neem has been reported to possess almost legendary medicinal, insecticidal, insect repellent, antifeedant, growth regulator, nematicidal and antifungal properties and is a highly explored tree among the present flora of pesticidal importance (Lim and Dale, 1994). In the field of agriculture among all the plants exploited so far for their bioefficacy against pests, neem tree occupies foremost status. Its different parts have been used by agriculturists either for the purpose of pesticide or fertilizer and have been evaluated for possessing pesticidal properties against a wide range of agricultural pests under both storage and field ecosystems. The results of these evaluations/tests are mentioned below.

NEEM LEAVES

Fresh neem leaves and also their dried powder have been tested against storage and field insect pests and found to exhibit biological activities against a wide range of insect pests as described below.

Biological Activity of Neem Leaves Against Storage Insects

Dried neem leaves admixed with cacao beans protected the beans and showed repellent activity against storage insects (Fry and Sons, 1938). Krishnamurthi and Rao (1950) reported its dried leaf powder to protect jola seed from insect attack in storage and it was also found to show insecticidal activity against almond moth, *Ephestia cautella*. These leaves admixed with wheat grains or placed in 6–8 cm layers protected the grains from the infestation of *S. oryzae*, *S. cerealella* and *R. dominica* (Pruthy and Singh, 1950). Neem leaves admixed with wheat grains also protected them from the attack of the red flour beetle, *Tribolium castaneum* (Atwal and Sandhu, 1970). But Prakash et al. (1979b) found these leaves ineffective to protect paddy grains against *S. cerealella*, *R. dominica* and *S. oryzae* in coastal climatic conditions of Orissa. Dried neem leaves powder admixed with the grains was reported as a traditional practice by the farmers to protect their stored paddy (Golob and Webley, 1980). Jilani and Su (1983) and Jaipal et al. (1983) reported neem leaf extract as repellent and checked penetration of *R. dominica* and *S. cerealella* into wheat stored in gunny bags treated with the extract. Ahmed (1984), however, reported neem leaves to protect rice and wheat grains stored in gunny bags against grain boring insects. Muda (1984) reported traditional use of dried neem leaves admixed with stored paddy or a thick layer of 20–30 cm leaves between the bags and floor to protect stored paddy from the insects. Neem leaf aqueous extract reduced fecundity and fertility of the rice moth, *C. cephalonica* in treated wheat (Pandey et al., 1985b). Leaf extract of neem in benzene showed repellency to the pulse beetle, *Callosobruchus chinensis* and reduced its oviposition in treated greengram (Pandey et al., 1986). Similarly, neem leaf powder admixed with cowpea seed showed toxicity to the pulse beetle, *C. chinensis* in storage (Yadav and Bhatnagar, 1987). Neem leaf powder (2%) admixed with sorghum seed also gave absolute grain protection against the infestation of *R. dominica* for 6 months in storage (Ramesh Babu and Hussaini, 1990b). Kumar and Mehla (1993) found 4% neem leaves to protect stored milled rice against infestation of *S. cerealella* and *R. dominica* under controlled conditions. Dried neem leaf powder admixed with sorghum @ 5% w/w reduced the adult emergence of *C. cephalonica* by 52–56% and also showed 40–45% larval mortality (Srivastava and Bhatt, 1995).

Biological Activity of Neem Leaves Against Field Insects

Leaf extract of neem was reported to show insecticidal (contact poison) activity against the larvae of lucerne weevil, *Hypera postica* (Chopra, 1928). Holman (1940) found neem leaf extract spray to protect crops against the locusts' attack. *Lipaphis erysimi*, an aphid, could not attack the crop sprayed with alcoholic extract of neem leaf (Atwal and Pajni, 1964). Crude neem leaf extract also inhibited the development and caused morphological alterations in Mexican bean beetle, *Epilachna varivestis*; diamond back moth, *Plutella maculipennis*; and beet leaf bug, *Piesma quadratum* when fed through foliar spray and topical application. Neem leaf extract showed antifeedant activity to the locust (Steets, 1975). Similarly, neem leaf extract showed toxicity and also affected the moulting and vitellogenesis in 3rd, 4th and 5th instar nymphs of *Dysdercus cingulatus* (Abraham and Ambika, 1979). Chandramohan and Sivasubramanian (1987) found aqueous neem leaf extract to show protection against the groundnut leaf miner, *Aproaerema modicella* on the groundnut leaves.

Roonwal (1953) reported that the locusts did not feed on neem leaves. Similarly, the desert locust, *Schistocerca gregaria* avoided feeding on neem leaves. Neem leaf extract failed to protect germinating cowpea seeds in rabi crop against surface grasshopper, *Chrotogonus trachypterus* but was effective against the larvae of cabbage butterfly, *Pieris brassicae* (Sandhu and Singh, 1975b). Though earlier Gupta (1973) found neem leaves to attract adults of sugarcane root borers, *Holotrichia serrata*, *H. consanguinea* and *H. insularias* and facilitated their killing by hand when fresh neem twigs were planted before sun set in the infested sugarcane fields. Aqueous and acetone extracts showed antifeedant activity against desert locust, *S. gregaria* when sprayed on the cabbage (Narayan et al., 1980). Neem leaf extract was reported to suppress the development of the larvae and pupae of broad bean leaf miner, *Liriomyza trifolii* on treated onion crop (Fagoonee and Toory, 1984). Chari and Muralidhran (1985) found 10% neem leaf extract to significantly control the castor semi-looper, *Achoea janata* when sprayed on the crop. Singh and Sharma (1986) reported 1 and 2% neem leaf extract in water to show antifeedancy to the cabbage and cauliflower aphid, *Brevicoryne brassicae* in the field and in pot experiments. Venkatesan (1987) reported toxicity in its leaf extract when tested on cotton against *Aphis gossypii*. Pathak and Krishna (1987) reported neem leaf odour to decline the hatchability of the cotton bollworm, *Earias vittella*. Kareem et al. (1989a) found neem leaf bitters (NLB) to significantly reduce the oviposition and development of the green leafhopper, *Nephotettix virescens* and the brown planthopper, *Nilaparvata lugens* when sprayed on rice seedlings and also applied by seedling root dip method. Koshiya and Ghelani (1990) found 5% neem leaf extract to show strong antifeedancy to the cotton leaf armyworm, *Spodoptera litura* larvae when sprayed on groundnut leaves. Reddy et al. (1990) and Chitra et al. (1991a) also found neem leaf extract (1.0%) spray on potted brinjal to reduce population level of the grubs of spotted leaf beetle, *Henosepilachna vigintioctopunctata* up to 95%. However, 5% neem leaf extract was found to be a promising crop protectant against infestation of Gujrat hairy caterpillar, *Amsacta moorei* and showed larvicidal activity when tested under field conditions (Patel et al., 1990a,b). Koshiya and Chaiya (1991) also reported 1.5% neem leaf extract to cause 100 per cent ovipositional deterrency to the adults of *S. litura* in a laboratory test. Sachan and Katti (1993) found 5% neem leaf extract significantly controlled *Helicoverpa armigera* on the treated chickpea plants. Bajpai and Sharma (1993) also reported neem leaf extract to reduce the oviposition of *Chilo partellus* by 73% over control when females were fed on the treated leaves of the host plants under net-house conditions.

NEEM SEED/KERNEL EXTRACT

Among the different plant parts of the neem its seeds/kernel extracts have been exploited to a great extent to find biological activity of various nature against a wide range of insect pests of agricultural importance. Neem seed/kernel crude extracts in different solvents like water, alcohol, methalnol, ethyl alcohol, petroleum, ether, acetone, benezene, ethyl ether and chloroform and also their dilutions have been reported to show antifeedant, repellent, insecticidal, growth inhibiting, juvenile hormone mimics, oviposition inhibiting, crop protectant and grain protectant activities against the insect pests in storage and in field agro-ecosystems as mentioned in Tables 1 and 2 respectively.

NEEM OIL

Neem oil extracted from its seed/kernel has also been evaluated against a wide range of insect species of agricultural importance in the field as well as in storage ecosystems. Neem oil has been reported to show antifeedant, repellent, insecticidal, grain protectant and growth inhibiting activities against these insects. For the last two decades neem oil has been exploited to a great extent especially against the insect pest infesting rice crop; against the brown planthopper, *Nilaparvata lugens*; rice green leaf hopper, *Nephotettix virescens*; rice leaffolder, *Cnaphlocrosis medinalis*; and whitebacked plant hopper, *Sogatella furcifera* under natural and controlled conditions of their infestations.

Table 1. Effectiveness of neem seed/kernel extract against storage insect pests

Biological activity	Insect	Results of the test	Reference
Repellent	*Sitophilus oryae, S. granarius*	Protected wheat grains treated with its aqueous extract.	Quadri, 1973
Grain protectant	*Callosobruchus chinensis*	Kernel extract admixed with the redgram seeds protected them from this beetle.	Balasubramaniam, 1977
Antifeedant	*Sitotroga cerealella, Rhyzopertha dominica, S. oryzae*	Protected stored paddy admixed with its aqueous extract.	Prakash et al., 1978–1987
Antifeedant	*R. dominica, S. oryzae*	Protected treated paddy grains in a laboratory test.	Prakash et al., 1980
Juvenile hormone mimics	*Ephestia kuehniella*	Caused mortality and affected the larval and pupal development when topically applied to 3rd instar larvae.	Sharma et al., 1990
Grain protectant	*R. dominica, S. cerealella*	Neem extract @ 0.5% significantly protected stored rice against these insects.	Devi and Mohandas, 1982
Insecticidal	*S. oryzae*	Neem seed extract showed toxicity to the weevil in stored maize.	Ivbijaro, 1983b
Grain protectant	*C. chinensis*	Kernel powder admixed with pigeonpea seeds protected them from beetle attack.	Radhakrishnan et al., 1984

Table 1. Effectiveness of neem seed/kernel extract against storage insect pests (Continued)

Biological activity	Insect	Results of the test	Reference
Grain protectant	*General storage insects*	Bags treated with 2% extract protected stored paddy.	Muda, 1984
Antifeedant	*S. oryzae*	Extract protected wheat grains against this weevil in a laboratory test.	Rout, 1986
Ovipositional deterrent	*Corcyra cephalonica*	Reduced its fecundity and fertility in wheat admixed with neem extract.	Pandey et al., 1985b
Antifeedant	*C. cephalonica*	Neem kernel suspension admixed with broken rice showed feeding deterrency to this moth.	Devraj Urs and Srilatha, 1990

But because of its oily nature having large droplet size, non-availability of suitable spraying equipment at farmers' levels and phyto toxicity at higher doses it did not show encouraging results for its commercial applications. However, research work carried out on the evaluation of biological activities of neem oil is mentioned below.

Biological Activity of Neem Oil Against Storage Insects

Neem oil for its bio-efficacy as a grain protectant against stored grain insect pests evaluated under both controlled and natural conditions of insect infestation by several workers is presented in Table 3.

Biological Activity of Neem Oil Against Field Insect Pests

Similarly, neem oil has been evaluated against a wide range of crop field insect pests either by spraying it on the seed or by using seedling and crop

Table 2. Effectiveness of neem seed/kernel extract against insect pests in field conditions

Biological activity	Insect	Results of the test	Reference
Antifeedant	*Locusta migratoria*	Neem seed extract showed feeding deterrency.	Pradhan et al., 1963
Repellent	*Euproctis lunata*	Protected castor leaves sprayed with its 0.5% aqueous extract.	Mane, 1968
Antifeedant	*Locusta migratoria*	Protected maize, cabbage and sorghum plants sprayed with its extract.	Pradhan and Jotwani, 1968a,b
Antifeedant	*Amsacta moorei*	Extract sprayed on sunn-hemp protected it from insect attack.	Patel et al., 1968
Repellent	*Euproctis lunata*	Aqueous, alcohol and acetone extracts sprayed on castor plant repelled this insect.	Babu and Beri, 1969
Antifeedant	*Rhopalosiphum nymphae, S. gregaria*	Insects did not feed on the plants treated with alcoholic extract of neem seed.	Goyal, 1971
Ecdysone analogue	*Antestiopsis orbitalis*	When the bug was fed on the treated coffee leaves with the extract, it showed morphological changes during its development.	Leuschner, 1972
Antifeedant	Aphids and jassids	Seed suspension 1–2% sprayed on brinjal crop protected it from insect attack.	Rajanasari and Nair, 1972

Table 2. Effectiveness of neem seed/kernel extract against insect pests in field conditions (Continued)

Biological activity	Insect	Results of the test	Reference
Antifeedant	*Spodoptera litura*	Tobacco seedlings treated by 1–5% extract suspensions were not attacked by this insect.	Anonymous, 1972
Antifeedant	*Chrotogonus trachypterus*	Treated with 2% suspension did not allow the insect to feed on germinating crops.	Sandhu and Singh, 1975
Antifeedant	*Leptinotarsa decemlineata*	Larvae did not feed on the potato leaves sprayed with the extract.	Steets, 1976
Antifeedant	*Euproctis fraterna, Nephantis serinopa*	Castor and coconut leaves treated with the extract were not attacked by the respective insects' larvae.	Kareem et al., 1976
Antifeedant	*Boarmia selenaria*	Caterpillars did not feed on host plant sprayed with extract.	Meisner et al., 1976
Antifeedant	*Locusta migratoria, Schistocerca gregaria*	Kernel extract showed feeding deterrency to the locust.	Tilak, 1977
Insecticidal, ovicidal	*Schistocerca gregaria*	Kernel suspension showed toxicity and reduced hatchability of the eggs.	Singh and Singh, 1987a,b
Antifeedant	*Spodoptera frugiperda*	Host plant sprayed with the extract was not fed on by the caterpillars.	Warthen et al., 1978
Insecticidal	*Spodoptera litura*	Kernel suspension sprayed on tobacco protected the crop.	Joshi et al., 1978

Table 2. Effectiveness of neem seed/kernel extract against insect pests in field conditions (Continued)

Biological activity	Insect	Results of the test	Reference
Antifeedant	*Popillia japonica*	Beetle did not feed on host plant sprayed with the extract.	Ladd et al., 1978
Antifeedant	*Madurasia obscurella*	Mung bean sprayed with neem kernel suspension avoided feeding of the pest.	Yadav et al., 1979
Insecticidal and juvenile hormone mimic activity	*Dysdercus cingulatus*	Bug that fed on the treated leaves of cotton with extract showed mortality and morphogenetic changes during its development.	Abraham and Ambika, 1979
Repellent	*Spodoptera litura*	Neem extract showed ovipositional deterrency to the insect.	Joshi and Sitaramaiah, 1979
Growth disrupter	*Epilachna varivestis, Apis millifera*	Insects fed with extract treated diets showed disrupted growth during their development.	Rembold et al., 1980b
Juvenile hormone mimetic activity and toxicity	*Apis millifera*	Extract caused mortality and affected larval and pupal ecdysis when applied topically to the last larval stage.	Sharma et al., 1980
Histopatho-logical alternator	*Epilachna varivestis*	When extract incorporated with its food was fed to the insect it showed histo-pathological changes in ovary.	Schulz, 1981
Juvenile hormone mimic activity	*Epilachna varivestis*	Produced abnormal adults and caused mortality to the larvae when fed through its diet.	Ascher and Sell, 1981

Table 2. Effectiveness of neem seed/kernel extract against insect pests in field conditions (Continued)

Biological activity	Insect	Results of the test	Reference
Antifeedant	*Leptinotarsa decemlineata*	Neem extract sprayed on the potato crop restricted feeding of the larvae and adult beetle.	Britski, 1982
Antifeedant	*Phyllotreta striolata*	Aqueous extract (0.4%) showed feeding deterrency to the beetle on the radish plant in a net-house test.	Meisner and Mitchell, 1982
Growth inhibitor	*E. varivestis, Pieris brassicae, Leptinotarsa decemlineata*	Extract 100 mg/kg caused mortality and inhibited their growth.	Feuerhake and Schmutterer, 1982
Juvenile hormone mimetic activity	*Ceratitis capitata*	Affected the metamorphosis and development when the fly was fed on diet incorporated with its ethanolic extract.	Steffens and Schmutterer, 1982
Ovipositional inhibitor	*Spodoptera littoralis*	Eggs treated with its 2% kernel suspension reduced 50% egg laying.	El-Sayed, 1982–1983b
Insecticidal	*Spodoptera littoralis*	Neem kernel suspension caused 100% mortality to 2nd and 3rd instar larvae when fed on treated leaves using 0.2 to 0.5% suspension.	El-Sayed, 1982–1983a

Table 2. Effectiveness of neem seed/kernel extract against insect pests in field conditions (Continued)

Biological activity	Insect	Results of the test	Reference
Ovipositional inhibitor and insecticidal	*Liriomyza sativae, L. trifolii*	In greenhouse, extract caused larval mortality in both species and inhibited oviposition of *L. trifolii* when sprayed on Henderson bush lima bean.	Webb et al., 1983
Insecticidal	*Melanagromyza obtusa, Heliothis armigera, Exelastis atomosa*	Protected arhar crop form these pod borers when sprayed with the extract.	Thakre et al., 1983
Toxicity and juvenile hormone mimic activity	*Crocidolomia* spp.	Caused mortality to its larvae and also affected moulting to cause poor adult emergence and abnormal pupae.	Fagoone, 1983
Juvenile hormone mimic activity	*Mythimna separata, Cnaphlocrocis medinalis*	Methanolic extract fractions showed developmental abnormalities and mortality to the larvae.	Schmutterer et al., 1983
Antifeedant	*Locusta* sp., *Stonoxys* sp.	Host plant sprayed with the extract was not attacked by these insects.	Omollo, 1983
Growth inhibitor	*Dysdercus koenigii*	Crude extract inhibited the development and growth of the 4th instar larvae.	Jaipal et al., 1983

Table 2. Effectiveness of neem seed/kernel extract against insect pests in field conditions (Continued)

Biological activity	Insect	Results of the test	Reference
Antifeedant and growth inhibitor	*Spodoptera littoralis*	Larvae did not feed on host leaves treated with methanolic extract but affected its development.	Meisner et al., 1983
Growth and oviposition inhibitor	*Schistocerca gregaria*	Extract inhibited the hatching of the eggs.	Bhanotar and Srivastava, 1984
Antifeedant	*Melanagromyza obtusa*	Ethanolic extract sprayed on pigeonpea crop controlled the fly.	Singh et al., 1985
Antifeedant	*Spodoptera litura*	Kernel suspension (0.5%) sprayed on the tobacco crops protected it from this insect.	Joshi et al., 1984
Insecticidal	*Acyrthosiphum pisum, Aphis fabae*	Seed extract was toxic to the aphids when sprayed on the host plants.	Schauer, 1984
Antifeedant	*Nephantis serinopa*	Seed extract showed antifeedancy to the insect on the coconut seedlings.	Kareem, 1984
Synergist and toxicity	*Plutella xylostella*	Seed extract showed synergistic effect with piperonyl butoxide when 4th instar larvae were fed on the treated leaf disc.	Lange and Feuerhake, 1984
Insecticidal	*Heliothis armigera*	Kernel extract was toxic to control this borer on the Bengal gram and red gram.	Srivastava et al., 1984; Kumar and Sangappa, 1984

Table 2. Effectiveness of neem seed/kernel extract against insect pests in field conditions (Continued)

Biological activity	Insect	Results of the test	Reference
Antifeedant	*S. littoralis*	Extract showed antifeedancy to the larvae in a laboratory test.	Rajasekaran and Kumarswami, 1985b,c
Antifeedant	*Achoea janata*	Kernel suspension (2%) sprayed on the castor controlled the pest effectively.	Chari and Muralidharan, 1985
Insecticidal	*Heliothis armigera*	Ethanol extract protected chickpea crop against this borer.	Singh et al., 1985
Insecticidal	*Maruca testulatis*	Kernel extract protected the pigeonpea crop from this insect.	Singh et al., 1985
Repellent, toxicity	*Schistocerca gergaria*	Kernel extract (0.4% crude) showed toxicity and also reduced its oviposition when fed through the diet.	Singh, 1985
Antifeedant	*Achoea janata*	2% kernel extract showed phago-deterrency to this insect.	Chari and Muralidhran, 1985
Antifeedant	*Pieris brassicae*	Seed extract showed antifeedancy to 5th instar larvae when fed through its diet.	Zhang and Chiu, 1985
Antifeedant	*Pelopedas maithias, Henosepilachna vigintioctopunctata, Pericalia ricini, Psalis panulata*	Seed extract sprayed on the brinjal protected the crop from these insects.	Dhandapani et al., 1985

Table 2. Effectiveness of neem seed/kernel extract against insect pests in field conditions (Continued)

Biological activity	Insect	Results of the test	Reference
Insecticidal	*Liriomyza trifolii*	Extract applied as soil drench to chrysanthemum crop caused mortality to its last larval instars and pupae.	Hiran et al., 1985
Insecticidal	*Spodoptera litura*	Kernel suspension sprayed on tobacco seedlings showed toxicity to the test insect.	Chari et al., 1985
Growth inhibitor	*Nephotettix virescens*	Paddy seed treatment with extract (10–100 mg/kg) reduced the growth and development of the insect.	Heyde et al., 1985
Insecticidal	*Heliothis armigera*	Kernel suspension (5%) was toxic to the larvae and pupae of the borer when sprayed on the host plant.	Rao and Srivastava, 1985; Gohokar et al., 1985
Insecticidal	*Lapaphis erysimi*	Seed extract was toxic to mustard aphid when sprayed on the plant.	Sharma et al., 1986
Insecticidal	*Spodptera litura*	Kernel suspension 0.5% sprayed on the tobacco plant and 2.0% in its nursery effectively protected the crop.	Ram Prasad et al., 1986
Growth inhibitor	*Bombyx mori*	Doses of extract @ 2 and 4 μg/larvae greatly affected its development when applied topically.	Srinivas, 1986

Table 2. Effectiveness of neem seed/kernel extract against insect pests in field conditions (Continued)

Biological activity	Insect	Results of the test	Reference
Insecticidal	*Heliothis armigera*	Kernel extract (5%) protected chickpea from this insect in mini-plot experiments under field conditions.	Siddappaji et al., 1986
Antifeedant and repellent	*Spodoptera litura*	Kernel suspension 2–3% showed antifeedancy and repellency to this insect when sprayed on tobacco nursery.	Joshi, 1986; Joshi et al., 1984
Insecticidal	*Oreseolia oryzae, Nephotettix* spp., *Cnaphalocrosis medinalis*	Application of neem coated urea in soil reduced the incidences of the insects.	David, 1986
Antifeedant	*Bervicoryne brassicae*	Kernel extract at 0.2 and 0.4% sprayed on cabbage and cauliflower crops protected the crops from aphid attack.	Singh and Sharma, 1986
Synergist	*S. gregaria*	Seed extract enhanced the bio-efficacy of fenitrothion and dichlorvos in combined sprays to control the locust.	Singh and Singh, 1987c
Insecticidal	*Calocoris angustatus*	Seed extract 2% effectively reduced its population.	Sharma and Leuschner, 1987
Insecticidal	*Bemisia tabaci*	Kernel extract 2% sprayed on blackgram reduced the transmission of yellow mosaic virus (YMV) and its vector whitefly.	Mariappan et al., 1987

Table 2. Effectiveness of neem seed/kernel extract against insect pests in field conditions (Continued)

Biological activity	Insect	Results of the test	Reference
Insecticidal	*Lipaphis erysimi*	Reduced the population of the aphid, a vector for yellow mosaic virus of black gram when kernel suspension was sprayed.	Mariappan et al., 1987
Repellent	*Nephotettix virescens*	5% aqueous kernel extract reduced oviposition when used as root-dip of rice seedling for 24 hrs.	Kareem et al., 1987a
Antifeedant	*Nilaparvata lugens*	Pre-sowing paddy seed treatment with 10% neem kernel extract reduced insect population and gave better seedling growth.	Kareem et al., 1987b
Insecticidal, oviposition inhibitor, repellent	*Schistocerca gregaria*	Kernel suspension inhibited oviposition and showed insecticidal activity when tested alone and also in combinations with insecticides.	Brajendra 1987, Brajendra and Singh, 1987a,b, 1988b
Growth and development inhibitor	*Sogatella furcifera, Nilaparvata lugens, Nephotettix virescens*	Root soaking of rice seedling in neem kernel extract or its sprays on the plant or soil incorporation inhibited the growth and development of the pests.	Saxena et al., 1987
Insecticidal	*Holotrichia* sp.	Kernel extract application in rice reduced incidences of the insects.	Rajamani et al., 1987

Table 2. Effectiveness of neem seed/kernel extract against insect pests in field conditions (Continued)

Biological activity	Insect	Results of the test	Reference
Antifeedant	*Nilaparvata lugens, Cnaphalocrosis medinalis*	Kernel extract checked the insect incidence in rice pots sprayed by ULV sprayer.	Rajasekaran et al., 1987a
Crop protectant	*Aulacophora foveicollis*	Seed extract in methanol protected the pumpkin crop from its attack.	Gujar and Mehrotra, 1988
Antifeedant	*Heliothis armigera*	Aqueous seed extract (5%) spray reduced the borer damage in the black gram.	Rajasekaran et al., 1987b
Antifeedant, oviposition inhibitor	*Spodoptera litura*	Effectively controlled *S. litura* in tobacco nurseries when 2% aqueous neem kernel suspension was spread 3 to 5 times at 10-day intervals.	Joshi and Ram Prasad, 1975; Joshi et al., 1978; Sitaramaiah et al., 1979; Ram Prasad et al., 1987
Oviposition inhibitor	*Phthorimaea operculella*	5% formulation of seed extract reduced oviposition of the moth.	Shelke et al., 1987
Growth inhibitor	*Nilaparvata lugens, Nephotettix virescens*	Reduced nymphal development and protected young rice seedlings when seeds were soaked in the extract.	Kareem et al., 1988a,b
Antifeedant, insecticidal	*Lipaphis erysimi*	Kernel extract sprayed on mustard protected the crop from this aphid.	Singh et al., 1988
Antifeedant	*Nephotettix virescens*	Seed extract sprayed on rice reduced the survival of the hoppers.	Narsimhan and Mariappan, 1988

Table 2. Effectiveness of neem seed/kernel extract against insect pests in field conditions (Continued)

Biological activity	Insect	Results of the test	Reference
Insecticidal	*Sylepta* sp., *Maruca* sp., *Heliothis* sp.	Seed bitters and kernel extracts both reduced the leaf and pod damage in mungbean due to these pests.	Kareem et al., 1988a,b
Oviposition inhibitor, growth inhibitor	*Nilaparvata lugens, Nephotettix virescens*	Seed bitters 0.25% reduced egg laying, development and adult emergence.	Kareem et al., 1989a
Insecticidal	*Nephotettix virescens*	Kernel extract admixed with carbofuran reduced rice tungro virus infection and development of the test insect.	Kareem et al., 1989b
Growth inhibitor	*Nephotettix virescens, Nilparavata lugens*	Kernel extract 2% checked adult emergence of these insects when rice seedling germinated from the treated seeds.	Kareem et al., 1989c
Antifeedant	*Apogonia blanchardi*	Aqueous and ethanolic extracts of seed (1 and 2.5%) checked feeding of its adult beetles on treated rose twigs.	Doharey and Singh, 1989
Repellent	*Nephotettix virescens*	Neem extract was more effective repellent than neem oil.	Songkittisuntron, 1989

Table 2. Effectiveness of neem seed/kernel extract against insect pests in field conditions (Continued)

Biological activity	Insect	Results of the test	Reference
Antifeedant	*Cnaphlocrosis medinalis*	Kernel extract (5%) and 0.16% teepol spray reduced its populations significantly.	Mohan and Gopalan, 1990
Insecticidal	*Helicoverpa armigera*	Kernel suspension (3%) sprayed on gram reduced its larval population at par with 0.05% malathion.	Butani and Mittal, 1990
Insecticidal	*Helicoverpa armigera*	5% kernel extract was as toxic as 0.07% endosulfan when tested by spraying on chickpea and pigeonpea.	Sachan and Lal, 1990
Insecticidal	*Spodoptera litura*	2% kernel extract sprayed on tobacco in nursery controlled the pest.	Joshi et al., 1990
Insecticidal	*Amsacta moorei*	5% seed extract sprayed on the crop controlled the caterpillar.	Patel et al., 1990b
Insecticidal	*Phyllocnistis citrella*	2% seed extract protected citrus crop.	Dharajothi et al., 1990a
Crop protectant	*Cnaphlocrosis medinalis, Hieroglyphus banian*	5% kernel extract showed superior rice plant protection when applied as high and low volume applications as compared to ULV application.	Dhaliwal et al., 1990a

Table 2. Effectiveness of neem seed/kernel extract against insect pests in field conditions (Continued)

Biological activity	Insect	Results of the test	Reference
Antifeedant	*Schistocerca gregaria*	Seed kernel extract in water at 0.003% cons. showed complete feeding deterrency.	Mishra and Singh, 1992
Growth inhibitor	*Schistocerca gregaria*	Aqueous kernel suspension at 0.5 and 1.0% admixed with sand disrupted embryonic development and also reduced survival of its hatchlings.	Singh and Singh, 1992
Antifeedant	*Leptinotarsa decemlineata*	Kernel extract reduced feeding rate of the insect.	Kaethner, 1992
Crop protectant	*Helicoverpa armigera*	5% kernel water extract protected the chickpea crop and yielded at par with neem oil treatment.	Sinha, 1993
Crop protectant	*Megalurothrips sjostedti*	Seed extract at 3% cons. reduced infestation of the thrips on cowpea when applied as foliar sprays twice using high volume application.	Saxena, 1993
Insecticidal	*Helicoverpa armigera*	Seed kernel extract (1%) caused 90–95% larval mortality.	Natrajan and Shanthi, 1994
Antifeedant	*Patialus techomella*	Kernel extracts at 0.5% cons. showed antifeedancy to 1st, 2nd and 3rd instar larvae.	Sundaraj, 1994

Table 2. Effectiveness of neem seed/kernel extract against insect pests in field conditions (Continued)

Biological activity	Insect	Results of the test	Reference
Crop protectant	*Hydrellia philippina*	Seed kernel extract (5%) reduced insect incidences when sprayed 21 days after rice transplanting.	Bhatia et al., 1994
Insecticidal	*Scirpophaga incertulas*	Kernal extracts (5%) significantly reduced the white ear heads.	Dash et al., 1995

treatments. The research work on the bio-efficacy of neem oil against the field insect pests is presented in Table 4.

NEEM KERNEL POWDER

Neem kernel powder has been intensively evaluated as grain protectant against stored grain insect pests of cereals, pulses and oil seeds by admixing it with the commodity. This powder was reported to possess antifeedant, repellent, ovipositional inhibitor and insecticidal properties and ultimately it protected the grains from infestation of insects in storage. The results of evaluation of neem kernel powder against storage insects under both natural and control conditions of their infestation are presented in Table 5.

However, against field insect pests, neem kernel powder has been evaluated either by spraying its suspension on the crop or by incorporating this powder with the soil. In the soil, neem kernel powder has been reported to kill soil fauna of insect pests of agricultural importance and also in a few cases it has shown systemic activity and reduced insect pest populations in the crop after sowing/transplanting. The results of evaluation of neem kernel powder against the field insect pests are presented in Table 6.

NEEM CAKE

De-oiled cake of neem kernel or seed has also been tested against a variety of insects in storage as well as in field agro-ecosystems. Neem cake has also been found to possess antifeedant, repellent, insecticidal and oviposition inhibitor properties against insect pests of agricultural importance. Research work carried out on exploitation of neem cake as a botanical pesticide against storage insects is presented in Table 7.

Table 3. Effectiveness of neem oil against insect pests of stored grains

Biological activity	Insect	Results of the test	Reference
Antifeedant	*Callosobruchus chinensis*	Bengal gram treated with 2% neem oil was not infested by the beetle for 6 months.	Anonymous, 1973–1974
Antifeedant	*Tribolium confusum*	Wheat grains treated with 2% oil were free from insect infestation.	Anonymous, 1973–1974
Repellent	*Rhyzopertha dominica*	Neem oil impregnated on gunny bags at the rate of 1100 mg/m^2 repelled this grain borer.	Savitri, 1975
Grain protectant	*Callosobruchus* spp.	Neem oil (1%) treatment to black gram protected the grains up to 10 months.	Jayraj, 1975
Grain protectant	*Bruchus* spp.	Neem oil (1%) treatment of pulse grains gave negligible grain protection against these beetles.	Talati, 1976
Grain protectant	*Sitotroga cerealella, Rhyzopertha dominica, Callosobruchus* spp.	Neem oil seed treatment of jowar, wheat, maize, peas, Bengal gram gave promising grain protection against these insects.	Ketkar et al., 1976
Grain protectant	*C. chinensis*	Protected redgram grain in storage against this beetle.	Sangappa, 1977
Grain protectant	*R. dominica, S. cerealella*	Population of these insects was checked in neem oil treated paddy.	Prakash et al., 1980

Table 3. Effectiveness of neem oil against insect pests of stored grains (Continued)

Biological activity	Insect	Results of the test	Reference
Grain protectant	*Callosobruchus maculatus*	Cowpea grains treated with neem oil @ 8 ml/kg were protected against the attack of this beetle.	Pereira, 1983
Grain protectant	*C. chinensis*	Caused 100% mortality to the adults and grubs and protected the greengram.	Ali et al., 1983
Repellent	*S. cerealella*	Neem oil treatment of grains inhibited the oviposition and reduced population build-up of the moth.	Verma et al., 1983
Repellent	*C. chinensis*	1% w/w neem oil admixed with cowpea seeds showed repellency to this beetle and also reduced its oviposition and adult emergence.	Naik and Dumbre, 1984
Ovipositional deterrent	*Corcyra cephalonica*	Neem oil admixed with wheat grains reduced oviposition and hatching of its eggs.	Pandey et al., 1985b
Reproduction inhibitor	*C. cephalonica*	Neem oil checked reproduction of this moth in the treated milled rice.	Pathak and Krishna, 1985
Grain protectant	*S. cerealella, R. dominica, S. oryzae*	Paddy grains treated with neem oil @ 1% w/w were not damaged by these insects in farm godowns.	Prakash et al., 1978–1987

Table 3. Effectiveness of neem oil against insect pests of stored grains (Continued)

Biological activity	Insect	Results of the test	Reference
Grain protectant	*C. maculatus*	Neem oil admixed with cowpea grains protected them against attack of the beetle.	Naik and Dumbre, 1985
Insecticidal	*C. chinensis*	Neem oil application to greengram @ 0.25% w/w checked insect multiplication and protected grains in storage.	Sujatha and Punnaiah, 1985
Ovicidal	*C. chinensis*, *C. maculatus*, *C. analis*	Neem oil showed ovicidal activity against these pulse beetles.	Yadav, 1985
Growth and development inhibitor	*C. cephalonica*	Neem oil was tested to be a potential inhibitor of its development and population growth.	Prathak and Krishna, 1985
Repellent	*S. cerealella*	Neem oil treatment in paddy seed inhibited the oviposition of the moth in a laboratory test.	Prakash et al., 1986
Oviposition inhibitor and grain protectant	*C. chinensis*	Neem oil treatment of the chickpea seeds @ 4 mg/100 grams showed inhibition in the egg laying of the test beetle and resulted in 100% grain protection against the beetle up to 5 months.	Das, 1986; Das and Karim, 1986

Table 3. Effectiveness of neem oil against insect pests of stored grains (Continued)

Biological activity	Insect	Results of the test	Reference
Repellent	*T. castaneum*	Neem oil (500 and 1000 ppm) repelled the beetle in treated milled rice.	Jilani et al., 1987
Repellent	*C. chinensis*	Pre-storage seed treatment with neem oil @ 10 ml/kg reduced its oviposition.	Babu et al., 1989
Repellent	*C. chinensis*	Neem oil showed highest olfactory repellency in reducing the oviposition.	Khaire et al., 1989
Grain protectant	*S. cerealella, R. dominica*	Neem oil treatment to paddy grains reduced their populations and protected the grains in storage.	Prakash et al., 1989
Grain protectant	*S. oryzae*	Neem oil (1%) admixed with rice protected the grains from infestation of this weevil.	Prakash et al., 1990c
Grain protectant	*C. chinensis*	Neem oil (1%) admixed with the peas' grains protected them in storage.	Kumari et al., 1990
Grain protectant	*R. dominica*	Neem oil (1%) admixed with stored paddy showed highest grain protection.	Prakash et al., 1991
Grain protectant	*C. maculatus, C. phaseoli, Zabrotes subfasciatus*	Arrested the adult emergence and also showed ovicidal activity.	Yadav, 1993

Table 4. Effectiveness of neem oil against field insect pests

Biological activity	Insect	Results of the test	Reference
Insecticidal	*Aphis gossypii, Urentius hystricellus*	Neem oil sprayed on the brinjal crop killed these pests.	Cherian and Gopala, 1944
Antifeedant, growth inhibitor	*Pieris rapae*	Neem oil showed antifeedancy and growth inhibitory activity to the larvae of this worm.	Atwal and Pajni, 1964
Insecticidal	*Atharigoni-avoria soccata*	Oil sprayed on the sorghum was toxic to the fly.	Anonymous, 1973–1974
Repellent	*Plutella xylostella*	Neem oil sprayed on okra crop showed repellency to this pest.	Anonymous, 1974
Repellent	*Earias vittella*	Neem oil sprayed on the cotton plant @ 10 1/ha reduced its incidences.	Thangavel et al., 1975
Antifeedant	*Hydrellia phillipina*	Neem oil sprayed on the baddy crop significantly protected the crop against this pest.	Murthy, 1975
Repellent	*Plutella xylostella*	Neem oil sprayed on grapes reduced incidences and repelled this moth.	Kalam, 1976
Antifeedant	*Schistocerca gregaria*	Neem oil sprayed on the cabbage crop showed feeding deterrency.	Attri and Singh, 1977b
Repellent	*Nilaparvata lugens*	Neem oil showed strong repellency to the hopper and affected its fecundity.	Balasubrahmanian, 1979; Saxena et al., 1979

Table 4. Effectiveness of neem oil against field insect pests (Continued)

Biological activity	Insect	Results of the test	Reference
Antifeedant	*N. lugens*	Neem oil sprayed on rice crop reduced its incidences and ragged stunt virus transmission through this vector.	Saxena et al., 1981a
Antifeedant	*Cnaphalocrosis medinalis*	Neem oil sprayed on rice crop reduced insect incidence.	Saxena et al., 1981b
Antifeedant	*Nephotettix virescens*	Neem oil 10% spray reduced its life span.	Mariappan et al., 1982a
Insecticidal	*Nephotettix virescens*	Neem oil sprayed on the rice plant reduced incidences of RTV transmitted by the insect and lowered hopper population.	Mariappan and Saxena, 1983
Antifeedant	*Nilaparvata lugens*	Neem oil reduced feeding of 2nd and 3rd instar nymphs.	Chiu et al., 1983
Insecticidal (Systemic)	*Maliarpha seperatella, Sesamia calamistis, Diapsis macro-phthalma, D. apicalis*	Neem oil sprayed at early rice vegetative stages protected the crop from the attack of these stem borers.	Ho and Kibuca, 1983
Ovipositional deterrent	*Dacus cucurbitae*	Neem oil sprayed on the pumpkin inhibited egg laying of the fly.	Singh and Srivastava, 1983
Antifeedant, growth inhibitor	*S. furcifera, N. lugens, N. virescens*	Neem oil sprayed on the rice plant showed antifeedant activity and inhibited growth and development of these insects.	Saxena et al., 1984a

Table 4. Effectiveness of neem oil against field insect pests (Continued)

Biological activity	Insect	Results of the test	Reference
Insecticidal	*Cyrtorhinus lividepennis*	Neem oil sprayed on the rice plant caused toxicity to the bug.	Saxena et al., 1984b
Antifeedant	*Nephotettix virescens*	Neem oil solution (5%) sprayed on the rice plant showed antifeedancy to the insect.	Heyde et al., 1984
Insecticidal	*Heliothis armigera*	Neem oil was toxic to this insect and protected Bengal gram crop in the field test.	Kumar and Sangappa, 1984
Insecticidal	*Hydrellia philippina, Nephotettix virescens, Nilaparvata lugens*	Neem oil coated urea application in the rice soil reduced incidences of the pest.	Krishnaiah and Kalode, 1984
Insecticidal	*Heliothis armigera*	Neem oil protected the sorghum and Bengal gram from this borer when sprayed in field conditions.	Rao et al., 1984; Babu and Rajesekaran, 1984
Antifeedant	*Amsacta moorei*	Neem oil sprayed on the groundnut crop showed antifeedancy to this insect.	Verma and Singh, 1985
Antifeedant, growth inhibitor	*Leptocorisa oritorius*	Deformed the nymphs and the adults of the bug when fed on neem oil treated host/food material.	Saxena et al., 1985

Table 4. Effectiveness of neem oil against field insect pests (Continued)

Biological activity	Insect	Results of the test	Reference
Antifeedant, insecticidal	*Nilaparvata lugens*	Rice plant sprayed with 5% neem oil suspension reduced the survival of hopper and incidences of the ragged stunt virus.	Saxena and Khan, 1985a
Antifeedant	*Nephotettix virescens*	Neem oil odour disturbed its feeding on the rice plants kept in an arena permeated with odour and reduced its longevity.	Saxena and Khan, 1985
Insecticidal	*Schistocerca gregaria*	Neem oil showed toxicity to this locust when sprayed on the maize crop.	Gujar and Mehrotra, 1985
Antifeedant	*Schistocerca gregaria*	Neem oil topical application to 5th instar nymphs reduced its feeding on the maize crop.	Gujar and Mehrotra, 1985
Growth inhibitor	*Nephotettix virescens*	Neem oil (0.5 to 5%) significantly retarded the growth and development of the insect.	Heyde et al., 1985
Antifeedant	*Heliothis armigera*	Neem oil reduced the population of the borer on treated chickpea plants.	Rao and Srivastava, 1985
Insecticidal	*Henosepilachna vigintiocto-punctata*	Neem oil protected eggplant crop from this beetle when sprayed on the foliage.	Tewari and Moorthy, 1985

Table 4. Effectiveness of neem oil against field insect pests (Continued)

Biological activity	Insect	Results of the test	Reference
Antifeedant, ovipositional repellent	*Spodoptera litura*	Neem oil emulsion (10%) sprayed on tobacco nursery protected the crop from this pest.	Joshi, 1986
Antifeedant, synergist	*Spodoptera litura*	Neem oil plus sesame oil (94:6 ratio) under laboratory tests showed antifeedancy.	Rajasekaran and Kumarswami, 1985b
Insecticidal	*Melanaphis sacchari*	Neem oil sprayed on the sorghum plant controlled the aphid.	Srivastava and Parmar, 1985
Antifeedant	*Heliothis armigera*	Neem oil was not effective to produce higher yield over control in chickpea and pigeonpea but showed antifeedant activity.	Singh et al., 1985; Sharma and Dahya, 1986; Sinha and Mehrotra, 1988
Repellent	*Brevicoryne brassicae*	Neem oil (0.4 and 0.8%) sprayed on cauliflower and cabbage showed repellency to the aphid.	Singh and Sharma, 1986
Ovicidal	*Earias fabia*	Neem oil vapour affected the hatchability of the eggs of test insect when female insects were exposed.	Pathak and Krishna, 1986

Table 4. Effectiveness of neem oil against field insect pests (Continued)

Biological activity	Insect	Results of the test	Reference
Insecticidal	*Dysdercus superstitiosus* adults and larvae of *Acraea eponina, Ootheca mutabilis, Riptortus dentipes*	Neem oil tested against these insects showed toxicity using its 0.1 to 2% concentration under controlled conditions.	Olaifa et al., 1987
Insecticidal	*Calocoris angustatus*	Neem oil sprayed on the sorghum protected the crop from this bug infestation.	Rao and Azam, 1987
Antifeedant	*Pericallia ricini*	Neem oil showed antifeedancy to the insect.	Mala, 1987
Antifeedant and gustatory repellent	*S. litura*	Distillates of its oil showed antifeedant activity to 3rd instar larvae.	Koul, 1987
Antifeedant	*Aulacophora foveicollis*	Neem oil showed highest degree of antifeedancy to the beetle under laboratory test.	Anonymous, 1987
Insecticidal	*Nilaparvata lugens*	Neem oil sprayed on the rice crop showed insecticidal activity to this hopper.	Velusamy et al., 1987
Growth inhibitor	*Crocidolomia binotalis*	Neem oil caused physico-chemical disorder during growth and development of the insect in the laboratory test.	Patnaik et al., 1987a

Table 4. Effectiveness of neem oil against field insect pests (Continued)

Biological activity	Insect	Results of the test	Reference
Insecticidal	*Athalia proxima*	Neem oil (0.4%) declined the population of the fly and also caused mortality to its larvae at 3% concentration.	Patnaik et al., 1987b
Crop protectant	*Aulacophora foveicollis*	Neem oil sprayed on the pumpkin protected the crop from insect infestation.	Gujar and Mehrotra, 1988
Insecticidal	*Stenchaeto-thrips biformis*	Neem oil (2%) sprayed on rice seedlings in the nursery and in the transplanted rice checked the thrips attack.	Pillai and Ponniah, 1988; Madhusudhan and Gopalan, 1988
Crop protectant	*Earias vittella*	Neem oil (2%) sprayed on okra protected its fruits from this fruit borer and increased the yield.	Sardana and Kumar, 1989
Insecticidal	*Nilaparvata lugens, Sogatella furcifera*	Neem oil (1%) application on the rice plants reduced the emergence of these insects in a net-house experiment.	Ramaraju and Sundarababu, 1989
Courtship inhibitor	*Nilaparvata ligens*	Males failed to copulate with treated females due to their continuous abdomen movement and thus prolonged premating period.	Saxena et al., 1989

Table 4. Effectiveness of neem oil against field insect pests (Continued)

Biological activity	Insect	Results of the test	Reference
Insecticidal repellent	*Maconellicoccus hirsutus*	Neem oil (2%) sprayed on the grape twigs showed repellency to this bug. Neem oil also caused mortality up to 91% to its nymphs when tested under controlled conditions using 1% spray directly on the pumpkin plants.	Abraham and Tandon, 1990
Insecticidal	*Orseolia oryzae*	Neem oil (3%) sprayed on the rice plants at 10, 25, 40 and 55 days after transplanting significantly reduced infestation of the gall midge.	Samlo et al., 1990
Antifeedant, growth inhibitor	*Dysdercus koenigii*	Topical application of neem oil (0.1 to 1.0%) to last nymphal instar showed toxicity and growth regulatory effects and caused morphogenetic activity during development of the bug.	Gujar and Mehrotra, 1990
Insecticidal	*Cnaphalocrosis medinalis, Scirpophaga (Tryporyza) incertulas*	Neem oil (1 to 4%) mixed with 0.4% teepol sprayed on the rice plant reduced the incidences of the pests.	Singh et al., 1990

Table 4. Effectiveness of neem oil against field insect pests (Continued)

Biological activity	Insect	Results of the test	Reference
Insecticidal	*Phyllocnistic citrella*	Neem oil (2 and 4%) sprayed on the young citrus trees and 15-day intervals checked the pest.	Dhara Jothi et al., 1990a
Insecticidal	*Toxoptera citricidus*	Neem oil (2%) reduced its population on lime.	Dhara Jothi et al., 1990b
Antifeedant	*Chilo partellus*	Neem oil (2 to 10%) reduced survival and feeding of the larvae.	Sharma and Bhatnagar, 1990
Crop protectant	*Earias vittella*	Neem oil (0.5%) and deltamethrin (0.09%) reduced yield loss in cotton.	Samjuthniravelu, 1990
Growth inhibitor	*Bemisia tobaci*	Neem oil (1%) suppressed growth and nymphal development of the fly when it was spread on the cotton.	Natarajan and Sundarmurthy, 1990
Insecticidal	*Lipaphis erysimi*	Neem oil (1.5%) showed 100% mortality to the aphids.	Mani et al., 1990
Grain protectant	*Leptocorisa acuta*	Neem oil (5%) significantly protected paddy grains on the caged plant.	Gupta et al., 1990
Insecticidal	*Chelonus blackburni*	Neem oil (3 to 5%) did not kill the parasite but significantly reduced the population of the tuber moth (host).	Shelke et al., 1990

Table 4. Effectiveness of neem oil against field insect pests (Continued)

Biological activity	Insect	Results of the test	Reference
Ovipositional deterrent	*Spodoptera litura*	Neem oil (1.5%) showed 100% ovipositional deterrency.	Koshiya and Chaiya, 1991
Insect growth disruptant	*Nephotettix virescens, Hydrellia philippina, Dicladispa armigera, Scirpophaga incertulas, Orseolia oryzae*	Neem oil (1%) disrupted the growth of 1st instar nymphs of *N. virescens* and affected the oviposition of other insects at higher oil concentrations when tested under controlled conditions.	Krishnaiah and Kalode, 1991
Plant protectant	*Cnaphlocrosis medinalis, Hieroglyphus banian*	Neem oil (2%) at high and low volume applications was more effective than Ultra Low Volume (ULV) application and also reduced the damage caused by insects on the treated rice plants.	Mohan et al., 1991
Crop protectant	Pest complex of the okra	Neem oil (1%) effectively protected the crop.	Rao et al., 1991
Crop protectant	*Helicoverpa armigera*	Neem oil (5%) protected chickpea crop at par with endosulfan.	Sinha, 1993
Crop protectant	*Scirpophaga incertulus*	Neem oil (4%) reduced white ear heads incidences in paddy fields.	Dhaliwal et al., 1993

Table 4. Effectiveness of neem oil against field insect pests (Continued)

Biological activity	Insect	Results of the test	Reference
Crop protectant	*Nilaparvata lugens, Sogatella furcifera*	Neem oil treatment @ 7.5 kg/ha effectively controlled pests' incidences in rice.	Sontakke, 1993
Crop protectant	*Hydrellia philippina*	Neem oil 3% sprays 21 days after transplanting in paddy fields reduced the damage caused by this insect.	Bhatia et al., 1994
Crop protectant	Scirpophaga incertulas	Neem oil (1.5%) combined with monocrotophos (0.05%) treatment effectively reduced dead hearts in rice at vegetative stage.	

Under field conditions neem cake has also been used by the farmers in India as a fertilizer. A number of research workers tested neem cake by incorporating it in the soil against insect pests of agricultural importance. Neem cake was found to kill or minimize the soil fauna of insect pests as well as to reduce the population of crop insect pests due to its systemic insecticidal action. The results of evaluation of neem cake as insecticide, antifeedant and repellent under field conditions are presented in Table 8.

FORMULATIONS OF NEEM PRODUCTS

Several formulations of neem products, viz., repelin, welgro, margosan-O, vermidin, neemrich, neemark, neemta, neemgaurd, nemidin, nemol, nemicidine, margocide CK, margocide OK, achook etc., have also been formulated by different agro-chemical industries and tested to have pesticidal properties against insect pests of agricultural importance. Some of these formulations have also been recommended and registered for their application in agricultural pest management programmes for the farmers. Neem products formulations, which have been tested and found effective, are described as follows:

(i) **Repelin**: This is also known as RD-9. It is formulated using a mixture of oils of neem, kharanja, castor, mahuwa and gingilly. It contains about 300 ppm of azardirachtin in its natural form and is formulated in two forms as

Table 5. Effectiveness of neem kernel powder (NKP) against insects infesting stored grains

Activity	Insect	Results of the test	Reference
Grain protectant	*R. dominica, S. oryzae, T. castaneum*	Protected wheat grains against storage insects when admixed with NKP.	Jotwani and Sircar, 1965
Repellent	*C. maculatus*	Protected grains of mung, Bengal gram, cowpea and peas against the beetle.	Jotwani and Sircar, 1967
Antifeedant	*C. maculatus, R. dominica*	Protected wheat, mung, Bengal gram, cowpea and grams when admixed @ 2% w/w with NKP for 6–12 months against these insects.	Pradhan and Jotwani, 1968b
Grain protectant	*Trogoderma granarium*	Protected wheat grains against this beetle when admixed with this powder.	Saramma and Verma, 1971
Grain protectant	*C. cephalonica, S. oryzae*	Impregnated gunny bags with NKP (2%) protected grains from cross infestation.	Chachoria et al., 1971
Oviposition inhibitor	*C. maculatus, C. chinensis*	NKP admixed with grains inhibited the oviposition of the test beetles.	Yadav, 1973
Antifeedant	*Trogoderma granarium*	Protected wheat grains and showed antifeedancy.	Girish and Jain, 1974
Grain protectant	*S. oryzae, C. chinensis*	NKP admixed @ 1 and 2% w/w protected stored grains.	Majumdar, 1974
Grain protectant	*C. maculatus*	Protected jowar seeds for 6 months against this beetle when the seeds were admixed with NKP @ 2% w/w.	Despande et al., 1974

Table 5. Effectiveness of neem kernel powder (NKP) against insects infesting stored grains (Continued)

Activity	Insect	Results of the test	Reference
Grain protectant	Storage insects in general	NKP admixed with paddy @ 2.0% w/w and also impregnated on the gunny bags protected the grains against storage insects for 6 months.	Nair, 1975
Grain protectant	*R. dominica*	Protected wheat seeds admixed with NKP against the pest.	Zag and Bhardwaj, 1976
Grain protectant	*S. oryzae*	NKP admixed with wheat @ 2% w/w protected the grains from this weevil attack for 3 months.	Satpathy, 1976
Grain protectant	*Bruchus* sp., *S. oryzae*	NKP admixed with cowpea and sorghum seeds @ 2% w/w absolutely protected grains against the pulse beetle.	Subramaniam, 1976
Grain protectant	*S. cerealella, R. dominica*	NKP admixed with rice grains @ 0.5–2% w/w protected grains against these pests.	Chellapa and Chelliah, 1976
Grain protectant	Storage insects in general	NKP admixed with grains reduced infestation of insects in stored wheat, jowar, maize and peas.	Ketkar et al., 1976
Grain protectant	*S. cerealella, R. dominica*	NKP admixed with paddy grains protected them from insects' attack.	Prakash et al., 1979a
Grain protectant	*R. dominica*	NKP @ 5% w/w effectively protected the wheat seed for 3 months.	Singh and Srivastava, 1980

Table 5. Effectiveness of neem kernel powder (NKP) against insects infesting stored grains (Continued)

Activity	Insect	Results of the test	Reference
Grain protectant	*T. granarium, R. dominica, S. oryzae*	NKP admixed with stored wheat protected the grains from these insects' attack.	Jotwani and Srivastava, 1981
Grain protectant	*S. oryzae, C. chinensis, C. maculatus*	NKP admixed with gram seed @ 1.5% w/w showed 100% mortality to the pulse beetles.	Ghosh et al., 1981
Grain protectant	*S. cerealella, R. dominica, S. oryzae*	Paddy grains admixed with NKP @ 2% w/w were protected under natural conditions of insects' infestation.	Prakash et al., 1982
Grain protectant	*C. chinensis*	Cowpea seeds admixed with NKP @ 0.5% w/w were protected for 4 months and @ 1% w/w for 8 months.	Sowunmi and Akinnusi, 1983
Antifeedant	*C. maculatus*	NKP admixed with cotton seeds @ 1% w/w reduced the progeny of the beetle.	Ivbijaro, 1983a
Repellent, insecticidal	*S. oryzae*	NKP admixed with maize seed @ 0.5–2.5 g/200g grains protected them for 6 months.	Ivbijaro, 1983b
Grain protectant	*C. maculatus*	NKP admixed with cowpea seeds @ 0.5 and 1% w/w controlled the beetle for 4 and 8 months respectively.	Sowunmi and Akinnusi, 1983
Grain protectant	*C. chinensis, C. maculatus*	NKP admixed with seeds of peas, black gram and grasspea protected them against the pulse beetles.	Yadav, 1984

Table 5. Effectiveness of neem kernel powder (NKP) against insects infesting stored grains (Continued)

Activity	Insect	Results of the test	Reference
Antifeedant	*C. chinensis*	NKP @ 5% w/w caused 69% mortality of the adults of this beetle.	Radhakrishnan et al., 1984
Grain protectant	*C. chinensis*	NKP admixed with greengram protected it from the beetle.	Mathur et al., 1985
Grain protectant	*T. granarium*	NKP admixed with wheat grains protected them from this beetle attack.	Singh and Singh, 1985; Singh and Kataria, 1986
Grain protectant	*E. cautella*	NKP admixed with the bajra grains protected them from this moth.	Lakhani and Patel, 1985
Grain protectant	*S. oryzae*	NKP reduced the population of the weevil when admixed with the wheat grains.	Rout, 1986
Antifeedant	*O. surinamensis, T. castaneum*	NKP admixed with rice @ 2% w/w showed antifeedancy to test insects.	Prakash and Rao, 1986
Antifeedant	*T. granarium*	1% NKP admixed with the wheat seeds showed antifeedancy to the insect.	Singh and Kataria, 1986
Grain protectant	*C. chinensis*	NKP (2%) admixed with greengram protected the grains.	Reddy and Reddy, 1987
Grain protectant	*R. dominica*	2% neem fruit powder gave above 90% grain protection for 6 months when admixed with the sorghum seeds.	Ramesh Babu and Husaini, 1990b
Grain protectant	*T. granarium*	NKP admixed with rice protected it from the beetle.	Mostafa, 1991
Grain protectant	*T. granarium*	2.5 and 5% NKP inhibited larval development and protected wheat.	Orounchi and Sharifi-zazdi, 1993

Table 6. Effectiveness of neem kernal powder (NKP) against field insect pests

Activity	Insect	Results of the test	Reference
Insecticidal	White grubs	NKP incorporated in the soil reduced the incidences of white grubs in groundnut.	Raodeo, 1973
Antifeedant	*Spodoptera litura*	3% NKP suspension in petriplates showed antifeedancy to the larvae of the insect.	Joshi and Ram Prasad, 1975
Antifeedant	*N. lugens, S. furcifera*	Decreased the secretion of honey dew of the insects and also showed antifeedancy.	Chiu et al., 1983
Antifeedant	*Pieris brassicae*	NKP suspension (0.4%) application protected cabbage leaves against the larvae of the pest.	Sandhu and Singh, 1975b
Insecticidal	White grubs	NKP incorporated with soil at the time of sowing protected crop from the white grubs' attack.	Patel, 1976
Insecticidal	Pest complex of mung bean	2% NKP dusting protected the crop significantly.	Yadav et al., 1979
Insecticidal (Synergist)	*Nephotettix virescens*	NKP in combination with carbofuran effectively controlled the population of rice green leafhopper below economic threshold level and reduced rice tungro virus incidences.	Kareem et al., 1988b
Growth inhibitor	*N. virescens*	3% NKP suspension inhibited growth of nymphs of this insect.	Krishnaiah and Kalode, 1990
Antifeedant	*Spodoptera litura*	2% neem seed kernel suspension inhibited the insect feeding in tobacco nurseries.	Murthy et al., 1993

Table 7. Effectiveness of de-oiled neem cake (DNC) against stored grain insect pests

Activity	Insect	Results of the test	Reference
Repellent	*S. cerealella, T. granarium*	DNC admixed with the wheat grains reduced insect attack.	Pruthy, 1937
Grain protectant	*C. chinensis*	DNC admixed with mung reduced grain damage due to this beetle.	Talati, 1976
Grain protectant	*C. chinensis, C. maculatus*	Black gram grains admixed with 1% DNC reduced insect damage to the pulse grains.	Prakash, 1982
Repellent	*S. oryzae*	Maize grains admixed with DNC reduced the oviposition.	Bowry et al., 1984
Grain protectant	*T. granarium*	DNC admixed with wheat flour inhibited insect development.	Singh and Singh, 1985
Grain protectant	*T. granarium*	Neem dust developed from DNC admixed with wheat grains @ 0.03% w/w were 100% protected.	Singh et al., 1993
Grain protectant	*R. dominica, S. oryzae, C. maculatus*	DNC powder admixed @ 1 g/kg of paddy, sorghum and pulse grains showed 100% protection.	Velumani et al., 1993

repelin-A and repelin-B. It has been found to repel a number of insect pests and protect the crops from their infestation. Against tobacco caterpillar, *Spodoptera litura* repelin was reported to be as effective as 2% neem seed kernel oil and controlled its infestation in tobacco crops in the nursery as well as in the fields (Subramaniam, 1988; Anonymous, 1989; Ramchandra Rao, 1990). Kareem et al. (1988) found this product as an effective formulation against diamond-back moth, *Plutella xylostella* when tested on broccoli plants. Repelin was found to show repellency to oviposition of *Chrysopa scelestes* (Yadav and Patel, 1990) and to pod borer, *Helicoverpa armigera* (Tirumala Rao and Venugopal Rao, 1990). Subba Rao (1990) also reported this product to show repellency to the rice pests, *Nilaparvata lugens* and *Sogatella furcifera*

Table 8. Effectiveness of de-oiled neem cake (DNC) against field insect pests

Activity	Insect	Results of the test	Reference
Repellent	Tobacco ground beetle	Protected tobacco crop from damage when neem cake was incorporated in the soil in pots @ 2% cake/plant.	Anonymous, 1961–1962
Repellent	*Schistocerca gregaria*	0.5% alcoholic extract of neem cake sprayed on the crops protected them against the locust.	Sinha and Gulathi, 1963
Insecticidal	Mustard aphids	Spray of non-fatty alcohol extract of neem cake was aphidicidal.	Singh and Gulathi, 1964
Insecticidal	*Rhopalosiphum nymphae, S. gregaria*	Extractives of pure isolates of neem cake were toxic to the insects under controlled conditions.	Goyal, 1971
Insecticidal	*Microtermes* sp.	Neem cake incorporated in the soil @ 3 kg/acre before sowing killed the white ants.	Dutta, 1974
Insecticidal	Soil inhabiting insects	Neem cake effectively controlled the population of these insects when incorporated in the soil.	Khound, 1975
Insecticidal	*Holotrichia insularis*	Neem cake application in soil protected chillis from this root grub.	Sachan and Pal, 1976
Insecticidal	*Holotrichia consanguinea*	Neem cake admixed in soil was toxic to the insect.	Nigam, 1977

Table 8. Effectiveness of de-oiled neem cake (DNC) against field insect pests (Continued)

Activity	Insect	Results of the test	Reference
Insecticidal	*Henosepilachna vigintiocto-punctata*	Neem cake powder showed toxicity to this beetle on the ridgeguard.	Borah and Saharia, 1982
Insecticidal	*Drosicha mangifera*	Soil application of neem cake controlled the infestation of mango mealy bug.	Tandon and Lal, 1980
Antifeedant	*Schistocerca gregaria*	Neem cake was toxic when tested against this locust.	Brajendra and Singh, 1982a
Insecticidal	*Cylas formicarius*	Neem cake was reported to protect the sweet potato from this weevil when tested in mini fields.	Rajamma, 1982
Antifeedant	*Nilaparvata lugens, Sogatella furcifera*	5% neem cake decreased the honey dew secretion and also showed antifeedant activity.	Chiu et al., 1983
Insecticidal (Systemic)	*Maliarpha seperatella, Sesamia calamistis, Diapsis macrophthalma, D. apicalis*	Neem cake application in soil inhibited the development and population build-up of these rice stem borers after transplantation.	Ho and Kibuka, 1983
Insecticidal	*Nilparvata lugenus*	Neem cake application in paddy fields could reduce the population of the pest.	Saxena et al., 1984b
Insecticidal	*Holotrichea consanguinea*	Neem cake incorporated in the soil controlled this pest in the groundnut.	Rao et al., 1984
Insecticidal	*Hydrellia philippina, Nephotettix* spp.	Neem cake coated urea reduced the incidences of the pests in rice.	David, 1986

Table 8. Effectiveness of de-oiled neem cake (DNC) against field insect pests (Continued)

Activity	Insect	Results of the test	Reference
Insecticidal	*Phyllocnistis citrella*	Neem cake extract showed insecticidal activity to this insect when sprayed on citrus crop.	Nagalingam and Savithri, 1980; Batra and Sandhu, 1981; Singh and Azam, 1986
Insecticidal	*Ophiomyia phaseoli, Madurasia obscurella*	Water extract of neem cake (10 g in 100 ml water) plus 3% neem oil emulsion sprayed 15 days after sowing of the urdbean controlled both pests.	Gunathilangaraj et al., 1987
Insecticidal	*Nilaparavata lugens, Sogatella furcifera*	Neem cake extract (5%) reduced the emergence of these insects when sprayed on the rice plants.	Ramaraju and Sundara Babu, 1989
Insecticidal	*Earias vittella*	DNC application @ 5 kg/plot reduced borer incidences in the okra crop.	Mallik and Lal, 1989
Insecticidal	*Cnaphlocrosis medinalis*	Neem cake (150 kg/ha) incorporated in the soil followed by 3% neem oil sprays at 10-day intervals effectively checked the incidences of leaffolder under field conditions.	Krishnaiah et al., 1990; Krishnaiah and Kalode, 1990
Insecticidal	*Nephotettix* sp., *S. furcifera*	Neem cake with urea was more effective to reduce the population and gave higher rice yield.	Viswanathan and Kandiannan, 1990

Table 8. Effectiveness of de-oiled neem cake (DNC) against field insect pests (Continued)

Activity	Insect	Results of the test	Reference
Insecticidal	*Spodoptera litura*	De-oiled neem cake extract (5%) sprayed on tobacco nursery significantly protected the crop from insect damage.	Murthy et al., 1990
Insecticidal	*Nymphula responsalis, N. enixalis, Nymphula* sp., *Cryptoblabes gnidiella*	DNC application in azolla field @ 250 kg/ha showed highest toxicity to the larvae and pupae of the pests.	Sasmal, 1991
Plant protectant	*Cnaphlocrosis medinalis, Hieroglyphus banian*	DNC extract (10%) applied as high and low volume applications showed superior rice crop protection than ULV application.	Mohan et al., 1991
Antifeedant	*Spodoptera litura*	1% DNC extract protected tobacco nurseries from insect feeding.	Murthy et al., 1993
Crop protectant	*Hydrellia philippina*	Neem cake @ 150 kg/ha applied as soil amendment reduced the damage caused by whorl maggots in paddy fields.	Bhatia et al., 1994

and also to protect rice crop from these planthoppers' attack. Rao et al. (1991) found repelin 1% spray alone to check the damage due to the pest complex in okra. Repelin was also reported to control brown planthopper, *Nilaparvata lugens* when tested on rice plants (Shukla et al., 1991). Pravakar and Rao (1993) reported repelin as 2.47 times more toxic to 2nd instar larvae of *S. litura* than annmet. Repelin-A (1%) and repelin-B (2%) both showed strong antifeedancy to the caterpillars of *S. litura* in tobacco nurseries (Chari et al., 1993). Dhaliwal et al. (1993) reported repelin (4 kg/ha) most effective against stem borers in the paddy fields. Repelin (2%) was found to be most effective to reduce the incidences of ber fruit fly, *Carpomyia vesuviana* when compared with other insecticides (Reddy et al., 1993). Sreedevi et al. (1993) reported RD-9 repelin as the most effective neem product compared to other formulations like neemicide, neknool and welgrow when tested against brinjal leaf beetle, *Henosepilachna vigintioctopunctata*.

(ii) **Margosan-O**: This is a formulation of neem oil fractions. Jilani and Su (1983) reported this product to show repellent and growth inhibitor activities to red flour beetle, *Tribolium castaneum*. However, this was also found to inhibit the growth of the cotton stainer, *Dysdercus koenigii* (Koul, 1988). Jacobson (1989) reported margosan-O as an effective formulation for controlling the insect pests of non-food crops especially in the nursery stage. Zehnder and Warthen (1990) found this product to effectively control Colorado potato beetle, *Leptinotarsa decemlineata* when tested on potato plants. Meisner et al. (1990) found margosan-O aqueous dilution to show toxicity to the larvae of notuid, *Earias insulana* when fed through the diets as well as sprayed on the cotton crop. Foliar and systemic treatments of its 2% solution sufficiently prevented the development of leafhopper, *Asymmetrasca decedens* on the cotton seedlings when tested under controlled conditions (Meisner et al., 1992).

(iii) **Neemark®**: This is also a neem seed oil formulation prepared by West Coast Herbochem Pvt. Ltd. Bombay, India and considered as active azardirachtin in the crude form. Yadav and Patel (1990) reported neemark to show gustatory repellency to *Chrysopa scelestes*. Ramchandra Rao (1990) also found repellency in neemark to *Spodoptera litura*, whereas neemark showed ovicidal activity against *Aleurolobus barodensis* (Patel et al., 1990b). Neemark as 1% spray alone did not significantly control the pest complex of okra; however, it showed superior results than other neem products like repelin or bisol (Rao et al., 1991). Patel et al. (1991) also reported neemark to show 86.67% repellency to the bean aphids on treated amrut gaurd. Its 2% concentration solution sprayed on the cotton leaves reduced the population of the sweet potato whitefly, *Bemisia tabaci* when tested under field conditions (Puri et al., 1991). Shukla et al. (1991) found neemark sprays to control white-backed planthopper, *Sogatella furcifera* in rice.

(iv) **Welgro**: This is a formulation of neem kernel powder found to show insecticidal activity when sprayed on tobacco crops against *Spodoptera litura* (Subramaniam, 1988). Kareem et al. (1988d) reported welgro spray on broccoli plants to check infestation of diamond-back moth, *Plutella xylostella*. Yadav and Patel (1990) reported this product to show ovipositional and gustatory repellency to *Chrysopa scelestes* when studied under laboratory conditions. Shukla et al. (1991) reported welgro sprays on rice plants to reduce damage due to white-backed planthopper, *Sogatella furcifera*. Nanda et al. (1993) found its 2% formulation as foliar spray to reduce the incidences of gall-midge and yellow stem borer in the paddy fields.

(v) **Neemrich**: This is a formulation of neem extract reported to protect the potato crop against the attack of *Phthorimaea operculella*. Sharma et al.

(1983) and Murthy et al. (1989) found neemrich (dichloroethane extract of neemrich) as effective as 2% neem seed kernel suspension against *S. litura* in the tobacco nursery. Sharma et al. (1984) also reported that treatment of a preferred ovipositional substrate, i.e., rough black muslin with 5 mg/cm^2 of neemrich-I, completely inhibited egg laying of the potato tuber moth, *P. operculella*. However, Yadav and Patel (1990) found it as a gustatory repellent to *Chrysopa scelestes*. Chari et al. (1993) reported neemrich-II (0.1%) to show 100% larval mortality to *S. litura* in tobacco nurseries 7 days after treatment.

(vi) **Nemol**: Neem seed extract in petroleum ether yielding to light yellow coloured oil after one month of extraction is called nemol. Chitra et al. (1987) reported 0.05% and 0.5% concentrations of nemol to show significant brinjal leaf protection against the 3rd instar grubs of the spotted beetle, *H. vigintio-ctopunctata*. Its 1000 and 500 ppm concentrations (W/V) showed absolute antifeedancy to the 3rd instar larvae of the castor semi-looper, *Achoea janata* (Purohit et al., 1989). Larval development was significantly inhibited when its 1000 ppm concentration was tested against *S. litura* and *H. armigera* (Nelson et al., 1993).

(vii) **Vemidin**: This formulation is extracted in Soxlet apparatus using ethanol to yield vemidin from the neem seed. Vemidin (0.05% and 0.5%) protected the brinjal leaf from 3rd instar grubs of *H. vigintioctopunctata* in a leaf-disc test (Chitra et al., 1987). Purohit et al. (1989) also found its 500 and 1000 ppm concentrations (W/V) to check the feeding of 3rd instar larvae of *A. janata* absolutely on the treated castor leaves. Vemidin (1000 ppm) also reduced adults' emergence of *S. litura* by 60% (Nelson et al., 1993).

(viii) **Nemidin**: Vemidin partitioned with ethyl acetate and water, the portion soluble in ethyl acetate concentrated and again partitioned of this concentrate/residue with mixture of aqueous methanol and petroleum ether (95:5), the aqueous methanol solution finally yielded a brown semi-solid residue on concentration called nemidin. Chitra et al. (1987) reported 0.5 to 0.05% concentrations of nemidin to protect the treated brinjal leaf from the attack of 3rd instar grubs of *H. vigintioctopunctata*. Its 500 and 1000 ppm concentrations (W/V) were reported to show absolute feeding inhibition to 3rd instar larvae of the castor semi-looper, *A. janata* (Purohit et al., 1989). Nelson et al. (1993) found nemidin at 1000 ppm concentration to inhibit larval development by 68% in the case of *S. litura*, *H. armigera* and *S. furcifera*.

(ix) **Achook®**: Godrej Soaps Limited (India) has developed a neem formulation under the registered trade name 'achook', which is a water soluble powder (WSP) that contains azardirachtin (0.03%), azardiradion, nimbocinol and ebinimbocinol as active principles. Achook showed strong antifeedancy and reduced the larval and pupal survival and also adult emergence depending upon the does in *Helicoverpa armigera* and *Earias vittella* (Raman et al., 1993). Prakash and Rao (1994) found its 0.5 and 1.0% concentrations to effectively control the populations of rice ear head bug, *Leptocorisa acuta* when sprayed 20 days after flowering of rice plants.

(x) **Other neem formulations**: Many other formulations of different fractions of neem seed extracts like nemicidin, margocide, neemgaurd, margosol, NK-100, Ne-pet-et, V-74, NS-58 and biosol have also been reported to show bio-pesticidal activity. Nemicidin has been reported to protect brinjal crop from attack of *E. vigintioctopunctata* (Chitra et al., 1987). Biosol was also reported to reduce incidences of rice leaffolder, *C. medinalis* when sprayed on rice plants (Anonymous, 1989b). Neemta-2100 and neemgaurd, the factions of neem seed extracts, were tested as biopesticide under greenhouse conditions (Krishnaiah et al., 1990). Shukla et al. (1991) reported margocide-CK and margocide-OK to significantly reduce the population of white-backed planthopper when sprayed on rice plants in field conditions. Jena and Dani (1994) found margocide-CK and margocide-OK 1% concentration sprayed on the potted rice plant to show 57 to 80% mortality to the 5th instar larvae of the brown planthopper, *Nilaparvata lugens*. Azardirachtin-rich fractions like Ne-pet-et, NK-100, V-74 and NS-58 were also found to show biological activities when tested against *Sogatella furcifera*, *Spodoptera litura* and *Helicoverpa armigera*. Oviposition was deterred by 89% in treatment of NS-58. NK-100 reduced insect feeding by 69%. Ne-pet-et (1000 ppm) inhibited larval development of *H. armigera* and *S. furcifera* by 68% (Nelson et al., 1993). Chitra and Reddy (1994) reported neemgaurd and margosol formulations @ 5 ml/l in combination with monocrotophos 0.05% as highly effective in controlling the groundnut leaf miner, *Aproaerema modicella*.

Active Components from Neem

From the neem seed extract and oil, a number of chemical components have been isolated and identified (Table 9).

But only a few chemicals, viz., azardirachtins, deacetyl-salannin, salannin, nimbin, epinimbin and meliantriol, are found to possess biological activities. During the process of evaluation of the biologically active components, these products were found to show insect repellent, antifeedant, growth inhibitor/regulator and insecticidal activities when tested against a wide range of insects under storage and field conditions. These active components are described as follows:

Azardirachtins

Azardirachtins is a group of C_{26} terpenoids that exhibit different biological activities and are structurally similar to the insect ecdysones. Some of them were found to act as antineoplastic and cytotoxic agents and others as pesticides and antifeedants (Kundu et al., 1985). Its 0.1 ppm dosage was reported to disrupt the growth of insects and act as antifeedant and sterilant (Jacobson, 1989). Zanno et al. (1975) reported on the structure of an insect phago-repellent when they gave the first complete structure for azardirachtin. Butterworth and Morgan (1986) first reported a substance from neem seeds, which they named azardirachtin, and used its feeding inhibitory effect for the control of the desert

Table 9. Chemical components isolated from neem, *A. indica*

Sl No.	Neem components	Source	Mol. formula	Mol. wt.
1.	Azadirachtanin A	L	$C_{32}H_{40}O_{11}$	600
2.	Azadirachtin	S	$C_{35}H_{44}O_{16}$	720
3.	3-Desacetyl-3-cinnamoyl	L	$C_{42}H_{48}O_{16}$	808
4.	22,23-Dihydro-23-methoxy	S	$C_{36}H_{48}O_{17}$	737
5.	1-Tigloyl-3-acetyl-1 1-methoxy azadirachtinin	B	$C_{36}H_{46}O_{16}$	734
6.	Azadirachtol	F	$C_{32}H_{46}O_{6}$	526
7.	3-Tigloylazadirachol	S	$C_{33}H_{42}O_{14}$	662
8.	Azadiradione	S	$C_{28}H_{34}O_{5}$	450
10.	7-Desacetyl-7-benzoyleposy	S	$C_{33}H_{36}O_{6}$	528
11.	7-Desacetyl-7-hydroxy	F	$C_{26}H_{32}O_{4}$	408
12.	1B,2B-Dieposy	S	$C_{28}H_{34}O_{7}$	436
14.	17-Epi-	S	$C_{28}H_{34}O_{5}$	450
15.	Epoxy-	S	$C_{28}H_{34}O_{6}$	466
16.	17B-Hydroxy	S	$C_{28}H_{34}O_{6}$	466
17.	1-Methoxy-1,2-dihydro	S	$C_{29}H_{38}O_{7}$	498
18.	Azadirone	S	$C_{28}H_{36}O_{4}$	436
19.	Gedunin	S	$C_{28}H_{36}O_{4}$	482
20.	7-Desacetyl	S	$C_{26}H_{32}O_{6}$	440
21.	7-Desacetyl-7-benzoyl	S	$C_{33}H_{36}O_{7}$	544
22.	Isoazadirolide	L	$C_{32}H_{42}O_{10}$	586
23.	Isonimbocinolide	L	$C_{32}H_{42}O_{9}$	570
24.	Isonimolicinolide	F	$C_{30}H_{38}O_{8}$	526
25.	Margosinolide	T	$C_{27}H_{32}O_{8}$	484
26.	Iso-	T	$C_{27}H_{32}O_{8}$	484
28.	Meliantriol	S	$C_{30}H_{50}O_{5}$	490
29.	Nimbandiol	S,L	$C_{26}H_{32}O_{7}$	456
30.	6-Acetyl-	S	$C_{28}H_{34}O_{8}$	498
31.	Nimbidinin	S	$C_{26}H_{34}O_{6}$	442
32.	Nimbin	SBW	$C_{30}H_{36}O_{9}$	540
33.	Desacetyl-	SBF	$C_{28}H_{34}O_{7}$	498
34.	4-Epi-	S	$C_{30}H_{36}O_{9}$	556
35.	Photo-oxidized	S	$C_{30}H_{36}O_{10}$	482
37.	6-Desacetyl	SLB	$C_{26}H_{32}O_{6}$	440
38.	Nimbiol	B	$C_{18}H_{24}O_{2}$	272
39.	Nimbocinone	L	$C_{30}H_{46}O_{4}$	470
40.	Nimbolide	L	$C_{27}H_{30}O_{7}$	466
41.	Nimbolin A	W	$C_{39}H_{46}O_{8}$	642
42.	Nimbolin B	W	$C_{39}H_{46}O_{10}$	674
44.	Nimilicinoic acid	F	$C_{26}H_{34}O_{6}$	442
45.	Nimolicinol	S	$C_{28}H_{34}O_{7}$	482

Table 9. Chemical components isolated from neem, *A. indica* (Continued)

Sl No.	Neem components	Source	Mol. formula	Mol. wt.
46.	Nimolinone	F	$C_{30}H_{44}O_3$	452
47.	Salannin	S	$C_{30}H_{44}O_3$	596
48.	3-Desacetyl-	S	$C_{32}H_{42}O_8$	554
49.	Photooxidized	S	$C_{34}H_{44}O_{10}$	612
50.	Salannol	S	$C_{32}H_{44}O_8$	556
52.	Sugiol	B	$C_{20}H_{28}O_2$	300
53.	7-Acetylnetrichilenone	S	$C_{28}H_{36}O_5$	452
54.	Vepinin	S	$C_{28}H_{36}O_5$	452
55.	Vilasinin	L	$C_{26}H_{36}O_5$	428
56.	3-Acetyl-7-tigloyllactone-	S	$C_{33}H_{46}O_8$	570
57.	1,3-Diacetyl-	S	$C_{30}H_{40}O_7$	512

S-Seeds, W-Wood, L-Leaves, B-Bark, F-Fruit
Source: The Chemistry of Neem Tree by P.S. Jones et al., In *Focus on Phytochemical Pesticides Vol. 1. The Neem Tree,* edited by M. Jacobson (1988): 19–45.

locust, *Schistocerca gregaria*. Further in another test with the pure compound, the remarkably low concentration of 5 gm/l gave complete inhibition of the feeding response. Biological activities of azardirachtin group components have been summarized in Table 11.

Besides its antifeedant activity, azardirachtin was also reported to have growth inhibitory effects (Butterworth and Morgan, 1971; Ouadri and Narsaiah, 1978; Ruscoe, 1972; Steets, 1975, 1976a; Steets and Schmutterer, 1975; Meisner et al., 1976; Ladd et al., 1978; Warthen, 1979). Schmutterer and Rembold (1980) and Rembold et al. (1986) found that neem compounds which did not deter the Mexican bean beetle, *Epilachna varivestis* from feeding were highly active insect growth inhibitors. Rembold (1989a,b,c) found a series of pure compounds, which were contained in the polar methanol soluble extract from neem seed and were not phago-repellents but in minute amounts disturbed metamorphosis of *Epilachna* and produced larval-pupal intermediates due to interference of these compounds with the insect hormonal system (Schmitterer and Rembold, 1990). The neem seed fractions also disrupted growth in other insects without inducing any feeding inhibition (Rembold et al., 1980b). The chemicals affected the insect hormone system directly, whereas antifeedants could indirectly cause developmental deviants and initiate hormonal effects through subsequent starvation by affecting some chemo-receptors (Slama, 1978).

The main insect growth inhibitor was identical with azardirachtin, a compound which until then had been treated as a phago-repellent (Rembold and Schmutterer, 1981). A short time later, it became clear that some of the unknown growth inhibitors were isomeric with the compound already known as 'azardirachtin' (Forster, 1983; Rembold et al., 1984a).

Table 10. MC_{50} values (ppm) for the natural isomeric azardirachtins, some of their chemical derivatives and salannin, as calculated from *Epilachna* bioassays

Azardirachtins	n	MC_{50}	Sign.
Azardirachtin A	40	1.66	ss
22,23-Dihydro-azardirachtin-A	6	1.22	w
11-Acetyl-azardirachtin A	13	8.68	s
3-Deacetyl-azardirachtin A	31	0.38	ss
23a-Ethoxy-22,23-dihydro-azardirachtin A	19	0.74	s
23b-Ethoxy 22,23-dihydro-azardirachtin A	7	0.52	s
Azardirachtin B	19	1.30	ss
22,23-Dihydro-azardirachtin B	27	0.28	ss
3-Detigloyl-3-(2-methylbutyryl)-22-23-dihydroazardirachtin B	17	0.45	ss
3-Detigloyl-azardirchtin B	13	0.08	s
23a-Ethoxy-22,23-dihydro-13,14-diepoxy-17-hydroxy-azardirachtin B	8	100	
Azardirachtin C	8	12.97	s
Azardirachtin D	15	1.57	ss
Azardirachtin E	25	0.57	ss
Azardirachtin F	18	1.15	s
Azardirachtin G	16	7.69	s
Salannin	4	100	

n = number of individual cage tests, s = statistical significance, ss = 99.9%, s = 98–99%, w = 95–99%.

Structure of the Azardirachtins

Azardirachtin A (Structure I, Figure 14) is the predominant growth inhibiting neem component. Its former structure as proposed by Zanno et al. (1975) has been reassigned by three laboratories (Turner et al., 1987; Belton et al., 1987; Kraus et al., 1987) and unequivocally gave the basis for a structural elucidation of the other isomeric azardirachtins by NMR spectroscopy.

Before 1983, azardirachtin was determined to be a mixture of several isomers due to their very similar chemical, physical and biological properties; therefore, it could be treated as a group of azardirachtin isomers, with azardirachtins A and B being the predominant ones (Forster, 1983; Rembold et al., 1984a). Further, at least five isomers have been purified and with the exception of one (azadirachtin C) a structure based on NMR data could be proposed. In addition, by cautious chemical modification of the mother compounds, 9 more isomers were prepared, and these 16 highly active insect growth inhibitors formed the basis for simpler structures with the same biological activity (Rembold, 1989a). The structures of these azardirachtins are discussed below.

Table 11. Effectiveness of active components of neem against field insect pests

Activity	Insect	Active component	Results of the test	Reference
Repellent	*Locusta migratoria*	Nimbidinin	Showed repellency to the insect on the treated cabbage leaves and the stripes with these components.	Pradhan et al., 1963
Antifeedant and growth inhibitor	*S. gregaria* (larvae), *Pieris brassicae* (eggs and larvae), *Dsydercus* sp. (eggs and larvae)	Azardirachtin	The component showed systemic action and antifeedant activity on the leaves.	Gill and Lewis, 1971
Antifeedant	*S. gregaria*	Azardirachtin	The component was a highly effective antifeedant against the locust in tests under controlled conditions.	Butterworth et al., 1972
Antifeedant and growth inhibitor	*Plutella xylostella* (larvae), *Pieris brassicae* (larvae), *Heliothis virescens* (larvae), *Dysdercus asiatus* (nymphs)	Azardirachtin	The component showed antifeedancy and growth inhibitory activities against the insects when fed through their diets.	Ruscoe, 1972
Antifeedant and growth inhibitor	*Plutella xylostella, Pieris brassicae, Dysdercus fasciatus, Heliothis virescens*	Azardirachtin	Deformed the pupae of *Pieris* and affected the growth and development of other insects and showed antifeedancy.	Hans-Jorg, 1972

Table 11. Effectiveness of active components of neem against field insect pests (Continued)

Activity	Insect	Active component	Results of the test	Reference
Antifeedant	*Epilachna varivestis*	Azardirachtin	Delayed the oviposition and reduced feeding of the beetle when fed through its diet.	Steets and Schmutterer, 1975
Antifeedant	*S. gregaria*	Neem oil fractions (N_1 and N_6)	Both fractions of neem oil showed antifeedancy to the locust.	Narayana et al., 1980
Antifeedant	*Acalymma vittatum*	Azardirachtin	Dipping cantaloupe leaves in 0.1% solution of the component showed antifeedancy to the beetle and protected the crop.	Pierie, 1981
Antifeedant and growth inhibitor	*Heliothis zea, S. gregaria, H. virescens*	Azardirachtin	Showed antifeedancy to *H. zea* and *S. gregaria* and inhibited the growth and development of *H. virescens.*	Kubo and Klocks, 1982
Antifeedant	*Phyllotreta striolata*	Azardirachtin	Showed antifeedancy to the beetle on the radish plant under laboratory conditions.	Meisner and Mitchell, 1992
Growth inhibitor	*Epilachna varivestis, Aphis mellifera*	Azardirachtin	4th instar larvae of *E. varivestis* and 3rd instar larvae of *A. mellifera* when fed on Azardirachtin mixed diets showed inhibition.	Rembold et al., 1982
Antifeedant	*Acalymma vittatum*	Azardirachtin, Salannin	Both components were found to be antifeedants against the beetle.	Reeds et al., 1982

Activity	Insect	Active component	Results of the test	Reference
Growth inhibitor	*Locusta migratoria*	Azaridirachtin	Caused morphogenetic effects and inhibited eclosion process in the last instar larvae.	Sieber and Rembold, 1983
Growth inhibitor	*Spodoptera litura*	Azardirachtin	Last instar larvae fed on the diet containing azardirachtin resulted to form larval-pupal intermediates and inhibited its growth.	Gujar and Mehrotra, 1983
Growth inhibitor, insecticidal	*Ostrinia furnacalis*	Azardirachtin	3rd and 4th instar larvae fed on artificial diet containing 20 ppm azardirachtin for 2 days inhibited pupation, produced black legs and brown spots on the thorax and caused larval mortality.	Chiu et al., 1984
Antifeedant and growth inhibitor	*Epilachna varivestis*	Salannin	Salannin disrupted the epidermal tissues, fat bodies and mitochondria formation and also showed antifeedancy to the beetle when fat through the diets.	Schmutterer, 1984

Table 11. Effectiveness of active components of neem against field insect pests (Continued)

Activity	Insect	Active component	Results of the test	Reference
Growth inhibitor, insecticidal	*Dysdercus koenigii*	Azardirachtin	The component prolonged development and caused deformities in the larvae and wingless adults developed when fed to its various stages and caused mortality when applied topically.	Koul, 1984
Antifeedant	*Heliothis virescens, Spodoptera eridania, Epilachna varivestis*	Azardirachtin	The component showed antifeedancy when fed to these insects through treated leaves.	Lidert et al., 1985
Growth inhibitor	*Spodoptera litura*	Azardirachtin	On treated castor leaves feeding of the young larvae was reduced and disrupted the larval growth.	Koul, 1985
Growth inhibitor	*Locusta migratoria migratorioides*	Azardirachtin	Injected component reduced the growth rate of 5th instar nymphs and caused death at dose of 80 μg/g body weight.	Mordrie-Luntg et al., 1985
Growth inhibitor	*Ostrinia furnacalis*	Azardirachtin	Disrupted larval development.	Shin Foon et al., 1985
Insecticidal	*Ostrinia nubilalis*	Azardirachtin	The component (0.0001%) sprayed on the host leaves caused 100% mortality to 3rd instar larvae.	Arnason et al., 1985

Activity	Insect	Active component	Results of the test	Reference
Antifeedant	*S. gregaria*	Azardirachtin	Showed antifeedancy against the insect.	Butterworth and Morgan, 1986
Growth inhibitor	*S. gregaria*	Azardirachtin	Caused morphogenetic changes during nymphal development.	Rao and Subrahmaniyam, 1986
Aphicidal	*Dactynotus carthami*	Azardirachtin and limonoids	Showed toxicity to the aphids.	Devakumar et al., 1986
Ecdysis inhibitor	*Heliothis virescens*	Diocetylazardir-achtin	Inhibited ecdysis during larval development.	Kubo et al., 1986
Growth inhibitor	*Spodoptera litura*	Azardirachtin	Caused developmental abnormalities and toxicity.	Rao and Subrahmaniyam, 1986
Growth inhibitor	*Bombyx mori*	Azardirachtin	Admixed with diet affected the development of the test insect.	Koul et al., 1987
Antifeedant	*Nephpotettix virescens*	Neem limonoides	Seedling root dip of rice (TN-1) in 2500 ppm concentration of the component for 25 seconds caused antifeedancy.	Saxena and Boncolin, 1988
Antifeedant	*Diacrisia oblique*	Thionenone	Showed antifeedant activity to this insect under a laboratory test.	Agrawal and Mall, 1988

Table 11. Effectiveness of active components of neem against field insect pests (Continued)

Activity	Insect	Active component	Results of the test	Reference
Antifeedant	*Diaphania indica, Schistocerca gregaria*	Azardirachtin	Under controlled conditions, this component showed feeding deterrency to the insects.	Chitra and Kandaswamy, 1988
Antifeedant	*Peridroma sancia*	Azardirachtin	Inhibited the feeding activity and the survival of this cutworm when fed through its diet.	Champagne et al., 1989
Enzyme inhibitor	*Spodoptera litura*	Azardirachtin	Injected azardirachtin (1 μg/g) to 6th instar affected its food consumption and digestive enzyme secretions resulting in poor nutritional utilization.	Ayyangar and Rao, 1990
Gustatory repellent	*Chrysopa scelestes*	Active azardirchtin (neemark)	Neemark found to be a repellent inhibiting its oviposition.	Yadav and Patel, 1990
Repellent	*Spodoptera litura*	Active azardirachtin (neemark)	Showed repellency and protected tobacco crop against this pest.	Ramachandra and Rao, 1990
Neuroendocrine inhibitor	*Spodoptera litura*	Azardirachtin	A single dose (1 μg/g) injection into larvae delayed accumulation of neuro-secretory material in median neuro-secretory cells.	Gupta and Rao, 1990

Activity	Insect	Active component	Results of the test	Reference
Growth inhibitor	*Corcyra cephalonica*	Azardirachtin	Inhibited larval development when fed on the treated diet.	Sharma, 1992
Probe reductant	*Rhopalosiphum padi, Sitobion avenae*	Azardirachtin	Less than 500 ppm topical application reduced probing activity of the aphids.	West et al., 1992
Serotoninergic inhibitor	*Locusta migratoria*	Azardirachtin	Interfered with the serotoninergic system of the somatogastric ganglia and blocked the mid gut peristalsis.	Dorn and Trumm, 1993
Ecdysteroid inhibitor	*Teleogryllus mitratus*	Azardirachtin	10 µl azardirachtin/insect administered to 1-day-old female cricket reduced the ecdysteroid level.	Strambi et al., 1993
Growth inhibitor	*Lipaphis erysimi*	AZT-VR-K (azardirachtin enriched acetone extract of neem seed kernel)	Treatment of 3rd instar @ 0.25 to 5.0 mg/nymph produced 16.7% intermediates and delayed development.	Bhathal and Singh, 1993
Ovicidal	*Dysdercus koenigii*	Thionemone	Toxicity to eggs of this bug.	Agrawal, 1993

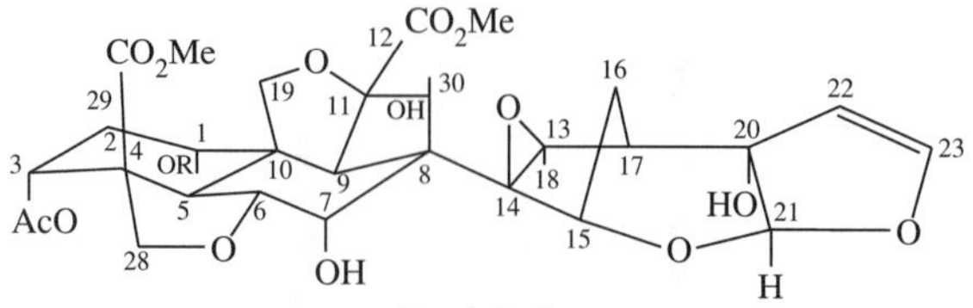

Figure 14. Azardirachtin A (structure I).

Azardirachtin B (Structure II, Figure 15) was isolated by following its growth inhibitory activity in the *Epilachna* bioassay (Rembold, 1987; Rembold et al., 1984a). Klenk et al. (1986) reported on the structure of 3 tigloylazard-irachtol and Kubo et al. (1984) reported on deacetylazardirachtinol. There were three structural differences, if azardirachtins A and B were compared with one another. In azardirachtin B, the tigloyl side chain was located at position 3, the molecule contained a free 1-hydroxy group and position 11 was reduced to the deoxy compound. Later a similar molecule was isolated from the neem seed and described as 3-deacetyl-11-desoxyazardirachtin, still having the tigloyl side chain in position 1 and a free 3-hydroxyl group (Bilton et al., 1987).

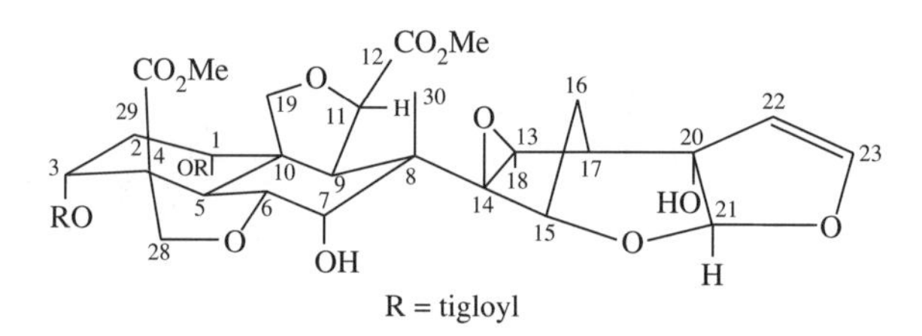

Figure 15. Azardirachtin (structure II).

Azardirachtin D (Structure III, Figure 16) differed from azardirachtin A by reduction of the ester group in position 4 to a methyl group. All the rest of the molecule was unchanged if compared with the main compound of this series 1-cinnamoly-3-feruloyl-11-hydroxymeliacarpin, which differed from azardirachtin D only in the substituents at positions 1 and 3. This component was isolated from *Melia azadarach* leaves (Kraus, 1986). Azardirachtin E (Structure IV, Figure 17) was the naturally occurring detigloylazardirachtin A (Rembold, 1989). Natural derivatives of azardirachtin B were the isomers F and G. In azardirachtin F (Structure V, Figure 18), the ether bridge in position 19 was reduced and opened by formation of a C-19 methyl group. Azardirachtin G (Structure VI, Figure 19) was an isomer of azardirachtin B with a double bond former instead of the epoxide ring and with a hydroxy group in position 17 (Rembold, 1989).

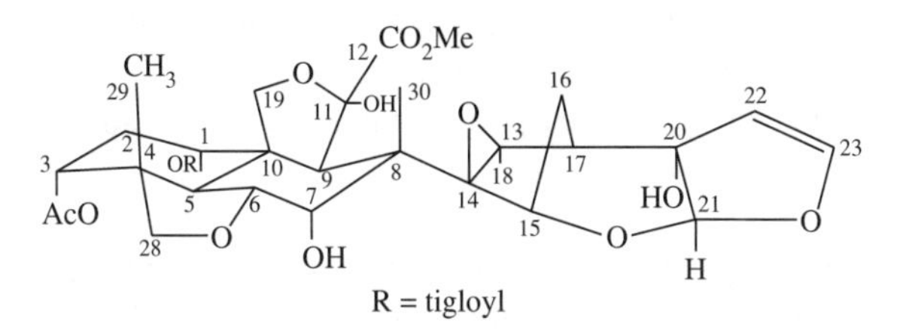

Figure 16. Azardirachtin (structure III).

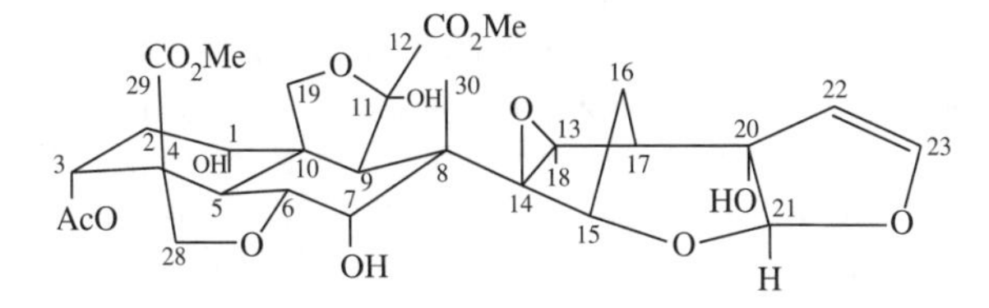

Figure 17. Azardirachtin (structure IV).

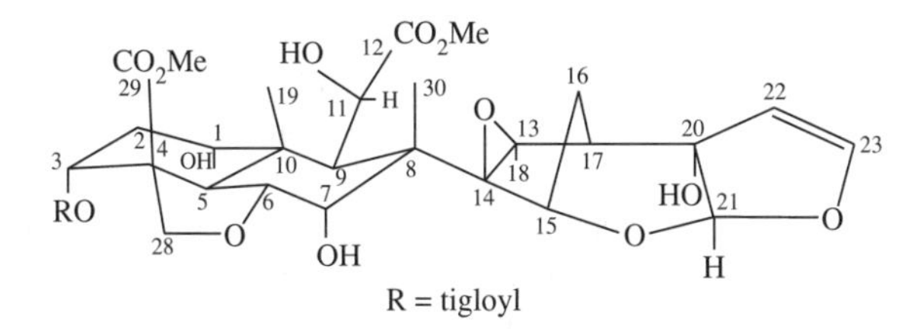

Figure 18. Azardirachtin (structure V).

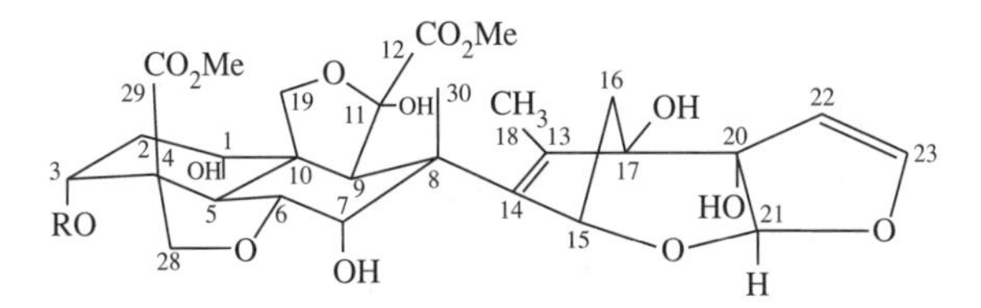

Figure 19. Azardirachtin (structure VI).

Forster (1987) degraded the azardirachtin A and B molecules and obtained a series of derivatives by minor modification of the respective mother compounds in addition to isolation of six active components from neem seed. 22,23-Dihydro-azardirachtins A and B were prepared by catalytic hydrogenation, and 11-acetyl-azardirachtin A by reaction of azardirachtin A with acetic anhydride (Butterworth et al., 1972).

Forster (1987) evaluated the biological activity of the 17 compounds presented in Table 11 and found that all these products with a biological activity at 10 ppm possessed several features in common. Although only 14 molecules with limited structural differences could be compared with each other, they all possessed some distinct properties in common which, if absent, abolish the growth inhibiting activity in the *Epilachna* bioassay. This can be explained as follows:

1. For the biological effect the type of substitution at the decalin rings A and B was important. Creation of a free 3-hydroxy group by deacetylation of azardirachtin A increased the MD_{50} value in growth inhibitory activity about 4-fold, and hydrolysis of the 3-tigloyl moiety from azardirchtin B increased 16-fold. But no activity difference between azardirchtin A and its three isomers D, E and F, with values between 1.15 and 2.80 ppm, was noticed because all possessed 3-hydroxy group substituted by either acetate or tiglate. However, highest activity was found with both of the 1- and 3-hydroxy groups being unsubstituted. Also, the ecdysteroid molecule possessed a decalin moiety and carried two hydroxy groups in the A ring. There was a fundamental difference between the ecdysteroid and the azardirachtin structures. However, the latter possessed rings A and B *trans*-connected and the ecdysteroids *cis*-connected. When compared to their molecular models, the two hydroxy groups in ring A were about the same distance from each other in the azardirachtin and in the ecdysteroid molecule.
2. The 22, 23-double bond of the dihydrofuran ring was present in all the naturally occurring azardirachtins, and could easily be removed either by hydrogenation or by addition of alcohol. In the latter case, biological activity was increased in all the molecules (Table 10). Comparison of the azardirachtin and ecdysteroid molecular models also indicated some similarities between the ecdysteroid side chain of eight carbon atoms and that of the four furan carbons in axardirachtin, due to the fact that elongation of this carbon skeleton by addition of an ethoxy group significantly increased biological activity.
3. Activity of azardirachtin B (1.30 ppm) and of its derivative azardirachtin F (1.15) was significantly higher than azardirachtin A (1.66) and especially of its 11-acetyl derivative (8.68). Thus structural variation at position 11 also affected the growth inhibitory activity of the azardirachtins. It appeared to be of importance for developing high biological activity that a group which occupied too much space at position 11 should not be present.
4. The most critical structural element responsible for any growth inhibitory activity of the azardirachtins seemed to be the 13,14-epoxy group, which became evident from a comparison of the two inactive compounds, which were free of the oxirane ring, i.e., 23a-ethoxy-22,23-dihydro-13,14-deepoxy-17-hydroxyazardirachtin B

and salannin. Further, biological activity was also reduced significantly in the two azardirachtins C (12.97 ppm) and G (7.69 ppm), when compared with MC_{50} values. The latter molecule had a ketal function at position C-21 and a free hydroxy group at C-7 as found in the other 12 active azardirachtins. Comparison of the azardirachtin G structure and its activity with 23a-ethoxy-22,23-dihydro-14,14-diepoxy-17-hydroxyazardirachtin B indicated a more complicated structural relationship.

Based on these observations, a reduced structure for the biologically active portion of the azardirachtin molecule could be proposed (Structure VII, Figure 20).

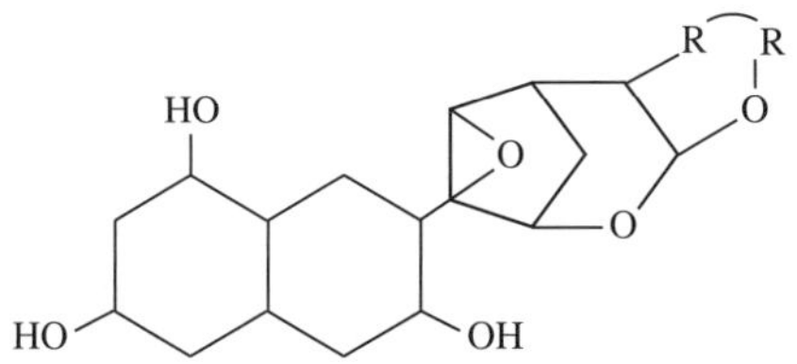

Figure 20. Azardirachtin (structure VII).

This reduced azardirachtin structure draws special attention to the seven isomeric azardirachtins, which have been isolated so far from neem seed and showed insect growth inhibition activity. Kraus (1986) and Bilton et al. (1987) reported that 3-deacetyl-11-desoxyazardirachtin and 1-cinnamoyl-3-feruloyl-11-hydroxymeliacarpin were also active as insect growth inhibitors. Based on the proposed minimal structure it could be stated that the decalin ring might be substituted at positions 1 and 3 hydroxy groups, either one or both of which were esterified or both were unsubstituted and the two free hydroxy groups yielded the highest biological activity; as epoxy group might be presented in correct steric distance from the two hydroxy groups and these were possibly of a hydroxy group in position 7 and/or a ketal function in position 21 and the reduced dihydrofuran ring or a side chain, again in correct steric distance from the epoxy group increased the growth regulating activity of the azardirachtin analogue.

For practical use in plant protection, although a synthetic compound of this structure does not seem to be economically feasible, it will be helpful for further studies on other simpler and biologically active compounds. Biological activity of azardirachtin substituents such as tiglate or acetale may be as a synergistic function in penetrating the lipophilic insect cuticle, may be some sort of slow release of the bio-active compound by ester hydrolysis in the gut or only in the target organ, or may be an alkylation reaction through splitting of the epoxy ring, which would explain the organ specificity and high biological activity of the azardirachtin molecule.

Mode of Azardirachtin Action

The antifeedant activity of azardirachtin was demonstrated against several insect species (Warthen, 1979; Jacobson, 1981, 1986; Jacobson et al., 1984; Schmutterer, 1984). However, as exemplified by the quantitative *Epilachna* bio-assay, the growth inhibitory effect was dose-dependent and diminished with reduced concentrations. Azardirachtin is known to inhibit feeding at much lower concentrations in hemimetabolous than in holometabolous insects. There are other examples showing higher effects as a phago-repellent for azardirachtin than as a growth inhibitor. Meisner et al. (1981) reported on the feeding response of *Spodoptera litura* and *Earias insulana* larvae, both of which showed a much higher inhibition with azardirachtin than with salannin. Kubo and Klocke (1982) tested the efficacy of azardirachtin as the phago-repellent and growth inhibitor in a cotton leaf and cotyledon-disk feeding assay. The larvae of lepidopteran such as *Heliothis zea*, *H. virescens*, *Spodoptera frugiperda* and *Pectinophora gossypiella* were inhibited from feeding at lower azardirachtin doses than those needed for growth inhibition. Due to its systemic action, the compound effectively protected the plant leaf and colyledon. The azardirachtin also showed the antagonistic effects under certain test situations and the growth inhibitory activity appeared only when the compound was fed before. However, antifeedant activity was effective due to the appearance of new foliage only for a short period of time, if no systemic effect was possible.

The pure azardirachtin and its derivatives as partially purified neem seed fractions were almost equally active against spinning and nonfeeding of *Ephestia kuehniella* and against feeding of *Apis mellifera* larvae, inducing delayed larval development in both the insects (Sharma et al., 1980). Similar results were found after feeding azardirachtin to the larvae of the Asiatic corn borer, *Ostrinia furnacalis* (Shin Foon et al., 1985).

Azardirachtin was also found to disrupt the development in some phytophagous insects (Ruscoe, 1972; Gill, 1972). Treatment of the insects and/or their food with the pure compound or with azardirachtin-containing extracts caused growth inhibition, malformation, mortality, and reduced fecundity (Schmutterer and Rembold, 1980; Steets, 1975, 1976a,b; Steets and Schmutterer, 1975; Redfern et al., 1981; Rembold and Czoppelt, 1981; Rembold et al., 1981). Similar morphogenetic defects could also be induced by synthetic hormone mimics and it was obvious that azardirachtin could function like the ecdysoids, which are known to be present in many plants. According to Kauser and Koolman (1984), azardirachtin reduced ponsasteron A-binding to ecdysteroid receptors in *Calliphora* epidermis *in vitro*. There was also no effect by azardirachtin on prothoracic gland secretion nor did it interfere with prothoracotropic hormone (PTTH) function. Koul et al. (1987) measured ecdyson secretion in 5th instar of *Bombyx mori* prothoracic glands, which were incubated in the presence and absence of PTTH, azardirachtin and mixtures of both. No marked difference was found when either prothoracic glands as controls, or activated with PTTH, were exposed to azardirachtin. These results proved that the compound neither exerted a direct, competitive effect on prothoracic gland secretion, nor did it block any PTTH receptor sites of the endocrine organ.

The effects of azardirachtin on the hormone system of the last larval instar of *Locusta migratoria* were studied and showed a typical dose-dependence effect on the responding larval instars (Sieber and Rembold, 1983). At a dose of 0.6 azardirachtin/g only 10% of the larval instars showed a reaction, whereas 2 μg/g elicited a maximal response and no larva was able to undergo ecdysis. The intermoult of azadirachtin-injected larvae varied between 8 and 60 days. The 4th and 5th instar larvae (control) lasted for 6 and 9 days respectively. Koul et al. (1987) also found similar results when azardirachtin was injected in 5th instar larvae of *Bombyx mori*, which was very effective in producing the pupal deformities and the larval growth inhibition at 1 and 2 μg/g respectively.

The inhibition of metamorphosis was indicated as an interaction of azardirachtin with the hormone system of the treated larvae. There was pronounced effect of azardirachtin on control of ecdysteroid titer as first demonstrated in 5th instar of *Locusta migratoria* (Rembold et al., 1981; Sieber, 1982; Sieber and Rembold, 1983). According to them this effect as an interference of the compound with the neuroendocrine system of the larvae, which was supported by histological studies, by increasing in paraldehyde-fuchsin stainable material in the neurosecretory cells of the Pars intercerebralis of azardirachtin treated last-instar locusts. Concomitantly, also with the ecdysteroid, the juvenile hormone synthesis was affected by azardirachtin and the dose dependent sharp increase of the larval response to the compound with an ED_{50} value around 1.5 μg/g was found to be due to the altered feeding behaviour, which resulted in high mortality possibly by influencing neural control centers (Rembold, 1981). Redfern et al. (1981) found persistent feeding inhibition in *Spodoptera frugiperda* larvae after transfer from an azardirachtin-treated to an untreated diet.

The effect on neural control centers was also indicated by changes in behaviour after azardirachtin treatment. Application of azardirachtins to the early larval stages of *Locusta* extended the duration of the larval stage to several weeks. Such larvae showed sexual behaviour like adults and flight pattern and formation of the flight muscle activity resembled the flight motor pattern of young locusts (Kutsch, 1985). The azardirachtin was extremely effective in regulating growth and behaviour in *Locusta* at low doses, and that feeding inhibition was not the primary cause for growth disruption. Inhibition of ecdysis in azardirachtin-treated larvae was due to interference with the hormonal control of moulting. The ecdysteroid titer in the treated larvae revealed a close relationship between endocrine conditions and morphogenetic processes. Shift or even complete disappearance of the moulting-hormone titer was the result of azardirachtin application (Sieber and Rembold, 1983). Such a shift of the ecdysone titer could affect feeding behaviour. Ecdysone administration at physiological concentrations (35 µg/5th instar *Locusta migratoria* nymph) reduced the live weight gain and all the treated insects failed to moult (Rao and Rembold, 1983), which demonstrated a relation between the actual hormone titer and feeding behaviour of the nymph. Therefore, not only a phago-deterrent could affect the hormone titer via starvation, but vice versa, the hormone titer could control feeding behaviour (Slama, 1978).

Dorn et al. (1986a) studied the effects of azardirachtin on the moulting cycle, endocrine system and ovaries in the last instar larvae of *Oncopeltus fasciatus* and supported the mutual and dose-dependent activity of this compound. The epidermis of the permanent larvae showed neither ecdysis nor apolysis; however, it was engaged in secretory activity, which might correspond to adult pro-cuticle secretion. These larvae showed an ecdysteroid peak, which was considerably delayed and was distinctly lower than in the controls. With high azardirachtin doses adult ovarian development began in some cases and resulted to chorionated eggs. Also neurosecretion appeared to be affected in these treated larvae and possibly the oviductal transformation from larval to adult form as well (Dorn et al., 1986b). Likewise, in the larvae of the tobacco hornworm, *Manduca sexta* growth and endocrine events were affected by azardirachtin (Schluter et al., 1985). According to them at low doses (0.5 µg/g) mainly defective pupae were formed, whereas with doses of 2 µg/g and higher mainly supernumerary larvae were formed. Most of these pupae failed to cast-off their exuvia and died after varying periods of survival. Here also the hemolymph ecdysteroid titers azardirachtin-treated last instar larvae showed the characteristic shift, which was reflected by delayed larval growth. The larvae which pupated usually failed to perform normal pupal development. The hormonal effects were reflected by changes in the histological pattern of the prothoracic glands (Bidmon, 1986). Similar results were described in 5th instar larvae of *Bombyx mori* (Koul et al., 1987) and *Epilachna* (Schmutterer and Rembold, 1980; Schluter, 1981). Thus it could be generalized that irreversibly, or at least for an extended period of time, azardirachtin blocked and sometimes changed developmental programs, as well as those which were normally expressed in the next instar only. Some examples were the adultoid characters of permanent larvae of *Locusta migratoria* (Shalon and Pener, 1984; Kutch, 1985), of *Oncopeltus fasciatus* (Dorn et al., 1984, 1986b) and of *Manduca sexta* (Schluter et al., 1985). However, differences in the reaction from azardirachtin treatment varied with insect species and some insects did not even react at all.

Inhibition of reproduction either by crude neem kernel extracts or by azardirachtin has long been known (Rembold and Sieber, 1981a,b; Koul 1984; Subrahmaniyam and Rao, 1986). As in the larvae, the main effect of azardirachtin was a change in the ecdysteroid titers (Redfern et al., 1981; Sieber and Rembold, 1983; Dorn et al., 1986b). Rembold et al. (1980a,b) studied the effects of azardirachtin into females of *Locusta migratoria* and observed that single injection of 10 µg/g azardirachtin into a female between 2 and 13 days resulted in ovaries with almost mature oocytes. Similar to the effects in the larva, there was a sensitive phase during oocyte development. Thus it was concluded that in *Locusta* females this was obviously the period of vitellogenesis, whose program could not be blocked by azardirachtin administration. Most of the treated locusts had no oviposition and only traces of ecdysteroid were present in the ovaries.

Similarly, azardirachtin when injected into newly hatched adults of *Oncopeltus fasciatus* affected their longevity, fecundity and hatchability of eggs from treated parents. There were marked differences between males and females, with higher mortality in the females and induction of impotence in the males with as little as 0.125 μg/insect, mainly due to mating failure, which could be the consequence of an azardirachtin-induced hormonal imbalance in the males (Dorn et al., 1986a). In *Schistocerca gregaria* injection of 2 μg/g of the compound into newly hatched females completely inhibited weight gain within the first 8 days. Azardirachtin delayed synthesis and release of neurosecretion from the A-type median neurosecretory cells of the brain, possibly leading to inhibition of ovarian development (Subrahmaniyam and Rao, 1986). Rembold et al. (1980a) and Subrahmaniyam et al. (1989) reported azardirachtin A as an insect growth inhibitor and hormonally controlled ovarian development in *L. migratoria* when injected at physiological dose. Under azardirachtin stress condition translocation and release of the neurosecretory proteins labeled with L-(35S)-systein in the corpus cardiacum was very poor. Thus it was concluded that azardirachtin influenced the release of trophic hormones from the corpus cardiacum leading to alterations in timing and titer of morphogenetic hormone pools.

According to Rembold (1981) and Bergot et al. (1981) juvenile hormone III was the only hormone analogue present in *Locusta migratoria*, which controlled synthesis of vitellogenin in the fat body and, as such, controlled egg maturation in the ovary and reduced synthesis of egg yolk protein. The ecdysteroids were synthesized in the ovary at the end of the egg maturation cycle and then released into the hemolymph. Under normal control conditions, juvenile hormone and ecdysterone concentrations were low in the hemolymph during previtellogenesis after adult emergence. From the 5th day onwards, the juvenile hormone titer increased steadily, reaching maximum at the end of egg maturation and then decreased steeply during oviposition. Due to the measurement of individuals it was possible to realize that, even in well-synchronized populations, there was a variation of the juvenile hormone peak, maximum in the range of plus or minus 1 day followed by a corresponding deviation in egg deposition. The juvenile hormone remained minimum during oviposition (Stay and Tobe, 1985) and generally the juvenile hormone titer increased during previtellogenesis, reaching its maximum during rapid oocyte growth. The juvenile hormone peak was therefore a sensitive marker for the female reproductive period. Its duration might be only a very few hours and the 24-h frame may therefore be too coarse. The same was true for the ecdysteroid titers. Here again, the hemolymph hormone titer was found to be closely connected with the progress of egg maturation. A steep increase of ecdysteroid titer began after the egg yolk protein was passed its maximum hemolymph titer and therefore an ecdysteroid peak became visible at the end of vitellogenesis. The two peaks of juvenile hormone and of ecdysone were more or less synchronous within the egg maturation cycles under normal conditions. Also the hemolymph peak of egg yolk protein coincided with the two hormone peaks (Rembold, 1989a).

Rembold (1989c) mentioned azardirachtins as a group of seven tetra nortriterpenoids, which were chemically similar to main compound azardirachtin-A. Azardirachtins interfered with neuro-endocrine regulation of juvenile and moulting hormone titers. Main cellular targets were the malpighian tubules and the corpus cardium of the insect. In the corpus cardium the azardirachtins reduced the turnover of neurosecretory materials. Consequently levels of the morphogenetic juvenile and moulting hormones were shifted and gradually decreased after azardirachtin injection. In this way metamorphosis of the juvenile insect was inhibited and reproduction of the adult was as well. Gupta and Rao (1990) also studied the effect of a single dose (1 μg/g) of injected azardirachtin into last instar larvae of *Spodoptera litura* on medium neurosecretory cells and corpora allata and found a delayed accumulation of neurosecretory materials in these tissues.

Application of azardirachtin could provide a better understanding of the correlation of these three components involved in egg maturation, i.e., juvenile hormone, ecdysteroids and vitellogenin.

Salannin

Salannin is another triterpenoid compound isolated from the neem seed extract and closely related to azardirachtins in its structure and functions (Figure 21). Its biological activity has been described in Table 11.

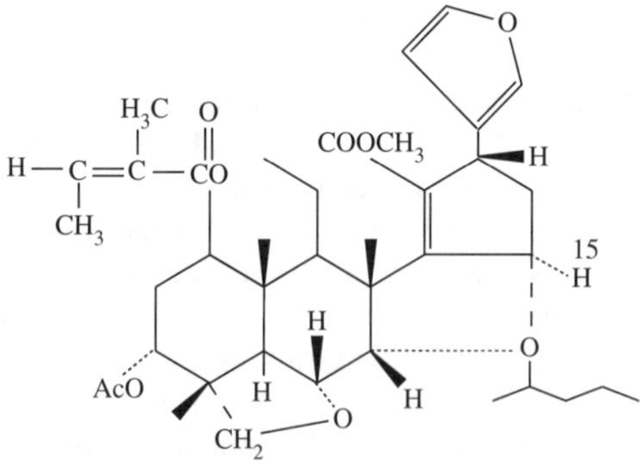

Figure 21. Salannin.

Deacetyl Azardirchtinol

This limonoid was found to be a protein ecdysis inhibitor when tested against *Heliothis virescens* (Kubo et al., 1986).

Nimbin and Nimbidinin

These are also structurally very similar triterpenoids to azardirachtin in their structure and functions. Their biological activity has also been mentioned in Table 11. Their chemical structures are represented in Figure 22 and Figure 23, respectively.

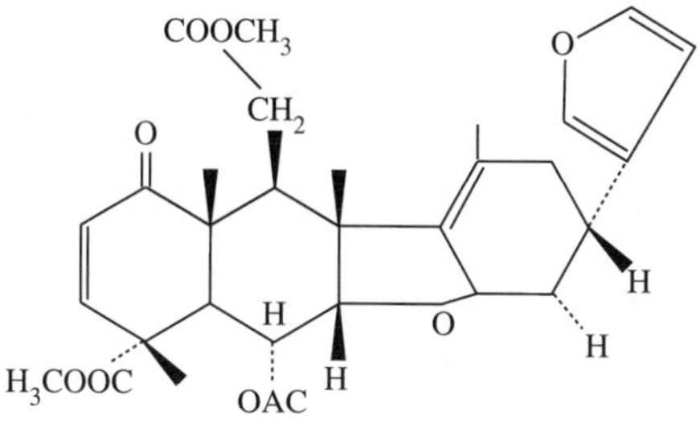

Figure 22. Nimbin.

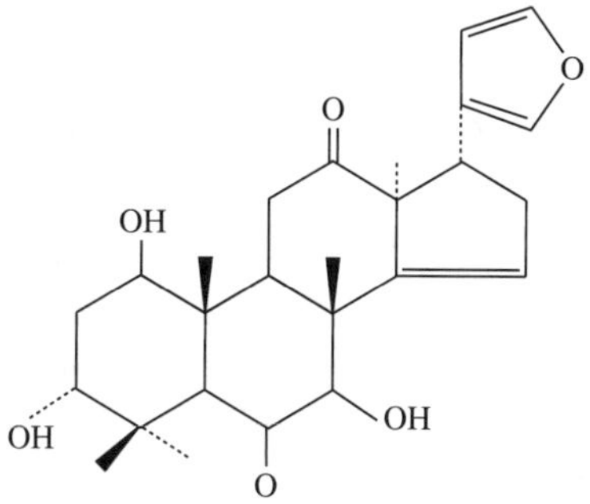

Figure 23. Nimbidinin.

Meliantriol

Meliantriol is triterpenoid alcohol, another active component isolated from neem seed oil. Biological activity of meliantriol is presented in Table 11 and its structure is represented in Figure 24.

112. *Balanites aegyptica* (Linn.) Delile. (Simaroubaceae)

(Syn.-*B. roxburghii* Blanch.)

Common name: Desert date or Inguli

This is a spiny evergreen perennial tree or shrub found in all the South East Asian countries in the tropics. In India, this plant is available in Rajasthan, Gujurat, Madhya Pradesh and South Indian States. Inguli is known to possess insecticidal property in its bark, root, fruit and seed extracts in water, alcohol and petroleum ether when tested under laboratory conditions against the melon

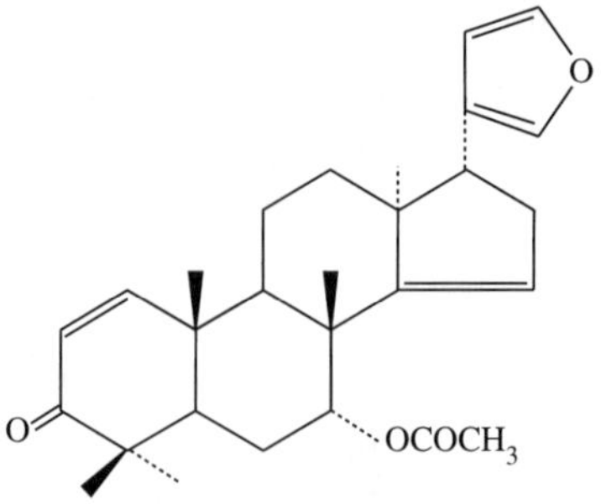

Figure 24. Meliantriol.

worm, *Diaphania hyalinata* and diamond-back moth, *Plutella xylostella* (Jacobson, 1958). Water and alcohol extracts of its bark and dry powder showed insecticidal activity against the aphids and the grasshoppers (McIndoo, 1983). Jain (1987) found an active component, a new 'saponin' (steroidal saponin), to show antifeedant activity to *Diacrisia obliqua* under controlled conditions when isolated from its seed extract.

113. *Bambusa vulgaris* Schrader ex. Wendland (Bambusaceae)

Common name: Bamboo tree or Feathery bamboo

This is a perennial shrub/tree found in tropical and sub-tropical zones. Bamboo tree is available in India, China, the Philippines, Indonesia, Bangladesh and Sri Lanka. The stem and branches of feathery bamboo were found to repel the insect pests of rice when placed in the corner of the paddy fields (Litsinger et al., 1978).

114. *Bania tolifolia* Linn. (Menispermaceae)

Common name: Hippe

This is a perennial shrub/small tree found in sub-tropical and temperate zones and available in India, China, USSR and West Africa. Its oil showed toxicity to the pulse bruchid, *Callosobruchus chinensis* when admixed with stored redgram (Sangappa, 1977).

115. *Barringtonia racemosa* (Linn.) Sperng. (Barringtoniace)

Common name Ejjool

This is a small perennial tree commonly found in tropical zones. Ejjool is distributed in South East Asia, Australia and West Africa. It is available in the riversides in the East and Western Ghats in India. Root, bark, stem, branch, fruit and seed extracts of this tree in water, alcohol and petroleum ether were reported to show insecticidal and antifeedant activities against the citrus aphid, *Aphis citricola* and the coffee tree aphid, *Toxoptera aurantii* (Jacobson, 1975).

116. *Bauhinia purpurea* Linn. (Caesalpiniaceae)

Common name: Camel's foot tree or Pink bauhinia
This is a perennial middle size tree indigenous to the lower slopes of the Himalayas and distributed in North India, Assam, Manipur, Khasi hills, Western Peninsula, Burma and Nepal. Water extract of its branch inhibited the growth and the development and finally reduced longevity of the cotton bug, *Dysdercus cingulatus* (Kareem, 1984).

117. *Bersama obyssinica* Fresen. (Melianthaceae)

Common name: Bersam
This is a perennial shrub/small tree found in temperate and sub-tropical zones. Its flower extract in ethyl ether was found to possess attractant property to the females of Mediterranean fruit fly, *Ceratitis capitata* and melon fly, *Dacus cucurbitae* (Keiser et al., 1975).

118. *Betula sollennis* A. Gray. (Betulaceae)

Common name: Birch genus
This is a perennial creeping shrub or weak tree found in temperate zones. Insecticidal property in its water and alcohol leaf extracts was reported against vinegar fruit fly, *Drosophila hydei* (Jacobson, 1958).

119. *Bidens pilosa* Linn. (Asteraceae)

Common name: Indian bidens
This is an annual aromatic shrub in tropical and temperate zones and commonly available in Indonesia, Nepal, Malaysia, Sri Lanka, India and China. Extracts of its whole plant, flower and meristem tissues in alcohol and water were reported to possess insecticidal and antifeedant properties against the milkweed bug, *Oncopeltus fasciatus* (Khan, 1982).

120. *Bigonia pearcei* Linn. (Bigoniaceae)

This is a small annual herb in tropical and sub-tropical zones. Insecticidal activity in its leaf extract in alcohol and water was observed against the diamond-back moth, *Plutella xylostella* (Jacobson, 1975).

121. *Blettia striata* Reichb. (Orchidaceae)

This is an annual or biennial herb found in temperate zones. Whole plant or its leaf extracts in water and alcohol were found to show antifeedant activity against the diamond-back moth, *Plutella xylostella* (Jacobson, 1975).

122. *Blumea eriantha* G. Don. (Asteraceae)

Common name: Nimurdi (morathi)/Blumea
This is an erect herbaceous tropical plant distributed in the Konkan, Eastern and Western Ghats and South India. Oil vapours of this plant reduced hatching of the eggs and reproductive potential of the cotton bollworm, *Earias vittella* in a saturated environment when tested under controlled laboratory conditions (Dongre and Rahalkar, 1982).

123. *Boehmerea cylindrica* (Linn.) Sw. (Urticaceae)

Common name: Rhea
This is an annual/perennial shrub/small tree found in sub-tropical and tropical zones. Stem, leaf and flower extracts of rhea in ethyl ether were found to possess insect attractant property to both the male and female melon fly, *Dacus cucurbitae* (Keiser et al., 1975).

124. *Bothrichloa intermedia* (R.Br.) A. Camus (Poaceae)

(Syn.-*Andropogon intermedius* R.Br.)
Common name: Sandhor
This is a tall grass found in semi-arid and arid zones. In India, it is available throughout the country. Its leaf and whole plant extracts in water were reported to show insecticidal activity against grasshoppers (Waespe and Hans, 1979).

125. *Bougainvillea spectabilis* Willd. (Nyctaginaceae)

Common name: Bougainvillea
This is a perennial wood scandent tree/or woody climber found in tropical India but a native to South America. Bougainvillea is known to possess insecticidal property against the rice weevil, *Sitophilus oryzae* (Jacobson, 1975). Its flower bracts powder was reported to show toxicity to the pulse beetles, *Callosobruchus chinensis* and *C. maculatus* under laboratory tests (Anonymous, 1980–1981).

126. *Bouvardia ternifolia* (Cav.) Schlecht. (Rubiaceae)

This is a perennial small tree or shrub found in tropical and temperate zones. Stem, leaf and flower extracts in ethyl ether showed attractancy to both the male and female of Mediterranean fruit fly, *Ceratitis capitata* and the melon fly, *Dacus cucurbitae* (Keiser et al., 1975).

127. *Brassica campestris* Linn. var. *sarson* Prain (Brassicaceae)

(Syn.-*B. campestris* Linn. var. *glauca* Duth. and Full., *B. campestris* Linn. var. *dichotome* Watt.

Common name: Yellow mustard or Indian colza

This is a seasonal herb cultivated for its oil in India and in several other tropical countries. Oil of mustard has been reported to protect the pulse grains against bruchids infestation in storage (Mummigatti and Raghunathan, 1977; Sangappa, 1977; Prakash, 1982). Ali et al. (1983) and Doharey et al. (1983, 1984a,b, 1987) reported its oil to be toxic to the pulse beetle, *Callosobruchus chinensis* when admixed with the greengram seed @ 1% w/w in storage and caused 100% mortality and protected the pulse seed in storage. Its oil also showed repellency to the pulse beetles, *C. chinensis* and *C. maculatus* and protected cowpea seeds admixed with the oil at the rate of 1% w/w (Naik and Dumbre, 1984). Begum and Quiniones (1990) and Singal and Singh (1990) found its oil @ 3 ml/kg treatment to greengram seeds a most effective protectant against infestation of *C. chinensis* and adversely affected hatching and development of the beetle.

128. *Brassica campestris* Linn. var. *taria* Duth. and Full. (Brassicaceae)

Common name: Indian rape or Rapeseed

This is an oil yielding seasonal herb grown in Uttar Pradesh, Rajastan, West Bengal and Punjab in India. Its seed extract was found to be toxic to the adults of the flour beetles, *Tribolium castaneum* and *T. confusum* (Edig and Davis, 1980). Its seed oil 1% was reported to show insecticidal activity against the pulse beetles, *Callosobruchus chinensis* and *C. maculatus* (Ali et al., 1983 and Doharey, 1990). Chauhan et al. (1987) found its seed extract to show toxicity to the adults and the larvae of the rice moth, *Corcyra cephalonica*. Rapeseed oil treatment at the rate of 3 ml/kg seeds of chickpea was also reported to inhibit the hatching of the eggs and development of *C. chinensis* and protected the grains from its infestations (Singhal and Singh, 1990).

Petterson (1976) reported an active component, 'sinigrin', in its undamaged plants, which was found to attract the mustard gall midge, *Dasyneura brassicae*.

129. *Brassica cernua* (Linn.) Juss. (Brassicaceae)

This is a seasonal herb found in tropical, sub-tropical, Mediterranean and temperate zones. This herb is reported to possess insecticidal property in its aqueous leaf and whole plant extracts against the vinegar fly, *Drosophila hydei* (Jacobson, 1958).

130. *Brassica hirta* Moench. (Brassicaceae)

(Syn. *B. alba* [Linn.] Rabenh., *Sinapis alba* Linn.)

Common name: White mustard

This is an annual or seasonal herb found in temperate, tropical and Mediterranean zones and available in India. Its oil was found to protect storage insects infesting stored pulses, especially the blackgram and the greengram, when admixed with stored grains (Prakash, 1982).

131. *Brassica juncea* (Linn.) Czern. & Coss. (Brassicaceae)

(Syn.-*B. juncea* Hook f. & Thomas, *B. Lanceolata* Lange, *Sinapis juncea* Linn.)

Common name: Indian mustard

This is a seasonal herb cultivated in North and Central India and in other tropical countries for its oil. Bowry et al. (1984) reported its oil cake to show repellency to the rice weevil, *S. oryzae* and to reduce its oviposition in stored maize grains admixed with the cake at the rate of 2.5 to 7.5% w/w. Verma and Singh (1985) found its seed extract to show antifeedant activity against the hairy caterpillar, *Amsacta moorei* on groundnut crop. Similarly, Singh and Vir (1987) and Babu et al. (1989) reported its oil to reduce the oviposition of the pulse beetles, *Callosobruchus maculatus* and *C. chinensis* when seeds were treated before storage. Choudhury (1990) and Kumari et al. (1990) also found 1% mustard oil to protect treated pea seeds in storage against the pulse beetle, *C. chinensis* when admixed with the seeds. Singh et al. (1993) and Ramzan (1994) further reported 3 mg/kg mustard oil admixed with the pulse grains to show no damage by *C. chinensis* and *C. maculatus* respectively in storage.

132. *Brassica oleracea* Linn. (Brassicaceae)

Common name: Wild cabbage

This is an annual or seasonal herb found in temperate zones. The roots and rhizome extracts of this herb are known to possess insecticidal property when tested against the vinegar fly, *Drosophila melanogaster* and the soldier fly, *Inopus rubriceps* (Jacobson, 1975).

133. *Brassica oleracea* Linn. var. *acephala* D.C. (Brassicaceae)

Common name: Borecole or Kale

This is an annual or seasonal herb found in tropical zones. In India, it is available in Assam, Kashmir, Maharastra and Gujarat and cultivated for its shoot and leaves used as vegetable. Root extract of this herb was reported to show insecticidal activity when tested against the vinegar fly, *D. melanogaster* and the black swallow-tail butterfly, *Papilio polyxenes* (Michael et al., 1985).

134. *Brassica oleracea* Linn. var. *capitata* Linn. (Brassicaceae)

Common name: Cabbage

This is a seasonal herb grown as a vegetable in tropical and sub-tropical zones. This is a native to Europe but now grown in North and Central States in India. Root extract of cabbage in water was reported to possess insecticidal property when tested against the vinegar fly, *D. melanogaster* (Jacobson, 1975).

135. *Brassica oleracea* Linn. var. *rapifera* Metz. (Brassicaceae)

(Syn.-*Brassica rapa* Linn.)

Common name: Turnip

This is an annual/seasonal branched herb, commonly cultivated for its swollen roots as a vegetable in Punjab and Uttar Pradesh in India and also in other tropical countries. Root extract of turnip was reported to possess insecticidal property when tested against the pea aphid, *Macrosiphum pisum* (Jacobson, 1975). An active component, '2-phenyl ethyl isothiocyanate', was isolated from edible parts of turnip and found to show toxicity to the insects but it showed no adverse effect on human health (Lichtenstein et al., 1962).

136. *Brongniartia benthamiana* Hemsl. (Fabaceae)

This is an annual/perennial herb found in tropical zones and commonly available in Central America. Its fruit and twig extracts in ethyl ether showed attractant activity to the females of Mediterranean fruit fly, *Ceratitis capitata* and the melon fly, *Dacus cucurbitae* (Keiser et al., 1975).

137. *Bryonia alba* Linn. (Cucurbitaceae)

Common name: Wild hop

This is an annual/perennial vine found in temperate zones. It is commonly available in vineyards in the middle of Europe. Its root extract was reported to show aphidicidal property when tested under controlled conditions (McIndoo, 1983).

138. *Buddleia lindleyana* Fort. (Loganiaceae)

Common name: Butterfly bush

This is a large perennial shrub native to China but grown in tropical zones as an ornamental shrub for its purple violet flowers. Water extract of its leaf was reported to show insecticidal activity against the black bean aphid, *Aphis rumicis* (Jacobson, 1975).

139. *Bupleurem fruticosum* (Linn.) Wall. (Umbelliferae)

This is an annual/perennial herb found in Europe, Australia and in temperate Asian countries. In India, it is available on northern hills from Kashmir to Khasi hills up to 1000 to 4000 m high. Phenyl propanoids isolated from ether extract of its leaf were found to show antifeedant activity when tested under controlled conditions against Colorado potato beetle, *Leptinotarsa decemlineata* and armyworm, *Mythimna unipunctata* (Muckenstrum et al., 1981).

140. *Butea monosperma* (Lamk.) Taubert. (Papilionaceae)

(Syn.-*B. frondosa* Koeining ex Roxb., *Erythrina monosperma* Kuntze.)
Common name: Flame of the forest
This is a small perennial deciduous tree available throughout India and also in other tropical countries. Extracts of this tree showed high juvenile hormone analogue activity when applied topically to the larvae of *Spodoptera litura* in the laboratory test. Further, this extract showed ovicidal activity to the test insect (Lalitha et al., 1981).

141. *Butyrospermum parkii* Kotaschy. (Sapotaceae)

Common name: Karite
This is a perennial tree or large shrub found in tropical Africa. Leaves of this plant showed promising grain protection when admixed with grains against the stored grain insect pests (Giles, 1964). Pereira (1983) reported its seed oil to protect cowpea seeds against the pulse beetle, *Callosobruchus chinensis* for 90 days when admixed with its oil at the rate of 5 ml/kg seeds.

142. *Buxus japonica* Linn. (Buxaceae)

Common name: Japanese buxus
This is a perennial shrub or woody climber found in sub-tropical and temperate zones. This is commonly available in Northern and Central America, Japan and tropical South Africa. Aqueous and alcohol extracts of its leaf and branch were found to possess insecticidal property when tested against the vinegar fly, *Drosophila hydei* under laboratory conditions (Jacobson, 1958).

143. *Buxus sempervirens* Linn. (Buxaceae)

(Syn.-*B. wallichiana* Baill.)
Common name: Common buxus or Box tree or Papari

This is a perennial small tree or woody climber found in temperate and subterranean zones. It is commonly available in dry chalky hills in Europe and in the West of Asia. Leaf extract of this plant showed repellent activity when tested against the Japanese beetle, *Popillia japonica* (McIndoo, 1983).

144. *Caesalpinia bonducella* Flem. (Caesalpiniaceae)

(Syn.-*Guilandina bonducella* Linn.)

Common name: Fever-nut or Nicker tree or Mohua bean

This is a perennial shrub or small tree found in tropical zones and available in Indian plains and the West Indies. Its nut powder was reported to protect leguminous and vegetable seeds against the common storage insects (Mookherjee et al., 1970) and also against the pulse beetles, *Callosobruchus* spp. (Anonymous, 1980–1981).

145. *Caesalpinia coriaria* (Jacq.) Willd. (Caesalpiniaceae)

Common name: Divi-Divi or American sumach

This is a perennial small tree or shrub found in tropical zones. American sumach is a native of the West Indies and South America but also available in South and North-West Indian States. The aqueous leaf extract of this plant was reported to show repellent activity when tested against the Japanese beetle, *Popillia japonica* (McIndoo, 1983).

146. *Caesalpinia pulcherrima* (Linn.) Sw. (Caesalpimaceae)

(Syn.-*Poinciana pulcherrima* Linn.)

Common name: Peacock flower or Barbados pride or Paradise flower or Gulmohar

This is a large perennial shrub or small ornamental tree found in tropical zones. This shrub is commonly available in North and Central India. Its root bark, leaf, flowers, dry powders and extracts in ether and alcohol were reported to show insecticidal and repellent activities against the rice weevil, *S. oryzae* (Jacobson, 1975; Prakash et al., 1978–1987).

147. *Caesalpinia sappan* Linn. (Caesalpiniaceae)

Common name: Sappan wood

This is a perennial shrub/small tree native to India and commonly grown as a hedge plant in West Bengal, Orissa, Andhra Pradesh, Kerala, Tamil Nadu and Karanataka. Its methanol and benzene leaf extracts were reported to reduce multiplication of Angoumois grain moth, *Sitotroga cerealella* (Prakash et al., 1978–1987).

148. *Calliandra calothyrsus* Benth. (Mimiosoidae)

This is a perennial, widely cultivated small tree/shrub found in tropical zones and commonly grown in Indonesia. From its leaf, seed and sap, a number of non-protein imino acids, viz., S-(β-carboxyethyl)-cystiene and cis-4-hydroxy pipecolic acid (Figure 25), were reported to show antifeedant activity against *Spodoptera frugiperda* and *Heliothis* sp. Non-protein imino acids are amino acids containing a heterocyclic nitrogen ring, which are largely confined to legumes. These compounds are derivatives of pipecolic acid, the higher homologue of proline and include acetylamino, 4-monohydroxy and 4-dighydroxy derivatives.

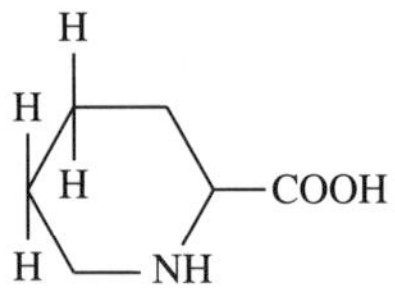

Figure 25. Cis-4-hydroxy pipecolic acid.

High concentrations above 5% of these components were found to be toxic to *S. frugiperda*. The chemical formula of S-(β-corboxyethyl)-cysteine is $HOOC-CH_2-CH_2-S-CH_2-NH_2CH-COOH$.

149. *Callicarpa japonica* Thunb. (Verbenaceae)

(Syn.-*C. mimurazaki* Sieb.)

This is an ornamental perennial shrub with pinkish white flowers and violet berries found in sub-tropical and temperate zones. This shrub is commonly available in Japan and China. Benzene extract of its leaf was reported to show antifeedant activity against the 3rd instar larvae of *Spodoptera litura* (Hosozawa et al., 1974b).

150. *Calophyllum inophyllum* Linn. (Clusiaceae)

Common name: Indian laurel or Laurelwood or Undi/Polang

This is a perennial evergreen tree found in tropical zones. Laurelwood tree is commonly distributed on the seacoast from Madagascar to Australia and the Pacific. In India, it is commonly available in Orissa, Karnataka, Maharastra and Andaman. Biological activities of its de-oiled cake, seed extracts and oil have been tested against the insect pests of stored products as well as insect pests of field crops. The results are as follows:

Biological Activity Against Field Insect Pests

Seed extract 1% and seed oil significantly inhibited the incidences of the yellow mosaic virus (YMC) when sprayed on blackgram and also reduced the population of its vector whitefly, *Bemisia tabaci* (Mariappan et al., 1987). Similarly, its 1% seed oil when sprayed on rice plant successfully reduced the survival and checked the population build up of the green leafhopper in rice, *Nephotettix virescens* and transmission of rice tungro virus (Narsimhan and Mariappan, 1988). Mariappan et al. (1988) reported its oil as an antifeedant when tested against the rice leaffolder, *Cnaphlocrosis medinalis* and the green leafhopper, *N. virescens* and also reduced their survival. Ramaraju and Sundara Babu (1989) found undi seed oil (1.0%) and 2% extract to significantly reduce the nymph and adult emergence of BPH, *Nilaparvata lugens* and WBPH, *Sogatella furcifera* in the rice crop. In greenhouse studies, the oil of its seed significantly reduced the incidences of *N. lugens* and *S. furcifera* and gave superior results than neem oil and also disrupted the growth of BPH at 5000 ppm and GLH at 2500 ppm (Krishnaiah and Kalode, 1990). Koshiya and Chaiya (1991) also reported its 1.5% oil to show 100% ovipositional deterrency to the adults of *Spodoptera litura*.

Biological Activity Against Storage Insect Pests

Naik and Dumbre (1984) reported its oil to show repellent activity against the pulse beetles, *Callosobruchus chinensis* and *C. maculatus* and it protected cowpea grains admixed with the oil at the rate of 1% w/w. Its seed oil also protected stored paddy against Angoumois grain moth, *Sitotroga cerealella* when the grains treated with its oil at the rate of 1% w/w were stored in the laboratory and godowns (Prakash et al., 1987, 1991).

151. *Calopogonium coeruleum* Jaq. (Leguminosae)

Common name: Jicana
This is a perennial vine found in sub-tropical and temperate zones and commonly available in North America. Its seed and pod extracts in water were reported to show insecticidal activity against the melon moth, *Diaphania hyalinata*, diamond-back moth, *Plutella xylostella* and fall armyworm, *Spodopterafrugiperda* (Jacobson, 1958).

152. *Calotropis gigantea* Ait. (Asclepiadaceae)

Common name: Crown plant or Madar or Aak
This is a perennial undershrub found in tropical and arid zones. Crown plant is commonly found in Arabian countries, Pakistan and Afghanistan. It is available throughout India. Its whole plant extract or extracts of its stem, branch, leaf and flower in water, alcohol, petroleum ether or kerosene were found to show antifeedant and insecticidal activities when tested against the rice weevil, *Sitophilus oryzae* (Jacobson, 1975) and the jute hairy caterpillar, *Diacrisia obliqua* and the leaf eating larvae (Jacobson, 1958). However, aqueous leaf extract sprayed on watermelon and vegetable crops avoided feeding

of the locusts (Bhatia, 1940). Bandara et al. (1987) reported methanol and dichloromethane extracts of its flower to show toxicity to the aphids. Pandey (1988) found its aqueous leaf extract to show toxicity to the sugarcane top shoot borer, *Tryporiza nivella* when tested under laboratory conditions. Petroleum ether extract of its leaf was also reported to reduce the grub and adult populations of the brinjal spotted leaf beetle, *Henosepilachna vigintioctopunctata* when tested under greenhouse conditions (Reddy et al., 1990).

153. *Calotropis procera* (Wild.) Dryand ex. Wt. Ai. (Asclepiadaceae)

(Syn.-*Asclepiad procera* Willd.)

Common name: Swallow wort or Akund or Madacer or White oak

This is an evergreen perennial shrub found in tropical and semi-arid zones. In India, it is commonly available in Western and Central States and used as a human poison. Extracts of its whole plant, stem, branch, leaf and flower in water, alcohol and petroleum ether were reported to show insecticidal activity when tested against the red pumpkin beetle, *Aulacophora foveicollis* and the cabbage butterfly, *Pieris brassicae* (Khanvilker, 1983). However, soil application of its leaves @ 9 t/ha was reported to show insecticidal activity against termite attack in the wheat crop (Hussain, 1928).

Against stored grain pests like *T. castaneum, R. dominca* and *T. granarium*, its aqueous leaf extract showed repellent activity in laboratory tests (Jilani and Malik, 1973). Sharma (1983) found its leaf extract to reduce the population of *R. dominica* in stored wheat when applied alone or in combination with malathion. Its flower extract was also reported to be toxic to 1st, 2nd, 3rd and 4th instar larvae of the lesser grain borer, *Rhyzopertha dominica* (Sharma, 1985). Yadav and Bhatnagar (1987) found its leaf powder to show toxicity to the pulse beetle, *C. chinensis* when admixed with cowpea seeds in storage at the rate of 2% w/w. Jacob and Sheila (1993) reported its leaf powder admixed with stored rice @ 2% w/w to protect it against *R. dominica* under controlled conditions.

154. *Camellia sinensis* (Linn.) O. Kuntze. (Theaceae)

(Syn.-*C. thea* Link., *C. theifera* Griff, *Thea sinensis* Linn.)

Common name: Tea

This is an evergreen perennial shrub cultivated in tropical and sub-tropical areas of China, Japan and USSR. In India, tea is grown in Assam, Darjiling, Nilgiris, Malabar, Dehradun and Kumaon hills. Its leaf extract in water or oil was reported to possess insecticidal and antifeedant properties against the sugarcane woolly aphid, *Ceratovacuna lanigera* (Jacobson, 1958), squash bug, *Anasa tristis* (McIndoo, 1983) and the termites (Jacobson, 1975).

155. *Canavalia maritima* (Aubl.) Thouars (Fabaceae)

Common name: Canavalia

This is a biennial or perennial vine found as a shore plant in tropical and sub-tropical zones. In India, it is grown in Tamil Nadu, Andhra Pradesh, West Bengal and Kerala. Its aqueous leaf extract was reported to possess insecticidal activity against the vinegar fly, *Drosophila hydei* (Jacobson, 1958).

156. *Canna indica* Linn. var. *orientalis* E. Rosc. (Cannaceae)

Common name: Indian shot

This is a perennial ornamental herb found in tropical zones and available in Malaysia, Sri Lanka, Bangladesh and India. It is commonly found in South Indian States. Stem, branch, leaf and flower extracts of this herb in water and alcohol were reported to possess insecticidal activity against the rice weevil, *Sitophilus oryzae* and the cotton stainer, *Dysdercus cingulatus* (Balasubramanian, 1982).

157. *Cannabis sativa* Linn. (Moraceae)

Common name: Marijuana or Hemp or Bhang

This is a tall aromatic annual/perennial herb found in tropical and sub-tropical zones. It is native of Central Asia but available in India, Bangladesh, Sri Lanka, Pakistan and Malaysia. Its whole plant and leaf extracts in water or alcohol were reported to show insecticidal or repellent activities when tested against the insect pests of agricultural importance (McIndoo, 1983). Khare (1972) found leaves of the bhang to protect stored wheat against *S. oryzae* when admixed with the grains. Similarly, dried leaves of this plant when admixed with stored rice at the rate of 2% w/w under a laboratory test gave 59% grain protection against *S. cerealella*, *R. dominica* and *S. oryzae* in controlled conditions (Prakash et al., 1978–1987). However, under natural storage conditions of insect infestation these leaves at 2% w/w dose failed to provide significant grain protection against the grain boring insects like *S. cerealella*, *S. oryzae* and *R. dominica* as well as against the infestation of other stored grain insect pests (Prakash et al., 1982).

158. *Canthium euroides* Lam. (Rubiaceae)

Common name: Arbutus unedo

This is a perennial tree available in tropical zones. Its leaf extract in water was reported to show insecticidal and antifeedant activities against Mexican bean beetle, *Epilachna varivestis* (Waespe and Hans, 1983).

An active component named 'unedoside' (Figure 26) was isolated and identified from its leaf extract and reported to show antifeedant activity against

E. varivestis and African armyworm, *Spodoptera exempta* when tested under laboratory conditions in a free choice test (Kubo and Nakanishi, 1978).

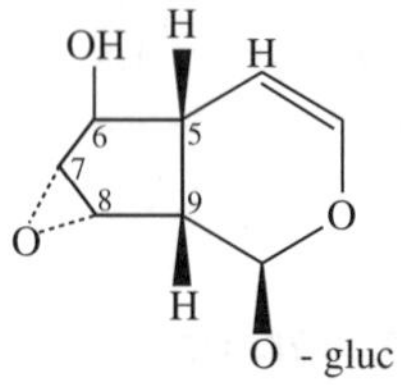

OH H H 6 5 7 8 9 O O H O - gluc

Figure 26. Unedoside.

159. *Capparis horrida* Linn. (Capporidaceae)

(Syn.-*C. zeylanica* Linn.)

Common name: Gitoran or Ardanda

This is a perennial thorny climbing shrub found in tropical zones and commonly available in South Indian States. Fruit extract of this shrub in water and its oil were reported to possess insecticidal property when tested against the rice armyworm, *Mythimna (Pseudoletia) unipunctata* (Nagarajah, 1983).

160. *Capsella bursa-pastoris* Medik. (Brassicaceae)

Common name: Shepherds purse

This is an annual herb found in temperate and sub-tropical zones and commonly available in Central USA. This plant extract was found to possess insecticidal property when tested against the Japanese beetle, *Popillia japonica* (McIndoo, 1983).

161. *Capsicum annum* Linn. (Solanaceae)

Common name: Red pepper or Chili

This is an annual/seasonal spicy herb grown in tropical and sub-tropical zones for its economic importance. This plant is native of the West Indies and tropical America but grown throughout India. Extracts of its fruiting body when applied topically or admixed with diets showed insecticidal activities against the corn weevil, *Sitophilus zeamais* (Javier, 1981), the cowpea bean weevil (Ponce De Leon, 1983) and certain flying insects (Omar, 1983). In Andhra Pradesh, dried fruits of red chilies are traditionally being used to control storage insects of milled rice and the pulses (Prakash et al., 1986).

162. *Capsicum fruitescens* Linn. (Solanaceae)

Common name: Giant green pepper or Pickle pepper or Goat pepper

This is an annual/seasonal herb grown in tropical and temperate zones and commonly cultivated in North India. Its aqueous fruit extract possessed insecticidal property against the rice weevil, *Sitophilus oryzae* and showed 100% mortality within 24 h when treated using 2% extract (Debkirtaniya et al., 1980b) and was also reported to show repellent activity against stored grain pests in general (Michael et al., 1985).

163. *Cardaria draba* Desv. (Brassicaceae)

This is an annual/biennial herb grown in Mediterranean and West Asian regions. Its whole plant extract in water was found to show insecticidal activity against the aphids of agricultural importance (McIndoo, 1983).

164. *Cardiocrinium cordatum* Endl. (Liliaceae)

This is a perennial herb grown in temperate zones and commonly found in the Himalayas and East Asian regions. Extracts of its leaf and stem were reported to show insecticidal activity against the vinegar fly, *Drosophila hydei* (Jacobson, 1958).

165. *Carex cliborum* Linn. (Cyperaceae)

This is an annual herb found in temperate zones. Its root extract in water was toxic to the vinegar fly, *Drosophila hydei* (Jacobson, 1958).

166. *Carex siderostica* var. *glabra* Cav. (Cyperaceae)

This is an annual herb found in temperate zones. Its stem and leaf extracts were toxic when tested against the vinegar fly, *Drosophila hydei* (Jacobson, 1958).

167. *Carthamus tinctorius* Linn. (Asteraceae)

Common name: Safflower (Kusum)

This is a seasonal herb native to India and cultivated for its oil seeds in Madhya Pradesh, Andhra Pradesh, Karanataka and Maharastra. Its seed oil was reported to show toxicity to the pulse beetles, *Callosobruchus chinesis* and *C. maculatus* when tested by admixing it with stored greengram (Anonymous, 1980–1981). Doharey et al. (1983, 1984a,b, 1987) also reported safflower oil to protect the greengram seeds in storage from the attack of *C. chinensis* and *C. maculatus* when admixed @ 1% w/w and it was found to show ovicidal activity and also to check their multiplication. Gupta et al. (1988) reported its oil as a most effective greengram protectant when seeds were coated @ 0.6% w/w and stored in jute bags for 6 months.

168. *Carum roxyburgharum* Benth. ex. Kurz. (Umbelleferae)

(Syn.-*Trachyspermum roxburgharum* [D.C.] Craib.)

Common name: Bishops weed

This is an annual herb found in tropical zones. This is commonly available in Karanataka, Andhra Pradesh and Orissa in India. Its seed powder was reported to show toxicity to the pulse beetles, *Callosobruchus chinesis* and *C. maculatus* (Anonymous, 1980–1981).

169. *Caryopteris divaricata* Bunge. (Vebrenaceae)

This is a perennial shrub found in temperate zones and commonly available in the Himalayas in India and Japan. One percent plant extract in benzene was reported to show antifeedant activity when tested against the 3rd instar larvae of *Spodoptera litura* (Hosozawa et al., 1974a).

170. *Cassia alata* Linn. (Caesalpiniaceae)

Common name: Ringworm cassia

This is a perennial ornamental shrub found in tropical and sub-tropical zones. In India, it is available in West Bengal and Tamil Nadu. Bark and leaf extracts of this plant were found to be toxic to the common black and red ants (Jacobson, 1975).

171. *Cassia artemisioides* Gaudichex D.C. (Caesalpiniaceae)

Common name: Wormwood senna

This is a perennial/annual bushy shrub found in the dry tropical and sub-tropical zones and a native to Australia. Flower extract of this plant in water was reported to show toxicity when tested against the rice weevil, *Sitophilus oryzae* (Jacobson, 1975).

172. *Cassia diclymobotiya* Collad. (Caesalpiniaceae)

This is a perennial shrub found in tropical and sub-tropical zones known to possess insecticidal property when its leaf aqueous and alcoholic extracts were tested against the aphids of agricultural importance (Worsley, 1934).

173. *Cassia fistula* Linn. (Caesalpiniaceae)

Common name: Golden shower or Indian laburnum or Purging cassia

This is an ornamental tree found in the tropics. It is available in Coastal Indian States. Its leaf extract in water was reported to show insecticidal activity when tested against the pulse beetle, *Callosobruchus chinensis* in controlled conditions (Satpathy, 1983) and termites (Jacobson, 1975). Jaipal et al. (1983) reported its crude seed extract to inhibit the metamorphosis of 5th instar larvae of the cotton stainer, *Dysdercus koenigii*. Its seed extract was also found to reduce population build up of *S. cereallella*, *S. oryzae* and protected the rice grains treated with the extract and stored under laboratory conditions (Prakash et al., 1989, 1990b,c, 1991).

174. *Cassia senna* Roxb. (Caesalpiniaceae)

(Syn.-*C. sophora* Linn., *C. esculenta* Roxb., *C. chinensis* Facq.)
Common name: Sophera senna
This is a perennial herb commonly grown in the gardens in India, China, Malaysia, the Philippines and East Africa. Glucosides extracted from the pod and leaf of this plant showed synergistic effect to enhance the toxicity of synthetic pyrethroids to the larvae of *Spodoptera littoralis* when combined with and tested under controlled conditions (Mesbatch et al., 1985).

175. *Cassia tora* Linn. (Caesalpiniaceae)

(Syn.-*C. obtusifolia* Linn.)
Common name: Coffee weed or Sickle weed
This is an annual herb or undershrub found in tropical, sub-tropical and temperate zones and commonly available in North India. Its whole plant aqueous extract was reported to possess insecticidal property against *Autographa brassica* when tested under controlled conditions (Khan, 1982). Petroleum ether extract of its seeds was found to be toxic on treated film in the paired petri dish against the red cotton bug, *Dysdercus koenigii* and then transferred to the okra leaf for feeding (Chandel et al., 1984).

176. *Cassine glauca* Von. Martins (Leguminosae)

This is a seasonal/annual herb found in tropical and sub-tropical zones and commonly available in South India. Its whole plant extract in acetone was an effective antifeedant against the castor semi-looper, *Achoea janata* (Despande et al., 1990).

177. *Castanea dentata* Mill. (Agaceae)

Common name: American chestnut
This is a perennial tree found in tropical and temperate zones. Its leaf aqueous extract showed repellency to the Japanese beetle, *Popillia japonica* (McIndoo, 1983).

178. *Catharanthus roseus* Leo. Don. (Apocynaceae)

(Syn.-*Lochnera roseus* [Linn.] Spach., *Vinca rosea* L.)

Common name: Rose periwinkle or Madagascar winkle

This is an ornamental small perennial herb found in the tropics. This is a native of the West Indies and distributed in all the South East Asian countries and East Africa. Rose periwinkle is available in Orissa and Tamil Nadu in India. Its leaf extract in water showed phago-deterrency against the larvae of *Spodoptera litura*. Larvae lost their weights and failed to pupate when fed on treated cotton leaves (Meisner et al., 1981).

Aqueous leaf extract of this herb was also reported to show insecticidal activity against the yellow stem borer of rice, *Scirpophaga incertulas* (Satpathy, 1983) and the cotton stainer, *Dysdercus cingulatus* (Balasubramanian, 1982). Its leaf extract was found to be antifeedant when sprayed on blackgram and fed to the blackgram pod borer, *Heliothis armigera* (Rajasekaran et al., 1987b). Root extract of this herb, however, was reported to be an antifeedant against *S. litura* (Kareem, 1984). Its 5% leaf extract caused higher larval mortality to the hairy caterpillar, *Amsacta moorei* than even neem leaf extract (Patel et al., 1990b).

Meisner et al. (1981) reported an active component, 'vinblastine', an alkaloid isolated from this herb, to show feeding deterrent activity against *S. littoralis*.

179. *Ceanothus americanus* Linn. (Rhambaceae)

Common name: Red root

This is a perennial shrub available in temperate zones and commonly found in North America. Its aqueous leaf and fruit extracts were reported to show repellent activity when tested against the Japanese beetle, *Popillia japonica* (McIndoo, 1983).

180. *Cecropia mexicana* Loeft. (Utricaceae)

This is a perennial tree found in the tropics. Its leaf extracts in water and alcohol were reported to show insecticidal activity when tested against the cotton leaf armyworm, *Spodoptera litura* (Jacobson, 1958).

181. *Cedrela odorata* Linn. (Meliaceae)

Common name: Spanish cedar

This is perennial tall tree found in temperate, sub-tropical zones. Ethanol extracts of its leaf, root, root-bark and twig showed antifeedant activity to the striped cucumber beetle, *Acalymma vittatum* when tested in a greenhouse test (Jacobson, 1989).

182. *Cedrus deodara* (Roxb. ex. Lanb) G. Don. (Pinaceae)

(Syn.-*Pinus deodara* Roxb., *Abies deodara* Linn.)

Common name: Himalayan cedar or Deodar

This is a perennial tall evergreen tree distributed in North West Himalayas from Kashmir to Garwal. Forests of deodar are found in Kullu, Kashmir, Chamba, Tehri-Garwal, Almorah and at other hill stations in India. Tare and Sharma (1988) reported its oil to show antifeedant activity against the castor semi-looper, *Achoea janata* and the potato tuber moth, *Phthorimaea operculella*.

183. *Celastrus angulata* Maxim. (Celastraceae)

Common name: Bittersweet

This is a perennial shrub or woody climber found in tropical and temperate zones. Root, bark and leaf extracts of this shrub in water and ether and also their dried powders were reported to show antifeedant, insecticidal and repellent properties when tested against a wide range of insect species, viz., European corn borer; the locusts of general category; the tent caterpillar, *Malacosoma disstria*; the willow leaf beetles (Jacobson 1958); the cabbage leaf beetle, *Colaphellus bowringi*; Hawaiian bed webworm, *Hymnea recurvalis*; the migratory locust, *Locusta migratoria*; and the cabbage beetle, *Phaedon brassicae* (McIndoo, 1983). Chiu (1989) referred its root bark powder to show antifeedant activity against the locust, *Locusta migratoria manilensis* and the larvae of the cabbage worm, *Pieris rapae*.

Zhang and Zhao (1983) reported its root bark powder admixed with rice grains at the rate of 0.5% w/w to inhibit the development of the weevils, *Sitophilus oryzae* and *S. zeamais*. Its root bark powder admixed with stored grains checked the multiplication of the corn weevil, *Sitophilus zeamais* and showed antifeedancy (Chiu, 1989). Seven sesquiterpene alkaloids and four sesquiterpene esters were isolated and identified from its root bark powder. The representative active components were maytansine, triptolide, maytoline and evanine. All these components contained a common mother nucleus 'agarofuran' (Figure 27).

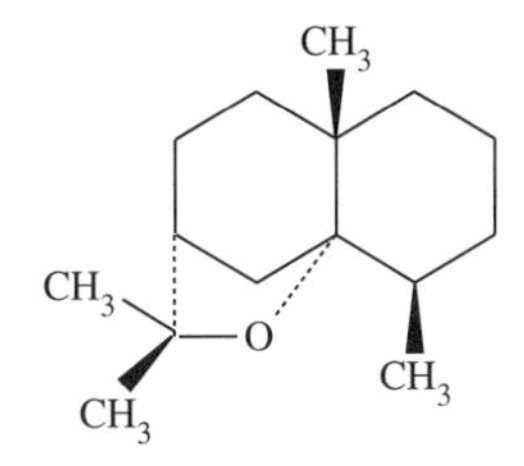

Figure 27. Dihydro-Beta-agarofuran.

The new sesquiterpene called 'celangalin' (Figure 28) as an agarofuran derivative was later identified as an active component, which showed antifeedant activity to *Ostrinea furnacalis* (Chiu, 1982).

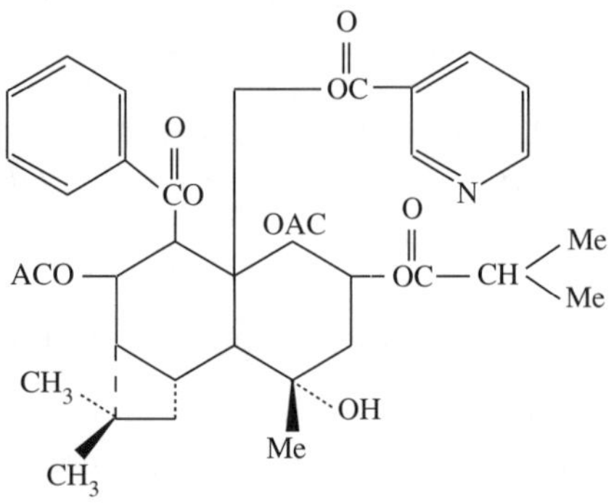

Figure 28. Celangalin.

184. *Celastrus articulatus* Linn. (Celastraceae)

Common name: Oriental bittersweet
This is a perennial woody climber available in tropical, sub-tropical and temperate regions. Its root and bark dried powders were reported to show insecticidal activity when tested against the rice yellow stem borer, *Scirpophaga incertulas* and June bug, *Anomala cupripes* (Chiu, 1982). From *Celestrus* sp. a number of toxic alkaloids, viz., wilforine, wilfordine, wilforgine and wilfortrine, were isolated and found to show insecticidal activity (Crosby, 1971).

185. *Celastrus glaucophyllus* Retd. et Wids. (Celastraceae)

This is a perennial woody climber found in tropical zones. Its root bark powder was found to show antifeedant activity against *L. migratoria manilensis* and *S. zeamais* and celangalin was reported to be an active component in the powder (Chiu, 1989).

186. *Centella asiatica* (Linn.) Urban (Apiaceae)

(Syn.-*Hydrocotyle asiatica* Linn.)
Common name: Asiatic pennywort or Brahmie
This is a prostate perennial herb or creeper found in tropical zones. In India, it is available in the marshy places rooting at its nodes. Its whole plant aqueous extract was reported to show insecticidal activity against the bean aphid, *Aphis fabae* and Mexican bean beetle, *Epilachna varivestis* (Jacobson, 1958).

187. *Centranthus macrosiphon* Boiss. (Velerianaceae)

Common name: Centranth

This is an annual herb with deep rose or white coloured flowers, cultivated in rockeries and native to Spain. Its seed oil was found to be an effective anti-feedant to the cotton boll weevil, *Anthonomus grandis* (Jacobson et al., 1981b).

188. *Centratherum anthelminticum* (Linn.) Kuntz. (Asteraceae)

Syn.-*Vernonia anthelmintica* [Linn.] Willd.)

This is an annual or perennial herb found in tropical and sub-tropical regions and distributed throughout India. Its seed extract in petroleum ether was reported to show toxicity to *Epilachna vigintioctopunctata* in a laboratory test (Chandel et al., 1987).

189. *Centrosema virginiana* Benth. (Papailionaceae)

Common name: Butterfly pea

This is a perennial herb found in tropical zones. Its whole plant aqueous extract was reported to show insecticidal activity when tested against the milkweed bug, *Oncopeltus fasciatus* (Heal et al., 1950).

190. *Cephalanthus occidentalis* Linn. (Rubiaceae)

This is a perennial shrub found in tropical and sub-tropical zones and commonly available in Asia and America. Its twig, leaf and fruit extracts in ethyl ether were reported to attract both the male and the female Mediterranean fruit fly, *Ceratitis capitata* (Keiser et al., 1975).

191. *Cephalotaxus harringtonia* Sieb. (Taxaceae)

Common name: Harrington plum yew

This is a perennial tree in tropical and temperate zones and commonly available in Northern USA. Its stem, branch and leaf aqueous extracts were found to show insecticidal activity against the vinegar fly, *Drosophila hydei* (Jacobson, 1958).

192. *Ceratonia siliqua* Linn. (Caesalpiniaceae)

Common name: Carob tree

This is a perennial tree native of Mediterranean zones but commonly grown in Punjab. Extract from carob pods was reported to show attractant activity against the saw toothed grain beetle, *Oryzaephilus surinamensis* (O'Donnell et al., 1983).

193. *Cestrum nocturnum* Linn. (Solanaceae)

Common name: Night jasmine
This is a perennial diffuse shrub native to the West Indies and found in tropical zones. It is distributed throughout India for its flower. Its leaf powder when admixed with wheat grains at the rate of 2% w/w showed promising grain protection against infestation of the rice moth, *Corcyra cephalonica* for a period of 2 months and significantly reduced the adults' emergence of the moth (Chander and Ahmed, 1986).

194. *Chamaecyparis formosensis* A. Murr. (Cupressaceae)

Common name: Japanese false cypress
This is a perennial tree found in sub-tropical and temperate zones. It is commonly available in Taiwan and Japan. Its leaf extract in ethanol was reported to show insecticidal activity against termites (Michael et al., 1985).

195. *Chamaecyparis lawsomania* (A. Murr.) Parl. (Cupressaceae)

Common name: Lawson's cypress
This is a perennial evergreen ornamental tree of hilly regions. Its aqueous seed extract was reported to show juvenile hormone analogue activity when tested against the mealworm, *Tenebrio molitor* (Jacobson, 1958).

196. *Chara globularis* Thullier. (Characeae)

This is an annual or perennial aquatic and semi-aquatic alga found in tropical and sub-tropical zones and commonly available in Eastern India. Jacobson and Pedersen (1983) reported insecticidal activity in the derivatives of 2-(methylthio)-1,3-propanedithion, an open chain analogue of the insecticidal compound, i.e., 4-(methylthio)-1,2-dithiolane and 5-(methylthio)-1,2,3-trithiane, isolated from this alga against the cotton stainer, *Dysdercus cingulatus* and the fruit fly, *Drosophila melanogaster*.

197. *Chenopodium ambrosioides* Linn. (Chenopodiaceae)

Common name: Wormwood seed or Mexican tea
This is an annual/perennial aromatic herb found in tropical and temperate zones. Worm seed is a native of the West Indies and South America but commonly available in West Bengal, Jammu and Kashmir and Maharastra in India. Its whole plant extract in water and root oil was reported to show insecticidal activity against the Japanese beetle, *Popillia japonica* (McIndoo, 1983) and antifeedant activity against the lesser grain borer, *Rhyzopertha dominica* (Malik and Mujtaba, 1984). Delobel and Malonga (1987) reported

its seed extract to show toxicity to the groundnut beetle, *Caryedon serratus* when tested in laboratory conditions and was found to reduce the survival of the beetle by 90–98%. Its seed extract also showed promising grain protection and reduced the population build up of Angoumois grain moth, *S. cerealla* in treated paddy grains stored under controlled conditions (Prakash et al., 1989).

198. *Chenopodium foetidum* Linn. (Chenopodiaceae)

Common name: Pigweed genus

This is an annual herb found in sub-tropical and temperate zones. Its whole plant aqueous extract was reported to show insecticidal activity when tested against the ants of agricultural importance (Usher, 1973).

199. *Chlorogalum pomeridianum* Kunth. (Liliaceae)

Common name: Soap plant

This is a perennial herb found in temperate and Mediterranean zones. Its aqueous bulb extract and also dried powder were reported to show insecticidal activity when tested against the melon worm, *Diaphania hyalinata* and the armyworm, *Pseudoletia unipunctata* (Jacobson, 1958).

200. *Chlorophora tinctoria* (Linn.) (Moraceae)

Common name: Yellow dye or Fustic

This is a perennial tree found in tropical zones. Its leaf and bark extracts in water were reported to show repellent activity when tested against the Japanese beetle, *Popillia japonica* (McIndoo, 1983).

201. *Chloroxylon sweitenia* D.C. (Flindersiacee)

This is a perennial, moderate size tree found in tropical regions and commonly available in dry deciduous forests throughout India and Sri Lanka. Its leaf extracts in petroleum ether, chloroform and methanol were found to show strong antifeedancy to the larvae of *Spodoptera litura* (Geetanjali et al., 1987). The active components, which showed insecticidal and antifeedant activities isolated from its leaf extract, were xylotenin (Figure 29), isopimpinellin, (Figure 30), bergaptan (Figure 31) and helientin (Figure 32) (Talapatra et al., 1968).

202. *Chromolaena odorata* (Linn.) King & Robinson (Asteraceae)

Common name: Crofton weed

This is a seasonal/annual herb found as a weed in the paddy fields in India, Indonesia, Malaysia, the Philippines, Thailand and Vietnam. Jacob and Sheila

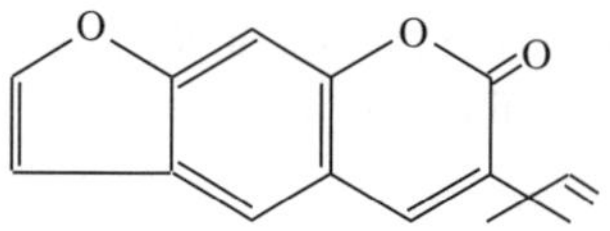

Figure 29. Xylotenin.

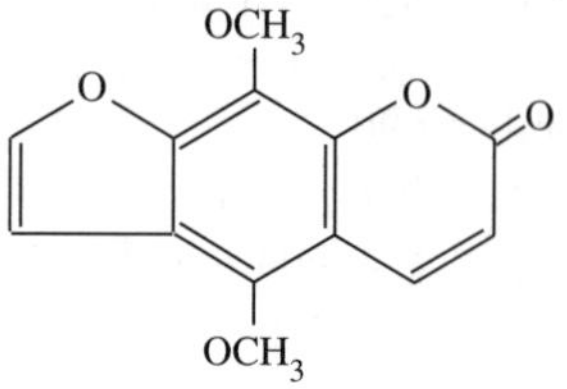

Figure 30. Isopimpinellin.

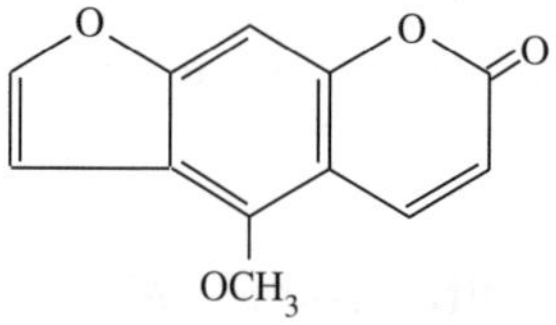

Figure 31. Bergaptan.

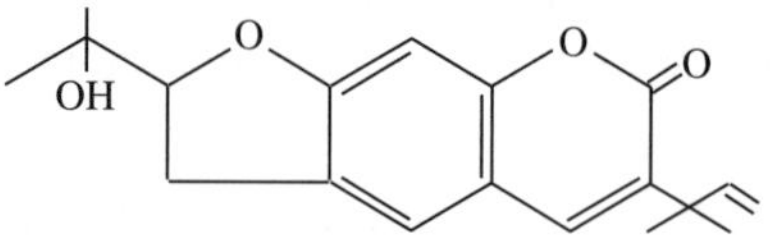

Figure 32. Helientin.

(1993) reported its leaves powder to protect stored rice from the attack of lesser grain borer, *Rhyzopertha dominica* when admixed @ 5% w/w and stored.

203–212. *Chrysanthemum* spp. (Asteraceae)

Common name: Pyrethrum plant/Marigold
Ten species of the pyrethrum plant are dealt with collectively as follows:

203. *C. balsamita* Linn.

(Syn.-*C. majus*)
Common name: Costmary chrysanthemum

204. *C. carneum Bieb.*

205. *C. cineraraefolium* (Thev.) Vis.

206. *C. coccineum* Willd.

(Syn.-C. roseum Adam.)
Common name: Painted daisy

207. *C. coronarium* Linn.

208. *C. frutiscens* Linn.

Common name: Paris daisy

209. *C. marshalli* Willd.

210. *C. partheneum* Linn.

Common name: Fever few

211. *C. segetum* Linn.

Common name: Corn marigold

212. *C. termutense* Linn.

Chrysanthemum, commonly known as the pyrethrum plant, is an annual/seasonal herb available in tropical and sub-tropical zones. This plant is grown in the USA, Japan, Kenya, Brazil, Congo, and Uganda. In India, it is cultivated in Jammu and Kashmir. Commercially pyrethrum represents the dried flowers of *C. cineraraefolium* (*Pyrethrum*) *cineraraefolium.* In the early eighteenth century, an American named Jumticoff discovered an effective insect powder used by the tribes of Caucasus, which was prepared from flower heads of certain species of pyrethrum plant. During the year 1828, it was introduced into Europe and later in 1876 in the United States and then Japan, Africa and South America. At the beginning of the nineteenth century, Dalmatia (Yugoslavia) and Japan became the principal pyrethrum producing countries. The earliest sources were the species of *Chrysanthemum roseum* and *C. carneum* and the more effective *C. cineraraefolium* was later discovered in Dalmatia as Dalmatian Insect Powder in 1840.

Of the ten species of *Chrysanthemum* exploited for their insecticidal property since 1851, only *C. cineraraefolium* was used for commercial purposes of the pyrethrum insecticide. The flowers of this *Chrysanthemum* species were ground into a very fine powder and produced as the Persian, Dalmatian and Japanese powders or Buchach etc.

This plant is known for possessing insecticidal, antifeedant and repellent properties in its flower, leaf and also whole plant extracts in water or their dry powders against a wide range of the insect pests of agricultural importance. The pyrethrum is commonly used to control the pests in stored products and also the pests of household importance. The pyrethrum aerosol sprays proved excellent home insecticides because of their safety and rapid action (Anonymous, 1974). However, a major disadvantage of the pyrethrum especially for use against the agricultural pests lies in lack of persistence due to its instability in the presence of light. The insects often recovered from exposure to sublethal doses of the pyrethrum, which means that the compound must be mixed with a small amount of other insecticides to ensure that the treated insects do not recover. One part of pyrethrum admixed with two parts of piperonyl butoxide was found to be seven times more toxic than the pyrethrum alone

(Cremlyn, 1978). Synergists are expensive but their use is justified with pyrethroids as they are found to enhance the multi-fold insecticidal activity. However, based on the structures and functions of the active components of natural pyrethrum, a number of stable pyrethroids have been synthesized and are found quite effective in controlling the insect pests to replace the use of natural pyrethrins (Elliott et al., 1978).

Natural pyrethrins tested for their bio-efficacy against a wide range of insects in storage and field ecosystems are as follows.

Biological Activity Against Storage Insects

Natural pyrethrins isolated from its flower head in the heavy white oil were reported to show insecticidal activity and successfully controlled the moth pests infesting stored grains (Potter, 1935). Atwal and Sandhu (1970) found natural pyrethrins protecting stored commodities from the attack of the flour beetle, *Tribolium castaneum* when admixed with the stored commodities. Dust of natural pyrethrins in stored wheat also successfully controlled lesser grain borer, *Rhyzopertha dominica* (Zag and Bhardwaj, 1976). The dusts and emulsifiable formulations of natural pyrethrins were also found to protect stored paddy grains from the attack of Angoumois grain moth, *S. cerealella* (Prakash and Pasalu, 1979; Prakash et al., 1980). Prakash et al. (1978–1987) reported 0.01% flower head extract of this plant in water using 2 litres/100 kg paddy to show 62% grain protection up to 120 days under natural conditions of storage in the farm godowns against major grain boring insects, viz., *S. cerealella*, *S. oryzae* and *R. dominica*. Ghosh et al. (1981) also reported *Chrysanthemum* flower powder to protect stored gram seeds against the pulse beetle, *C. chinensis* when admixed at the rate of 1.5% w/w and to cause 100% adult mortality.

Ahmed et al. (1976) tested various formulations of the selective ingredients of natural pyrethrins with piperonyl butoxide for prolonged toxicity against *T. castaneum* and found its emulsifiable concentrations (1:10) more effective in maintaining the residual toxicity. Natural pyrethrins performed better as synergists in controlling insect pests in bulk storage with piperonyl butoxide (Gillenwater and Burden, 1973; Anonymous, 1978b).

Biological Activity Against Field Insects

McIndoo (1924) found natural pyrethrins to kill the potato beetle. Charles (1929) reported Dalmatian and Californian powders from the pyrethrum plant as effective insecticides. Pyrethrum extract showed repellency to the codling moth (Yothers and Carlson, 1944). De Ong (1948) mentioned the original source of the Dalmatian insect powder, *C. cineraraefolium*, which effectively killed the cucumber beetles, *Diabrotica* spp., the cabbage worms and the aphids. Natural pyrethrum 'marc' showed 40–78% toxicity to adults and 1st stage nymphs of the red cotton bug, *D. cingulatus*; adults of the red pumpkin beetles, *Aulocaphora foveicollis*; *A. atripennis*; Indian honey bee, *Apis indica*; and the citrus pest, *Diaphornia citri* when tested under laboratory conditions (Mathur et al., 1961). Pyrethrum and piperonyl butoxide aerosols effectively controlled the aphids, *Myzus persicae* and *Macrosiphum rosae* when sprayed only once on the common horticultural crops (Jacobson and Crosby, 1971).

Mathews (1981) reported natural pyrethrins to control the imported cabbage worm, *Pieris rapae*. Under laboratory toxicity tests, Krishnamurthy (1982) reported natural pyrethrins to show toxicity when tested against the gall midge, *Aspodylia* spp.; the fruit borer, *Earias fabia*; and the eggplant fruit borer, *Leucinodes orbonalis*. Pandey et al. (1984) found its aqueous flower extract toxic to the sugarcane leaf hopper, *Pyrilla perpusilla*. Similarly, Tada and Chiba (1984) reported its flower extract from *C. coronarium* as an effective antifeedant when tested against 5th instar larvae of *Bombyx mori*. Leaf extract of *C. roseum* at 600 ppm also showed 47.2% juvenile hormone activity when tested against *Euproctis fraterna* and caused morphological deformities (Nalinasoundari et al., 1994).

Active Components

Pyrethrum extract obtained from the dried *Chrysanthemum* flowers by extraction with kerosene or ethyl dichloride and concentrated by vacuum distillation contains four main insecticidal components, which were collectively termed pyrethrins. Four active components, viz., pyrethrin-I, pyrethrin-II, cinerin-I, cinerin-II (Figure 33–36), were identified from the pyrethrum extracts (Jacobson and Crosby, 1971).

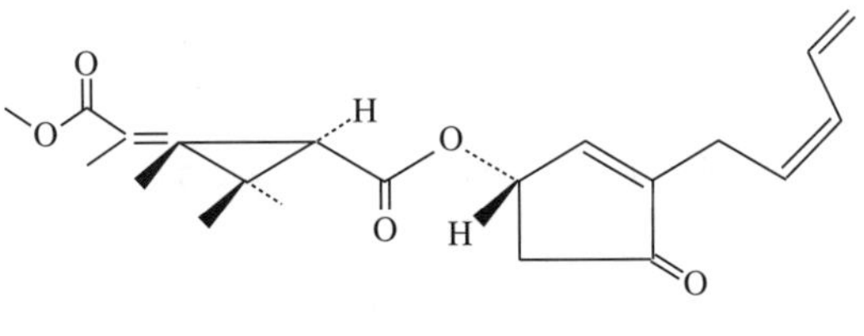

O O H O O H O

Figure 33. Pyrethrin-I.

O O H O O H O

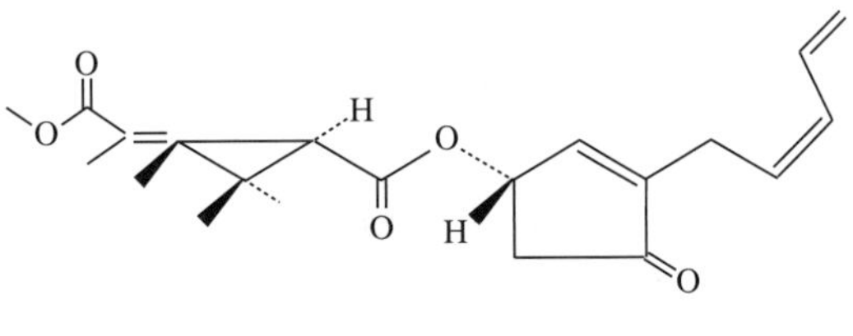

Figure 34. Pyrethrin-II.

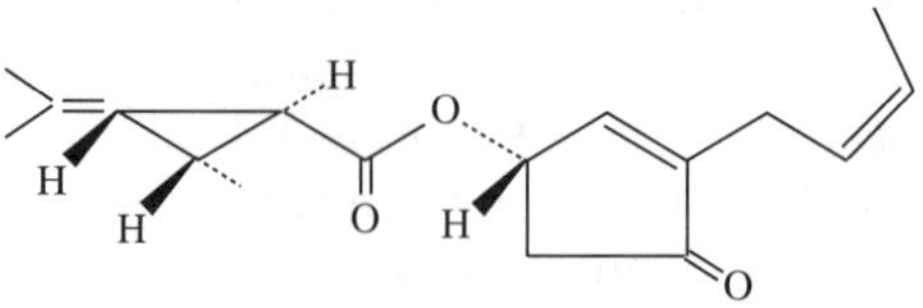

Figure 35. Cinerin-I.

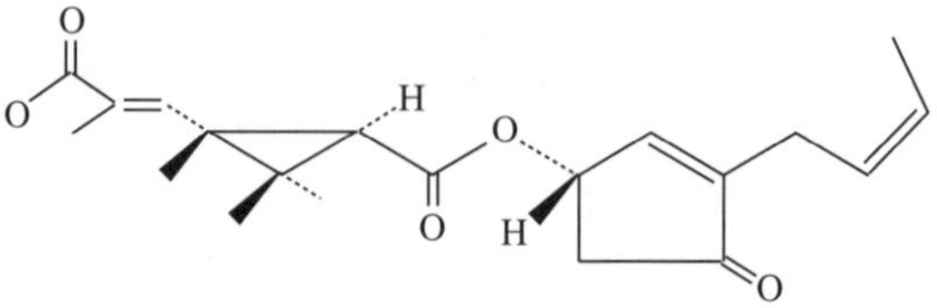

Figure 36. Cinerin-II.

These are the esters of two cyclopentenolones (6,R = CH = CH3 or CH3) called Pyrethrins and cyclopropane carboxylic acid (7,R = CH3 or COCH3) termed Cinerins. Martin (1964) and Srimmannarayan and Rao (1985), however, mentioned another two active components as jasmolin-I (Figure 37) and jasmolin-II (Figure 38) in this extract. Tada and Chiba (1984) reported an active component cis-spiro-enol ether[2-(2,4-hexandiynylidane)-1,6-dioxanspinol[4,4] non-3-ene] from *Chrysanthemum coronarium* as an effective antifeedant against 5th instar larvae of *Bombyx mori*. Head (1967) reported that from the extracts of mature pyrethrum flowers insecticidal constituents differed greatly with different clones. The ratio of pyrethrin-I and pyrethrin-II and the proportion of the six constituents were the characteristics of a particular clone. The proportion of cinerin, jasmolin and pyrethrin in the pyrethrin-I fraction was similar to that of the pyrethrin-II fraction and no relation between the relative proportion of cinerin, jasmolin and pyrethrin and ration of pyrethrin-I and pyrethrin-II was observed. Bestmann et al. (1987) reported essential oil containing pyrethrin-I isolated from *C. balsamita* to show insecticidal activity to the aphid, *Melapolophus dirhodum*.

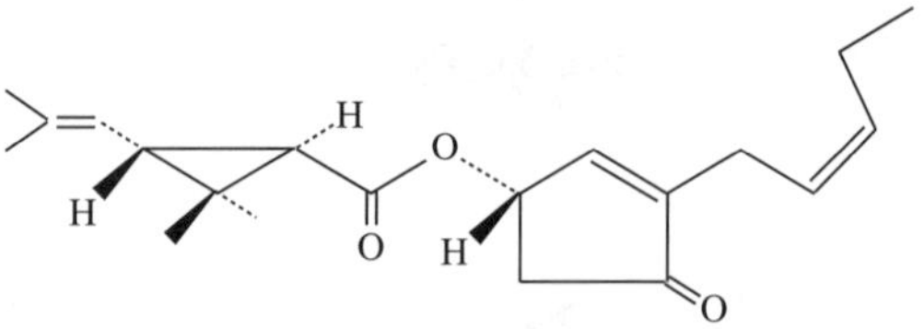

Figure 37. Jasmolin-I.

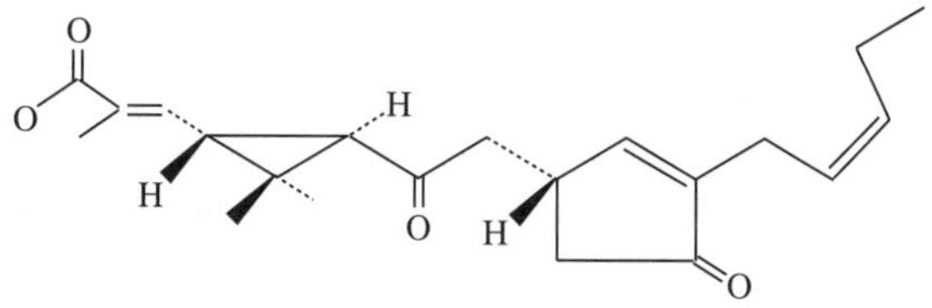

Figure 38. Jasmolin-II.

Based on biological activity of natural pyrethrum, a number of synthetic pyrethroids have been synthesized by alterations in their structures and configurations, which have photostability and great insecticidal potency. Allethrin was the first synthetic pyrethroid prepared by esterification of synthetic (±) chrysanthemic acid (7,R=CH_3) with alcohol allethrolone (6,R′=H). Allethrin had a strong insecticidal activity and removal of the keto group gave another insecticidal synthetic pyrethroid called bioallethrin. Synthetic pyrethroids have not been dealt with in this book.

213. *Chrysopsis villosa* (Nutt.) Ell. (Asteraceae)

This is an annual herb found in tropical and sub-tropical zones. Its whole plant aqueous extract was reported to show insecticidal activity when tested against the range land grasshopper, *Melanoplus femmurubrum* (Jacobson, 1975).

214. *Chrysosplenium flagelliferum* Linn. (Saxifragaceae)

This is an annual or biannual creeping/herb found in temperate zones. Water extracts of its leaf, root and branch were reported to show insecticidal activity against the vinegar fly, *Drosophila hydei* (Jacobson, 1958).

215. *Chrysosplenium yesoense* Linn. (Saxifragaceae)

This is an annual creeper found in temperate zones. Water extracts of its leaf, branch and flower were toxic to the vinegar fly, *Drosophila hydei* (Jacobson, 1958).

216. *Chrysothamnum nauseosus* Nutt. (Asteraceae)

This is an annual herb found in the United States. Essential oil of this plant and its polyphenols showed insecticidal and antifeedant activities when tested against Colorado potato beetle, *Leptinotarsa decemlineata* (Jermy et al., 1980).

217. *Cinchona calisaya* Wedd. (Rubiaceae)

Common name: Quinine or Peruvian bark

This is a perennial tree found in tropical and alpine regions. This is a native of South America but available in Nilgiri hills and Sikkim in India. Leaf and bark extracts of this tree in water and alcohol were reported to show insecticidal activity when tested against the melon worm, *Diaphania hyalinata* and the diamond-back moth, *Plutella xylostella* (Jacobson, 1958).

218. *Cinnamomum camphora* (Linn.) T. Nees and Eberm. (Lauraceae)

Common name: Camphor tree

This is a perennial tree found in sub-tropical and temperate zones. This is a native of China and Japan but available in Ootacamund in South India. Secretions (latex) of its leaf and bark extracts in water and methanol were reported to show insecticidal and repellent activities when tested against the aphids and the armyworm, *Mythimna (Pseudoletia) unipunctata* (McIndoo, 1983) and the screw worm, *Cochliomya homnivorax* (Michael et al., 1983).

219. *Cissus rhombifolia* Linn. (Vitidaceae)

Common name: Venezuela tree pine

This is a perennial vine found in tropical zones. Its branch extract showed toxicity and repellency to the diamond-back moth, *Plutella xylostella* (Jacobson, 1975).

220. *Citrullus colocynthis* (Linn.) Kuntze. (Cucurbitaceae)

(Syn.-*Cucumis colocynthis* Linn., *Colocynthis vulgaris* Schrod.)

Common name: Bitter apple or Colocynth

This is an annual/perennial trailing herb found in arid zones, native to Asia and Africa but available in Madhya Pradesh, Gujarat, North and South Indian States. Its root powder was reported to show toxicity to the pulse beetle, *Callosobruchus chinensis* (Puttarudriah and Bhatia, 1955). Its root, leaf and fruit extracts in water and also their dried powders were toxic to the white grub, *Holotrichia insularis* (Sachan and Pal, 1976) and also against the hairy caterpillar, *Euproctis fraterna* and the diamond-back moth, *Plutella xylostella* (McIndoo, 1983).

221. *Citrullus vulgaris* Schard. var. *fistulosus* (Stocks) Dutch & Full. (Cucurbitaceae)

(Syn.-*C. lantus* [Thunb.] Mansf. var. *fistulosus* [Stocks]-Mansf., *C. fistulosus* Stocks.)

Common name: Round gourd

This is an annual/perennial trailing herb found in tropical and sub-tropical zones. In India, it is cultivated mainly in Uttar Pradesh, Punjab, Haryana, Rajsthan and Bihar for its vegetable and pickles. Juices from its leaf and fruit pericarp were reported to inhibit the oviposition of cotton bollworm, *Earias vittella*, a pest of okra (Dongre and Rahalkar, 1984).

222. *Citrus aurantifolia* (Christm.) var. *limetta* Wight. and Arn. (Rutaceae)

(Syn.-*C. limettiodides* Tanaka, *C. medica* var. *limetta* Wright. and Arn., *C. lononia Linn.*)

Common name: Sweet lime or Lemon

This is a perennial shrub or small tree found in tropical and sub-tropical zones and cultivated in China, Malaysia, Japan, USA, USSR, Australia, the Philippines, etc. Sweet lime is commonly available in North and Central India. Extract and powder of dried citrus peels of sweet lime inhibited the multiplication of stored product insects (Su et al., 1972). Lemon oil was reported to be toxic to the pulse beetles, *Callosobruchus chinensis* and *C. maculatus* under laboratory tests (Su, 1976). Volatiles of this plant also inhibited the oviposition of the cotton leaf hopper, *Amrasca devastans* (Saxena and Basit, 1982).

Su and Horvat (1988) isolated and characterized four major components from insecticidally active lemon peel extract, viz., (i) 5,7-dimethoxy-2-h-1-benopyran-2-one (component A), (ii) 9-[3,7(dimethyl-2, 6-octadienly) oxy]-7 H-furo[3,2-g] [1] benzopyran-7-one (component B), (iii) 4-[3,7-(dimethyl-2,6-octadienyl)oxy]-7-H-furo [3,2-g][1] benzopyran-7-one (component C) and (iv) 5-[3,7-(dimethyl-2,6-octadienly)oxy]-7-methoxy-2H-1-benzopyran-2-one (component D). These components were tested for their toxicity to *S. oryze* and *C. maculatus* and the components B, C and D showed weak to moderate toxicity to both the test insects. The component A was non-toxic to *C. maculatus* but slightly toxic to *S. oryzae*.

223. *Citrus aurantium* Linn. (Rutaceae)

Common name: Sour orange or Seville orange

This is a perennial small tree/shrub distributed in temperate, tropical and sub-tropical zones. In India, this shrub is cultivated for its fruits as well as juice extractions in Himachal Pradesh, Punjab, Haryana, Delhi, Orissa and North Eastern States. Its fruit peels, dried powder and oil extracted from raw peels were evaluated for their biological activities and were reported to show insecticidal, antifeedant and repellent activities against a wide range of the insect pests of agricultural importance as mentioned below.

Biological Activity Against Storage Insects

Orange peel extract was reported to be insecticidal against the pulse beetle, *Callosobruchus phaseoli* (Taylor and Vickery, 1974). Milled rice grain treated with its aqueous extract at the rate of 20 ml/kg rice showed above 80% protection against the rice weevil, *S. oryzae* and the red flour beetle, *Tribolium castaneum* (Prakash et al., 1978–1987). Zhang and Zhao (1983) reported its volatile essences to show toxicity to the rice and corn weevils, *S. oryzae* and *S. zeamais* in a laboratory test and completely checked their multiplication. Don Pedro (1985) reported sun-dried powder of its fruit peels to show antifeedancy to the pulse beetle, *Callosobruchus maculatus* when admixed with stored cowpea @ 4% w/w. Its peel powder was also reported to reduce the grain damage and weight loss due to *S. oryzae* when admixed with the wheat grains (Rout, 1986). El-Sayeed and Abdel-Rahik (1986) found its peel oil to show ovicidal activity to the eggs of the pulse beetle, *C. maculatus* in stored cowpea seeds treated with its 1% formulation.

Biological Activity Against Field Insect Pests

Jacobson (1975) reported insecticidal activity in the oil extracted from its peels against the diamond-back moth, *Plutella xylostella*. Leaf extract in petroleum ether, however, showed juvenile-hormone mimetic activity against 5th instar larvae of the red cotton bug, *Dysdrcus cingulatus*. Volatiles from its peels were also found to inhibit the oviposition of the cotton leafhopper, *Amrasca devastans* (Saxena and Basit, 1982). Kareem (1984) reported its stem and leaf extracts to deform the juveniles during development and malformed the emerged adults of *D. cingulatus*.

Active Components

Limonin (Figure 39), an active component, was isolated from the sour orange and found to show insecticidal activity when tested against the pulse beetle, *Callosobruchus phaseoli* (Taylor and Vickery, 1974). 'Citronellol' [3.7-dimethy 1-6-octen-1-ol], another active component from this plant, was found to be an oviposition inhibitor when tested against the cotton leafhopper, *Amrasca devastans* (Saxena and Bait, 1982). Limonin, nomilin (Figure 40) and obacumone (Figure 41), the major limonoids isolated from its seeds, were reported to be active components possessing antifeedant property. Nomilin and obacumone were more active growth inhibitors even than the azardirachtin at 10 ppm concentration (Klocke and Kubo, 1982). Bentley et al. (1983) also reported antifeedant activity of limonin when tested against the 6th instar larvae of *Spodoptera exempta*. Suspensions of nomilin, however, showed antifeedant activity against the larvae of *S. frugiperda* on maize and *Trichoplusia ni* on kale as a host plant (Altieri, 1984).

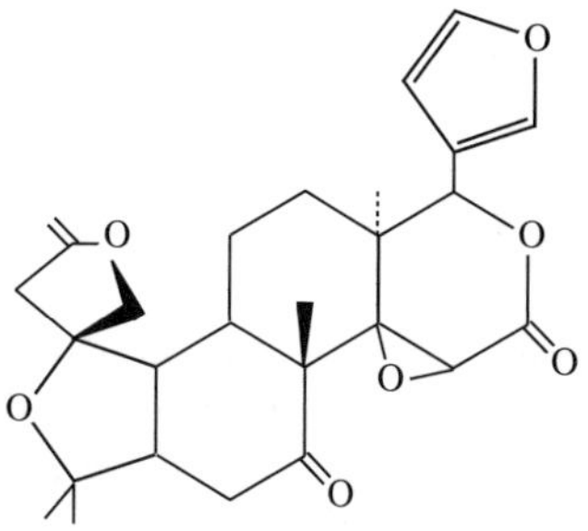

Figure 39. Limonin.

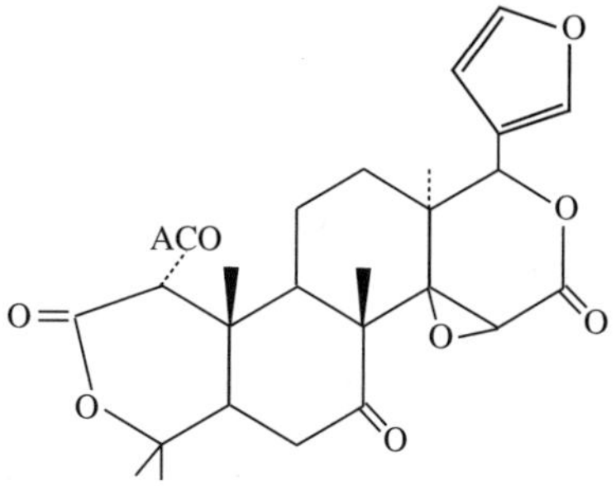

Figure 40. Nomilin.

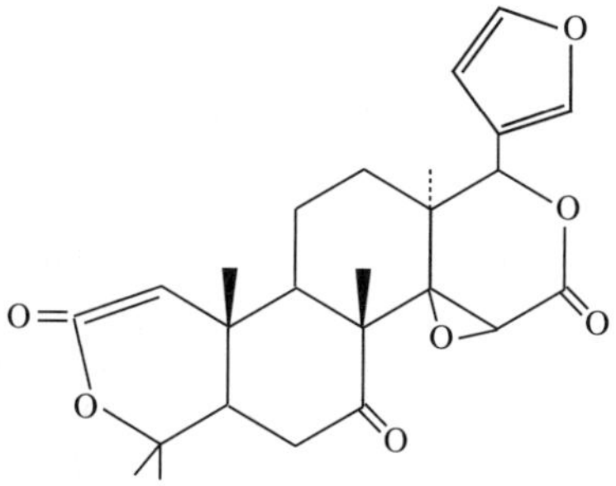

Figure 41. Obacumone.

224. *Citrus medica* Linn. (Rutaceae)

(Syn.-*C. tuberosa* Mill., *C. cedrata* Raf., *C. aurantium* var. *medica* Wight & Arn., *C. medica* var. *medica proper* Hook f.)

Common name: Citron

This is a small perennial tree found in tropical and sub-tropical zones. In India, it is available in mountainous Himalayas and Kumaun hills, Sikkim and Western Ghats. Oil of citron was reported to show repellent activity when tested against the bean aphid, *Aphis fabae* (McIndoo, 1983).

225. *Citrus paradisi* Macf. (Rutaceae)

(Syn.-*C. decumana* var. *Racemosa* Roem, *C. decumana* var. *paradisi* Nichols., *C. racemosa* Marc.)

Common name: Grapefruit

This is a perennial tree found in tropical and sub-tropical zones. In India, it is cultivated in Punjab, Uttar Pradesh, Madhya Pradesh and West Bengal. Powder of sun-dried peels of the grapefruit was reported to show antifeedancy to the pulse beetle, *Callosobruchus maculatus* when admixed with cowpea seeds in storage @ 4% w/w (DonPedro, 1985). El-Sayed and Abdel-Rahik (1986) reported its oil to show inhibition to oviposition of *C. maculatus* on the treated cowpea seeds using 0.73 to 1.0% formulations. El-Sayed et al. (1991) found its oil treatment with stored @ 0.25% w/w to significantly reduce the populations of wheat weevil, *Sitophilus granarius* and to cause 44% mortality 16 days after treatment.

Suspension of nomilin, a limonoid isolated from its seed, was found to show antifeedant activity against the larvae of *Spodoptera frugiperda* (Altieri et al., 1984). Jacobson (1989) also reported nomilin as an effective antifeedant against *Spodoptera frugiperda* and *Trichoplusia ni*. Limonin, another limonoid, was also registered a potent antifeedant against *Leptinotarsa decemlineata*.

226. *Citrus reticulata* (Linn.) Blanco. (Rutaceae)

(Syn.-*C. nobnilis* Andrews non Lour.)

Common name: Mandarin or Tangerine or Orange

This is a small perennial tree available in tropical and sub-tropical zones. In India, it is cultivated in Maharastra, Assam, Sikkim, Karnataka, Punjab, West Bengal, Jammu and Kashmir. Oil of its peel was reported to protect the husked rice grains from the paddy moth, *Sitotroga cerealella* when treated and stored under laboratory conditions and did not allow build-up of its populations (Prakash et al., 1989). Further, Prakash et al. (1990c) also found its powder of dried fruit peel to protect stored rice against the rice weevil, *Sitophilus oryzae*. Seed extract of orange in water when sprayed on the paddy grains of caged rice plant 20 days after flowering was reported to protect the grains from infestation of rice stink bug, *Leptocorisa acuta* and thus minimize the deterioration of the paddy grain quality in field conditions (Gupta et al., 1990).

227. *Citrus sinensis* (Linn.) Osbeck. (Rutaceae)

(Syn.-*C. aurantium* Linn. var *sinensis* Linn., *C. aurantium* sub. sp. *sinensis* [Gall.] Engl., *C. aurantium* sub. sp. *decumana* var. *sinensis* Thell., *Aurantium sinensis* Mill.)

Common name: Sweet orange

This is a perennial small tree found in tropical and sub-tropical regions. It is a native of China and now available in Tamil Nadu, Maharastra, Andhra Pradesh, Uttar Pradesh, Madhya Pradesh, Punjab and Rajasthan in India. Its fruit juice was reported to show attractant activity when tested against the leaf cutting ants (Jacobson, 1958). El-Sayed et al. (1991) reported oil of sweet orange to show 64% mortality to wheat weevil, *Sitophilus granarius* in stored treated grains @ 0.25% w/w.

228. *Clausena anisata* (Willd.) Hook. F. Ex. Benth. (Rutaceae)

Common name: Samanobere

This is a perennial tree found in tropical zones. Its aqueous leaf extract was reported to show antifeedant activity when tested against African armyworm, *Spodoptera exempta* (Gebreyens, 1980). Its root bark extract also showed antifeedant activity to *Spodoptera litura* and *Collosobruchus phaseoli* (Jacobson, 1989).

229. *Claytonia virginica* Linn. (Portulacaceae)

Common name: Virginia spring beauty

This is a perennial herb found in temperate regions. Its whole plant and leaf extracts in water were reported to show repellent activity when tested against the Japanese beetle, *Popillia japonica* (McIndoo, 1983).

230. *Cleistanthus collinus* (Roxb.) Benth. ex. Hook.f. (Euphorbiaceae)

(Syn.-*Clutia collina* Roxb.)

Common name: Karada

This is a perennial small tree found in the tropics. In India, it is commonly available in Orissa, Bihar and West Bengal. Leaf extract of karada was reported to show antifeedant activity when tested against the pulse beetle, *Callosbruchus chinensis* (Satpathy, 1983), insecticidal activity against the white ants (McIndoo, 1983) and also against the cotton leaf armyworm, *Spodoptera litura* (Krishnamurthy Rao, 1983).

Against stored grain insect pests, its leaf extract in methanol showed larvicidal activity against the saw toothed grain beetle, *O. surinamensis* under laboratory testing (Prakash and Rao, 1986) and insecticidal property against the rice weevil, *Sitophilus oryzae* (Prakash et al., 1978–1987) but against the rice moth, *Corcyra cephalonica* dried leaves of karada and its extract could not check its multiplication when admixed/treated with milled rice gains and used as diet for multiplication of the moth (Prusty et al., 1989).

From the crude extracts of bark in petroleum ether and chloroform, cleistanthin B was isolated as an active component and reported to show insecticidal and antifeedant activities (Jyotsna and Srimmannarayan, 1987).

231. *Clematis dioca* Linn. (Ranunculaceae)

Common name: Honduras fish poison

This is a perennial woody climbing shrub found in tropical regions and commonly available in the lower hills in Jamaica. Its aqueous leaf extract was reported to show insecticidal activity against the silkworm, *Bombyx mori* (McIndoo, 1983).

232. *Clematis ternifolia* Linn. (Ranunculaceae)

This is a perennial woody climbing shrub found in temperate regions. Water extracts of its branch, leaf and powdered seed were reported to show insecticidal activity to the vinegar fly, *Drosophila hydei* (Jacobson, 1958).

233. *Clematis virginian* Linn. (Ranunculaceae)

This is a perennial woody climber mainly found in temperate regions and commonly available in the United States. Its stem, leaf, flower and fruit extracts in ethyl ether showed attractancy to both sexes of the Mediterranean fruit fly, *Ceratitis capitata* (Keiser et al., 1975).

234. *Clematis vitalba* Linn. (Ranunculaceae)

Common name: Traveller's joy

This is a perennial woody climber found in temperate zones and commonly available in Central and South Europe. Water extracts of its branch, leaf and fruit showed repellency to the coleopteran pests in grain storage (McIndoo, 1983).

235. *Cleome visosa* Linn. (Cleomaceae)

Common name: Stickly cleome or Hurhur

This is an annual glandular pubescent herb found in tropical zones and commonly available in eastern India. Its leaf extract in water and also its juice were reported to show insect repellent and insecticidal activities when tested against the cotton leaf armyworm, *Spodoptera litura* (Krishnamurthy Rao, 1982).

236. *Clerodendrum calamitosum* Borneo. (Verbenaceae)

This is a perennial shrub available in tropical and sub-tropical regions. Its leaf and branch extracts in ether and water were reported to show antifeedant and also insecticidal activities to the 3rd instar larvae of the cotton leaf armyworm, *Spodoptera litura* (Waespe and Hans, 1979).

237. *Clerodendrum crytophyllum* Linn. (Verbenaceae)

This is a perennial shrub found in sub-tropical and temperate zones and commonly available in Japan and China. Its 1% leaf extract in ether was reported to be an antifeedant to the 3rd instar larvae of *Spodoptera litura* (Hosozawa et al., 1974b).

238. *Clerodendrum fragrans* (Vent.) Willd. (Verbenaceae)

(Syn.-*C. japonicum*)

Common name: Glory tree

This is a perennial sweet smelling and ornamental shrub found in temperate zones. This is a native of Japan but also available in China and North Indian States. Its leaf extract in ethyl ether was found to show attractancy to both sexes of the Mediterranean fruit fly, *Ceratitis capitata* (Keiser et al., 1975). However, earlier its 1% leaf extract in ether was reported to be antifeedant to the 3rd instar larvae of *Spodoptera litura* (Hosozawa et al., 1974b). Jha and Roychodhury (1988) also reported its leaf extract to show antifeedant activity against the rice weevil, *Sitophilus oryzae* in the treated grains.

239. *Clerodendrum imane* Vent. (Verbenaceae)

Common name: Banjui

This is a perennial shrub found in tropical zones. Its leaf and branch extracts in water were tested to be insecticidal in action against the mango hopper, *Idiocerus* sp. (McIndoo, 1983).

240. *Clerodendrum indicum* (Linn.) Kuntze. (Verbenaceae)

(Syn.-*Clerodendron sphonanthus* R.Br., *Siphonanthus indica* Linn.)

Common name: Tube flower or Turk's turban

This is a perennial tall shrub found in tropical zones. Its aqueous leaf extract showed repellent activity to the long headed flour beetle, *Latheticus oryzae* (Prakash et al., 1978–1987). Jha and Roychoudhury (1988) reported its leaf extract to show antifeedant activity against the rice weevil, *Sitophilus oryzae* in stored grains.

241. *Clerodendrum inerme* (Linn.) Gaertn. (Verbenaceae)

(Syn.-*Volkameria inerme* Linn.)

This is a perennial shrub found in tropical zones and commonly available in India. Its leaf extract in water was reported to show toxicity to the pulse beetles, *Callosobruchus chinensis* and *C. maculatus* under laboratory conditions (Anonymous, 1980–1981). Powdered leaves of this shrub admixed with the grains at the rate of 1, 2 and 5% w/w significantly reduced adult emergence of the rice moth, *Corcyra cephalonica* and gave promising grain protection (Chander and Ahmed, 1986). El-Ghar and El-Sheikh (1987) reported its leaf extract in petroleum ether to show toxicity and ovipositional deterrency to *C.*

chinensis. Further, its 5% aqueous leaf extract protected the field crop against hairy caterpillar, *Amsacta moorie* and significantly reduced its larval population (Patel et al., 1990a).

242. *Clerodendrum infortunatum* Linn. (Verbenaceae)

Common name: Bhant

This is a perennial shrub found in tropical zones and available throughout India. Its leaf, flower and whole plant extracts were reported to show repellent activity when tested against stored grain insect pests and found to be more effective to Angoumois grain moth, *S. cerealella* than other test insects (Abraham et al., 1972a). Rajamma (1982) reported its leaf extract to reduce the infestation of the sweet potato weevil, *Cylas formicarius* in the sweet potato crop.

The trans-declain unit of 'clerodendrin', a bitter active principle isolated from this shrub, was reported to show antifeedant activity against the cabbage butterfly, *Pieris brassicae* under a laboratory test (Geuskens et al., 1983).

243. *Clerodendrum myricordes* Linn. (Verbenaceae)

This is a perennial shrub found in tropical and sub-tropical zones. Its leaf extract in water was reported to show insecticidal activity when tested against African armyworm, *Spodoptera exempta* (Snyder, 1983).

244. *Clethra alnifolia* Bert. ex. Steud. (Clethraceae)

Common name: Vivania

This is a perennial shrub or small tree found in Asian and American tropical regions. Aqueous extracts of its leaf, stem and fruit were reported to show juvenile hormone analogue activity when tested against the milkweed bug, *Oncopeltus fasciatus* (Jacobson, 1958).

245. *Clibadium surinamensis* Allem ex. Linn. (Asteraceae)

This is an annual herb found in temperate zones. Its branch and leaf extract in water showed insecticidal activity to the silkworm, *Bombyx mori* (Jacobson, 1958).

246. *Cnidoscolus urens* Pohl. (Euphorbiaceae)

Common name: Spurge nettle

This is a perennial tree found in tropical zones. Its dried fruit powder was reported to be insecticidal in nature when tested against Hawaiian webworm, *Hymnea recurvalis* and the southern beet webworm, *Pachyzancla bipunctalis* (Jacobson, 1958).

247. *Cocculus tribolus* D.C. (Menisperimaceae)

This is a perennial shrub found in tropical and subtropical zones. Its aqueous leaf and root extracts were reported to possess insecticidal and antifeedant properties when tested against the pulse beetle, *Callosobruchus chinensis*; the vinegar fly, *Drosophila hydei*; the green leafhopper, *Nephotettix cincticeps*; and the cotton leaf armyworm, *Spodoptera litura* (Jacobson, 1975).

Two alkaloids, i.e., isoboldine and cocculolidine (Figure 42), were isolated from its extract, of which the former one showed repellency, whereas the latter one was reported to be toxic to the insects (Cosby, 1971).

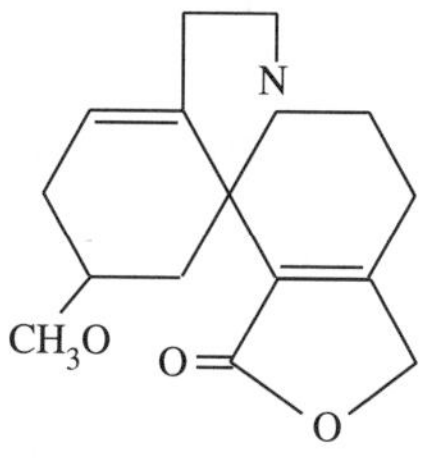

Figure 42. Cocculolidine.

248. *Cocos nucifera* Linn. (Palmaceae)

Common name: Coconut

This is a perennial tree found in tropical and sub-tropical coastal regions. Coconut oil sprayed/admixed with stored grains was reported to protect them from attack of storage insects, viz., the lesser grain borer, *R. dominica* and the rice weevil, *S. oryzae* (Fletcher, 1919). Similarly, its mint-oil spray was found to be toxic to sucking and chewing insects (Platon et al., 1970). Its edible oil was also reported to inhibit the development and multiplication of stored grain insect pests (Su et al., 1972). McIndoo (1983) also found coconut oil as a growth inhibitor and disinfestant to the rice weevil, *S. oryzae*. Mittal (1971), Mummigatti and Raghunathan (1977), Mazumder and Amla (1977), Verma and Pandey (1978), Golob and Webley (1980), Ali et al. (1983), and Doharey et al. (1983, 1984a,b, 1987) reported its oil to show insecticidal activity to the pulse beetle, *Callosobruchus chinensis* when admixed with greengram seed and caused 100% mortality to the adults and the eggs and ultimately protected the pulse seeds in storage. Coconut oil (1%) was found to show repellency to the pulse beetles, *C. chinensis* and *C. maculatus* and also inhibited their oviposition and adult emergence and finally protected greengram seeds in storage (Naik and Dumbre, 1984; Doharey et al., 1990). Jacob and Sheila (1990) also reported coconut oil to protect greengram seeds against infestation of *C. chinensis* when treated using 1 ml/kg seeds and caused above 60% mortality to the adults 3 days after treatment. But 100% cowpea seeds were protected when treated with coconut oil @ 2.5 ml/kg (El-Sayed et al., 1991).

Sujatha and Punnaiah (1984) and Singh et al. (1993) found treatment of 3 ml coconut oil with 1 kg of redgram grain to protect them from *C. chinensis* and recorded no damage.

249. *Coffea arabica* Linn. (Rubiaceae)

Common name: Coffee or Arabian coffee

This is an evergreen perennial shrub or small tree, native to Ethiopia and now commonly grown in Nilgiris, Karnataka and Travencore in India. Its twig, leaf and fruit extracts in ethyl ether were reported to attract both male and female of the Mediterranean fruit fly, *Ceratitis capitata* (Keiser et al., 1975).

250. *Coffea robusta* Linn. (Rubiaceae)

(Syn.-*C. canephora* Pierre Ex. Froaeh.)

Common name: Congo coffee

This is a robust evergreen perennial shrub or small tree native to Congo and now grown in Tamil Nadu, Karnataka and Travoncore in India. Its stem and leaf extracts in ethyl ether were found to be attractants to both male and female of the Mediterranean fruit fly, *Ceratitis capitata* and the melon fly, *Dacus cacurbitae* and the female of Oriental fruit fly, *Dacus dorsalis* (Keiser et al., 1975).

251. *Colchicum autumnale* Linn. (Liliaceae)

Common name: Meadow saffron

This is a perennial herb found in temperate and Mediterranean zones. It is commonly available in the moist rich pastures in many parts of England and Europe. Its root and seed aqueous extracts were reported to possess insecticidal and antifeedant properties when tested against the aphids (McIndoo, 1983) and also against the plum curculio, *Conotrachelus nenuphar*, Colorado potato beetle, *Leptinotarsa decemlineata* and the cotton leaf armyworm, *Spodoptera litura* (Jacobson, 1975).

252. *Coleus aromaticus* Lour. (Labiatae)

Common name: Country borage

This is an annual/perennial aromatic herb found in tropical zones and commonly available in the Coastal States of India. Its dried leaf powder and extract in benzene were reported to show grain protection against the infestation of storage insects, viz., the pulse beetle, *Callosobruchus chinensis* (Reddy and Reddy, 1987) and Angoumois grain moth, *Sitotroga cerealella* (Prakash et al., 1989).

253. *Comandra umbellata* Nutt. (Santalaceae)

Common name: Bastard toad flax

This is an annual herb found in temperate zones. Its whole plant extract in water was tested to be repellent to the Japanese beetle, *Popilia japonica* (McIndoo, 1983).

254. *Coniferous lumber* Juss. (Pinaceae)

This is a perennial tree found in temperate and sub-tropical zones. Its volatile oil contained bornyl acetate, α-terpenol, α-pinene and β-pinene as active components, which were reported to show toxicity to *Callidium vialaceum* (Karasev, 1976).

255. *Coniogramme japonica* Fee. (Polypodiaceae)

Common name: Bamboo fern

This is a fern found in temperate regions and native to Japan. Its aqueous leaf extract showed toxicity to the vinegar fruit fly, *Drosophila hydei* (Jacobson, 1958).

256. *Conringia orientalis* Adans. (Brassicaceae)

This is an annual herb found in the Mediterranean region. Its leaf, flower and fruit extracts were reported to show insecticidal activity when tested against the vinegar fly, *Drosophila melanogaster* (Michael et al., 1985).

257. *Consolida regalis* (D.C.) Opiz. (Ranunculaceae)

Common name: Forking larkspur

This is an annual herb found in temperate zones and commonly available in the United States. Its root, stem, leaf and seed extracts in ether were reported to show insecticidal activity against the cabbage worm, *Pieris brassicae* and the whitefly, *Bemisia tabaci* (McIndoo, 1983).

258. *Copaifera officinalis* Linn. (Caessalpiniaceae)

Common name: African copaifera

This is a perennial tree found in tropical America and African countries. Its oil was reported to be an attractant when tested against the fruit fly, *Ceratitis rosa* (McIndoo, 1983). Jacobson et al. (1987) found α-copaene, an active component isolated from its root oil, to attract Mediterranean fruit fly, *Ceratitis capitata* in a laboratory test.

259. *Corchorus capsularis* Linn. (Tiliaceae)

Common name: White jute or Jute
This is an annual shrub available in tropical regions. It is commonly grown in West Bengal, Assam, Bihar, Uttar Pradesh and Orissa in India. Mallick et al. (1980) reported its leaf extract to show antifeedant activity against the grey weevil, *Myllocerus discolor* when tested under controlled conditions. Its seed powder was also found to be toxic against the insect pests of rice (Zaman, 1983).

260. *Cordia dichotoma* Linn. (Borganiaceae)

Common name: Anonang (Philipp.)
This is a perennial shrub/small tree found in tropical zones. It is a native to the West Indies and commonly available in the Philippines and Indonesia. Its branches when placed in the paddy fields were observed to repel insect pests (Litsinger et al., 1978).

261. *Cordyline roxburgiana* Fabr. (Liliaceae)

Common name: Tingue (Philipp.)
This is a perennial shrub or tree found in tropical and sub-tropical zones and commonly available in the Philippines, Malaysia, and Indonesia. Its whole plant and leaf extracts were reported to minimize losses in paddy fields caused by insect pests (Litsinger et al., 1978).

262. *Coreopsis grandifolia* Linn. (Asteracease)

This is an annual/perennial herb found in tropical America and African countries. Its whole plant extract was reported to show repellent activity when tested against the Japanese beetle, *Popillia japonica* (McIndoo, 1983).

263. *Coriandrum sativum* Linn. (Apiacease)

Common name: Coriander
This is an annual aromatic herb found in temperate and tropical zones and commonly available in North Indian States. Its whole plant extract and oil were reported to show toxicity and growth inhibition to Colorado potato beetle, *Leptinotarsa decemlineata* (Mathews, 1981) and the cotton aphid, *Aphis gossypii* (McIndoo, 1983). Volatiles isolated from its whole plant extract, however, inhibited the oviposition of the leafhopper, *Amrasca devastans* (Saxena and Basit, 1982).

264. *Coriaria ruscifolia* sub. sp. *microphylla* (Poir.) L.E.Skog. (Coriariaceae)

Common name: Japanese coriaria
This is a perennial shrub found in the Mediterranean zone and available in Japan, Mexico and China. Its twig, leaf and fruit extracts in ethyl ether showed attractancy to both sexes of the Mediterranean fruit fly, *Ceratitis capitata* (Keiser et al., 1975).

265. *Corypha elata* Roxb. (Arecaceae)

Common name: Buri palm
This is a perennial tree found in tropical zones and commonly available in all the South East Asian countries. Its leaf extract was reported to show repellent activity against the rice field insects (Litsinger et al., 1978).

266. *Crassocephalum crepedioides* Benth. (Asteraceae)

This is an annual herb found in tropical regions and commonly available in Malaysia, China and the Philippines. Crude extract/juice of its leaf was reported to show insecticidal activity when tested against the cotton stainer, *Dysdercus cingulatus* (Carino, 1983). Further, its leaf and root extracts were also found to be toxic to the corn weevil, *Sitophilus zeamais* and the flour beetle, *Tribolium castaneum* and caused nearly 80% mortality (Carino, 1981).

267. *Crescentia cujecta* Linn. (Begoniaceae)

Common name: Calabash tree
This is a small perennial and ornamental tree native to tropical America and the West Indies and now grown in gardens in Orissa, Bihar and West Bengal in India. Powder of its fruit testas was reported to significantly reduce the population build-up of Angoumois grain moth, *Sitotroga cerealella* and finally protected the paddy grains when tested under controlled conditions (Behera, 1992).

268. *Crinum asiaticum* Linn. (Amaryllidaceae)

Common name: Poison bulb
This is a large bulbous herb found in tropical zones. It is cultivated as an ornamental plant and commonly available in Central and North India. Its leaf aqueous extract was reported to show repellency to several noxious insects (Nadkarni, 1954) and antifeedancy to the desert locust, *Schistocerca gregaria* (Singh and Pant, 1980).

269. *Crinum bulbispermum* Milne-Redhead & Schew. (Amaryllidaceae)

Common name: Sukhadarsan

This is an annual/perennial herb found in tropical zones and commonly available in North India. Its leaf aqueous extract was reported to show antifeedant activity to the desert locust, *Schistocerca gregaria* (Singh, 1974; Singh and Pant, 1980) and also to the mustard sawfly, *Athalia proxima* (Pandey et al., 1977).

270. *Crinum defixum* Ker-gaul (Amaryllidaceae)

This is an annual/perennial herb found in tropical zones and commonly available in Uttar Pradesh and Punjab in India. Sudhakar et al. (1978) and Pandey et al. (1979) reported its leaf extract in ethyl ether to show antifeedancy to the larvae of mustard sawfly, *Athalia proxima*.

271. *Crotalaria juncea* Linn. (Fabaceae)

Common name: Sunn-hemp

This is an annual tall herb/shrub cultivated in tropical and sub-tropical zones and commonly available in India. Its leaf, root and flower extracts in water and petroleum were toxic to the rice weevil, *S. oryzae* (Jacobson, 1975; Prakash et al., 1978–1987).

272. *Croton aromaticus* Linn. (Euphorbiaceae)

(Syn.-*C. lacciferus* Linn.)

This is a perennial aromatic shrub or small tree found in tropical and sub-tropical regions. The active component, '(-) hardwickiic acid', isolated from its dried root, was reported to show 62% mortality to the aphid, *Aphis craccivora* (Bandara et al., 1987). Further, the active components, i.e., 3-Ent-kauranoids and 2-oleananes, isolated from its leaf extract, were also reported to show toxicity to this aphid at 5 ppm concentration of the components' solutions. The components were structurally identified as Ent-kaur-15-en-3 β, 17-diol and Ent-15β, 16-epoxykauran-17-ol (Bandara et al., 1988).

273. *Croton californicus* Linn. (Euphorbiaceae)

This is an annual/perennial herb/small shrub found in tropical and sub-tropical zones. Water extract of its leaf was reported to show insecticidal activity against the aphids (Gunter and Jepson, 1982).

274. *Croton sparsiflorum* Marong. (Euphorbiaceae)

Common name: Croton/Jamalghota

This is an annual herb available in tropical and sub-tropical zones. Its plant extract was found to be the most toxic to, the larvae and adults of rice moth, *Coryra cephalonica* when compared with four other effective test products like *Acorus calamus*, *Annona squomosa*, *Brassica compestris* and *Melia azedarach* (Chauhan et al., 1987). Its seed oil was found to reduce the survival and longevity of the green leafhopper, *Niphotettix* spp. when sprayed on the rice plant (Narsimhan and Mariappan, 1988). Nigam et al. (1990) reported its plant extract to effectively minimize the adults and nymphal populations of the mustard aphid, *Lipaphis erysimi* under field conditions.

275. *Croton tiglium* Linn. (Euphorbiaceae)

Common name: Purging croton or Jamalghota

This is a perennial shrub/small tree found in tropical and sub-tropical zones. This plant is native of South East Asia and commonly available in Assam and West Bengal in India. Its seed extract in acetone was reported to possess insecticidal property against the silkworm, *Bombyx mori*; the cotton stainer, *Dysdercus koenigii* (Borle, 1982); the mulberry white caterpillar, *Rondotia nenciana* (McIndoo, 1983); and the beet armyworm, *Spodoptera exigua* (Yingchol, 1983). Crosby (1971) found its seed oil to show insecticidal activity and mentioned as its active component.

276. *Cuminum cyminum* Linn. (Umbelliferae)

Common name: Zeera/Fennel

This is a condimental herb grown in tropical regions. In India, it is available in the Central and Coastal States. An oil extract from its dry seeds, known as 'cumin oil' was found to be toxic to the cotton stainer, *Dysdercus cingulatus* when tested under controlled conditions (Purohit et al., 1983). Mohiuddin et al. (1987) reported repellency in its oil against the red flour beetle, *T. castaneum* under controlled conditions.

277. *Curcuma amada* Roxb. (Zingiberaceae)

Common name: Mango ginger

This is a perennial wild herb found in the sub-tropical and temperate zones and available in Chhota Nagpur, Eastern Himalayas and Western Ghats in India. Ahamad and Ahmed (1991) reported its rhizome powder to protect stored grain from infestations of rice weevil, *Sitophilus oryzae* and rice moth, *Corcyra cephalonica* when admixed @ 3% w/w and stored. This powder was most effective and showed 100% toxicity to the test insects.

278. *Curcuma longa* Koeining non-Linn. (Zingiberaceae)

Common name: Turmeric

This is an annual/seasonal herb grown as one of the condiments in tropical and temperate zones. In India, it is cultivated in Tamil Nadu, Andhra Pradesh, Bihar, Karnataka and Orissa. Turmeric powder was reported as a repellent and protected stored grains from the infestation of insects, viz., the rice weevil, *S. oryzae* (Jilani, 1980; Prakash et al., 1982; Jilani and Saxena, 1987; Jilani et al., 1987); the grain weevil, *Sitophilus granarium* (Jilani and Su, 1980, 1983); and the pulse beetle, *Callosobruchus chinensis* (Anonymous, 1980–1981; Mathur et al., 1985; Pronata, 1986). Its rhizome powder also protected stored rice against Angoumois grain moth, *S. cerealella* (Prakash et al., 1989); rice weevil, *Sitophilus oryzae* (Prakash et al., 1990c); and also against the lesser grain borer, *Rhyzopertha dominica* (Prakash et al., 1991) when admixed with grains. Vir (1990) reported its rhizome extract sprayed on the mothbean crop to provide 42.85% leaf protection against the attack of the mothbean defoliator, *Cyrtozemia cognata*. Ahamad and Ahmed (1991) found its rhizome powder to show 100% mortality to *S. oryzae* and *C. cephalonica* up to 45 days even at 1% w/w dose admixed with stored grains.

Two active components, viz., ar-turmerone [2-methyl-6-(4-methylphenyl)-2-hepten-4-one] and turmerone [2-methyl-6-(4-methyl-1, 4-cyclohexadien-1-yl)-2-hepten-4-one], were isolated, identified and evaluated as repellents against the flour beetle, *T. castaneum*. Turmerone was thermally unstable and at ambient temperature in the presence of air it was reported to yield to a dimer or the more stable ar-tumerone (Su et al., 1982).

279. *Curcuma zedoaria* Rosc. (Zingiberaceae)

Common name: Round zeodery

This is an annual herb grown in tropical and sub-tropical zones and available in Karnataka and Andhra Pradesh in India and distributed in South East Asia, Australia and Africa. Its seed and dried fruit powders were reported to show toxicity to the pulse beetles, *Callosobruchus chinensis* and *C. maculatus* when tested under laboratory conditions (Anonymous, 1980–1981). Ahamad and Ahmed (1991) found its rhizome powder to show 100% mortality to *S. oryzae* and *C. cephalonica* up to 45 days even at 1% w/w dose admixed with stored grains.

280. *Cymbopogon caesius* (Nees.) Stapf. (Poaceae)

(Sny.-*Andropogon caesius* Nees., *A. schoenanthus* sub. sp. *genuinus* var. *caesius* Hack.)

Common name: Ginger grass

This is a perennial grass found in tropical zones and distributed throughout South East Asian countries. This grass is commonly available in Travancore,

Bangalore and Tamil Nadu in India and in China. Zhang and Zhao (1983) reported its volatile essences to kill adults of the weevils, *Sitophilus oryzae* and *S. zeamais* when tested under controlled conditions.

281. *Cymbopogon citratus* (D.C.) Stapt. (Poaceae)

(Syn.-*Andropogon citratus* D.C., *A. citriodorum* Derf., *A. roxburghii* Nees)

Common name: Lemon grass

This is a tufted perennial grass found in tropical and sub-tropical zones. In India, this grass is available in Maharastra, Andhra Pradesh, Karnataka, Orissa and Kerala. Its leaf extract in water was reported to possess grain protectant property when tested against Angoumois grain moth, *Sitotroga cerealella* (Prakash et al., 1989) and also insecticidal property against the cotton aphid, *Aphis gossypii* (McIndoo, 1983). Devaraj Urs and Srilatha (1990) reported its plant extract to show repellent activity to the rice moth, *Corcyra cephalonica* when tested under controlled conditions. Gupta et al. (1990) found its leaf extract to protect paddy grains against rice stink bug, *Leptocorisa acuta* when sprayed on the caged rice plants and grains. Third instar larvae of *S. litura* treated with its oil either as isolated droplets or incorporated into test diets consisting of a castor meal made from green pods of peanut varieties M-13 and TMV-2 showed 100% mortality to the larvae (Rajapakse and Jayasena, 1991).

Banerji (1988) reported 'citral', a major constituent of its oil, to show weak juvenile hormone activity against the nymphs of *Dysdercus koenigii.*

282. *Cymbopogon marginatus* Spreng. (Poaceae)

Common name: Citronella

This is a perennial tropical grass commonly available in Central America and Africa. Its root extract was reported to show repellent activity against the moths and also observed to attract both male and female of the Oriental fruit fly, *Dacus dorsalis* (Jacobson, 1975).

283. *Cymbopogon martinni* (Roxb.) Watts. (Poaceae)

Common name: Palmarosa or Geranium oil plant

This is a perennial tropical grass distributed throughout South Asian countries and Africa and also available in India. Its essential (2%) oil was reported to show toxicity to the pulse beetles, *Callosobruchus chinensis* and *C. maculatus* (Srivastava et al., 1988a,b). Bhargawa (1994) found its oil admixed with food to reduce adult emergence and it showed feeding deterrency to rice moth, *Corcyra cephalonica.*

284. *Cymbopogon nardus* (Linn.) Rendle (Poaceae)

(Syn.-*Andropogon nardus* Linn., *Andropogon ampliflorus* Stend.)
Common name: Citronella grass
This is a robust perennial grass found in tropical and sub-tropical zones and distributed in the Philippines, Malaysia, Sri Lanka and East Africa. It is commonly available in South India. Oil extracted from its leaf was reported to reduce oviposition of leafhopper, *Amrasca devastans* (Saxena and Basit, 1982) and it also showed insecticidal activity to the screw worm, *Cochiliomyia hominivorax* (McIndoo, 1983).

285. *Cymbopogon wintercanus* Spreng. (Poaceae)

Common name: Java citronella
This is an annual/perennial herb found in tropical zones and commonly available in Central India. Essential oil (0.4%) extracted from its leaf extract was reported to show toxicity to the pulse beetles, *Callosobruchus chinensis* and *C. maculatus* (Srivastava et al., 1989). Bhargawa (1994) found its oil admixed with food to reduce adult emergence and it also showed antifeedancy to the rice moth, *Corcyra cephalonica*.

286. *Cynodon dactylon* (Linn.) Pers. (Gramineae)

(Syn.-*Panicum dactylon* L., *Digitaria dactylon* [L.] Scop., *Paspalum dactylon* [L.] Lamk., *Milium dactylon* [Linn.] Moench., *Chloris cynodon* Trin., *Caperiola dactylon* [L.] O. Kuntze.)
Common name: Bermuda grass or Bahama grass
This is a perennial grass found in tropical and sub-tropical zones and distributed in Japan, the Bahamas, Bermuda Islands, the Philippines and Coastal Australia. Bahama grass is also available in India. Verma and Singh (1985) reported its whole plant aqueous extract to show antifeedant activity against the sunn-hemp caterpillar, *Amsacta olbistriga*.

287. *Cynoglossum officinale* Linn. (Boraginaceae)

Common name: Hound's tongue
This is an annual/perennial herb found in temperate zones and distributed in the United States, Canada and European countries. Its root, leaf and branch extracts in water were reported to show growth inhibiting activity when tested against the vinegar fly, *Drosophila melanogaster* (Michael et al., 1985).

288. *Cyperus rotundus* Linn. (Cyperaceae)

(Syn.-*C. hexastachyos* Roxb.)
Common name: Nut grass
This is an annual/perennial sedge commonly available as weed in the agricul-

tural fields in tropical zones. It is distributed throughout India and also found in Sri Lanka, the Philippines, China, Malaysia, Pakistan and Indonesia. Its root and leaf extracts in water were reported to show antifeedant activity to the grasshoppers (Jacobson, 1958) and also to stored grain insect pests (Khan, 1982). Verma and Singh (1985) reported its aqueous leaf extract to show antifeedant activity against the sunn-hemp caterpillar, *Amsacta olbistriga*. Its shade dried stem extract in acetone was also found to be toxic to the brinjal leaf beetle, *Henosepilachna vigintioctopunctata* in a laboratory test (Chandel et al., 1987).

289. *Cyperus scariosus* Linn. (Cyperaceae)

Common name: Nut grass

This is an annual/perennial sedge commonly available as weed in the agricultural fields in tropical zones. Nut grass is distributed in India, the Philippines, Malaysia, Sri Lanka, Bangladesh, China, Australia and West Africa. Its leaf extract in water was found to be insecticidal to the larvae of *Utetheisa pulchella*, a pest of sunn-hemp and did not allow its larvae to develop (Pandey, 1978a,b).

290. *Cytisus laburnum* Linn. (Fabaceae)

Common name: Golden-chain

This is a perennial tree found in tropical and Mediterranean zones. Its leaf extract was reported to possess insecticidal property when tested against the aphids (Michael et al., 1985).

291. *Cytisus scoparius* Linn. (Fabaceae)

Common name: Scotch brown

This is a perennial tree or herb found in temperate and Mediterranean zones and commonly available in North America and European countries. Its meristemic tissue extract was reported to posses insecticidal property when tested against the aphids of agricultural importance and also against the cabbage white butterfly, *Pieris brassicae* (McIndoo, 1983).

292. *Dalbergia sisso* Roxb. (Fabaceae)

Common name: Sisso

This is a perennial tree found in tropical and sub-tropical zones. This is distributed in India, China, Malaysia, Indonesia, Sri Lanka, the Philippines and Pakistan. Its leaf and root extracts are known to possess larvicidal and antifeedant properties to different insects of agricultural importance. Its aqueous leaf extract was reported to show antifeedancy to the larvae of *Utetheisa pulchella*, a pest of sunn-hemp and it also affected its oviposition and hatching (Pandey, 1976). Further, its root extract did not allow the adults of *U. pulchella* to emerge when its larvae were fed on the treated leaves of *Crotolaria juncea* (Pandey et al., 1977). Two fractions from its root extract, one in methanol and another in acetone, were isolated and found to cause 100 and 57% mortality to the larvae of *U. pulchella* respectively (Pandey, 1978b). Its root extract in methanol was also found to show insecticidal activity to lepidopterans, viz., *Diacrisia obliqua*, *Spodoptera litura* and *Argyria cribraria* (Srivastava et al., 1979).

293. *Daphne odora* B. Don. (Thymelacaceae)

(Syn.-*D. cannabina* Wall., *D. papyracea* Wall.)
This is a perennial shrub or small tree found in Himalayan tract, Khashi and Nilagiri hills in India and also in Bhutan, Nepal and Burma. Three flavons isolated from the root and bark of this plant, i.e., daphnodrins A, B and C, were reported to show insecticidal activity when tested against *Spodoptera litura* and *Callosobruchus chinensis* (Inamori et al., 1987).

294. *Datura metal* Linn. (Solanaceae)

(Syn.-*D. alba* Nees, *D. fastuosa* Linn.)
Common name: Angel trumpet or Black datura
This is an annual/perennial herb or shrub available in tropical zones. Black datura is distributed in India, Bangladesh, the Philippines, Malaysia, Australia and East Africa. Its whole plant extract was found to show insecticidal activity against *Spodoptera litura* (Balasubrahmanian, 1982). Yadav and Bhatnagar (1987) reported its 2% leaf powder to show toxicity to the pulse beetle, *Callosobruchus chinensis* when admixed with cowpea seeds in storage. Devaraj Urs and Srilatha (1990) found its whole plant extract to show repellent activity against the rice moth, *Corcyra cephalonica* in a laboratory test.

295. *Datura stramonium* Linn. (Solanaceae)

(Syn.-*D. tatula*)
Common name: Jimson weed or Stramonium or Thorn apple or Datura
This is a perennial shrub found in temperate and sub-tropical zones and commonly available in North India, North America and Europe. Water extracts from its whole plant, branch, leaf, seed and flower were reported to possess insecticidal, antifeedant and nematicidal properties against the pests of agricultural importance. Its dried leaves admixed with stored grains were found to protect them from the infestation of insect pests (Giles, 1964). Its leaf extract in acetone was reported to show toxicity to 5th instar nymphs of *Dysdercus*

cingulatus and 3rd instar larvae of *Spodoptera litura* and *Pericallia ricini* (Rajendran and Goplan, 1979). Its whole plant extract was reported to be toxic to the fruit borer, *Earias vitella* when tested on potted okra crop (Anonymous, 1986). Similarly, Kareem (1984) found its whole plant extract possessing insecticidal property against *D. cingulatus*. Balasubrahmanian (1989) also reported its leaf extract possessing insecticidal property against *S. litura*.

296. *Datura suaveolens* Humb. & Bonpl. ex. Willd. (Solanaceae)

(Syn.-*Brugmansia suaveolens* [Humb. & Bonpl. ex. Willd.] Bercht & Presl.

This is a perennial shrub found in tropical zones and commonly available in Gujrat and Maharastra in India. Patel et al. (1990b) reported its aqueous leaf extract to show significant mortality to the larvae of Gujrat hairy caterpillar, *Amsacta moorei* in a bio-efficacy test under field conditions.

297. *Delonix regia* (Boj.) Raf. (Caesalpiniaceae)

(Syn.-*Poinciana regia* Boj. ex. Hook)

Common name: Peacock flower or Royal poinciana

This is a perennial tree found in the tropics and commonly available in coastal India and other South East Asian countries. Its flower extract in water was reported to show insecticidal activity against leaf eating caterpillars and beetles (Jacobson, 1975) and also antifeedant activity to the larvae and pupae of the pulse beetle, *Callosobruchus maculatus* (Candrakantha, 1988).

298. *Delphinium brunonianum* Royle. (Ranunculaceae)

Common name: Musk larkspur

This is an annual herb available in sub-tropical and temperate zones. It is native to China and distributed in the West Himalayas. Its branch, leaf and seed extracts in water and methanol were found to show repellency to the melon worm, *Diaphania hyalinata*; the cross striped cabbage worm, *Evergestis rimosalis*; European corn borer, *Ostrinia nubilalis*; the cabbage looper, *Trichoplusia ni*; the bean leafroller, *Uranus proteus* (Jacobson, 1975); and also to grasshoppers and locusts in general (McIndoo, 1983).

299. *Delphinium delavaji* Linn. (Ranunculaceae)

This is an annual herb found in temperate zones and commonly available in North America. Its flower and seed extracts were reported to be insecticidal to Mexican bean beetle, *Epilachna varivestis* (Jacobson, 1958).

300. *Delphinium grandiflorum* Linn. (Ranunculaceae)

Common name: Bouquet larkspur
This is a perennial herb found in temperate zones and commonly available in Canada and the United States. Its branch, leaf, flower, fruit and seed extracts were reported to be insecticidal and antifeedant to the diamond-back moth, *Plutella xylostella* (Jacobson, 1975).

301. *Delphinium laxiflorum* D.C. (Ranunculaceae)

This is an annual herb available in temperate zones. Its branch, leaf and seed extracts in water were found to show insecticidal activity against the black and red ants of agricultural importance (Jacobson, 1958).

302. *Delphinium Staphysagria* Linn. D.C. (Ranunculaceae)

Common name: Satvesacre
This is a biennial herb found in temperate zones and commonly available in the USA and Europe. Its branch, leaf and seed extracts in water and petroleum ether were reported to be insecticidal to grasshoppers (McIndoo, 1983). However, earlier its seed oil and alkaloids were found to be insecticidal (Davidson, 1929).

303. *Derris chinensis* Lour. (Fabaceae)

Common name: Chinese tuba-root
This is a perennial shrub found in tropical zones and distributed in India, China, Taiwan, Pakistan and USSR. Its root extract was reported to possess a poisonous alkaloid, 'rotenone', which was found to be toxic to the aphids and other soft bodied insects of agricultural importance (Nagai, 1902).

304. *Derris elliptica* (Wall.) Benth. (Fabaceae)

Common name: Derris or Tuba-root
This is a perennial shrub/woody climber available in tropical and sub-tropical regions. Derris is native to East Asia. In fact 'tuba-root' is a collective term used for various kinds of plant materials used to obtain fish poison or to aid hunting. These plants were called *toeba* or *tuba* in Malaya and found to be the members of the family Leguminosae. Among them the genus *Derris* is the most effective against fish and insects. Derris root has long been used as an insecticide. As early as 1848, Oxley suggested that tuba-root was effective against the leaf eating caterpillars. Hooker (1877) recorded that tuba-root had been used by Chinese in Singapore for preparation of an insecticide.

The main source of tuba-root is the root of *Derris elliptica*. The root contains mainly the *Derris* resin (25% rotenone) and related components called rotenoids. Numerous reports are available with regard to the action of rotenoids against a variety of insect species. These are summarized fully and concisely in several reviews (Brown, 1951; Shephard, 1951; Feinstein and Jacobson, 1953; Negherbon, 1959) but there was a little data on this subject with respect to recently isolated rotenoids such as munduserone (Figure 53), dolineone (Figure 54), etc.

Rotenone is extremely potent against many insects. Its action and that of other potent plant origin insecticides are shown in Table 12. The results indicated that rotenone was no exception to the rule that most of the insecticides varied in toxicant action with different species. The rotenone acted as a contact insecticide and also as a stomach poison of slow action.

Biological Activity Against Storage Insects

McIndoo et al. (1919) reported rotenone as an effective insecticide to control the pea beetle, *Bruchus pisonum*. Against the rice weevil, *Sitophilus oryzae* seed powder of *D. elliptica* was found to protect stored wheat when admixed with the grains (Krishnamurthi and Rao, 1950). Puttarudriah and Bhatia (1955) reported its root powder to be toxic to the pulse beetle, *Callosobruchus (Bruchus) chinensis*. Fukami et al. (1959) found toxic effects of 35 rotenone analogues against the azuki bean weevil, *Callosobruchus chinensis*. Similarly, toxic effect of rotenone was also reported against *Tenebrio molitor* when applied topically (Negherbon, 1959).

Biological Activity Against Field Crop Insects

Its root and leaf extracts as rotenone mixture were reported to be toxic to California red scale, *Aonidiella aurantii* (Cressman and Brodbent, 1943); the mustard web worm, *Crocidolomia binotalis*; the diamond-back moth, *Plutella interpunctella* and *Idiocerus* sp. (Puttarudriah and Bhatia, 1955); to Mediterranean fruit fly, *Ceratitis capitata* (Loke, 1983); and also to the cotton leaf armyworm, *S. litura* (Puttarudriah and Bhatia, 1955; Krishnamurthy, 1982).

Active Components

Thirteen rotenoids have been reported so far occurring in the higher plants as shown in Table 13. Rotenone was the foremost rotenoid to be isolated and explored chemically for its use and synthesis. The word rotenone is derived from *Derris chinensis*, commonly called 'gyoto', grown in Formosa (presently Taiwan) and also called 'roten' by the natives because it was shown to be a keton and thus called 'rotenone'. In 1916, Ishikawa isolated a compound showing the same melting point from *Derris elliptica* and named it tubotoxin and in 1923 both the compounds rotenone and tubotoxin were confirmed identical and the correct molecular formula $C_{23}H_{22}O$ was proposed. Later rotenone was also isolated from various species belonging to family Leguminosae (Jacobson and Crosby, 1971).

Table 12. Comparative toxicity of *Derris* resin (25% rotenone) and other compounds to several insect species

	Amount (mg/g) needed to produce mortality							
	Rotenone		Pyrethrins		Na arsenate		Nicotine	
	Route	100%	50%	100%	50%	100%	50%	100%
Anara tristis	T	2600	7	26	--	--	350	1250
	I	25	10	25	20	40	200	350
Bombyx mori (Larva)	T	0.7	--	0.4	--	--	4	8
Ceratomia catalpae (Larva)	T	5	2	6	--	--	100	200
	I	6	4	6	20	30	80	150
Oncopeltus fasciatus	T	60	8	28	--	--	190	450
Periplaneta americana	T	2000	6.5	12	250	1300	650	1300
	I	13	6	11	45	70	100	200
Popillia japonica	T	60	40	130	850	1700	650	1000
	I	110	40	110	50	100	400	900
Tenebrio molitor	T	75	35	100	--	--	3200	4400

T = Topical contact, I = Injection

Source: Jacobson and Crosby, 1971

Table 13. Occurrence of rotenoids and isoflavonoids in the plants

Plant	Rotenoid	Isoflavonoid
Derris malaccensis	Rotenone, Sumatrol, Duguelin, Toxicarol, Elliptone, Malaccol	Toxicarol isoflavone -- --
Mellettia auriculata	--	Aurmilline
Mundulia sericea	Munduserone	Mundulone, Munetone
Neorautanenia pseudopachyrrhiza	Dolineone	Neotenone, Nepseudin
Pachyrrhizus erosus	Rotenone, Pachyrrhizone	Pachyrrhizin, Erosnin
Tephrosia purpuria var. *maxima*	-- -- --	Maxima substance-C, Hydroxy-2′-methoxy-3-7-4-methylene-dioxy-isoflavone,Purpuranin-A, Purpuranin-B
Tephrosia villosa	Villosin, 12-a-dehydro isosumarol	

Source: Fukani and Kakajima, 1971, Srimmannarayan and Rao, 1985

Jacobson and Crosby (1971) mentioned that there were four independent schools of chemists and they simultaneously proposed an identical structure of rotenone giving the basis for the structural elucidation of other rotenoids that followed. Every rotenoid has a four-fused ring system, i.e., chromano-chromanone (Figure 43) of 6a, 12a-dihydro-6H-rotoxen-12-one, named from the parent heterocyclic system rotoxene. The trivial names for rotenoids and their degradation products are represented in Figures 44–56. Thus rotenone (Figure 44) was 6a, 12a, 4, 5-tetrahydro-2, 3-dimethoxy-5-isopropenylfurano-3,2,2,9(-hydrotozen-12-one).

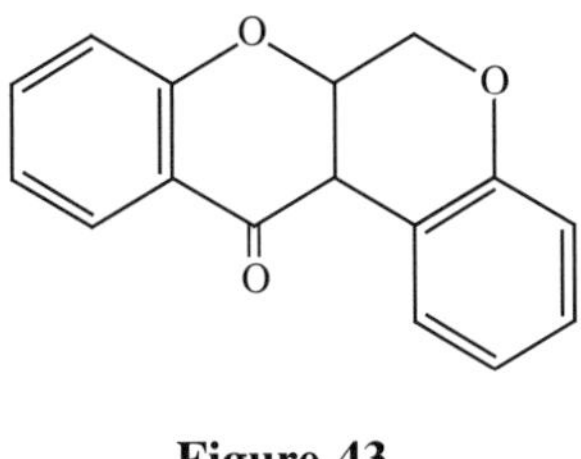

Figure 43. Chromanochromanone.

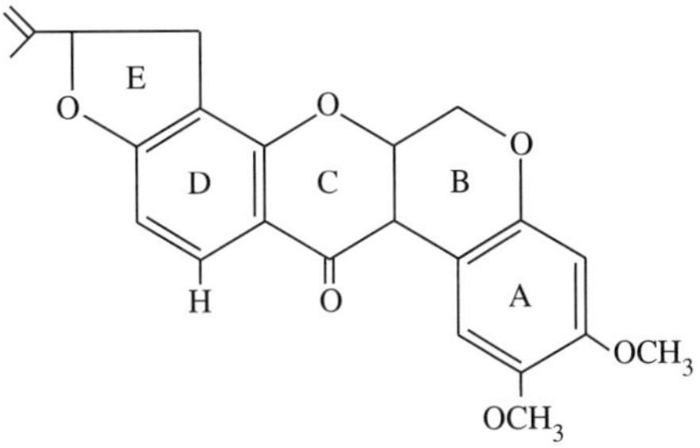

Figure 44. Rotenone.

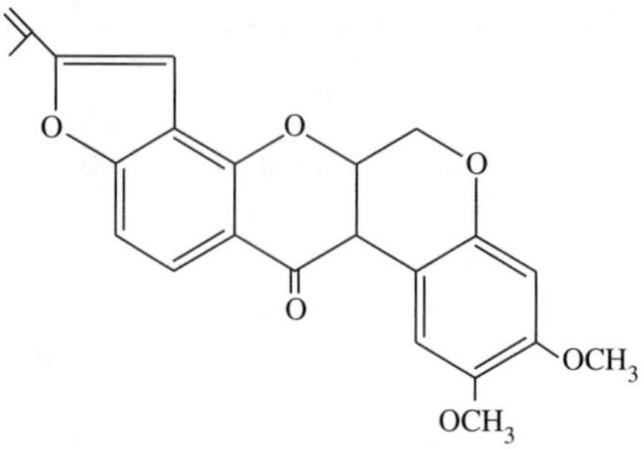

Figure 45. Isorotenone.

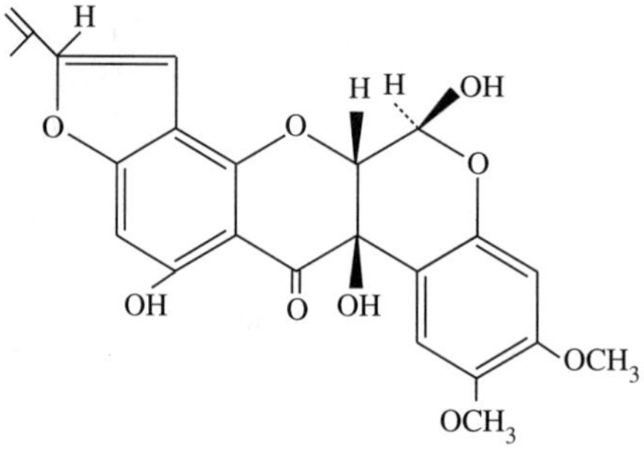

Figure 46. Villosin.

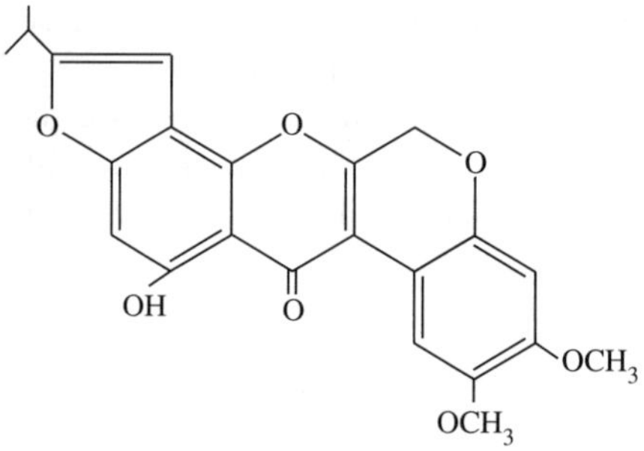

Figure 47. 6a, 12a-Dihydroisosumatrol.

The inhibition of the electron transport chain appeared to arise from the binding of rotenone to a component of the chain, but $NADH_2$ dehydrogenase was not inhibited (Corbett, 1974). The symptoms of insects poisoned by rotenone differed from those produced by neurotoxic insecticides and were characterized by reduction of oxygen-consumption, depressed respiration and eventual paralysis (Martin, 1964).

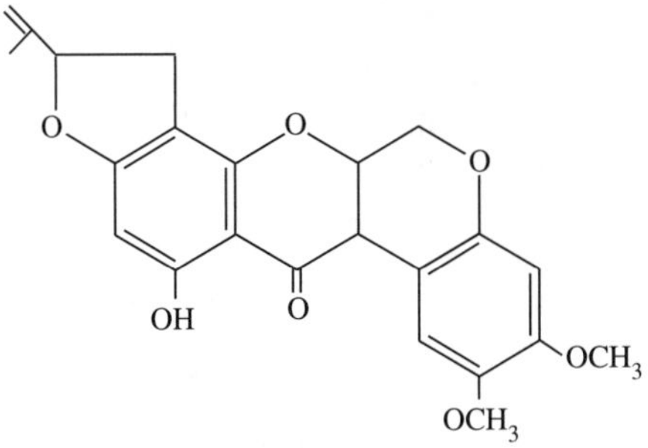

Figure 48. Sumatrol.

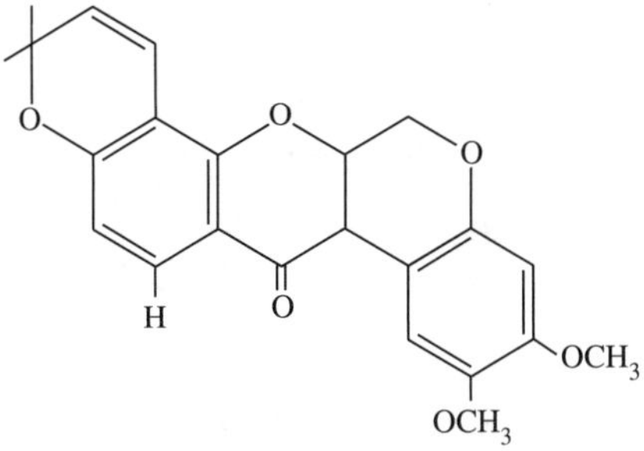

Figure 49. Duguelin.

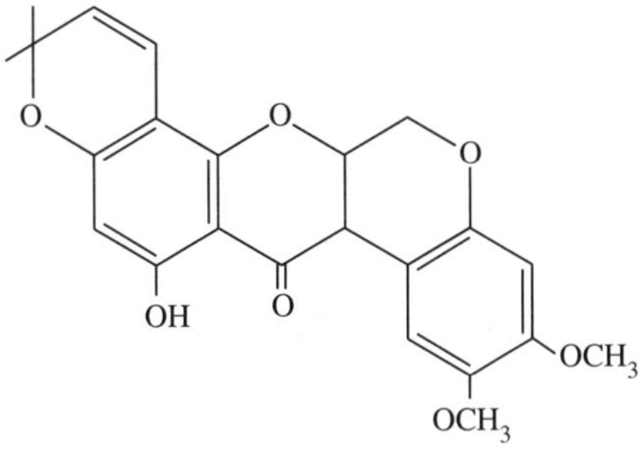

Figure 50. Toxicarol.

305. ***Derris ferruginea*** (Roxb.) Benth. (Fabaceae)

(Syn.-*Pongamia ferruginea* Wall., *Robinia ferruginea* Roxb.)
This is a perennial woody climber found in temperate and sub-tropical regions and distributed in the Eastern Himalayas and Assam in India, China and Nepal. Its root powder was reported to possess insecticidal property against Angoumois grain moth, *Sitotroga cerealella* (Prakash et al., 1978–1987).

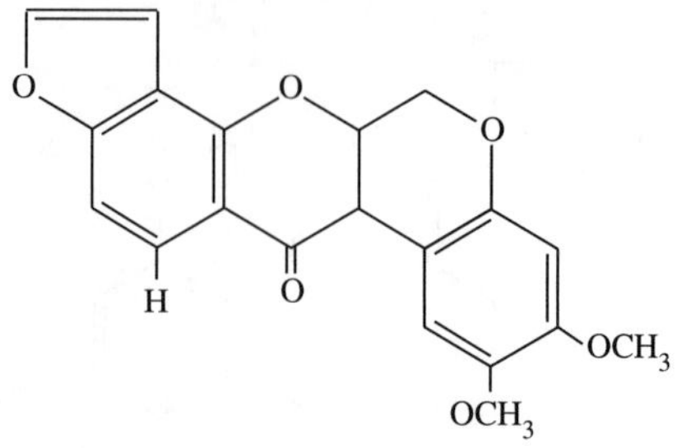

Figure 51. Elliptone.

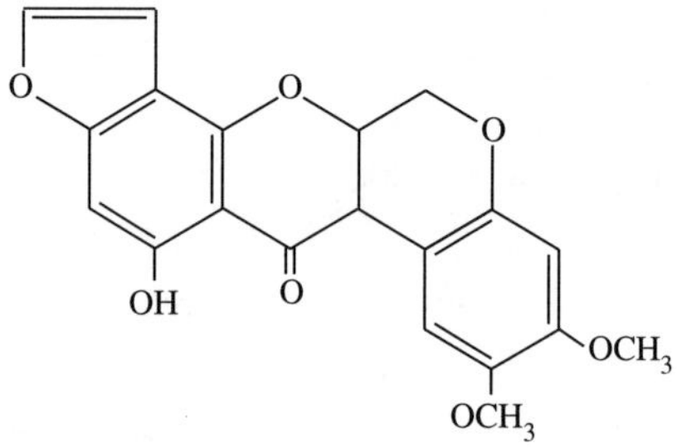

Figure 52. Malaccol.

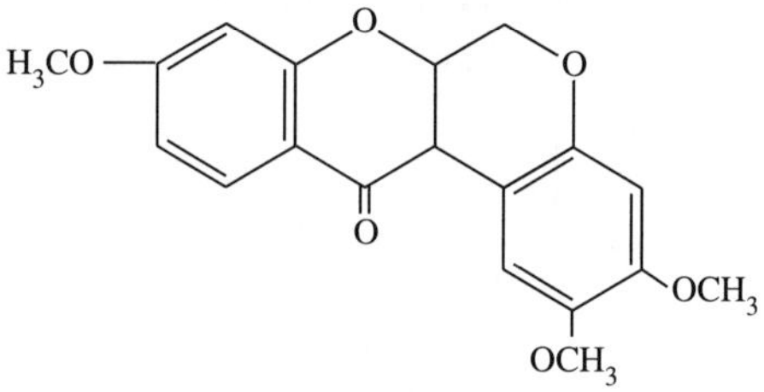

Figure 53. Munduserone.

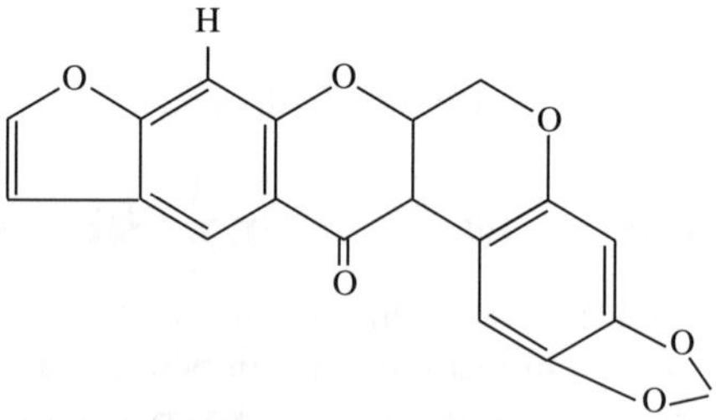

Figure 54. Dolineone.

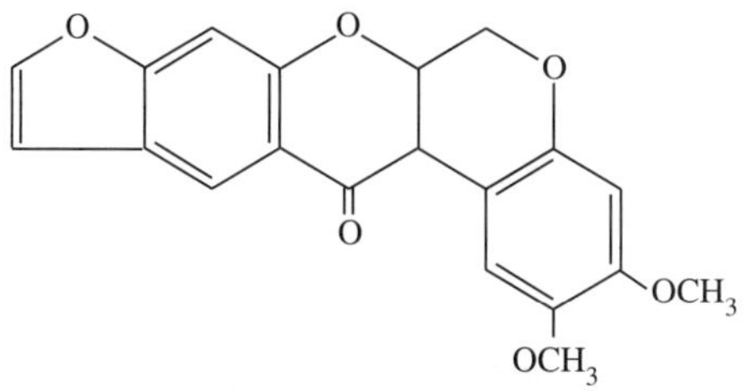

Figure 55. Erosone.

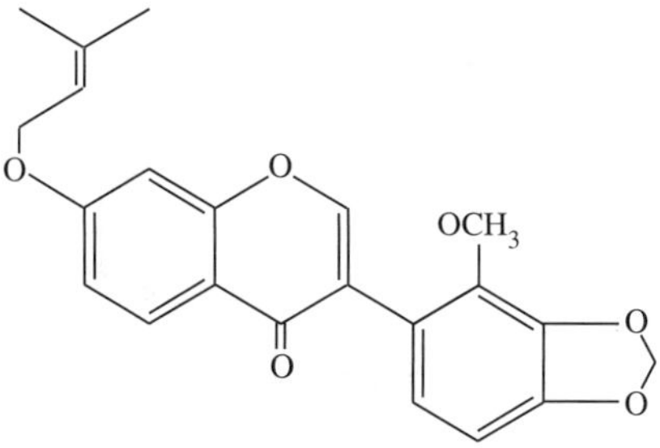

Figure 56. Maxima substance-C.

306. *Derris fordii* Lour. (Fabaceae)

This is a perennial woody climber found in temperate and sub-tropical zones and available in North America. Its root extract in acetone was reported to show insecticidal activity when tested against the aphid, *Aphis fabae* (Jacobson, 1975).

307. *Derris malaccensis* Prain. (Fabaceae)

This is a perennial woody climber/vine found in tropical regions but is a native of Malaysia. Its root powder and extract in water were reported to show insecticidal and repellent activities against the citrus aphid, *Aphis citri*; the cabbage leaf webworm, *Crocidolomia binotalis*; the spotted ladybird beetle, *Hinosepliachna sparsa*; and the diamond-back moth, *Plutella xylostella* (Crooker, 1983). A number of rotenoids, viz., rotenone (Figure 44), sumatrol (Figure 48), duguelin (Figure 49), toxicarol (Figure 50), elliptone (Figure 51) and malaccol (Figure 52) and an iso-flavonoid called toxicarol isoflavone (Figure 61), were isolated from this plant, showed insecticidal activity and are mentioned in Table 13. A number of other isoflavonoids isolated from other plants and found to show biological activity are also mentioned in Figures 56 to 61.

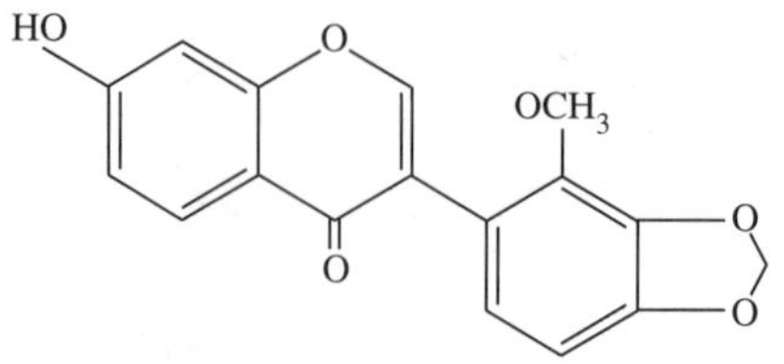

HO O OCH3 O O O

Figure 57. Hydroxy-2, methoxy-3,-4 methoxylene-dioxy-isoflavone.

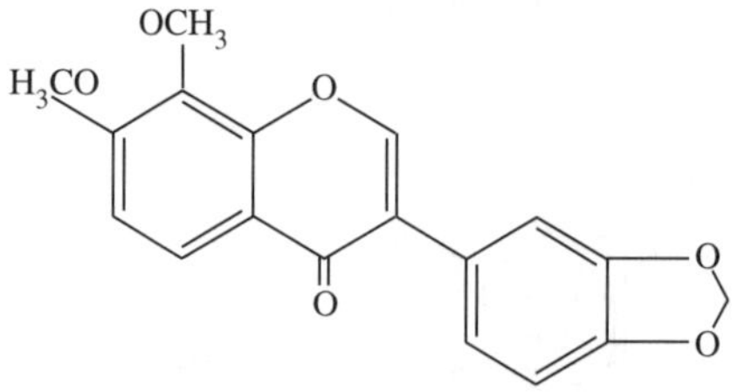

OCH3 H3CO O O O O

Figure 58. Purpuranin-A.

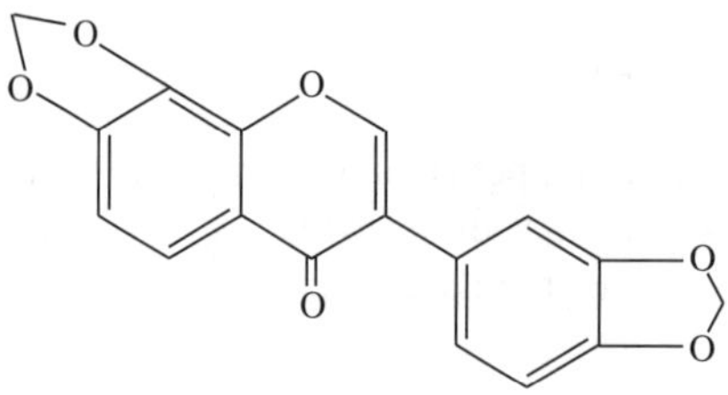

O O O O O O

Figure 59. Purpuranin-B.

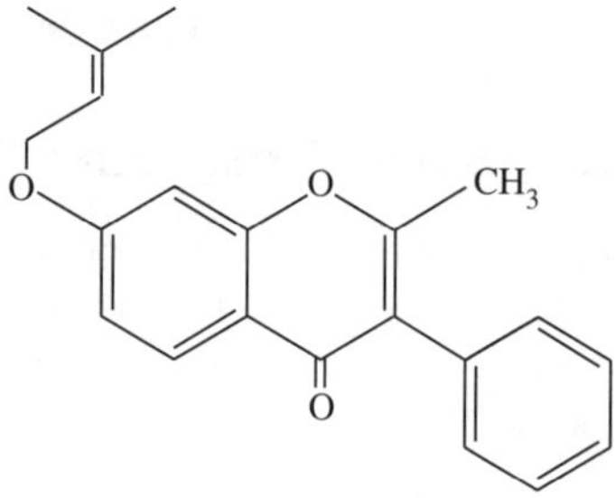

O O CH3 O

Figure 60. 7-Dimethyl-allyoxy, 2-methyl-isoflavone.

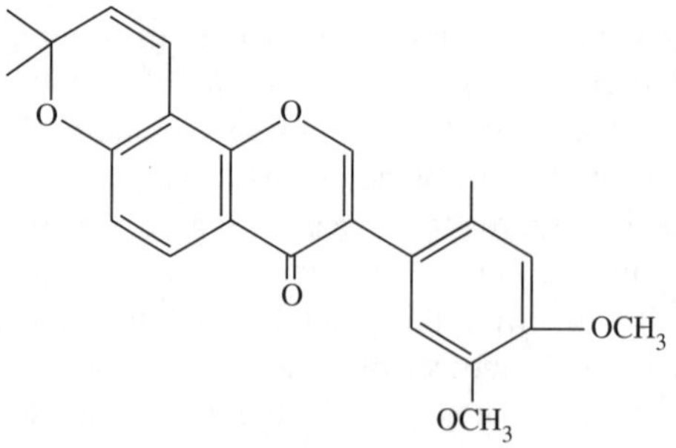

O O O OCH3 OCH3

Figure 61. Toxicarol isoflavone.

308. *Derris malaccensis* var. *sarwakenis* Lour. (Fabaceae)

This is a perennial tree or woody climber found in sub-tropical regions and available in China and Korea. Its root powder was reported to possess insecticidal property against the bean aphid, *Aphis fabae* (Jacobson, 1958) and also against Mediterranean fruit fly, *Ceratitis capitata* and the fruit fly, *Dacus* sp. (Loke, 1983).

309. *Derris philippinensis* Lour. (Fabaceae)

This is a perennial tree found in tropical regions and available in the Philippines. Its aqueous root extract was reported to possess insecticidal property when tested against the diamond-back moth, *Plutella xylostella* (Mohganoy and Morallo-Rajessus, 1975).

310. *Derris polyantha* (Roxb.) Benth. (Fabaceae)

This is a perennial tree or woody climber found in temperate and sub-tropical regions. Aqueous extract of its root was reported to show insecticidal activity against the bean aphid, *Aphis fabae* (Jacobson, 1958).

311. *Derris robusta* Benth. (Fabaceae)

This is a perennial deciduous hardy tree available in tropical zones. In India, it is found from Himalayan to Kumaon hills and Assam. The active component 'robustic acid' from its root extract was found to show antifeedant activity to the 4th instar larvae of *Spodoptera litura* (Srimmannarayan and Rao, 1985).

312. *Derris scandens* (Roxb.) Benth. (Fabaceae)

This is a perennial deciduous tree found in tropical zones and commonly available in the forests of Assam, Bengal, Uttar Pradesh and South India. A natural coumarin, 3-aryl-4-hydroxy coumarin, called lonchocarpic acid (Figure 62), and also its methyl ether 4,4-dio-methyl lonchocarpic acid (Figure 63) isolated from its root extract showed high antifeedant activity when tested against the 4th instar larvae of *Spodoptera litura* (Srimmannarayan and Rao, 1985).

313. *Derris trifoliata* Lour. (Fabaceae)

(Syn.-*D. uliginosa* Benth., *Pongamia uliginosa* D.C., *Roninia uliginosa* Willd.)

This is a perennial woody climber found in the tropical zone. Extracts from its branch and root were reported to be insecticidal against the aphids and the silkworm, *Bombyx mori* (McIndoo and Sievers, 1924).

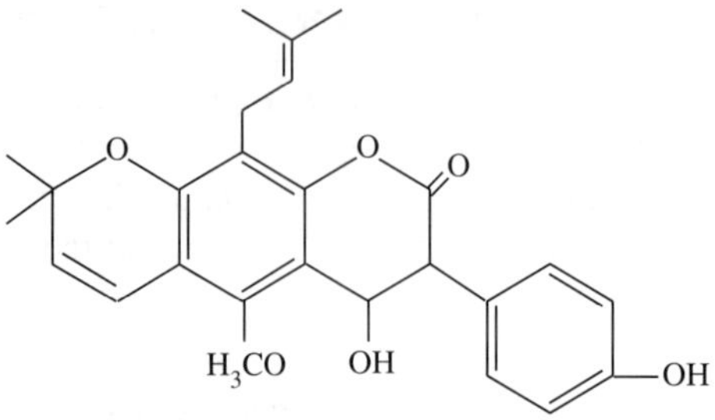

Figure 62. Lonchocarpic acid.

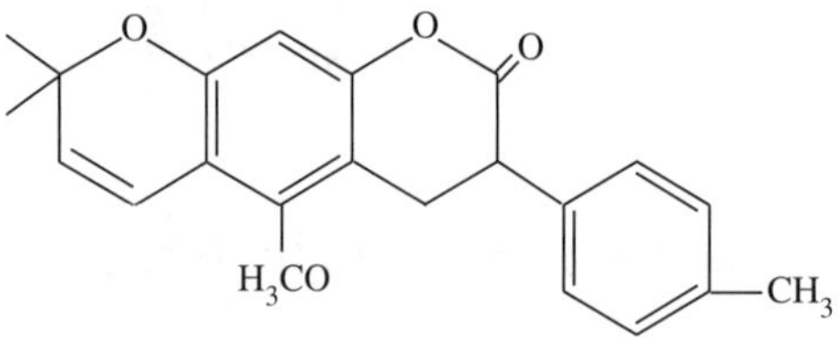

Figure 63. 4,4-Diomethyl lonchocarpic acid.

314. *Dichapetalum ruhlandi* Thou. (Dichapetalaceae)

This is a perennial woody climber found in tropical and sub-tropical regions and commonly available in China. Its leaf aqueous extract was reported to possess insecticidal property against the aphids (Michael et al., 1985).

315. *Digitalis ambigua* Linn. (Scrophulariaceae)

Common name: Yellow foxglove

This is a perennial herb found in temperate and Mediterranean zones. Its branch, leaf and flower extracts were reported to show insecticidal activity when tested against the aphids (McIndoo, 1983).

316. *Digitalis lonata* Ehrh. (Scrophulariaceae)

Common name: Grecian foxglove

This is a perennial herb found in temperate and Mediterranean zones. This is a native of Europe and now cultivated in Jammu and Kashmir in India. Its leaf extract in ether was found to show insecticidal activity against termites (Jacobson, 1975).

317. *Digitalis purpurea* Linn. (Scrophulariaceae)

(Syn.-*D. tomentosa* Link. & Hoffmagg.)

Common name: Common foxglove

This is a perennial/biennial woody climber found in temperate zones. This is a native of Western Europe and also available in hilly tracts in North India. Water extract of its leaf was reported to show insecticidal and antifeedant activities against aphids and termites (Jacobson, 1958; McIndoo, 1983).

318. *Dillenia indica* Linn. (Dilleniaceae)

(Syn.-*D. spiciosa* Thunb.)

Common name: Elephant apple

This is a perennial middle size tree found in tropical, sub-tropical and temperate zones. In India, this tree is available in the sub-Himalayan tracts, West Bengal, Madhya Pradesh, Assam and also South Indian States. Alcoholic extracts of its root, branch and leaf were reported to show insecticidal activity against the cabbage leaf webworm, *Crocidolomia binotalis*; the hair caterpillar, *Euproctis fraterna*; and the cotton leaf armyworm, *Spodoptera litura* (McIndoo, 1983).

319. *Dioscorea hispida* Linn. (Dioscoreaceae)

This is a perennial woody climber found in tropical and sub-tropical regions. Its aqueous extract was reported to be toxic to the aphids of agricultural importance (Jacobson, 1975).

320. *Dioscorea piscatroum* Linn. (Dioscoreaceae)

This is a perennial vine found in tropical and sub-tropical zones. Its aqueous root extract was reported to show insecticidal activity against the limacodid, *Parasa herbifera* (McIndoo, 1983).

321. *Diospyros chloroxylon* Roxb. (Ebenaceae)

This is a perennial tree found in tropical zones and commonly available in Pakala forest in Andhra Pradesh, India. Its wood extract in methanol was found to show absolute feeding deterrency to the 3rd instar larvae of the castor semi-looper, *Achoea janata* when treated castor leaves with its 500 and 1000 ppm concentrations were fed to this insect (Purohit et al., 1989).

322. *Diospyros discolor* Willd. (Ebenaceae)

This is a perennial tree found in sub-tropical and temperate regions. Its twig, leaf and bark extracts in ethyl ether were found to attract the females of Mediterranean fruit fly, *Ceratitis capitata* (Keiser et al., 1975).

323. *Diospyros montana* Cl. (Ebenaceae)

(Syn.-*D. cardifolia* Roxb.)

Common name: Date plum

This is a perennial shrub or small tree found in tropical zones. Water and alcoholic extracts of its leaf and fruits were reported to show antifeedant and insecticidal activities when tested against the croton caterpillar, *Achoea janata* and the jute hairy caterpillar, *Diacrisia obliqua* (McIndoo, 1983).

324. *Diospyros virginiana* Linn. (Ebenaceae)

Common name: American ebony or Persimmon

This is a perennial tree in tropical and sub-tropical zones and distributed in South East Asian countries, Australia, the United States, and West Africa. The active component, viz., 7-methyljuglone (5-hydroxy-7-methyl-1,4-naphthalendione), isolated from its wood extract was reported to show insecticidal activity when tested against the termite, *Reticulitermes flavipes* (Carter et al., 1975).

325. *Dipterocarpus turbinatus* Gaertn. (Dipterocarpeceae)

(Syn.-*D. indicus* Bedd. *D. levis* Mam.)

Common name: Gurjun

This is a perennial large evergreen tree found in tropical zones and distributed in China, Malaysia, Indonesia, Pakistan, Bhutan and India. Gurjan is available in Western Ghats in India. Oil of gurjun was reported to show repellent activity against the bamboo insects (McIndoo, 1983).

326. *Dolichos buchani* Linn. (Fabaceae)

This is an annual herb found in tropical regions and available in South Asian countries. Its root extracts in water, alcohol and petroleum ether showed antifeedant and insecticidal activities when tested against the red flour beetle, *Tribolium castaneum* and the milkweed bug, *Oncopeltus fasciatus* (Heal et al., 1950).

327. *Dorycnium rectum* (Linn.) Serigne (Fabaceae)

Common name: Psoralea

This is an annual herb found in tropical and sub-tropical zones. Its stem, leaf and flower extracts in ethyl ether showed attractancy to both the males and females of Mediterranean fruit fly, *Ceratitis capitata* (Keiser et al., 1975).

328. *Dracocephalum moldavica* Linn. (Labiatae)

This is an annual herb found in tropical zones and available in the East Indian States. Its oil showed toxicity to the cotton aphid, *Aphis gossypii* (McIndoo, 1983).

329. *Drymaria pachyphylla* Willd. ex. Roem & Schult. (Caryophyllaceae)

This is an annual/perennial herb found in temperate and sub-tropical zones. Its whole plant extracts in water and petroleum ether showed insecticidal activity to milkweed bug, *Oncopeltus fasciatus* (Heal et al., 1950).

330. *Dryopteris fillix-mas* (Linn.) Schoott. (Polypodiaceae)

Common name: Male fern

This is a cosmopolitan terrestrial fern. Its root powder and extract in acetone were found to be insecticidal to the bean aphid, *Aphis fabae* (McIndoo, 1983) and the armyworm, *Pseudoletia unipunctata* (Jacobson, 1958).

331. *Duboisia hopwoodii* Krast. (Solanaceae)

Common name: Pleurothallis

This is a perennial shrub found in tropical regions. Its leaf extract was toxic to the bean aphid, *Aphis fabae* (McIndoo, 1983) and mealy bugs (Michael et al., 1985).

Two forms of d and dl- nornicotine were isolated from this plant as toxic/insecticidal active components. Nor-nicotine (β-pyridyl-α-pyrolidine) was found to be more toxic than nicotine when tested against *Aphis fabae*. Brown (1951) and Schmeltz (1971) reported two alkaloids, i.e., nicotine and anabasine, as active components to show toxicity to a number of soft bodied insects like the aphids and the jassids.

332. *Echinacea angustifolia* D.C. (Asteraceae)

Common name: American cornflower

This is an annual herb found in temperate zones. Its root extract in petroleum ether showed insecticidal activity to the mealworm, *Tenerbrio molitor* and an active component, i.e., echinolone ($C_{24}H_{24}O_2$), with chemical structure [dextrorotatory (E)-10-dydroxy-4, 10-dimethyl-4, 11-dodecadien-2-one] was isolated from its root extract and found to be an effective juvenile hormone mimic against *T. molitor* (Jacobson et al., 1975a). Echinacein ($C_{16}H_{25}NO$) (Figure 64), an isobutylamide, as an active component was isolated from its root extract and found to show toxicity to *T. molitor* (Jacobson and Crosby, 1971; Jacobson, 1989).

$$CH_3CH=CHCH=CHCH=CH(CH_2)_2CH=CHCONH\ CH_2\ CH\ (CH_3)_3$$

Figure 64. Echinacein.

333. *Echinacea pallida* (Natt.) Britton (Asteraceae)

Common name: American cornflower

This is an annual herb found in temperate and sub-tropical zones and available in Central America. From its root extract an active component, 'echinacein', was also isolated and found to show insecticidal activity to a wide range of insects of agricultural importance (Jacobson and Crosby, 1971).

334. *Echinochloa crus-galli* (Linn.) var. *oryzicola* (Poaceae)

Common name: Barnyard grass

This is an annual grass growing as rice weed in almost all the rice growing countries. Its whole plant extract was reported to show antifeedant activity against the rice brown planthopper, *Nilaparvata lugens*. An active component was isolated from its whole plant extracts and identified as 'transaconitic acid' [trans-1-propene-1,2,3-tricarboxylic acid], which showed antifeedant activity against *N. lugens* (Kim et al., 1976).

335. *Eclipta alba* Hussk. (Asteraceae)

Common name: Morchand

This is an annual herb/weed commonly available in the lowland rice fields and also in their irrigating channels in all the tropical countries. Its whole plant extract was reported to show ovicidal activity when tested under laboratory conditions against the eggs of Angoumois grain moth, *Sitotroga cerealella* (Prakash et al., 1979a), but under natural conditions of insect infestation in farm godowns, this extract did not show promising grain protection (Prakash et al., 1978–1987). Rao and Prakasa Rao (1979) reported its shoot and root extracts to show antifeedant activity against the rice brown planthopper, *Nilaparvata lugens* in mini-plot experiments under field conditions.

336. *Eclipta prostata* Linn. (Asteraceae)

This is an annual herb found in tropical zones and distributed throughout India. Topical application of its leaf extract showed less than 30% toxicity to the red flour beetle, *Tribolium castaneum* and its root extract was reported to be toxic to the maize weevil, *Sitophilus zeamais*. Further, its flower extract was also reported to show toxicity to the cotton stainer, *Dysdercus koenigii* (Carino, 1981).

337. *Ehretia canarensis* (Roxb.) Genth. (Ehretiaceae)

This is a perennial tree commonly available in Nilagiri hills and also in Deccan mountains in India. Extract of this plant was reported to show moderate antifeedancy (60%) when tested against the jute hairy caterpillar, *Spilosoma obliqua* (Tripathi et al., 1987).

338. *Eichornia crassipes* Mart. (Pontederiaceae)

Common name: Water hyacinth

This is a perennial aquatic weed commonly found in tropical and sub-tropical ponds and commonly available in India. Water extract of its leaf was reported to exhibit insect growth regulating activity when tested against the development of nymphal instars of *Dysdercus similis* (Sita and Thakur, 1984).

339. *Elaeis guineensis* Jacq. (Palmaceae)

Common name: Palm or Oil palm tree

This is a perennial small tree found in tropical coastal zones. It is commonly grown for its oil in Kerala, Karnataka, Tamil Nadu and Indian Archipelago. Its oil when admixed with greengram and cowpea seeds showed mortality to the adults and checked multiplication of the pulse beetles, *Callosobruchus chinensis* and *C. maculatus* and finally protected the pulse grains in storage (Mittal, 1971; Varma and Pandey, 1978; Golob and Webley, 1980; Ali et al., 1983; Peirie, 1983; Naik and Dumbre, 1984). Palm oil (2.5%) treatment was reported to check infestation of *C. chinensis* in stored treated greengram (Sujatha and Punnaiah, 1985) and blackgram (Sujatha and Punnaiah, 1984). Jacob and Sheila (1990) also found palm oil treatment to the greengram @ ` ml/kg seeds to protect them against infestation of *C. chinensis* and showed above 60% mortality to the adults 3 days after treatments.

340. *Elephantopus elatus* Bertol. (Asteraceae)

This is an annual/perennial herb found in tropical and sub-tropical zones and commonly available in South America. Its root, stem, leaf and fruit extracts in ethyl ether showed attractancy to both males and females of the Mediterranean fruit fly, *Ceratitis capitata* (Keiser et al., 1975).

341. *Elephantopus scaber* Linn. (Asteraceae)

Common name: Elephant's foot

This is an annual herb grown in tropical regions and distributed in the Philippines, Japan, the United States, and India. Topical application of its leaf extract was found toxic to the red flour beetle, *Tribolium castaneum*; to the cotton stainer, *Dysdercus cingulatus*; and to the maize weevil, *Sitophilus zeamais* (Carino, 1981; Michael et al., 1985).

342. *Elephantopus tomentosus* Hbk. (Asteraceae)

This is an annual herb found in tropical zones and available in Japan and the Philippines. Topical application of its leaf extract was reported to be toxic to the red flour beetle, *Tribolium castaneum*. Its flower and leaf extracts were found to show insecticidal activity to the cotton stainer, *Dysdercus cingulatus* and root extract was toxic to maize weevil, *Sitophilus zeamais* when tested under laboratory conditions (Carino, 1981).

343. *Elsholtzia iriostachya* Benth. (Labiaceae)

This is a perennial shrub or undershrub found in the Himalayan tracts from Jammu and Kashmir to Sikkim and Khashi Hills in India. Essential oil of this plant was reported to show insect growth retardant activity against vinegar fly, *Drosophila melanogaster* and red flour beetle, *Tribolium castaneum* when tested incorporating its oil in their diets and it inhibited pupal formation and adult emergence of both the insects (Bhattacharya and Bordoloi, 1986).

344. *Elymus canadensis* Linn. (Poaceae)

Common name: Canada wild rye
This is a perennial grass found in temperate regions and commonly available in the United States and Canada. Its whole plant extract was reported to be insecticidal to the red-legged grasshopper, *Melanoplus femmurubrum* (Jacobson, 1975).

345. *Embelia ribes* Burm.f. (Myrsinaceae)

Common name: Baibarang
This is a perennial shrub found in tropical and sub-tropical zones and distributed throughout India and other South East Asian countries. Its aqueous extract was reported to show toxicity to the pulse beetles, *Callosobruchus chinensis* and *C. maculatus* (Anonymous 1980–1981). Similarly, its seed extracts in petroleum ether, benzene and alcohol protected cowpea seeds against *C. maculatus* when tested in a laboratory test (Bhaduri et al., 1985). Chander and Ahmed (1987a,b) also reported its seed extract in ethyl acetate at 0.1% concentration to show up to 76% and 67% mortality to *Tribolium castaneum* and *R. dominica* respectively in the treated wheat grains.

Chander and Ahmed (1985) tested the efficacy of natural embelin, a possible active component found in the seed extract of this shrub, and it showed toxicity to *T. castaneum*.

346. *Emilia tuberosa* Cass. (Asteraceae)

This is an annual herb found in tropical regions. Its leaf extract in water was reported to show repellent activity against the leaf eating caterpillars (Jacobson, 1958).

347. *Encelia farinosa* Gan. (Asteraceae)

Common name: Desert sunflower

This is an annual/perennial herb found in tropical and sub-tropical zones and commonly available in India, Pakistan and Afghanistan. Chloroform leaf extract of this herb showed feeding deterrency to the larvae of *Heliothis zea* (Wisdom et al., 1983). Its whole plant extract was found to show insecticidal and anti-feedant activities to the larvae of noctuid pests, *Peridroma saucia* and *Plusia gramma*. The active components in its whole plant extract were reported to be chromenes and benzofurans (Isman and Prokasch, 1985). Isman (1988) reported encecalin, an acetylchromene, a major active component to show toxicity to the grasshoppers and antifeedancy to the cutworm, *Peridroma saucia*.

348. *Entada polystachia* Adams. (Mimosaceae)

Common name: Paris rosa

This is a perennial shrub found in tropical regions. Its whole plant aqueous extract was reported to show insecticidal activity when tested against the cotton leaf worm, *Alabama argillacea* (Jacobson, 1958).

349. *Enterolobium cylocarpum* Mart. (Mimosaceae)

Common name: Elephant's ear

This is a perennial tree found in tropical zones and commonly available in West Africa. Its root, bark and fruit extracts in water, alcohol or petroleum ether and also their powders were reported to show insecticidal and antifeedant activities when tested against the boll weevil, *Anthonomus grandis*; the southern beet webworm, *Pachyzancla bipunctalis*; the cotton leaf worm, *Alabama argillacea* (Jacobson, 1958); and the milkweed bug, *Oncopeltus fasciatus* (Heal et al., 1950).

350. *Enterolobium saman* Prain. ex. King (Mimosaceae)

(Syn.-*Samanea saman* [Jacq.] Merr., *Pithecolobium saman* Benth.)

Common name: Rain tree or Monkey pod

This is a perennial tree found in tropical zones and available in the Philippines, Korea and Japan. Its branches and leaves propagated in the rice fields were found to show repellent activity against the general field pests of rice (Litsinger et al., 1978).

351. *Equisetum arvense* Linn. (Eauisetaceae)

Common name: Horse tail

This is a perennial herb found in tropical zones. Nicotine, an alkaloid, was isolated from this herb as an active component and reported to show insecticidal activity to a wide range of the insects of agricultural importance (Schmeltz, 1971).

352. *Eranthis hyemalis* Salish. (Ranunculaceae)

Common name: Winter aconite

This is a perennial herb found in tropical regions. Its bulb/corn dried powder and also extract were reported to show insecticidal activity to the diamond-back moth, *Plutella xylostella* (Jacobson, 1958).

353. *Eremocarpus setigerus* Lindl. (Euphorbiaceae)

Common name: Dove weed

This is an annual weed found in semi-arid zones and available in Andhra Pradesh and Madhya Pradesh in India. Its whole plant aqueous extract was reported to show insecticidal activity to the cross striped cabbage worm, *Evergestis rimosalis* (Jacobson, 1958). An active component, 'tillinoside', isolated from the plant extract was found to show feeding deterrency to the insects of agricultural importance (Bajaj et al., 1986).

354. *Eriangea cordifolia* Bail. (Asteraceae)

This is a perennial herb found in sub-tropical and temperate zones and commonly available in USSR and China. From its whole plant extract an active component named 'eriancorin' (Figure 65) was isolated and reported to show antifeedant activity to African armyworm, *Spodoptera exempta* and Mexican bean beetle, *Epilachna varivestis* when tested under laboratory conditions (Kubo and Nakanishi, 1978).

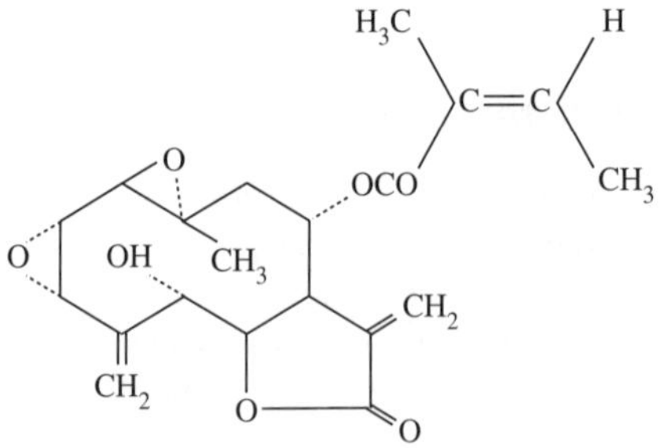

Figure 65. Eriancorin.

355. *Erigeron affinis* D.C. (Euphorbiaceae)

This is an annual weed found in sub-tropical and semi-arid zones. An active component named 'affinin' (Figure 66), an unsaturated isobutylamide

($C_{14}H_{25}NO$), was isolated from its root extract and reported to show insecticidal activity against Indian meal moth, *Tenebrio molitor* and also against a wide range of the insects of agricultural importance (Jacobson and Crosby, 1971).

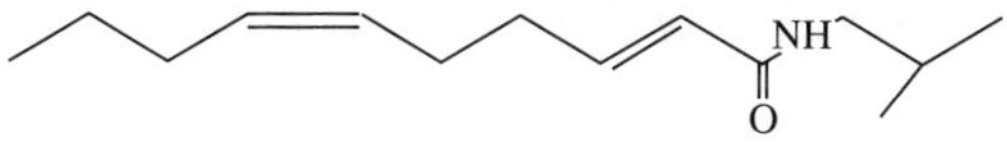

NH
O

Figure 66. Affinin.

356. *Erigeron bellibiastrum* Linn. (Asteraceae)

This is an annual herb found in temperate regions and commonly available in the United States and Canada. Its whole plant aqueous extract and also dried powder were reported to be toxic to the melon worm, *Diaphania hyalinata* and the southern beet web worm, *Pachyzancla bipunctalis* (Jacobson, 1958).

357. *Erigeron canadensis* Linn. (Asteraceae)

Common name: Horse weed or Canada fleabane

This is an annual highly branched herb found in tropical and sub-tropical regions. In India, it is available in Punjab, Upper Gangatic plain and Western Ghats in the Himalayas. Its whole plant and leaf extracts were reported to show toxicity to the melon worm, *Diaphania hyalinata* (Jacobson, 1958) and the Japanese beetle, *Popillia japonica* (McIndoo, 1983).

358. *Erigeron repens* Linn. (Asteraceae)

This is an annual herb found in temperate regions and available in North America and Europe. Its whole plant extract was reported to be toxic to European corn borer, *Ostrinia nubilalis* (Jacobson, 1958).

359. *Eriobotrya japonica* (Thunb.) Lindl. (Rosacee)

(Syn.-*Mespilus japonicus* Thunb., *Photinia japonica* Gray)

Common name: Japan medlar

This is a perennial small tree found in tropical, sub-tropical and temperate regions. This is a native of China but now available in Himachal Pradesh, Uttar Pradesh and Punjab in India. Its flower extract in water was reported to be toxic to the vinegar fly, *Drosophila* sp. (Jacobson, 1958).

360. *Eriogonum annuum* Nutt. (Polygonaceae)

This is an annual or perennial herb found in tropical, sub-tropical and temperate zones. Its root, stem, leaf, flower and fruit extracts in ethyl ether were reported to show attractancy to both male and female of the melon fly, *Dacus cucurbitae* (Keiser et al., 1975).

361. *Erlangea cordifolia* Sch. (Asteraceae)

This is an annual herb found in tropical regions and commonly available in Africa. Its whole plant extract was reported to show repellent activity to the corn earworm, *Heliothis armigera* (Waespe and Hans, 1983).

362. *Eruca sativa* Mull. (Cruciferae)

(Syn.-*E. vesicaria* [Linn.] Cav. sub sp. *sativa* [Mill.] Thel.)

Common name: Roquette or Rocket salad

This is a seasonal herb native to South Europe and now grown in North India. Cropping of Indian mustard, *Brassica juncea* with this herb received a lesser number of colonizing aphids, *Lipaphis erysimi* than those which were farther. It appeared that volatile allelo-chemicals from *E. sativa* made *B. juncea* plants inapparent to *L. erysimi* to a limited extent (Dilawari and Dhaliwal, 1992). Singh et al. (1993) found its 3 ml oil treatment to 1 kg of the redgram grains protected them from the pulse beetle, *C. chinensis* in storage.

363. *Eruca vesicaria* (Linn.) Cav. var. *sativa* (Mill.) Thell. (Brassicaceae)

(Syn.-*E. sativca* Mill.)

Common name: Rocket salad or Roquette or Taramira

This is an annual herb found in tropical and sub-tropical regions. It is a native of South Europe but now available in North India. Its seed oil was reported to control the rice pest, *Sogatella longifurcifera*; the rice dephacid, *Sogata striatus*; *Perkinsiella insigns*; and *Toya attenuata* (Khan and Khan, 1985). Doharey et al. (1983, 1984a,b, 1987) reported taramira oil to protect gram when admixed with the seeds @ 1% w/w against the pulse beetles, *C. maculatus* and *C. chinensis* in the controlled conditions of storage.

364. *Erythrinia indica* Lam. (Papilionaceae)

Common name: Coral tree

This is an annual/perennial ornamental shrub grown for its bright cinnamon red flowers in tropical zones and commonly available in India, Sri Lanka and Malaysia. Its seed extract was reported to be toxic to the larvae and the pupae of the pulse beetle, *Callosobruchus maculatus* and it showed antifeedant property to the adult beetles (Chandrakantha, 1988).

365. *Erythronium denscanis* Linn. (Liliaceae)

Common name: Dog tooth violet
This is a perennial herb found in sub-tropical and temperate zones. Its bulb and leaf extracts were reported to show toxicity to the vinegar fruit fly, *Drosophila hydei* (Jacobson, 1958).

366. *Erythrophleum guineense* G. Donn. (Leguminosae)

This is an annual/perennial herb found in tropical zones and commonly available in Sierra Leone. Its dried leaves admixed with the stored grains were reported to protect them from insects' infestation (Giles, 1964).

367. *Erythrophleum suaveolans* Afzel. ex. G. Donn. (Leguminosae)

Common name: Red water tree
This is a perennial tree found in the tropics. Its wood, leaf and bark extracts in water were reported to protect grains from insects' infestation in storage (Jacobson, 1975).

368. *Erythroxylon coca* Lam. (Erythroxyleaceae)

Common name: Peru coca
This is a perennial shrub or small tree found in tropical zones. It is commonly cultivated on the Andes of Peru from 600 to 2700 m above the sea. Anabasine, an alkaloid isolated as its active component, was reported to show insecticidal activity against a wide range of insects of agricultural importance (Schmeltz, 1971).

369. *Eucalyptus* sp. (n.m.) (Myrtaceae)

This is a perennial tree found in tropical and sub-tropical regions. Its leaf extract showed repellency to the woolly apple aphid, *Esiosoma lanigerum* and the screw worm, *Cochlomyia hominivorax* (McIndoo, 1983). However, a juveno-mimetic activity in its stem and leaf extracts in petroleum ether was reported against the 5th instar larvae of the cotton stainer, *Dysdercus cingulatus* (Rajendran and Gopalan, 1978; Kareem, 1984). *Eucalyptus* powder admixed with rice grains at the rate of 1% w/w effectively reduced the populations of the paddy moth, *Sitotroga cerealella* and checked the cross infestation of the lesser grain borer, *Rhyzopertha dominica* in laboratory test (Dakshinamurthy, 1988). Devraj Urs and Srilatha (1990) found its oil to show highest repellency among 5 other products evaluated against the rice moth, *Corcyra cephalonica* in a laboratory test. Similarly, Vir (1990) reported its leaf extract to greatly impair the fecundity of the pulse beetle, *Callosobrurchus chinensis*.

370. *Eucalyptus camaldulensis* L'Herit. (Myrtaceae)

This is a perennial tree found in sub-tropical and temperate regions. Its leaf extract showed promising control of outbreak of red locust, *Nomadocris septemfaciata* (Robertson, 1958). Jacobson (1975) reported a juvenile hormone analogue activity in its leaf, stem and fruit extracts when tested against the milkweed bug, *Oncopeltus fasciatus*. An active component 1, 8 cineole called eucalyptol (Figure 67) isolated from its leaf extract was found to show juvenile hormone mimic activity against the insects.

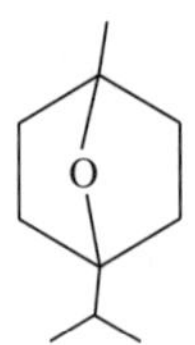

Figure 67. Eucalyptol.

371. *Eucalyptus citriodora* Hook. (Myrtaceae)

(Syn.-*E. maculata* Hook. var. *citriodora* Bailey)

Common name: Lemon scented eucalyptus

This is a perennial tall tree found in tropical regions and commonly available in Punjab, Uttar Pradesh and Andhra Pradesh in India. Its oil and volatiles of the oil were reported to inhibit the oviposition of the cotton leafhopper, *Amrasca devastans* (Saxena and Basit, 1982).

372. *Eucalyptus globulus* Labill. (Myrtaceae)

Common name: Fever tree or Tasmanian blue-gum

This is a perennial ornamental tree found in sub-tropical and Mediterranean zones. It is a native to Australia but commonly available in the hilly tracts in India. Its leaf extract in ether was reported to show insecticidal and antifeedant activities to the aphids (Mathews, 1981). Its dried leaf powder spread over stored potato as 2.5 cm layer was reported to protect potato promisingly for 6 months against the potato tuber moth, *Phthorimaea operculella* (Lal, 1987). Srivastava et al. (1988) reported its oil to show toxicity to the pulse beetles, *Callosobruchus maculatus* and *C. chinensis* and it protected pulse seeds from these beetles in storage when its oil was admixed with the seeds. Reddy et al. (1990) and Chitra et al. (1991) reported petroleum ether extract of its leaf to significantly reduce the population of the brinjal spotted leaf beetle, *Henosepilachna vigintioctopunctata* when sprayed on the potted plants.

373. *Eucalyptus naudiniana* Dum. (Myrtaceae)

(Syn.-*E. diglupta* Desf.)

This is a perennial tree found in tropical and sub-tropical zones. Crude extracts of its wood and bark were reported to inhibit the metamorphosis and development of the 5th instar larvae of the cotton stainer, *Dysdercus koenigii* and were suggested for its use in the control of this pest (Jaipal et al., 1983).

374. *Eucalyptus rostrata* Schlecht. (Myrtaceae)

This is a perennial tree found in tropical zones and available in Central and North Indian States. Its leaf extract and oil odour were reported to decline the hatchability of the eggs of cotton bollworm, *Earias vittella* (Pathak and Krishna, 1987).

375. *Eugenia aromaticum* (Linn.) Merr. and Perr. (Myrtaceae)

(Syn.-*E. carophyllata* Thunb.)

This is a perennial tree found in tropical and sub-tropical zones. Its flower bud/cloves powder admixed with the grains was reported to check the multiplication of storage insects and it was 'eugenol', an active component in the powder, which showed repellency to the corn weevil, *Sitophilus zeamais* when tested under controlled conditions (Hassanali and Lwande, 1989).

376. *Euonymus eropaea* Linn. (Celastraceae)

Common name: European spindle tree

This is a perennial tree found in tropical and sub-tropical zones. Its fruit extracts in water and petroleum ether and also its dried powder and latex were found to show antifeedancy to the saw toothed grain beetle, *Oryzaephilus surinamensis* and insecticidal activity against the chrysanthemum aphid, *Macrosiphoniella sanborni* (Jacobson, 1958). Crosby (1971) reported its fruit extract possessing biologically active alkaloids, i.e., 'wilforine' and 'wilfordine', as the active components to show insecticidal property.

377. *Euonymus japonica* Linn. (Celastraceae)

Common name: Japanese spindle tree

This is a perennial shrub or small tree found in Mediterranean zones. Its leaf extract showed repellency to the diamond-back moth, *Plutella xylostella* (Heal et al., 1950).

378. *Eupatorium* sp. (Asteraceae)

This is an annual herb found in tropical and sub-tropical zones and commonly available in North and Central India and Pakistan. Its whole plant extract in petroleum ether was reported to inhibit the growth and development of the cotton stainer, *Dysdercus cingulatus* (Kareem, 1984). Chockalingam (1986) found its leaf extract to protect tobacco crop against *Spodoptera litura* in the mini plot experiments. Saradamma (1988) reported its benzene extract to protect the brinjal crop against the attack of *Henosepilachna vigintioctopunctata*.

379. *Eupatorium hyssopifolium* Linn. (Asteraceae)

This is an annual/perennial herb found in temperate zones. Its leaf and flower aqueous extracts were reported to show toxicity to the Japanese beetle, *Popillia japonica* (McIndoo, 1983).

380. *Eupatorium japonicum* Linn. (Asteraceae)

This is an annual herb found in temperate zones. Its leaf extract in petroleum ether was reported to show insecticidal activity when tested against the vinegar fly, *Drosophila melanogaster* (Fagoonee, 1982).

381. *Eupatorium odratum* Linn. (Asteraceae)

This is an annual/perennial herb found in tropical regions and commonly available in South India. Its leaf extract in water was reported to reduce the infestation of the sweet potato weevil, *Cylas formicarius* (Rajamma, 1982).

382. *Eupatorium perfoliatum* Bulb. (Asteraceae)

This is a perennial herb found in temperate zones and commonly available in the United States and Canada. Its leaf extract in water was reported to show insecticidal activity against the vinegar fly, *Drosophilla melanogaster* (Michael et al., 1985).

383. *Eupatorium quadrangularae* Linn. (Asteraceae)

This is an annual/perennial herb found in tropical and sub-tropical regions and available in India, Malaysia, China, Korea and Vietnam. Chloroform extract of its leaf showed repellency to the leaf-cutting ant, *Atta cephalotes*. Quadragolide and three other known active components of sesquiterpene lactone group, viz., 4-desoxy-8-epi-ivangustin, isoalantolactone and diplophyllolide, were isolated and also reported to show repellent activity to *A. cephalotes* (Srimmannarayan and Rao, 1985; Michael et al., 1985).

384. *Eupatorium trapezoideum* Linn. (Asteraceae)

This is an annual herb found in tropical regions and commonly available in North India. The active components, as six derivative compounds of cadines isolated from its leaf extract were reported to exhibit antifeedant activity to the larvae of the silkworm, *Bombyx mori* (Shukla et al., 1986).

385. *Euphorbia adenochlora* Linn. (Euphorbiaceae)

This is a perennial herb found in tropical zones. Its branch and leaf aqueous extracts were reported to show insecticidal activity to the vinegar fruit fly, *Drosophila hydei* (Jacobson, 1958).

386. *Euphorbia antiquorum* Linn. (Euphorbiaceae)

This is a perennial herb or small shrub found in tropical regions. Its latex and leaf extracts were reported to possess insecticidal property against the dipteran pests of agricultural importance (McIndoo, 1983).

387. *Euphorbia hirta* Linn. (Euphorbiaceae)

Common name: Golindria or Sun-spurge
This is a perennial herb found in tropical regions. In India, it is available in the Himalayan tract and Punjab. Its leaf extract in water was reported to show insecticidal activity to the New Guinea sugarcane weevil, *Rhabdoscelus obscurus* (Litsinger et al., 1978).

388. *Euphorbia khasiana* Linn. (Euphorbiaceae)

This is an annual/perennial herb found in temperate zones and commonly available in North America and Canada. Its dried root powder was reported to be toxic to the Mexican bean beetle, *Epilachna varivestis* (Jacobson, 1958).

389. *Euphorbia maculata* Linn. (Euphorbiaceae)

This is an annual/perennial herb found in temperate zones and commonly available in North America and Europe. Its branch and leaf aqueous extracts were reported to show insecticidal activity to the milkweed bug, *Oncopeltus fasciatus* (Jacobson, 1958).

390. *Euphorbia pulcherrima* (Willd.) ex. Klotz. (Euphorbiaceae)

(Syn.-*Poinsettia pulcherrima* [Willd. ex. Klotz.] R. Grah.)
Common name: Poinsettia
This is a perennial shrub found in tropical zones. Poinsettia is a native to Central America and now available in East India. Its leaf and flower extracts

in water were reported to show insecticidal activity to the rice weevil, *Sitophilus oryzae* under controlled conditions (Jacobson, 1975; Prakash et al., 1978–1987). Prakash et al. (1989) also found its leaf extract to protect paddy against Angoumois grain moth, *S. cerealella* when treated and stored under natural conditions of infestation.

391. *Euphorbia royleana* Bosid. (Euphorbiaceae)

This is a perennial tree found in tropical regions and distributed in India, Sri Lanka, Malaysia, the Philippines and South China. Its fresh leaf extract in ether was reported to show antifeedant activity to the mustard sawfly, *Athalia proxima* (Pandey et al., 1977; Sudhakar et al., 1978), whereas its leaf extract in water showed toxicity to the adults and juveniles of this fly (Hameed, 1982). Its leaf extract in ether was also found to be larvicidal to *A. proxima* (Pandey et al., 1979). Sharma et al. (1992) reported its green leaf extract in 50% and 100% crude formulations to show significant antifeedancy to the jute stem weevil, *Apion corchori*; the jute semi-looper, *Anomis sabulifera*; and the 3rd instar larvae of Bihar hairy caterpillar, *Spilosoma obliqua*.

392. *Euphoribia tirucalli* Linn. (Euphorbiaceae)

Common name: Milk bush or Indian tree spurge

This is a perennial succulent spineless tree found in the tropics and native to Africa. Its meristematic tissue and branch extracts in water were reported to show insecticidal and repellent activities against the citrus aphid, *Aphis citri* on the treated host leaves (McIndoo, 1983).

393. *Evodia rutaecapa* Hook. & Thoms. (Rutaccae)

This is a perennial tree found in temperate zones. From its ripe fruits and root bark five indo-quinazoline alkaloids were isolated and reported to show antifeedant activity to the silkworm, *Bombyx mori* (Jacobson, 1989).

394. *Fagara chalybea* Engl. (Rutacene)

This is a perennial tree found in tropical zones and commonly available in East Africa. The alkaloid, N-methyl of linidersine and several benz (C)-phenanthridine alkaloids isolated from its root were reported to show antifeedant activity against the beet armyworm, *Spodoptera exiqua* and Mexican bean beetle, *Epilachna varivestis* (Jacobson, 1989).

395. *Fagara holstii* Fabr. (Rutaceae)

This is a perennial tree found in tropical zones and commonly available in East Africa. The alkaloid, N-methyl of linidersine and several benz (C)-phenanthridine akaloids were obtained from its root bark and reported to show antifeedancy to the beet armyworm, *Spodoptera exiqua* and Mexican bean beetle, *Epilachna varivestis* (Jacobson, 1989).

396. *Fagara microphylla* Desf. (Rutaceae)

This is a perennial small evergreen tree found in tropical and sub-tropical hills in West African countries. Its bark extract incorporated with artificial diets of the cotton pests, *Pectinophora gossypiella*, *Heliothis zea*, *H. virescens* and *Spodoptera frugiperda* was reported to show toxicity and also insect growth regulating activity when tested under controlled conditions and '5-isobutylamide', an active component, was found to be responsible for the biological activities (Kubo et al., 1984b).

397. *Fagus sylvatica* Linn. (Fagaceae)

Common name: Beech

This is a perennial tree found in tropical and temperate zones. In India, it is commonly available in Kulu Valley and Nilgiris. Its leaf extract in water was reported to be an antifeedant to gypsy moth, *Portheria dispar* when tested under controlled conditions (Jacobson, 1975).

398. *Ferula assafoetida* Linn. (Apiaceae)

Common name: Asafoetida

This is a perennial herb grown in tropical and sub-tropical zones. It is commonly available in Punjab and Kashmir in India. Its whole plant extract was found to be repellent to the cornfield ants (McIndoo, 1983). Tiwari et al. (1988) reported its 2% dust to protect coriander against the cigarette beetle, *Lasioderma serricorne* when admixed with the grains.

399. *Ficus caria* Linn. (Moraceae)

Common name: Fig

This is a small tree or bush found in tropical and sub-tropical zones. In India, it is available in Uttar Pradesh, Andhra Pradesh, Punjab, Maharastra and Rajasthan but a native of the Mediterranean region. Its leaf extract in benzene was reported to show antifeedant activity against the cotton leaf armyworm, *Spodoptera litura* when tested under controlled conditions (Hosozawa et al., 1974b).

400. *Foeniculum vullgatre* Mill. (Apiaceae)

(Syn.-*F. foeniculum* Karst., *F. officinale* All. S.; *Anithum foeniculum* Linn.)

Common name: Common fennel

This is a seasonal/annual aromatic herb found in the Mediterranean region. In India, it is commonly available in Punjab, Assam, Maharastra and Gujrat. Its aqueous leaf extract was found to show repellency to the screw worm, *Cochliomyia hominivorax* (McIndoo, 1983).

401. *Ganoderma lucidum* (Curtis) P. Krast (Polyporaceae)

Common name: Poisonous mushroom

This is a tropical mushroom available in coastal states of India. Its fruiting body extract in petroleum ether was reported to possess insecticidal property when tested under controlled conditions against the rice weevil, *Sitophilus oryzae* (Jacobson, 1975; Prakash et al., 1978–1987).

402. *Garcinia morella* Desr. (Guttiferae)

(Syn.-*G. gutta* Wight, *Mangostana morella* Garetn.)

Common name: Ceylon gamboge

This is a perennial tree found in Khasi hill and Western Ghats in India and Sri Lanka. Its leaf extract was reported to show synergistic activity by enhancing the toxicity of malathion against the red flour beetle, *T. castaneum* when tested under controlled conditions (Parmar and Dutta, 1982).

403. *Gardenia jasminoides* Ellis. (Rubiaceae)

(Syn.-*G. florida* Linn., *G. augusta* Merrill.)

Common name: Cape jasmin

This is a perennial ornamental shrub found in tropical zones and native to China. It is commonly available in Orissa in India. Water extract of its leaf was reported to show repellent activity against the red flour beetle, *Tribolium castaneum*; the lesser grain borer, *Rhyzopertha dominica*; and the khapra beetle, *Trogoderma granarium* (Jilani and Malik, 1973).

404. *Gaura coccinea* Linn. (Onagraceae)

This is a perennial herb found in sub-tropical and temperate zones. Water extract of its branch was reported to show insecticidal activity against the vinegar fly, *D. melanogaster* (Michael et al., 1985).

405. *Gendarussa vulgaris* Nees. (Acanthaceae)

This is a perennial shrub found in tropical zones and commonly available in the gardens in India, Sri Lanka, Indonesia and Malaysia. Its dried root and stem powders were reported to show insecticidal activity against the pulse beetle, *Callosobruchus chinensis* (Puttarudriah and Bhatia, 1955).

406. *Geranium eriostemon* var. *onoei* Linn. (Germiaceae)

This is a perennial herb found in temperate zones and commonly available in the United States and Canada. Its leaf and branch aqueous extracts were reported to be toxic to the vinegar fruit fly, *Drosophila hydei* (Jacobson, 1958).

407. *Gingiber officinale* Rosc. (Gingeberaceae)

Common name: Ginger

This is an annual/seasonal herb grown in tropical and sub-tropical zones and commonly found in India, Sri Lanka, Malaysia, Indonesia as a condiment. Its dry rhizome powder admixed with wheat grains was reported to reduce grain damage and their weight loss caused by storage insects (Rout, 1986).

408. *Giliricidia sepium* (Jacq.) Walp. (Fabaceae)

(Syn.-*G. maculata* [Hb. & K.] Steud., *Lonchocarpus maculatus* D.C.)

Common name: Madre tree or Spotted giliricida

This is a perennial/ornamental small tree found in tropical zones and native to tropical America but also available in South East Asian countries and the Philippines. In India, it is available in Tamil Nadu, Kerala, Karnataka and Maharashtra. Its whole plant, root, bark, leaf, fruit and seed extracts in petroleum ether and also their dry powders were reported to possess insecticidal and antifeedant properties against the yellow woolly bears, *Diacrisia virginica*; the southern armyworm, *Spodoptera eridania*; and the cabbage looper, *Trichoplusia ni* (Jacobson, 1958, 1975). Its branches planted in the paddy fields were reported to repel the caseworms and the whorl maggots by reducing their infestation in the crop (Litsinger et al., 1978). Morallo-Rajessus (1984) found its leaf extract to control the tobacco budworm, *Helicoverba armigera* when sprayed on the tobacco plant.

409. *Ginkgo biloba* Linn. (Ginkgoaceae)

Common name: Maiden hair tree

This is a perennial ornamental tree found in temperate zones and commonly available in North America and Europe. Its root and leaf extracts in water were reported to show insecticidal and repellent activities against European corn borer, *Ostrinia nubilalis* (Jacobson, 1975).

410. *Glycine max* (Linn.) Merr. (Fabaceae)

Common name: Soybean

This is an annual or seasonal bushy herb grown in tropical and sub-tropical zones. It is commonly available in South East Asian countries, tropical America and China. Water extracts of its branch and leaf were reported to show insecticidal activity to the sugarcane woolly aphid, *Ceratovacuna lanigera* and the codling moth, *Laspeyresia pomonella* (Khan, 1982). Oca et al. (1978) found its purified oil to reduce the survival and progeny of Angoumois grain moth, *S. cerealella* and the rice weevil, *S. oryzae* when stored grains/seeds of maize, wheat and sorghum were treated with its oil at the rate of 5 and 8 ml/kg and gave complete grain protection up to 120 days, without any adverse effect on their viability. Its oil was also reported to enhance toxicity of pyrethriod insecticides to the tobacco budworm, *Helicoverba armigera* when admixed with the insecticides (Ochou et al., 1986). Crude and refined soybean oil showed 100% adult mortality within 24 hrs and pigeonpea grains were protected against the pulse beetle, *Callosobruchus chinensis* in storage when grains were treated at the rate of 0.5 ml/100 g (Singh et al., 1988). Singh et al. (1993) also reported its oil to protect gram seeds from *C. chinensis* infestation when treated @ 3 ml/kg seed.

411. *Glycosmis calamus* Corr. (Rutaceae)

This is an evergreen perennial shrub found in sub-tropical and tropical regions and commonly available in North Indian States like Himachal Pradesh and Uttar Pradesh. Dichloromethane and methanol extracts of its root bark were reported to show toxicity to the aphids of agricultural importance (Bandara et al., 1987).

412. *Glycosmis pentaphyll* Corr. (Rutaceae)

(Syn.-*G. chylocarpa* W & A, *G. aborea* D.C., *G. retzii* Roem, *Limonia pentaphylla* Retz., *L. arborea* Roxb., *Myxospermum chlocarpum* Roem.)
Common name: Wild lemon

This is a perennial evergreen shrub commonly found in tropical and subtropical Himalayas, Sikkim, Assam, Burma, Sri Lanka and Malaysia. Deshpande et al. (1990) reported acetone extract of its leaf to show insecticidal activity when tested against the armyworm, *Mythimna separata*.

413. *Glycyrrhiza glabra* Linn. (Fabaceae)

Common name: Spanish liquorice

This is a perennial or annual herb found in tropical and Mediterranean zones and commonly available in Jammu and Kashmir and South India. Its root content, 'glycyrrihizin' (a glycoside), is considered 50 times sweeter than sugar. However, extracts from its leaf and stem were reported to show insecticidal activity against termites (Jacobson, 1975).

414. *Glycyrrhiza lepidota* Linn. (Fabaceae)

Common name: Liquorice

This is a perennial or annual herb found in temperate and sub-tropical zones and commonly available in South China, India and the Philippines. Its seed extract in water was reported to be toxic to the vinegar fly, *D. melanogaster* (Michael et al., 1985).

415. *Gossypium aroboreum* Linn. (Malvaceae)

(Syn.-*G. nanking* Meyen., *G. indicum* Tod., *G. neglectum* Tod.)

Common name: Tree cotton

This is an annual/biennial shrub grown in tropical and semi-arid zones and commonly available in Egypt and Syria. Its seed oil was reported to show insecticidal activity against the pulse beetles, *Callosobruchus chinensis* and *C. maculatus* (Shaaya and Ikan, 1970).

416. *Gossypium hirsutum* Linn. (Malvaceae)

(Syn.-*G. mexicanum* Tod., *G. religiosum* Linn.)

Common name: Cotton

This is a biennial/seasonal shrub grown in tropical zones and distributed in all the South East Asian countries, China, Israel, the Philippines and Egypt. It is commonly cultivated in Uttar Pradesh, Punjab, Madhya Pradesh, Tamil Nadu and Rajasthan in India. Its seed oil was reported to protect stored bean seed against the bruchid, *Zabrostes subfasciatus* when applied at the rates of 1–5 ml/kg seed and it gave complete protection to grains for 75 days (Schoonhoven, 1978). Similarly, stored maize, sorghum and wheat grains treated with its oil at the rates of 5 and 8 ml/kg were also found free from the infestation of Angoumois grain moth, *S. cerealella* and the rice weevil, *S. oryzae* up to 120 days without affecting their viability (Oca et al., 1978). Further, its seed oil showed insecticidal activity against field insect pests, i.e., oyster-shell scale, *Lepidosphes glomi* and Mexican mealy bug, *Phenococcus gossypii* (Jacobson, 1958). A new insecticidal C_{25}-terpenoid 'helicocid-H' was isolated from its seed oil and extract and reported to show toxicity to the tobacco budworm, *Heliothis virescens* (Robert et al., 1978). Its oil was also reported to enhance insecticidal toxicity to the tobacco budworm, *Heliothis virescens* (Robert et al., 1978). Its oil was reported to enhance toxicity to the pyrethroid insecticide when admixed with pyrethroid and tested against the larvae of tobacco budworm, *Helicoverpa armigera* (Ochou et al., 1986) and the adults of the boll weevil, *Anthonomus grandis* and also to tobacco budworm, *Heliothis virescens* (Wolfenbarger and Guerra, 1986). Butler et al. (1988) evaluated cotton seed oil for the control of the sweet potato whitefly, *Bemisia tabaci* in greenhouse studies in Phoenix, USA and established its potentiality. Further, Broza et al. (1988) tested cotton seed oil in field conditions on cotton in Israel during 1987 using 18-row inflatable boom sprayer and mistblower sprayer for its application and registered promising control of *B. tabaci*. Similarly, two formulations of cotton seed oil, i.e., flyteck-1® and a local formulation MAU-CSO, were found to reduce the number of adults of *B. tabaci* on the treated cotton leaves up to 7 days and gave better results than even neem seed oil formation, i.e., Neemark® (Puri et al., 1991).

417. *Grindelia humilis* Willd. (Asteraceae)

This is a perennial herb found in temperate and Mediterranean zones. Its whole plant and leaf extracts were toxic to the aphids (Waespe and Hans, 1979).

418. *Guaiacum officinale* Linn. (Zygophyllaceae)

This is a perennial tree found in tropical zones and native to the West Indies. Its wood extracts in water and alcohol showed repellency to the termites, *Cryptotermis brevis* and *Retculitermis flavipes* (Jacobson, 1975).

419. *Guarea rusbyi* Buch. (Meliaceae)

Common name: Cocillana

This is a perennial tree found in tropical and alpine regions. Its dried bark powder showed repellency to the Japanese beetle, *Popillia japonica* (McIndoo, 1983).

420. *Gymnema sylvestres* (Retz.) Schult. (Asclepiadaceae)

Common name: Merasingi

This is a perennial woody climber found in tropical zones. It is commonly available in India. Gymnemic acids, the secondary plant substances isolated from its leaf extract, showed antifeedancy to *Spodoptera eridania* (Granich et al., 1974).

421. *Gymnocladus dioicus* (L.) K. Koch. (Caesalpiniaceae)

Common name: Kentucky coffee tree

This is a perennial tree found in tropical and temperate zones. Its branch, seed and leaf extracts in water were reported to show antifeedant and repellent activities to the Japanese beetle, *Popillia japonica* and the southern armyworm, *Spodoptera eridania* (Waespe and Hans, 1979).

422. *Gynandropsis gynandra* D.C. (Cleomaceae)

This is a perennial herb found in tropical and sub-tropical zones and commonly available in North Indian States. Its seed extract in petroleum ether was reported to show toxicity to the spotted leaf beetle, *Henosepilachna vigintio-ctopunctata* in a laboratory test (Chandel et al., 1987).

423. *Gynandropsis pentaphylla* D.C. (Cleomaceae)

This is a perennial herb found in tropical and sub-tropical zones and distributed in all South East Asian countries. It is commonly available in Andhra Pradesh in India. Its seed extract in petroleum ether reduced the oviposition of the brown planthopper, *Nilaparvata lugens* when sprayed on the rice plant (Reddy and Urs, 1988).

424. *Gypsophila paniculata* Linn. (Caryophyllaceae)

Common name: Baby's breath

This is a perennial shrub found in temperate zones. Its whole plant and root extracts in water and also their dried powders were reported to show insect growth inhibiting activity to the vinegar fly, *Drosophila melanogaster* (Michael et al., 1985).

425. *Haematoxylon campechianum* Linn. (Caesalpiniaceae)

Common name: Log wood or Campeachy tree

This is an ornamental perennial small tree or shrub found in tropical zones and native to Central America and the West Indies. Its leaf extract in water was reported to show repellent activity to the Japanese beetle, *Popillia japonica* (McIndoo, 1983).

426. *Halesia carolina* Loefl. (Styraeaceae)

Common name: Silver bell tree

This is a perennial tree found in temperate zones and distributed in North America and Europe. Its bark, leaf and branch extracts in petroleum ether were reported to show antifeedant and insecticidal activities against the Japanese beetle, *Popillia japonica* (McIndoo, 1983).

427. *Haplophyllum buxbaumii* (Poir) G. Don. (Rutaceae)

This is a perennial shrub found in tropical, sub-tropical and temperate zones. Its root, stem and leaf extracts in petroleum ether were reported to show attractancy to both male and female of the Mediterranean fruit fly, *Ceratitis capitata* (Keiser et al., 1975).

428. *Haplophyton cimicidum* D.C. (Apocynaceae)

Common name: Cockroach plant
This is a perennial shrub found in tropical, sub-tropical and semi-arid zones and commonly available in Central America and Australia. Its whole plant, bark, leaf and branch extracts in water, alcohol and ether and also their dried powders were reported to show insecticidal activity against the striped blister beetle, *Epicauta vittata*; Mexican bean beetle, *Epilachna varivestis*; the codling moth, *Laspeyresia pomonella*; European corn borer, *Ostrinia nubilalis*; the imported cabbage worm, *Pieris rapae*; and the southern armyworm, *Spodoptera eridania* (Jacobson, 1958). However, earlier these extracts were reported to show toxicity to Mexican fruit fly, *Anastrepha vittata* (Feinstain, 1952). According to him, it was as toxic as pyrethrum to the squash bug, *Anara tristis*. Crosby (1971) reported two alkaloids, i.e., eburnamine ($C_{19}H_{24}N_2O$) (Figure 68) and haplophytine ($C_{37}H_{40}N_4O_7$) (Figure 69), to possess insecticidal property in their bark and leaf dried powders.

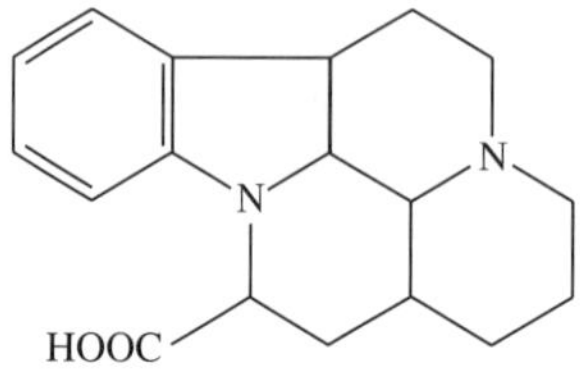

Figure 68. Eburnamine.

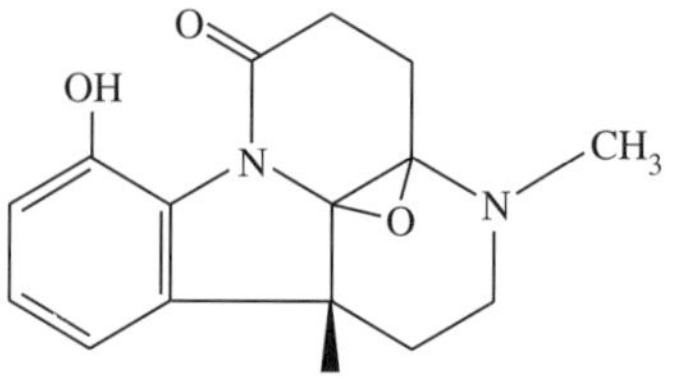

Figure 69. Haplophytine.

429. *Hardwickia manni* Roxb. (Leguminosae)

This is a perennial tree found in tropical regions. Its branch extract in water was reported to show insecticidal activity against termites (Khan, 1982).

430. *Harrisonia abyssinica* R.Br. (Simaroubaceae)

This is a perennial shrub found in tropical and sub-tropical regions and native to East Africa. Kubo et al. (1986) and Kubo and Nakanishi (1978) isolated an active component, 'harrisonin' (Figure 70), from its root and bark extract, which was found to show antifeedant activity to African armyworm, *Spodoptera exempta*. Hassanali et al. (1983) reported another active component called 'pedonin' from the crude extract of its root bark in methanol and reported it to show potent antifeedancy to *Eldane saccharina* and *Maruca testulatis*.

Blade et al. (1987) isolated a new chromene, '2-hydroxy methyl-alloptaeroxylin', from its root bark extract and reported it to show antifeedancy when tested against *S. exempta*.

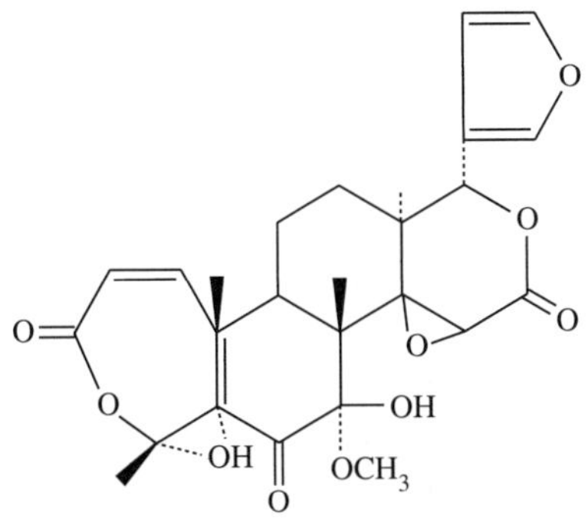

Figure 70. Harrisonin.

431. *Hedera helix* Linn. (Araliaceae)

Common name: English ivy

This is a perennial woody climber found in tropical and temperate zones. Its leaf, flower, fruit and seed extracts in water and alcohol showed insecticidal and antifeedant activities when tested against the diamond-back moth, *Plutella xylostella* (Jacobson, 1975).

432. *Helenium aromaticus* Linn. (Asteraceae)

This is an annual/perennial weed or herb found in temperate and sub-tropical zones. Treatment of 'helenalin', an active component isolated from this plant, was reported to protect packaging materials against infestation of wheat weevil, *Sitophilus granarius* and lesser grain borer, *Rhyzopertha dominica* when tested under artificial infestations of the test insects (Bloszyk et al., 1990).

433. *Helenium mexicanum* Mill. (Asteraceae)

Common name: Yerba leda pulga

This is an annual/perennial herb found in tropical, sub-tropical and temperate zones. Its root, leaf and flower extracts and also their dried powders were reported to show repellent and insecticidal activities to the codling moth, *Laspeyresia pomonella* and the southern beet webworm, *Pachyzancla bipunctalis* in a test under controlled conditions (Jacobson, 1958).

434. *Helianthemum lavendulifolium* Mill. (Castaceae)

This is a perennial herb found in tropical and sub-tropical zones. Its stem, leaf and flower extracts in ethyl ether showed attractancy to the male and female of Mediterranean fruit fly, *Ceratitis capitata* (Keiser et al., 1975).

435. *Helianthus annus* Linn. (Asteraceae)

Common name: Sunflower

This is a seasonal/annual herb found in tropical and sub-tropical zones and commonly grown in India as an oil seed crop. Sunflower is distributed in all South East Asian countries, China, Australia and Central America. Sunflower oil was reported to inhibit the multiplication of the pulse beetles, *Callosobruchus chinensis* and *C. maculatus* in the greengram and blackgram seeds admixed with the oil @ 0.3–1.0% w/w (Mummigatti and Raghunathan, 1977; Sangappa, 1977; Anonymous, 1980–1981). Doharey et al. (1983, 1984a,b, 1987) also reported its seed oil to protect greengram in storage against the pulse beetles, *C. maculatus* and *C. chinensis* when admixed @ 1% w/w and found it to show ovicidal activity and also checked their multiplication. Singh et al. (1993) found its seed oil to protect gram seeds from *C. chinensis* infestation when treated @ 3 ml/kg seed.

436. *Helianthus petiolaris* Linn. (Asteraceae)

This is an annual herb found in tropical and temperate regions and available in Central America and Canada. Its whole plant aqueous extract was found to be toxic to the red legged grasshopper, *Melanoplus femmurubrum* (Jacobson, 1975).

437. *Helichrysum hookeri* Mill. (Asteraceae)

This is an annual herb found in tropical zones and commonly available in South Indian States. Its whole plant extract and oil were reported to be toxic to the cotton stainer, *Dysdercus cingulatus* (Balasubramaniam, 1982).

438. *Heliopsis longipes* (A. Gray) Blake. (Asteraceae)

Common name: Chilcuan

This is a perennial herb found in temperate regions and distributed in Canada and the United States. Its root extracts in water, alcohol and ether were reported to be toxic to the bean weevil, *Acanthoscelides obtectus* (Jacobson, 1975); the melon worm, *Diaphania hyalinata*; the southern beet web worm, *Pachyzanclabipunctalis* (Jacobson, 1958); and also the codling moth, *Laspeyresia pomonella* (Feinstein, 1952). Its active component was reported to be an insecticidal amide called 'affinin' [N-isobutyl-2,6,8-decatrienamide] with chemical structure $CH_3CH= CH.CH = CH(CH_2)_2CH = CH.NH.CH_2CH(CH_3)_2$ (Jacobson and Crosby, 1971; Jacobson, 1975).

439. *Heliopsis scabra* Dunal. (Asteraceae)

This is a perennial herb found in temperate zones and distributed in North Asia and Europe. Dried powders of its root and branch and also their extracts in ether were reported to be toxic to the armyworm, *Pseudoletia unipunctata* (Jacobson, 1958). The active component isolated from its ethanolic root extract was identified as an isobutylamide ($C_{22}H_{35}NO$), 'scabrin', and reported to be more toxic than pyrethrins (Jacobson, 1952, 1971). Another active component, an isobutylamide 'heliopsin' ($C_{22}H_{33}NO$), was also isolated from its root extract and was reported to show insecticidal activity similar to natural pyrethrins (Jacobson and Crosby, 1971).

440. *Helonias bullate* Linn. (Liliaceae)

Common name: Swamp pink
This is a perennial herb found in temperate zones. Its aqueous leaf extract was reported to be repellent to the Japanese beetle, *Popillia japonica* (McIndoo, 1983).

441. *Hepatica bobilis* var. *nipponica* Gaj. (Ranunculaceae)

Common name: Liver leaf genus
This is a perennial herb found in temperate zones. Its branch, leaf and flower extracts in water were toxic to the vinegar fruit fly, *Drosophila hydei* (Jacobson, 1958).

442. *Hephrolepsois exalitata* Schot. (Davalliaceae)

This is a perennial shrub found in tropical zones and distributed in South East Asian countries. Its leaf extract in water was reported to inhibit the development of the red cotton bug, *Dysdercus cingulatus* (Kareem, 1984).

443. *Hibiscus rosa-sinensis* Linn. (Malvaceae)

Common name: Rose of China or Chinese hibiscus or Shoe flower
This is a perennial ornamental shrub or small tree grown in tropical and subtropical zones and commonly available in East Indian States. Its flower extract was found to be toxic to the rice weevil, *Sitophilus oryzae* (Jacobson, 1975). Prakash et al. (1978–1987) reported its aqueous flower extract @ 10 ml per 100 g in milled rice to give promising grain protection against the rice weevil, *S. oryzae*.

444. *Hibiscus syriacus* Linn. (Malvaceae)

Common name: Rose of Sharon or Shrubby althaea
This is a perennial shrub found in temperate and tropical regions and commonly available throughout India. Its flower extract in acetone was reported to be repellent to the boll weevil, *Anthonomus grandis* (Jacobson, 1975).

445. *Hieracium japonicum* Linn. (Asteraceae)

Common name: Hawk weed genus

This is a perennial herb found in temperate zones and distributed in Canada and the United States. Its root, branch and leaf extracts in water were reported to be toxic to the vinegar fruit fly, *Drosophila hydei* (Jacobson, 1958).

446. *Hieracium pratense* Linn. (Asteraceae)

This is a perennial herb found in sub-tropical zones. Its whole plant extract in water showed repellency to the Japanese beetle, *Popillia japonica* (McIndoo, 1983).

447. *Hippeastrum hybridum* Linn. (Amaryllidaceae)

This is an annual/perennial herb in tropical and sub-tropical zones and commonly available in North and Central India. Mid-rib sap of its leaf was reported to show antifeedant activity to the desert locust, *Schistocerca gregaria* (Singh and Pant, 1980).

448. *Hordeum vulgare* Linn. (Graminaceae)

Common name: Barley

This is a seasonal crop grown in tropical zones and distributed in South East Asian countries, South and East Africa and Australia. Two active chemicals, 'hordenine' and 'gramine' isolated from aqueous extract of the whole plant, were reported to be toxic to the stupid grasshopper, *Melanoplus bivittatus* (Harley and Thorsteinson, 1967).

449. *Humulus lupulus* Linn. (Moraceae)

Common name: Common hop

This is a perennial vine/herb found in temperate zones and native to Europe. It is also grown in Jammu and Kashmir and Himachal Pradesh in India. Its leaf, root and bark extracts in water were reported to show toxicity and repellency to the screw worm, *Cochliomyia hominivorax* (McIndoo, 1983) and also to the melon worm, *Diaphania hyalinata* and the southern armyworm, *Spodoptera eridania* (Jacobson, 1958).

450. *Hunteria eburnea* Pichon. (Apocynaceae)

This is a perennial tree found in sub-tropical and temperate zones. Crosby (1971) reported an alkaloid as active component called 'isoburanamine' ($C_{19}H_{24}N_2O$) in its leaf and seed extracts to show toxicity to Mexican bean beetle, *E. varivestis;* Colorado potato beetle, *Leptinotarsa decemlineata;* and the codling moth, *Laspeyresia pomonella.*

451. *Hura crepitans* Linn. (Euphorbiaceae)

Common name: Money's dinner bell or Sandbox tree

This is a perennial tree found in tropical zones and native to tropical America. Its branch, juice/aqueous extract was toxic to the southern armyworm, *Spodoptera eridania* (Jacobson, 1958) and also to *Aphis spirecola* (McIndoo, 1983).

452. *Hydrocarpus alpina* Wight. (Flacourtiacea)

This is a perennial large tree found in tropical and sub-tropical zones. Reddy and Urs (1988) reported its seed extract to reduce oviposition of the brown planthopper, *Nilaparvata lugens* when sprayed on the rice plant.

453. *Hydrocarpus laurifolia* (Dennstr.) Slkeumer (Flacourtiaceae)

(Syn.-*H. wightiana* Blume)

Common name: Morotti or Maravitti tree

This is a perennial tree found in tropical zones. Its fruit and seed extracts showed repellency to the coconut rhinoceros beetle, *Xyloryctes jamaicensis* (Jacobson, 1975; McIndoo, 1983).

454. *Hymenocallis littoralis* Salibs. (Amaryllidaceae)

Common name: Lilies

This is a perennial bulbous herb found in North India. Its leaf, fruit, root and bulb extracts having alkaloids showed antifeedancy to the desert locust, *Schistocerca gregaria* on the treated cabbage leaves (Singh and Pant, 1980).

455. *Hyocyanus niger* Linn. (Solanaceae)

Common name: Black honbane

This is a biennial/perennial herb found in tropical and Mediterranean zones. Its leaf, flower and branch extracts in water were toxic to aphids (McIndoo, 1983).

456. *Hypericum hookerianum* Linn. (Hypericaceae)

This is a perennial herb found in temperate and sub-tropical zones. Its whole plant aqueous extract was reported to be toxic to the red cotton bug, *Dysdercus cingulatus* and the cotton leaf armyworm, *Spodoptera litura* (Kareem, 1984).

457. *Hyptis spicigera* Jacq. (Labiatae)

This is an annual herb found in the tropics. Its seed extract and oil were reported to show repellent activity to the stored grain insect pests (Giles, 1964; McIndoo, 1983).

458. *Hyptis suavelobens* (Linn.) Poit. (Labiatae)

Common name: English basil

This is an aromatic annual herb found in North India. Its aqueous leaf extract was toxic to *Lipaphis erysimi* on the treated cabbage leaf (Roy and Pande, 1991).

459. *Intsia bijuga* A. Gray (Leguminosae)

This is an annual/perennial herb found in sub-tropical and tropical regions. Its seed oil showed repellency to *T. castaneum* (Mohiuddin et al., 1987).

460. *Inula helenium* Linn. (Asteraceae)

This is an annual/perennial herb found in sub-tropical and temperate zones. From its whole plant extracts the active components like sesquiterpene lactone and alantolactone (Figure 71) were isolated and showed antifeedancy against *Tribolium confusum*. Iso-alantolactone (Figure 72), another active component in this shrub, altered the feeding of *Sitophilus granarius* and *Trogoderma granarium* (Jacobson, 1989).

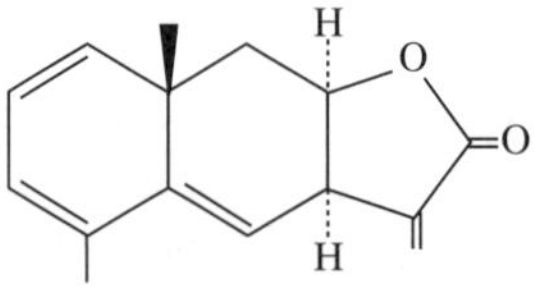

Figure 71. Alantolactone.

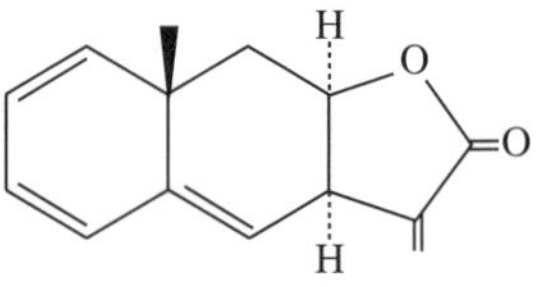

Figure 72. Iso-alantolactone.

461. *Inula racemosa* Hook. (Asteraceae)

This is an annual/perennial herb found in sub-tropical and temperate zones. Alantalides isolated from its roots showed feeding deterrency to the hairy caterpillar, *Spilosoma obliqua* when tested by leaf-disc dipping method. Solid alantolide (a mixture of alantolactone and iso-alantolactone) proved significantly superior even over coumarin (2H-1-benzopyron-2-one) (Figure 73) treatment and showed 70–87% feeding deterrency at 5% concentration (Bhathal et al., 1993).

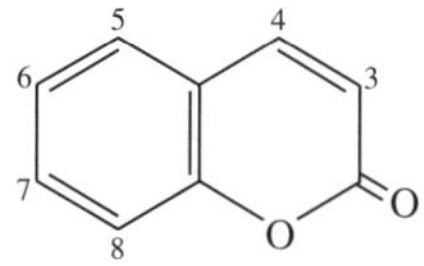

Figure 73. Coumarin.

462. *Ipomea cornea* Linn. (Convolvulaceae)

Common name: Morning glory genus

This is a perennial vine/creeper found in tropical zones and commonly available in India. Pandey and Singh (1976) reported its rhizome extract and dried powder to show toxicity to the pulse beetles, *Callosobruchus chinensis* and *C. maculatus*. Its leaf and flower extracts in benzene showed repellency to these beetles and reduced their oviposition and multiplication in stored greengram (Pandey et al., 1986). Bhaduri et al. (1985) found its leaf extracts in petroleum ether, benzene and alcohol to show promising grain protection to cowpea seeds against *C. maculatus*. Its leaf extracts in water and acetone and also dried powder were found to be toxic to *C. chinensis* (Hameed, 1983). Prakash et al. (1978–1987) also reported its leaf extract to protect stored milled rice from the saw toothed grain beetle, *Oryzaephilus surinamensis* and the red flour beetle, *Tribolium castaneum* when treated with the extract and stored in the farm godown. Its aqueous leaf extract also showed 90% repellency to the bean aphids on the treated amrut gaurd (Patel et al., 1991).

463. *Ipomea nil* Linn. (Convolvulaceae)

(Syn.-*I. headeracea* Clarke. [non-Jacq.] *Convolvulus nil* Linn.)

This is an annual/perennial vine found in sub-tropical regions. Its dried seed powder was reported to be toxic to aphids (Jacobson, 1958).

464. *Ipomea palmata* Forsk. (Convolvulaceae)

(Syn.-*I. cairica* [L.] Sweet., *Convolvulus cairicus* Linn.)

Common name: Railway creeper

This is a perennial climbing shrub found in tropical and sub-tropical zones and grown as an ornamental plant in India. Its seed extract in petroleum ether showed toxicity and ovipositional deterrency to the pulse beetle, *Callosobruchus chinensis* on the treated cowpea seeds (El-Ghar and El-Sheikh, 1987).

465. *Ipomea pundurata* (Linn.) G. Meyer (Convolvulaceae)

This is a perennial vine found in sub-tropical and temperate zones. Its rhizome, stem and leaf extracts in ethyl ether were reported to show attractancy to both the male and female of the melon fly, *Dacus cucurbitae* (Keiser et al., 1975).

466. *Ipomea purpurea* (Linn.) Roth. (Convolvulaceae)

Common name: Common morning glory or Tall morning glory

This is an annual vine found in tropical, sub-tropical and temperate zones. Its branches and leaf aqueous extracts showed toxicity to aphids (McIndoo, 1983).

467. *Ipomea quamoclit* Linn. (Convolvulaceae)

(Syn.-*Quamoclit pennata* [Desr.] Boj., *Q. vulgaris* Choisy, *Q. Quamoclit* Brit. Bron., *Convolvulus pennatus* Ders.)

Common name: Cypress vine

This is an annual vine found in tropical zones. Its whole plant aqueous extract and also dried powder were reported to be toxic to aphids (Jacobson, 1958).

468. *Iris douglasiana* Linn. (Iridaceae)

This is a perennial herb found in temperate and tropical zones. Its leaf, stem and fruit extracts in water showed juvenile hormone analogue activity to the milkweed bug, *Oncopeltus fasciatus* (Jacobson et al., 1975).

469. *Isocoma wrightii* Nutt. (Asteraceae)

Common name: Rayless golden rod

This is an annual/perennial herb found in temperate and sub-tropical zones and commonly available in North America. Volatiles of its oil and methanol-water partition fraction at 200 and 500 ppm concentrations completely inhibited feeding of the larvae of the fall armyworm, *Spodoptera frugiperda* and also increased the period of its life cycle and reduced the adults' emergence when tested under controlled conditions (Zalkow et al., 1979).

470. *Isodon shikokianus* Schrad. Ex. Benth. (Lamiaceae)

This is an annual/perennial shrub found in temperate and sub-tropical zones. Its leaf aqueous extract showed antifeedant activity to African armyworm, *Spodoptera exempta* (Waespe and Hans, 1979).

471. *Iva axillaris* Pursh. (Asteraceae)

This is an annual/perennial herb found in tropical, sub-tropical and temperate zones. Its stem, leaf and inflorescence extracts in ethyl ether showed attractancy to the male and the female of the Oriental fruit fly, *Dacus dorsalis* and the female of Mediterranean fruit fly, *Ceratitis capitata* (Keiser et al., 1975).

472. *Jacaranda mimosifolia* D. Don. (Bigoniaceae)

This is a perennial tree found in tropical, sub-tropical and temperate zones. Oil extracted from its dried seed was reported to show antifeedant activity when tested against the cotton boll weevil, *Anthonomus grandis* (Jacobson, et al., 1981b).

473. *Jacaranda obtusifolia* var. *rhombifolia* D. Don. (Bignoniaceae)

This is a perennial tree found in tropical regions. Its sap or latex was reported to be insecticidal in nature when tested against the leaf cutting ants (Jacobson, 1958).

474. *Jacquinia aristata* Mutis ex. Linn. (Thesphrastaceae)

This is a perennial shrub found in tropical, sub-tropical and semi-arid zones. Its root and fruit extracts in water and also their dried powders were reported to be toxic to the melon worm, *Diaphania hyalinata* and the diamond-back moth, *Plutella xylostella* (Jacobson, 1958).

475. *Jasminum arborescens* Roxb. (Oleaceae)

(Syn.-*J. roxburghianum* Wall.)

Common name: Tree jasmine

This is a perennial shrub found in tropical and sub-tropical zones. Its leaf extract in petroleum ether was reported to be toxic to the rice weevil, *Sitophilus oryzae* (Jacobson, 1975).

476. *Jatropha curcas* Linn. (Euphorbiaceae)

Common name: Barbados nut or Physic nut

This is a perennial tree found in tropical regions and commonly available in India. Its wood, leaf, fruit and seed extracts in water were reported to be toxic to the leaf cutting larvae (Jacobson, 1958). Shelke et al. (1987) reported its seed extract to reduce oviposition of the potato tuber moth, *Phthorimaea operculella* by 32% when tested under laboratory conditions using 3% formulation.

477. *Jatropha gossypifolia* Linn. (Euphorbiacea)

This is a perennial ornamental shrub found in tropical zones and distributed in all South East Asian countries. Its leaf and seed extracts were reported to be toxic to the red flour beetle, *Tribolium castaneum* (Chatterjee et al., 1980b).

478. *Juglans nigra* Linn. (Juglandaceae)

Common name: Black walnut

This is a perennial tree found in temperate zones and commonly available in Europe and North America. Its leaf and fruit extracts in ethyl ether were reported to show attractancy to the male and female of the melon fly, *Dacus cucurbitae* and the female of Mediterranean fruit fly, *Ceratitis capitata* (Keiser et al., 1975). However, its leaf and fruit extracts in water were found to be insecticidal and repellent against the striped cucumber beetle, *Acalymma vittatum* (McIndoo, 1983) and also to the smaller European elm bark beetle, *Scolytus multifasiatus* (Waespe and Hans, 1979).

479. *Juniperus verginiana* Linn. (Cuppressaceae)

Common name: Red cedar or Pencil cedar or Cedarwood

This is a perennial sturdy ornamental tree native to North America but also available in Himachal Pradesh in India. Its wood or steam distillation yields 1–3% of volatile oil, known as cedarwood oil, which is used in insecticides, perfumery, soaps etc. Cedarwood oil was found to protect stored wheat up to 60 days from rice weevil, *Sitophilus oryzae* and lesser grain borer, *Rhyzopertha dominica* when grains were treated at 100 ppm concentration of its oil solution (Sighamony et al., 1986).

480. *Justicia betonica* Linn. (Acanthaceae)

This is a perennial undershrub found in tropical regions and distributed in Malaysia, Indonesia, Australia, the Philippines, South China and Central America. This shrub is available throughout India. Its dried leaf powder admixed with the wheat grains @ 1, 2 and 5% w/w significantly reduced the adult emergence of the rice moth, *Corcyra cephalonica* in a laboratory test (Chander and Ahmed, 1986).

481. *Justicia gendarussa* L.f. (Acanthaceae)

(Syn.-*Gendarussa vulgaris* Nees.)

Common name: Water willow genus

This is an evergreen perennial shrub found in tropical and sub-tropical zones. Its root, branch and leaf extracts in water and also their dried powders were toxic to the pulse beetle, *Callosobruchus chinensis*; the hairy caterpillar, *Euproctis fraterna*; and the cotton leaf armyworm, *Spodoptera litura* (McIndoo, 1983).

482. *Justicia tranquebariensis* Linn. (Acanthaceae)

This is a perennial undershrub found in tropical and sub-tropical zones. Macre and Towers (1984) reported antifeedant activity of liganins isolated from this plant. Chitra et al. (1995) found 'lariciresinol' (Figure 74), an active component isolated from this plant extract, as a potent antifeedant to 2nd stage grubs of *H. viginitioctopuntata*.

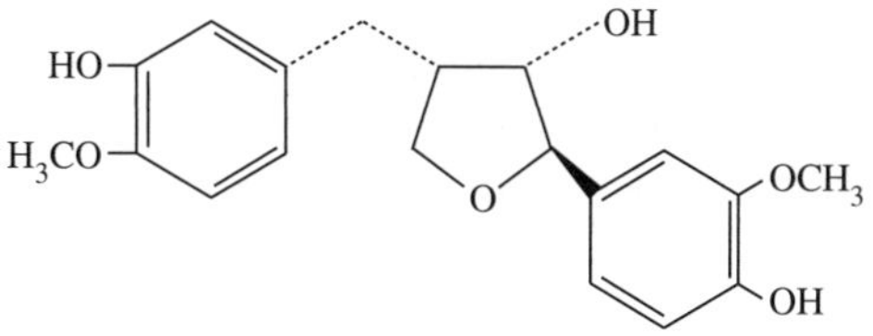

Figure 74. Lariciresinol.

483. *Kallistroemia maxima* (Linn.) Hook & Arn. (Zygophyllaceae)

This is a perennial shrub found in temperate zones. Its root, stem, leaf, flower and fruit extracts in ethyl ether showed attractancy to Mediterranean fruit fly, *Ceratitis capitata* and to the female of the melon fly, *Dacus dorsalis* (Keiser et al., 1975).

484. *Kalmia latifolia* Linn. (Ericaceae)

Common name: Mountain laurel

This is a perennial shrub found in tropical and sub-tropical zones and available in Saudi Arabia, Egypt and Israel. A synthetic fraction of ethanolic residues of its leaf extract showed antifeedant activity to the larvae of gypsy moth, *Lymantria dispar*. The active components 'lyoniol-A' and gryanotoxin-X, VIII, II and III isolated from its leaf extract were found to be the effective antifeedants. In addition, kalmitoxin-I, II, III, IV, V and VI were also reported as active components but kalmitoxin-I was a potent antifeedant (El-Naggar et al., 1980).

485. *Khaya myasica* A. Juss. (Meliaceae)

Common name: Red mahogany or Lice seeds

This is a perennial tree found in tropical zones. Its seeds' oil and extract in water showed insecticidal and also repellent activities against termites (Usher, 1973).

486. *Kleinhovia hospita* Linn. (Sterculiaceae)

This is a perennial tree found in tropical regions. Its leaf and branch extracts in water were reported to show repellant activity to the insects when fresh twigs were planted in the rice fields (Litsinger et al., 1978).

487. *Lachnanthes caroliniana* (Lam.) Wilb. (Haemodoraceae)

Common name: Red root or Paint root

This is a perennial herb found in tropical and sub-tropical zones. Its whole plant extract showed repellency to the Japanese beetle, *Popillia japonica* (McIndoo, 1983).

488. *Lactuca sativa* Linn. (Asteraceae)

(Syn.-*L. scariola* Linn. var. *sativa* C.B. Clarke)

Common name: Garden lettuce

This is a biennial/annual shrub found in temperate zones and distributed in North America and Europe. Its dried leaf powder was reported to be toxic to the tomato horn worm, *Manduca sexta* (Jacobson, 1975) and the cabbage worm, *Pieris rapae* (McIndoo, 1983).

489. *Lantana camara* Linn. (Verbenaceae)

(Syn.-*L. aculeata* [L.] Moldenke, *L. aculeata* Linn.)

Common name: Common lantana

This is a perennial shrub found in tropical and sub-tropical zones and distributed in South East Asian countries, China, Australia and Africa.

Biological Activity Against Storage Pests

Dried leaf powder of common lantana was reported to show antifeedant activity when tested against the rice weevil, *Sitophilus oryzae* (Morallo-Rajessus, 1982) whereas its wood extract showed toxicity to the corn weevil, *Sitophilus granarius* (Morallo-Rajessus, 1984). Its leaf and flower extracts showed antifeedancy (Pandey et al., 1986) and toxicity (Saxena et al., 1992) when tested against *Callosbruchus chinensis* in stored greengram.

Biological Activity Against Field Pests

Its leaf extract in water was reported to be toxic to the black bean aphid, *Aphis rumicis* and repellency to the diamond-back moth, *Plutella xylostella*

(Jacobson, 1975). However, Pandey et al. (1977) and Hameed (1983) found its leaf extractives in ether to show antifeedant activity to 3rd instar larvae of the mustard sawfly, *Athalia proxima*. Its oil was reported to be repellent to honey bees (Attari and Singh, 1977a). Further, its dried leaf powder showed insecticidal, antifeedant and repellent activities against the Asian corn borer, *Ostrinia furnacalis* (Morallo-Rajessus, 1982) and against turnip aphid, *Hyadaphis erysimi* (Yingchol, 1983). Jaipal et al. (1983) reported its wood extract to inhibit the metamorphosis of 5th instar larvae of the cotton stainer, *Dysdercus koenigii*. Topical application of its flower extract, however, was found to be toxic to the rice brown planthopper, *Nilaparvata lugens* (Morallo-Rajessus, 1984). Lal (1987) reported its dried leaf powder spread over the stored potato to show significant protection to the potato against infestation of potato tuber moth, *Phthorimaea operculella*. Reddy et al. (1990) reported its petroleum ether extract to reduce the population of the brinjal spotted leaf beetle, *Henosepilachna vigintioctopunctata* significantly when sprayed on the potted plants. Similarly, its leaf extract showed toxicity to 2nd instar larvae of the hairy caterpillar, *Amsacta moorei* (Patel et al., 1990b). Sharma et al. (1992) also reported its crude leaf extract in water to show significant antifeedancy to the 3rd instar larvae of jute semi-looper, *Anomis sabulifera* and 4th instar larvae of Bihar hairy caterpillar, *Spilosoma obliqua*. Gopakumar et al. (1993) found 1% aqueous suspension of the stem extract of this weed to show growth inhibitory activity when tested against 4th instar larvae of silkmoth, *Eupterote undata*.

490. *Lantana rugosa* Thunb. (Verbenaceae)

(Syn.-*L. salvifolia* Jacq.)
This is an annual/perennial shrub found in tropical regions and commonly available in West Africa and Central India. Its whole plant extract was reported to protect stored grain against the insect pests in storage (Giles, 1964). Its fresh leaf extracts showed larvicidal and antifeedant activities to the larvae of the mustard sawfly, *Athalia proxima* (Sudhakar et al., 1978; Pandey et al., 1979).

491. *Laserpitium archangelica* Jacq. (Umbelliferae)

This is an annual/perennial herb found in temperate and sub-tropical zones and commonly available in Europe. From its plant extracts, sesquiterpene lactones were isolated and found to show antifeedant activity against the stored grain pests, viz., *S. granarius*, *T. granarium* and *T. castaneum*. 'Trilobolide', an active component isolated from the plant extract, showed highest antifeedancy to these insects when tested under controlled conditions (Nawort et al., 1983).

492. *Laserpitium siler* Linn. (Umbelliferae)

This is an annual/perennial herb found in temperate regions in Europe. Eight sesquiterpene lactones were isolated from its whole plant extract and reported to show antifeedant activity against stored grain insects, *T. castaneum*, *T. granarium* and *S. granarius* when tested under controlled conditions. Trilobolide showed the highest degree of antifeedancy to these insects (Nawort et al., 1983).

493. *Laser trilobum* Desf. (Umbelliferae)

This is an annual/perennial herb or small shrub found in dry and stony plains in Europe. Nawort et al. (1983) reported deterrent activity of eight sesquiterpene lactones isolated from this plant to the stored grain pests, viz., the adults of *S. granarius*, *T. granarium* and the adults and the larvae of *T. castaneum* in a laboratory test. The highest antifeedant activity was recorded in trilobolide.

494. *Laurus nobilis* Linn. (Lauraceae)

Common name: Sweet laurel or Bay laurel

This is an evergreen bush or small tree found in Mediterranean basins, Asia Minor, Greece, Italy, Spain and Portugal. Its aqueous leaf extract was reported to possess repellent property against the red flour beetle, *Tribolium castaneum*. The active components, viz., benzeldehyde, piperidine and geranion, isolated from this tree were found to show repellency to *T. castaneum*. Muckenstrum et al. (1981) isolated another active component, i.e., methyl euginol, a phenyl propanoid, from its ether extract and reported it to show a high degree of antifeedancy when tested against *Leptinotarsa decemlineata* and *Mythimna unipunctata*. Another active component, 'alpha-pinine', isolated from its leaf extract was also found to show attractancy to *T. castaneum* (Saim and Clifton, 1986). Shaaya et al. (1991) reported fumigant toxicity of its essential oil to the lesser grain borer, *Rhyzopertha dominica* having 'terpinen-4-ol' as an active component.

495. *Lavandula angustifolia* Mill. (Lamiaceae)

(Syn.-*L. officinalis* Chaix, *L. vera* D.C., *L. spica* Linn.)

Common name: Lavender

This is an annual small shrub found in temperate zones and native to South Europe. It is commonly available in Jammu and Kashmir in India. Oil and extracts of its leaf and flower were reported to show repellent activity against the cotton aphid, *Aphis gossypii* (McIndoo, 1983).

496. *Lavandula gibsoni* Lam. (Labiatae)

This is a small perennial shrub found in sub-tropical and Mediterranean zones and distributed in North India, Nepal and South China. Its leaf extract in acetone was reported to show insecticidal activity against the cotton stainer, *Dysdercus koenigii*; the potato tuber moth, *Phthorimaea operculella*; and the red flour beetle, *Tribolium castaneum* (Sharma et al., 1981).

497. *Leonotis leonurus* (Pers.) R.Br. (Lamiaceae)

Common name: Lion's ear

This is a perennial herb or small shrub found in tropical zones. Its root extract in water was reported to show antifeedant activity against the milkweed bug, *Oncopeltus fasciatus* (Hael et al., 1950).

498. *Leonotis neptaefolia* (Linn.) W. Ait. (Lamiaceae)

Common name: Lion's ear

This is a perennial herb found in tropical and temperate zones and commonly available in North India and Pakistan. Its leaf and seed dried powder were reported to be toxic to the bean leaf beetle, *Cerotoma trifurcata* and the melon worm, *Diaphnia hyalinata* (Jacobson, 1958) and also to the cotton stainer, *Dysdercus flavidus* (Khan, 1982).

499. *Lepidium ruderale* Linn. (Brassicaceae)

Common name: Wild pepper grass

This is an annual herb found in tropical and temperate zones. Water extract of its whole plant was reported to be toxic to the aphids of agricultural importance (McIndoo, 1983).

500. *Lepidium virginicum* Linn. (Brassicaceae)

This is an annual herb found in sub-tropical and temperate zones. Its root, stem, leaf and flower extracts in ethyl ether showed attractancy to the female Mediterranean fruit fly, *Ceratitis capitata* (Keiser et al., 1975).

501. *Letharia vulpina* (Lichen)

This is a lichen found in sub-tropical and temperate regions. Solutions of vulpinic acid isolated from this lichen, coated onto the leaves of sprouted broccoli (*Brassica oleracea* var. *italica*), showed antifeedant activity when fed to the striped armyworm, *Spodoptera ornithogalli* (Slansky, 1979).

502. *Leucas aspera* (Willd.) Spreg. (Lamiaceae)

Common name: Dandha-kalas

This is an annual/perennial herb found in tropical regions and distributed in South India, Sri Lanka, Malaysia, Indonesia and Australia. Its branch, leaf, flower and fruit extracts in either showed juvenile-mimetic activity to the cotton stainer, *Dysdercus cingulatus* (Rajendran and Gopalan, 1978) and also showed insecticidal activity to this bug (Balasubramanian, 1982).

503. *Leucothoe grayana* D. Don. (Ericaceae)

This is a perennial shrub found in temperate regions. Its leaf extract in water and also dried powder were reported to be toxic to the rice weevil, *Sitophilus oryzae* and the vinegar fruit fly, *Drosophila hydei* (Jacobson, 1958) and also to the cabbage beetle, *Phaedon brassicae* (McIndoo, 1983).

504. *Leucothoe keiskei* D. Don. (Ericaceae)

This is a perennial tree found in temperate zones. Its branch, leaf and flower extracts in water were reported to be toxic to the vinegar fruit fly, *Drosophila hydei* (Jacobson, 1958).

505. *Liatris punctata* Gaertn. ex. Schreb. (Asteraceae)

This is a perennial herb found in temperate zones and commonly available in North America and Europe. Its whole plant extract was reported to show insecticidal activity to the red legged grasshopper, *Melanoplus femmurubrum* (Jacobson, 1975).

506. *Libanotis ugonsis* Hill. (Apiaceae)

This is a perennial herb found in temperate zones. Its root, leaf and branch extracts in water were reported to be toxic to the vinegar fruit fly, *Drosophila hydei* (Jacobson, 1958).

507. *Licaria salicifolia* (Sw.) Kostermi (Lauraceae)

This is a perennial tree found in sub-tropical regions. Its twig and leaf extracts in ethyl ether showed attractancy to both male and female of the Oriental fruit fly, *Dacus dorsalis* and female of the Mediterranean fruit fly, *Ceratitis capitata* (Keiser et al., 1975).

508. *Ligustrum obturifolium* Linn. (Oleaceae)

This is a perennial shrub in temperate and sub-tropical zones. Its leaf extract in water was reported to be toxic to the vinegar fruit fly, *Drosophila hydei* (Jacobson, 1958).

509. *Lindenbergia grandiflora* Lehm. (Scrophulariaceae)

This is an annual/perennial diffuse herb found in tropical regions and available throughout India on the walls and the banks in marshy places. Its leaf extract was reported to show strong antifeedant activity to the jute hairy caterpillar, *Spilosoma obliqua* when tested under controlled laboratory conditions (Tripathi et al., 1987).

510. *Linum usitatissimum* Linn. (Linaceae)

Common name: Flax or Linseed

This is an annual/seasonal shrub grown in tropical zones. It is commonly cultivated as oil-seed crop in India. Its oil cake was reported to show repellency to the rice weevil, *Sitophilus oryzae* and reduced its oviposition (Bowry et al., 1984). Double refined linseed oil acted as surface coating agent for endosulfan encapsulated formulations and increased efficacy of the chemical against the sorghum stalk borer, *Chilo partellus* (Makherjee et al., 1984; Srivastava and Saxena, 1986). Trivedi (1987) reported its oil as a carrier for pyrethrin dusts when tested against the lesser grain borer, *R. dominica* and the red flour beetle, *T. castaneum* and found it to enhance the effectiveness of the pyrethrin. Linseed oil 1% admixed with peas protected the seeds from the attack of the pulse beetle, *Callosobruchus chinensis* (Kumari et al., 1990).

511. *Lippia adoensis* Rich. (Verbenaceae)

This is a perennial herb/shrub found in tropical and sub-tropical regions and commonly available in West Africa and India. Its oil was reported to be toxic to the adults of *Dysdercus superstiosus* and the larvae of *Acrea eponina*, *Ootheca mutabilis* and *Riptortus dentipes* (Olaifa et al., 1987).

512. *Lippia berlandieria* (Linn.) Kunth. (Verbenaceae)

Common name: Oregano

This is an annual/perennial herb found in sub-tropical zones. Its leaf extract in water attracted the striped cucumber beetle, *Acalymma vittata* (Mathews, 1980).

513. *Lippia geminata* Hb. and Kunth. (Verbenaceae)

(Syn.-*L. alba* [Mill.] N.E.Br.ex. Britton & Willson)
Common name: Wild sage

This is an aromatic annual/perennial herb found in tropical and sub-tropical zones and distributed in all South East Asian countries and West Africa. It is available in Orissa, West Bengal, Nilgiri Hills and Bihar in India. Its aqueous leaf extract and dried powder were reported to show insect repellent activity and protected stored paddy against infestations of Angoumois grain moth, *S. cerealella*; the lesser grain borer, *R. dominica*; and the rice weevil, *S. oryzae* (Prakash and Rao, 1984; Prakash et al., 1987, 1978–1987, 1989). Prusty et al. (1989) also found its leaf powder to inhibit the larval and pupal populations of the rice moth, *Corcyra cephalonica* when admixed with rice grain in storage. Prakash et al. (1991) found its leaf extract in methanol to protect stored paddy grains against *R. dominica*.

514. *Lippia stoechadifolia* Kunth. (Verbenaceae)

Common name: Puleo

This is a perennial herb found in tropical, sub-tropical and temperate zones. Its aqueous leaf extract was reported to protect the stored paddy grains when treated with the crude extract against grain boring insects (Prakash et al., 1978–1987). From the leaf extract of this plant, an active component possessing insecticidal activity was isolated and identified as a monoterpenoid called 'pulegone 1,2-epoxide', i.e., [6-methyl-3-(1-methyl ethylidene)-7-oxabicyclo (4,10) heptan-2-one]. This component was reported to be neurotoxic to the vinegar fly, *D. melanogaster* and also to *Gromphadorhina portentosa* and found to inhibit the synthesis of acetylcholinestrases (Grundy and Still, 1985).

515. *Litsea cubeba* (Lour.) Pers. (Lauraceae)

(Syn.-*L. citrata* Bl.)

This is a perennial aromatic shrub or small tree distributed in the eastern Himalayas and Khasi hills in India. Its volatile essences using 0.5 ml/petri dish were found to kill the adults of storage weevils, *Sitophilus oryzae* and *S. zeamais* and completely suppressed their populations in a laboratory test (Zhang and Zhao, 1983).

516. *Lobelia inflata* Linn. (Campanulaceae)

Common name: Indian tobacco

This is an annual herb found in temperate and sub-tropical regions and native to USA and Canada. In India, it is cultivated in the eastern Himalayas, Nilgiris and Kerala. Its leaf extract in water was toxic to the bean aphid, *Aphis fabae* (McIndoo, 1983).

517. *Lolium pernae* Linn. (Poaceae)

(Syn.-*L. brasilianum* Nees.)
Common name: Rye grass

This is a perennial grass found in sub-tropical and temperate zones. In India, it is available in Sikkim, Kashmir and Nilgiris and used as fodder. Its whole plant extract was reported to possess an alkaloid named 'halostachine', which was found to show insecticidal activity against the locusts (Aassen et al., 1969).

518. *Lonchocarpus* sp. (Fabaceae)

Common name: Cube

This is a perennial tree found in tropical zones. Its aqueous root extract and dried powder showed insecticidal activity against silkworm, *Bombyx mori* and Colorado potato beetle, *Leptinotarsa decemlineata* (McIndoo and Sievers, 1924). The active component in its root was reported as 'rotenone' (De-Ong, 1948).

519. *Lonchocarpus chrysophyllum* Kunth. (Fabaceae)

Common name: Black haiari

This is a perennial tree found in tropical zones. Its root extract in water and dried powder were reported to be insecticidal to the black bean aphid, *Aphis rumicis* and the winter moth, *Cheimatobia bruma* (Michael et al., 1985).

520. *Lonchocarpus deniflorus* Kunth. (Fabaceae)

Common name: Haiari

This is a perennial vine found in tropical zones. Its root powder was reported to be toxic to the bean aphid, *Aphis rumicis* and the winter moth, *Cheimatobia bruma* (Michael et al., 1985) and also to the storage insects, *S. cerealella*, *S. oryzae* and *R. dominica* (Prakash et al., 1978–1987).

521. *Lonchocarpus nicou* (Aubl.) D.C. (Fabaceae)

Common name: Timbo

This is a perennial tree found in Peru. Its root extract was reported to be toxic to aphids (De-Ong, 1948). The active component in its root extract was found to be 'rotenone' (Martin, 1964).

522. *Lonchocarpus urucu* Killip and Smith. (Fabaceae)

Common name: Timbo

This is a perennial tree found in tropical zones. Its root extract was reported to be toxic to the milkweed bug, *Oncopeltus fasciatus* (Hael et al., 1950). The rotenone was found to be an active component from its root extract (Martin, 1964).

523. *Lophotocarpus calycinus* Durand. (Alismataceae)

Common name: Swamp potato genus

This is an aquatic and also semi-aquatic annual/perennial herb found in arctic zones. Its whole plant aqueous extract was reported toxic to the milkweed bug, *Oncopeltus fasciatus* (Hael et al., 1950).

524. *Lotus crassifolius* (Benth.) Greene var. *subglaber* (Ottley) C.L. Hitche (Fabaceae)

This is a perennial herb found in temperate zones. Its root extract in ethyl ether attracted the adults of Mediterranean fruit fly, *Ceratitis capitata* (Keiser et al., 1975).

525. *Lotus pediculatus* Linn. (Fabaceae)

Common name: Lotus major

This is a perennial herb found in temperate zones and commonly available in Europe and North America. Its whole plant extract or juice showed insecticidal and repellent activities against the grass grub, *Castelytra zealanica* (Sutherland, 1983).

526. *Luffa acutangula* (Linn.) Roxb. (Cucurbitaceae)

(Syn.-*Cucumis acutangulus* Linn.)

Common name: Ribbed/ridged gourd or Angled luffa or Vegetable sponge or Dishcloth loofah

This is an annual vine found in tropical and temperate zones. Its flowers and seed extracts in alcohol and petroleum ether were reported to be toxic to the rice weevil, *Sitophilus oryzae* when tested under controlled conditions (Jacobson, 1975).

527. *Luffa aegyptica* Mill. (Cucurbitaceae)

(Syn.-*L. cylindrica* Auct. pl. non.M.J. Roem., *Momordica cylindrica* Linn.)

Common name: Dishcloth gourd or Smooth loofah or Vegetable sponge

This is an annual climbing vine found in tropical Asia and commonly cultivated in India. Its dried leaf powder admixed with the stored grains was reported to protect them from the infestation of lesser grain borer, *Rhyzopertha dominica*; rice weevil, *Sitophilus oryzae*; and red flour beetle, *Tribolium castaneum* under controlled conditions (Giles, 1964).

528. *Lupinus argenteus* Linn. (Fabaceae)

This is a perennial herb found in temperate zones. Its leaf, flower, seed and whole plant extracts in water and their dried powder showed insecticidal and insect growth regulating activities against the vinegar fly, *Drosophila melanogaster* (Michael et al., 1985).

529. *Lycium halimifolium* Linn. (Caryophillaceae)

Common name: Matrimony vine

This is a perennial shrub found in temperate and Mediterranean zones. Its whole plant dried powder was reported to be insecticidal when tested against the aphids of agricultural importance (Michael et al., 1985).

530. *Lycopersicon esculentum* Mill. (Solanaceae)

(Syn.-*L. lycopersicum* [Linn.] Karst., *Solanum lycopersicum* Linn.)

Common name: Tomato

This is an annual or seasonal herb grown as vegetable in tropical, subtropical, temperate and semi-arid zones and widely cultivated now all over the world. Its whole plant, leaf, fruit and branch extracts in water were reported to be insecticidal to the cabbage butterfly, *Pieris brassicae* and to flies and bees (McIndoo and Sievers, 1924) and also to Colorado potato beetle, *Leptinotarsa decemlineata* (Jacobson, 1975). Saxena and Basit (1982) reported its leaf extract and volatiles from the extracts as an oviposition inhibitor to the leafhopper, *Amrasca devastans*. Cooke-Stinson (1986) isolated an active component called '2-tridecanone' from its leaf, which showed repellent activity to the grain weevils, *S. granarius*, *S. oryzae* and *S. zeamais* in the treated wheat grains stored in test jars under controlled conditions and also found it to reduce their oviposition.

531. *Lycopersicon hirsutum f. glabratus* Mill. (Solanaceae)

Common name: Wild tomato

This is an annual herb found in temperate and tropical zones. Dimock et al. (1982) reported toxicity of the analogous '2-tridecanone', a naturally occurring toxicant in its leaf extract to the maize borer, *Heliopthis zea*.

532. *Lycopersicon pimpinellifolium* (Juslen.) Mill. (Solanaceae)

Common name: Currant tomato

This is an annual herb found in tropical, temperate and alpine zones and a native of Paris. Its whole plant extract in water was reported to be toxic to the potato beetle (Jacobson, 1975).

533. *Lycopodium clavatrum* Linn. (Lycopodiaceae)

Common name: Club moss

This is an annual/perennial moss found in the mountainous heaths and the moors all over Europe. An alkaloid 'nicotine' isolated from its whole plant extract was reported to show toxicity to a wide range of insect pests of agricultural importance (Schmeltz, 1971).

534. *Lycoris radiata* L'Herit (Ericaceae)

Common name: Spider lily

This is a perennial bulbous herb found in temperate zones. Its root and branch extracts in petroleum ether were reported to be toxic to the vinegar fruit fly, *Drosophila hydei* (Jacobson, 1958).

535. *Lysimachia hummularia* Linn. (Primulaceae)

Common name: Money-wort

This is a perennial vine/herb found in temperate zones. Its leaf and flower extracts in alcohol were reported to show insecticidal activity to stored grain insect pests (McIndoo, 1983).

536. *Lysimachia mauritiana* (Linn.) Mill. (Primulaceae)

This is an annual herb found in temperate zones. Its root, branch and leaf extracts in water were reported to show toxicity to the vinegar fruit fly, *Drosophila hydei* (Jacobson, 1958).

537. *Macleaya cordata* R. Br. (Papaveraceae)

Common name: Pink poppy

This is a perennial tree found in tropical and temperate zones. Its leaf extract in water showed toxicity to the vinegar fruit fly, *Drosophila hydei* (Jacobson, 1958).

538. *Madhuca butyracea* (Roxb.) Macb. (Sapotaceae)

(Syn.-*Diploknema butyracea* [Roxb.] H.J. Lam.; *Bassia butyraceae*)

Common name: Butter tree or Bassia fat or Indian roxb

This is a perennial large tree found in tropical and sub-tropical regions and commonly available in sub-himalayan tract, Kumaun hills, Sikkim and also in South India. Its seed extracts in ethanol, butanol and water at 0.5% concentration showed toxicity to the pulse beetle, *Callosobruchus chinensis* in treated greengram seed in storage (Lalitha et al., 1988).

539. *Madhuca indica* J.F. Gmel. (Sapotaceae)

(Syn.-*Bassia latifolia* Roxb., *M. latifolia* [Roxb.] Macb.)

Common name: Mahua

This is a perennial tree found in tropical and sub-tropical regions and distributed in all the South East Asian countries and Central America. Mahua is commonly available in Central and South India.

Biological Activity Against Storage Insects

Its oil when admixed with grains was reported to protect them from infestation of storage insects, viz., *R. dominica* and *S. oryzae* (Fletcher, 1919). Its stem and leaf dusts were found to be toxic to the pulse beetle, *Callosobruchus chinensis* (Puttarudriah and Bhatia, 1955). However, its oil cake admixed with the stored paddy effectively protected it from insect damage (Prakash et al., 1978–1987). Its leaf, bark, branch and seed extracts in water were also reported to show insecticidal and repellent activities against the pulse beetle, *C. chinensis* (McIndoo, 1983). Ali et al. (1983) found mahua oil to show toxicity to the pulse beetle, *C. chinensis* when admixed with greengram seeds @ 1 ml/100 g in storage. Bowry et al. (1984) reported its oil cake admixed with stored maize to reduce oviposition of the rice weevil, *Sitophilus oryzae*. Kumari et al. (1990) also found its 1% oil to protect pea seeds from the pulse beetle, *C. chinensis*.

Biological Activity Against Field Crop Insects

Sachan and Pal (1976) reported its cake application in the soil to reduce the incidences of the white grub, *Holotrichia insularia* in chilies. Its bark and seed extracts in water were found to be toxic to the cabbage leaf borer, *Crocidolomia binotalis*; the hairy caterpillar, *Euproctis fraterna*; and also to the common cutworm, *Spodoptera litura* (McIndoo, 1983).

Seed kernel extracts of mahua sprayed on blackgram crop reduced the incidences of the whitefly, *Bemisia tabaci* and the yellow mosaic virus (YMV) transmitted by this fly (Mariappan et al., 1987). Its seed oil was also reported to reduce the survival of the green leafhopper, *Nephotettix virescens*; the rice leaffolder, *Cnaphlocrosis medinalis*; and the rice tungro virus when sprayed on the rice leaves (Narsimhan and Mariappan, 1988; Mariappan et al., 1988). Ramaraju and Sunder Babu (1982) reported its 1% oil and 2% seed extract to reduce the emergence of the brown planthopper, *Nilaparvata lugens* and the white backed planthopper, *Sogatella furcifera* in rice. Its 2% oil in 3 sprays given at 15-day intervals significantly minimized the infestation of the citrus leaf miner, *Phyllocnistis citrella* (Dhara et al., 1990a) and reduced the population of citrus aphid, *Toxoptera citricidus* (Dhara et al., 1990b).

540. *Magnolia virginian* Linn. (Magnoliaceae)

Common name: Sweet flag
This is a perennial shrub found in temperate zones and distributed in Europe and North India. Its leaf extract in water was reported to be repellent to the Japanese beetle, *Popillia japonica* (McIndoo, 1983).

541. *Maianthemum canadense* Weber. (Liliaceae)

Common name: Wild lily of the valley
This is a perennial shrub found in temperate zones. Its leaf extract in water was reported to show repellent activity to the Japanese beetle, *Popillia japonica* (McIndoo, 1983).

542. *Mammea americana* Linn. (Clusiaceae)

Common name: Mammey apple
This is a perennial tree found in tropical and sub-tropical zones. It is a native of the West Indies but available in Central America and India. Its seed powder and also root bark, branch, leaf and fruit extracts in water and alcohol were reported to show insecticidal and repellent activities when tested against the rice weevil, *S. oryzae*; the gulf white cabbage worm, *Ascia monuste*; the bean beetle, *Cerotoma ruficornis*; the aphid, *Macrosiphum sonchi*; the southern beet worm, *Pachyzancla biupunctalis*; the imported cabbage worm, *Pieris rapae*; the armyworm, *Pseudoletia unipunctata*; and the southern armyworm, *Spodoptera eridania* (Jacobson, 1958). Crosby (1971) reported an active component called 'mammein' ($C_{22}H_{28}O$) (Figure 75) in its ground seed powder extract and found it to show insecticidal activity against American armyworm, *Spodoptera frugiperda* and the melon worm, *Diaphania hyalineata*.

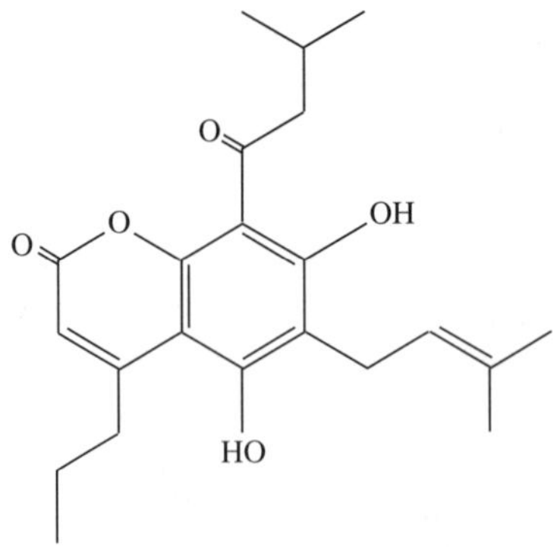

O O O OH HO

Figure 75. Mammein.

543. *Mangifera indica* Linn. (Ancistrocladaceae)

Common name: Mango
This is a perennial fruit tree found in tropical and sub-tropical regions and available throughout India and Pakistan. Its leaf powder was reported to show repellency to stored grain insects (Zaman, 1983).

544. *Markhamia stipulata* Seem. ex. Baill. (Begoniaceae)

This is a perennial small tree/shrub found in tropical zones. Its leaf extract in ethanol was reported to be insecticidal when tested against the termites of agricultural importance (Jacobson, 1975).

545. *Maytenus rigida* Molin. (Celastraceae)

This is a perennial shrub found in tropical zones and commonly available in Brazil. An active component isolated from its root extracts was reported to show a high degree of antifeedant activity when tested against *Pieris brassicae* and *Locusta migratoria* (Monache et al., 1984).

546. *Medicago sativa* Linn. (Fabaceae)

Common name: Alfalfa or Lucerne
This is a perennial herb found in temperate and semi-arid zones. Its whole plant extract in ether was reported to show insecticidal and antifeedant properties when tested against Mexican grasshopper, *Melanoplus mexicana* (Jacobson, 1975) and also against the grass grub, *Castelytra zealancia* (Sutherland, 1983). Pandey and Faruqui (1990) found repellency to the lucerne weevil, *Hyptera postica* in the crude petroleum ether extracts of *M. sativca*, *M. intertexta*, *M. arabica*, *M. scutellata* and *M. rugosa* up to 1000 micrograms strength but at lower doses of 10 and 100 micrograms these extracts showed attractancy to these weevils.

547. *Melaleuca leucadendron* Linn. (Myrtaceae)

Common name: Australian indomel
This is a perennial tree found in Australia and Pacific regions. From its leaf extract a feeding deterrent, (E,S)-nerolidol, was isolated and tested against the larvae of *Lymantria dispar* (Doskotch et al., 1980).

548. *Melia azedarach* Linn. (Meliaceae)

Common name: China-berry or Pride of India or Persian lilac or Dharek
This is a perennial tree found in tropical and sub-tropical zones distributed in South East Asian countries. Its leaf, bark, branch, fruit and seed extracts and

kernel powder and also its oil cake were reported to show repellent, antifeedant, insecticidal and growth inhibiting activities against a wide range of insects in both the storage and the field agro-ecosystems. Research work on bio-pesticidal properties presented for both ecosystems as follows:

Biological Activity Against Storage Insects

Dried leaf powder of dharek was reported to protect the stored wheat grains and found to act as an antifeedant and also repellent to Angoumois grain moth, *Sitotroga cerealella*; the lesser grain borer, *Rhyzopertha dominica*; the rice weevil, *Sitophilus oryzae*; the wheat weevil, *S. granarius*; and khapra beetle, *Trogoderma granarium* (Pruthy and Singh, 1950; Atwal and Sandhu, 1970; Saramma and Verma, 1971; Zag and Bhardwaj, 1976; Teotia and Pandey, 1978). Similarly, its fruit and seed powders were reported to protect stored wheat and rice grains from the infestation of storage insects when admixed with the grains (Atwal and Pajni, 1964; Atwal and Sandhu, 1970; Saramma and Verma, 1971; Jilani and Malik, 1973; Teotia and Pandey, 1978; Hameed, 1982). Kumar and Mehla (1993) found its 2% leaf power admixed with basmati rice to protect it in storage against insects.

Biological Activity Against Field Crop Pests

The biological activity in terms of insecticidal, repellent, antifeedant and growth regulating properties of dharek leaf and drup are given in Table 14 against a number of insects of agricultural importance.

Active Components

From its leaf extract, an active component as one of the derivatives of 'paraisin' was isolated. This component was not destroyed even at 37°C and also possessed no adverse effects on mammals. Later in 1946, another active component, 'meliantin', was isolated from its leaf extract and reported to show anti-locust activity (Chauvin, 1946). From its seed oil an active component, 'meliantriol-I', was also isolated and found to show biological activity (Lavie et al., 1967). Similarly, an active component, 'azardirachtin', was also isolated from its leaf and fruit extract and reported to show antifeedant activity against the locusts (Morgan and Thornston, 1973). Kubo and Klocke (1981) reported 'azardirachtin' from its fruit extract and found it to inhibit the larval molting of *Bombyx mori*, *Pectinophora gossypiella*, *Heliothis virescens* and *Spodoptera frugiperda*. Chiu (1989) isolated from its seed kernel another active component, i.e., 1-cinnamoyl-3-feruloyl-11 hydroxy meliacarpin, and found it to inhibit the development of the 3rd instar nymphs of the lychee stink bug, *Tessaratoma papilosa* and deformed 23.3% of the nymphs with partially developed prothorax and abnormal antennae. This tree contains several limonoids

Table 14. Biological activity of dharek leaf

Activity	Plant part/formulation	Test insect	Reference
Insecticidal	Drup extract	*Bombyx mori*	McIndoo and Sievers, 1924
Insecticidal	Leaf extract	*Brevicoryne brassicae*	Strakhantzer et al., 1937
Antifeedant	Leaf extract	Locusts of watermelon and bajra plants	Bhatia, 1940
Repellent	Leaf extract	*Medicago arborea*	Sergent, 1944
Repellent	Drup cake	Termites	Feinstain, 1952
Insecticidal	Drup extract	*Pieris brassicae* (Larvae)	Atwal and Pajni, 1964
Insecticidal	Drup extract	Larvae of *Pieris brassicae* and adults of *Aulacophora foveicollis*	Pajni, 1964
Repellent	Drup extract	Nymphs and adults of *Chrotogonus trachypterus*	Pajni, 1964
Antifeedant	Drup crude extract	Locusts	Shapan-Gabrieth, 1965
Antifeedant and growth regulator	Leaf extract	Larvae of *Heliothis zea* and *Spodoptera litura*	McMilan et al., 1965
Antifeedant	Fruit extracts	*Chrotogonus trachypterus*	Sandhu and Singh, 1975b
Juvenile hormone mimic activity	Fruit extract in methanol (Limonoids)	*Bombyx mori*, *Pectinophora gossypiella*, *Heliothis virescens*, *Spodoptera frugiperda*	Kubo and Klocke, 1981
Antifeedant repellent	Bark, leaf and seed extracts	*Pieris rapae*, *Spodoptera littoralis*, *Phyllocnistic citrella*, *Diaphornia citri*	Chiu, 1982
Antifeedant	Leaf and seed extracts	*Spodoptera frugiperda*, *Heliothis virescens*	Epino and Saxena, 1982

Table 14. Biological activity of dharek leaf (Continued)

Activity	Plant part/formulation	Test insect	Reference
Antifeedant, insecticidal	Seed oil (0.04%)	*Scirpophaga incertulas, Sogatella furcifera, Nilaparvata lugens*	Su et al., 1983
Antifeedant	Leaf and seed extracts	*Scirpophaga incertulas*	Wen, 1983
Antifeedant	Petroleum ether extract of seed kernel	Nymphs of *Nilaparvata lugens*	Chiu et al., 1983
Antifeedant	Leaf and seed extracts	Locusts of wheat crops	Omollo, 1983
Antifeedant	Leaf and seed extracts	*Myzus pericae*	Fagoonee, 1983
Antifeedant	Leaf and seed extracts	*Pectinophora gossypiella*	Kubo, 1983
Juvenile hormone mimic	Crude extracts of its bark and wood	*Dysdercus koenigii* larvae	Jaipal et al., 1983
Insecticidal	Leaf extractive	*Corcyra cephalonica*	Chauhan et al., 1987
Insecticidal	Leaf and fruit extracts	*Pieris rapae*	Zhang and Chiu, 1985
Synergist	Seed oil with diazinon	*Orseolia oryzae*	Chiu, 1989

which because of their higher mammalian toxicity have not been exploited adequately for future use (Jacobson, 1989).

549. *Melia toosendan* (Linn.) Juss. (Meliaceae)

Common name: China-berry

This is a perennial tree found in temperate and sub-tropical zones and commonly available in China, Europe and USSR. Its seed oil and extract in ether were reported to show insecticidal, repellent and antifeedant activities when tested against Asiatic citrus psyllid, *Diaphornia citri*; the white veined rice armyworm, *Leucania veralba*; the rice brown planthopper, *Nilaparvata lugens*; the rice gall midge, *Orseolia oryzae*; Asiatic corn borer, *Ostrinia furnacalis*; citrus leaf miner, *Phyllocnistic citrella*; the imported cabbage worm, *Pieris rapae*; the rice stem borer, *Scripophaga incertulas*; and the rice noctuid,

Spodoptera abyssina (Chiu, 1982). Chiu et al. (1983) also reported its seed extract in petroleum ether to show antifeedant activity to *Nilaparvata lugens*.

An active component, 'toosendanin', was isolated from its seed extract and reported to show antifeedancy to the cabbage worm, *Pieris rapae* when its emulsified solution (0.3% concentration) was sprayed on the cabbage seedlings (Zhang and Chiu, 1985). Shi et al. (1986) also reported toosendanin as a strong antifeedant which was responded by maxillary chemoreceptors of *Mythimna separata*.

550. *Melia volkensii* Linn. (Meliaceae)

This is a perennial tree found in temperate regions and distributed in North India and Europe. Its seed extract was reported to show antifeedant activity against *Schistocerca gregaria* when aqueous solution (2%) was sprayed on the host plant. The feeding inhibition rendered a poor relative growth, prolonged intermolt period and caused high mortality to the test insect (Mwangi, 1982; Omollo, 1983).

551. *Melinis minutiflora* P. Beavu. (Poaceae)

Common name: Molasses grass

This is a perennial grass found in tropical and sub-tropical zones. It is a native of Africa but now grown in India. Its whole plant expressed juice was found to show repellency to the aphids and sugarcane leafhopper, *Pyrilla purpusilla*. Its viscous oily secretion of cumin odour was observed to be an active component possessing repellent activity against this hopper (McIndoo, 1983).

552. *Menispermum canadense* Linn. (Menispermaceae)

Common name: Yellow parilla

This is a perennial woody climber found in temperate zones. Its leaf, branch and seed extracts in water were found to show toxicity and growth inhibiting activities when tested against vinegar fly, *Drosophila melanogaster* (Michael et al., 1985).

553. *Menispermum cocculus* Linn. (Menispermaceae)

(Syn.-*Amamirta cocculus* [L.] Wight. & Arn., *A. paniculata* Colebr.)

Common name: Fish berry or Lewant nut

This is a perennial woody climber found in temperate zones. Its leaf, fruit and seed aqueous extracts showed insecticidal and growth regulating activities against the hairy caterpillar, *Euproctisfraterna* and mango hopper, *Idiocerus* sp. (McIndoo, 1983).

554. *Mentha avensis* Linn. (Lamiaceae)

Common name: Japanese mint

This is an annual/seasonal herb found in tropical, sub-tropical and temperate zones and commonly available in North India. Its leaf essential oil was reported to be toxic and also exhibited fumigant activity to the pulse beetles, *C. chinensis* and *C. maculatus* and ultimately protected pulse grains in storage (Srivastava et al., 1989).

555. *Mentha ciliata* Ehrh. (Lamiaceae)

This is an annual/seasonal aromatic herb found in tropical and sub-tropical zones. It is commonly grown in India as a condiment. Its oil was reported to show fumigant activity when tested against the pulse beetle, *C. chinensis* (Mishra et al., 1981). Its leaf powder was also found to show toxicity to the adults of *C. chinensis* when admixed with chickpea grains in storage. An active component in its oil distillate was reported to be 'linonool' which was found to possess insecticidal property (Mishra et al., 1984).

556. *Mentha longifolia* (Linn.) Huds. (Lamiaceae)

(Syn.-*M. sylvestris* Linn.)

Common name: Horse-mint

This is an annual/perennial aromatic herb found in tropical and Mediterranean zones. Its leaf extract was reported to show repellent activity to the cotton aphid, *Aphis gossypii* (McIndoo, 1983).

557. *Mentha piperita* Linn. (Lamiaceae)

Common name: Peppermint

This is an annual aromatic herb found in tropical zones and commonly available in Indian plains, Europe, Egypt and South America. Its seed oil was reported to show fumigant activity to pulse beetle, *Callosobruchus chinensis* (Mishra et al., 1981) and also to red flour beetle, *Tribolium castaneum* (Mishra and Jitender, 1983; Shaaya et al., 1991). Its leaf powder showed toxicity to the pulse beetle, *C. chinensis* when admixed with stored chickpea (Mishra et al., 1984). An active component, 'menthol', isolated from its oil distillates, was found to show toxicity to *C. chinensis* (Mishra et al., 1984).

558. *Mentha pulegium* Linn. (Lamiaceae)

Common name: Penny-royal

This is an annual/perennial herb found in tropical zones. This herb is native to Europe and Western Asia but now grows in Jammu and Kashmir in India. Its oil at 5000 ppm dilution showed absolute inhibition in feeding of the larvae of *Spodoptera frugiperda*, and also reduced adults' emergence when admixed with the diets (Zalkow et al., 1979).

559. *Mentha spicata* (L.) Huds. (Lamiaceae)

(Syn.-*M. viridis* Linn.)
Common name: Spearmint or Wild-mint
This is a perennial/annual herb found in temperate, tropical and sub-tropical zones and available throughout India, especially in sandy areas and also in other neighbouring countries. Its dried leaves admixed with the wheat grain protected it from the infestation of storage insects (Quadri and Rao, 1977; Anonymous, 1979; Prakash et al., 1982). Mishra et al. (1981) reported its oil to show repellent activity to the pulse beetle, *Callosobruchus chinensis*. However, Srivastava et al. (1988a,b) reported its oil to show toxicity to *C. chinensis* and *C. maculatus* when admixed with the stored pulses. McIndoo (1983) reported its leaf extract as a repellent to screw worm, *Cochliomya hominivorax*. Mishra et al. (1984) found 'carvone', an active component in its oil distillates and also in leaf powder, which showed toxicity to the adults of *C. chinensis* when admixed with stored chickpea grains.

560. *Mesua ferrea* Linn. (Clusiaceae)

Common name: Iron weed or Nagakesar or Nageswar Champa
This is a perennial tree found in tropical zones and distributed in South East Asia. Iron weed is commonly available in eastern Himalayas, Assam, West Bengal, Western Ghats, Travancore and Andaman Islands in India. Its leaf and flower extracts were reported to show growth inhibiting activity when tested against the cotton stainer, *Dysdercus koenigii* (Krishnamurthy Rao, 1982).

561. *Millettia auriculata* Baker ex. Brandis (Fabaceae)

(Syn.-*M. extensa* Benth. ex. Baker)
This is a perennial woody climber/small tree found in tropical, sub-tropical and alpine regions and available in Central India. Its root and branch extracts in water were reported to show repellency to the cattle grubs (McIndoo, 1983). An active component, a natural isoflavone called 'aurmillone' (Figure 76), was isolated from its root extract and observed to show antifeedant activity to *Spodoptera litura* (Srimmannarayana and Rao, 1985).

562. *Millettia pachycarpa* Bench. (Fabaceae)

Common name: Fish poison climber
This is a perennial vine found in tropical and sub-tropical zones. Its root, fruit and branch extracts in water and also their dried powders were mentioned possessing insecticidal and repellent properties when tested against the bean

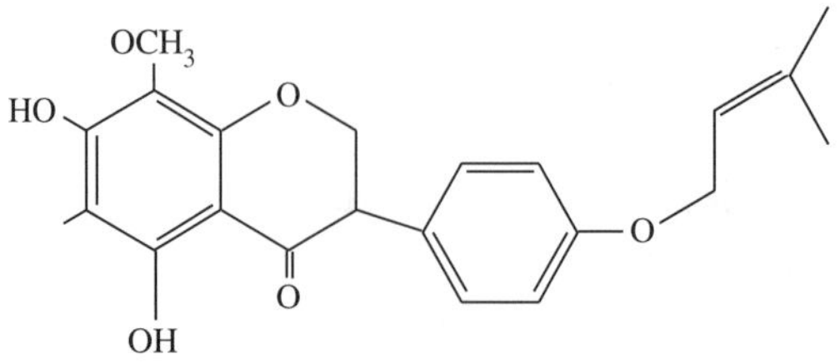

Figure 76. Aurmillone.

aphid, *Aphis fabae*; the silkworm, *Bombyx mori*; and Mexican bean beetle, *Epilachna varivestis* (Michael et al., 1985).

563. *Millettia reticulata* Wight. & Arn. (Fabaceae)

This is a perennial woody climber found in temperate zones and available in North America. Its leaf extract was reported to show insecticidal activity when tested against the cabbage leaf beetle, *Colaphellus bowringi* (McIndoo, 1983).

564. *Momordica charantia* Linn. (Cucurbitaceae)

Common name: Bitter-gourd

This is an annual/seasonal vine found in tropical and sub-tropical zones and available throughout India and Sri Lanka. Its whole plant extracts in water and alcohol were reported to show insecticidal and antifeedant activities when tested against the mustard sawfly, *Athalia proxima* (Kumar et al., 1979). Mohiuddin et al. (1987) reported repellency in its oil against the red flour beetle, *T. castaneum*. Its crude alcoholic leaf extract was also found to show feeding deterrency when tested against red pumpkin beetle, *Aulacophora foveicollis* (Mehta and Sandhu, 1992).

Okabe et al. (1980) isolated bitter principles (momordicosides) from its seeds and characterized them as glycosides of cucurbit-5-ene-3, 22, 23, 24, 25 pentanol. Metcalf et al. (1980) also reported 'cucurbitacins' tetracyclic terpenoids to act as kairomones for the cucumber beetle. Chandravadena and Pal (1983) established that triterpenoid glycoside isolated by them was different from momordicosides A and B and at 2 mg or more doses of this triterpenoid glycosides inhibited the feeding of *A. foveicollis*. A terpenoid called 'momordicine-II' (23-O-β-glucopyranoside of 3, 7, 23-trihydroxycucurbita-5, 25-dien-19 al) was isolated as an active component from this plant extract and found to show feeding deterrency to the red pumpkin beetle, *Aulacophora foveicollis* (Chandravadena, 1987).

565. *Momordica cochinchinensis* Spreng. (Cucurbitaceae)

This is an annual climbing herb found in tropical and temperate regions. Oils extracted from its dried seed powder were reported to show antifeedant activities against the cotton boll weevil, *Anthonomus grandis* (Jacobson and Crystal, 1971).

566. *Momordica foetida* Linn. (Cucurbitaceae)

This is an annual vine found in tropical zones. Its leaf and whole plant aqueous extracts were reported to be toxic to the moths, the weevils and the ants of agricultural importance (Jacobson, 1975).

567. *Monodora tenuifolia* Dual. (Annonaceae)

This is a perennial tree found in temperate forests. Its seed oil was reported to show insecticidal activity when tested against the adult *Dysdercus superstitiosus* and the larvae of *Acrea eponina*, *Ootheca mutabilis* and *Riptortus dentipes* under controlled conditions (Olaifa et al., 1987).

568. *Morinda pterygosperma* Gaertn.

(Syn.-*M. oleifera* Lamk., *Guilandina moringa* Linn.)

This is a perennial small tree found in tropical zones and available in Eastern India and Bangladesh. Its aqueous leaf extract was reported to be grain protectant when tested against the rice weevil, *Sitophilus oryzae* in stored wheat (Rout, 1986).

569. *Morinda royoc* Linn. (Rubiaceae)

This is a perennial small tree found in temperate zones. Its leaf extracts in ethyl ether attracted and Mediterranean fruit fly, *Ceratitis capitata* (Keiser et al., 1975).

570. *Mucuna cochinensis* (Lour.) A. Cheval (Leguminosae)

(Syn.-*M. nivia* D.C.)

Common name: Lyon bean or Velvet bean

This is an annual/biennial vine herb grown in tropical and sub-tropical zones. Extract of its mature leaf showed antifeedancy to the larvae of the jute hairy caterpillar, *Spilosoma* (*Diacrisia*) *obliqua* (Premchanda, 1989). The active components from its leaf extract were found to be tetrahydro-isoquinolines and a large amount of phenolic substances, which showed antifeedant activity against *S. obliqua* (Premchanda, 1989).

571. *Mucuna mustisiana* D.C. (Leguminosae)

This is an annual vine found in tropical and sub-tropical zones. Its whole plant extract was toxic to southern armyworm, *Spodoptera eridania* (Michael et al., 1985).

572. *Mundulea suberosa* Benth. (Leguminosae)

(Syn.-*M. sericea* [Willd.] Cheval.)

Common name: Lilac silky laburnum or Sweet lane

This is a perennial small tree found in tropical regions and available in Central and South India. Its seed powder was reported to protect wheat grains from infestation of the rice weevil, *Sitophilus oryzae* (Krishnamurthi and Rao, 1950). Its stem and bark powder were also reported to be toxic to insect species, viz., the common cutworm, *Spodoptera litura*; the mustard webworm, *Crocidolomia binotalis*; and the pulse beetle, *Callosobruchus chinensis* and *Idiocerus* sp. (Puttarudriah and Bhatia, 1955). However, aqueous extracts of its bark and seed were toxic to *Aphis rumicis*, *Coccus viridis* and *Leptinotarsa decemlineata* (McIndoo, 1983). Rotenone, an active component in its roots, and another active component, a natural isoflavone called 'mundulone' (Figure 77), were also isolated from seed extract and showed antifeedancy to the larvae of *Spodoptera litura* under controlled conditions (Srimmannarayan and Rao, 1985).

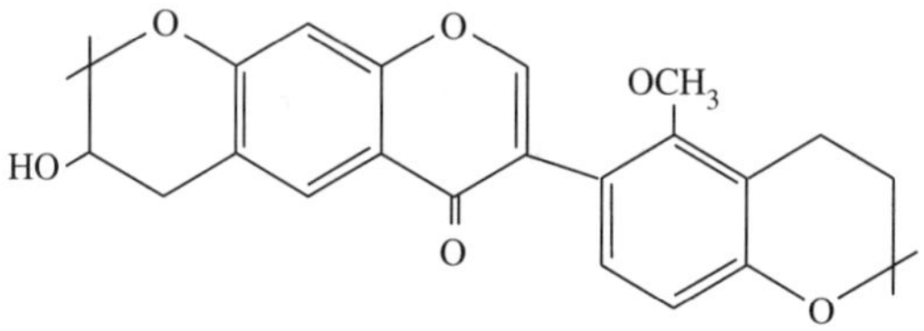

Figure 77. Mundulone.

573. *Musa ferra* Linn. (Musaceae)

This is an annual herb found in tropical zones and commonly available in Central and South India, Sri Lanka and Malaysia. Its leaf extract in petroleum ether showed repellency to the cotton stainer, *Dysdercus koenigii* (Krishnamurthy Rao, 1982).

574. *Neorantanenia fisitolia* Schnz. (Fabaceae)

This is an annual tropical herb found to show insecticidal activity in its aqueous root extract against the black bean aphid, *Aphis rumicis* (Michael et al., 1985).

575. *Neorantanenia pseudopachyrhiza* Schnz. (Fabaceae)

This is an annual tropical herb found to show insecticidal activity in its aqueous root extract against the black bean aphid, *Aphis rumicis*. An active component, 'pachyrhizin' ($C_{19}H_{12}O$), was reported in its root extract to possess insecticidal property (Crosby, 1971).

576. *Nepeta cataria* Linn. (Lamiaceae)

Common name: Catnip or Catmint

This is an annual aromatic herb found in tropical zones. In India, it is available in Jammu and Kashmir but is native to Europe. Its root, leaf and flower extracts in water and dried powder of its flower were reported to show insecticidal and also antifeedant activities when tested against the imported cabbage worm, *Pieris rapae* and Colorado potato beetle, *Leptinotarsa decemlineata* (Mathews, 1981).

Essential oil of *Nepeta* spp. contained diasteroisomers of nepetalactone, which showed repellency to ants (Velasco-Negueruela et al., 1981).

577. *Nepeta subsessilis* Linn. (Lamiaceae)

Common name: Catnip genus

This is an annual herb found in temperate and semi-arid zones. Its root, branch and leaf extracts in water were toxic to the vinegar fly, *Drosophila hydei* (Khan, 1982).

578. *Nephrolepis cingulatus* Schoot. (Polypodiaceae)

This is an annual/perennial fern found in tropical and sub-tropical zones. Its whole plant extract in ethanol was reported to show insecticidal activity when tested against the cotton stainer, *Dysdercus cingulatus* (Gunasena, 1983).

579. *Nephrolepis exaltata* Shoot. (Polypodiaceae)

This is a perennial herb found in tropical zones and commonly available in South India. Its leaf extract in petroleum ether showed juveno-mimetic activity to 5th instar larvae of the cotton stainer, *Dysdercus cingulatus* and caused morphogenetic changes in adults during their development (Rajendran and Gopalan, 1978).

580. *Nerium indicum* Mill. (Apocynaceae)

(Syn.-*N. odourum* Soland., *N. oleander* Blanco)

Common name: Oleander or Kaner

This is a perennial/biennial evergreen shrub found in tropical, sub-tropical, temperate and arid zones but native to the Mediterranean region. Oleander is distributed in North India, Nepal, Bhutan, South China, Indonesia and Australia. Its dried leaf powder was reported to protect wheat from infestation of the red flour beetle, *Tribolium castaneum* when admixed with the stored grains (Atwal and Sandhu, 1970). Whole plant extract or extracts of its root, branch, leaf and flower in water, ether and alcohol were toxic to the vinegar fly, *Drosophila hydei* and the rice weevil, *Sitophilus oryzae* (Jacobson, 1975).

Pandey and Singh (1976) also found its aqueous leaf extract to show toxicity to *Callosobruchus chinensis* and *C. maculatus*.

581. *Nicandra physaloides* (Linn.) Goertn. (Solanaceae)

Common name: Apple of Peru or Peruvian ground cherry

This is an annual herb found in tropical and sub-tropical zones. It is available in hilly tracts of the Himalayas and West Deccan Peninsula in India. Its leaf extract in water showed insecticidal and antifeedant activities against the tomato hornworm, *Manduca sexta* (Jacobson, 1975); Colorado potato beetle, *Leptinotarsa decemlineata* (Britski, 1982); Mexican bean beetle, *Epilachna varivestis* (Waespe and Hans, 1979); and the green house whitefly, *Trialeyodes vaporariorum* (McIndoo, 1983).

An active component, 'nicablin A' (Figure 78), a steroid isolated from its leaf extract, showed antifeedancy against *Epilachna varivestis* (Ascher et al., 1981; Ascher, 1983).

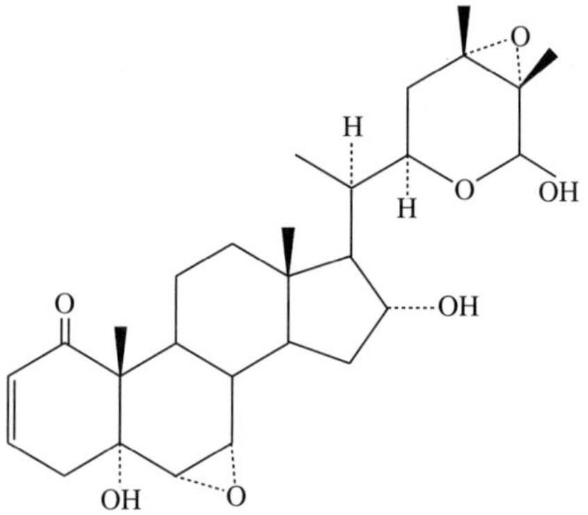

Figure 78. Nicablin A.

582. *Nicotiana benthanian* Domin. (Solanaccae)

This is an annual or seasonal herb found in temperate and sub-tropical zones. Its liquid and sticky exudates were reported to be toxic to the tobacco horn caterpillar, *Manduca sexta* when tested under controlled conditions. In addition, exudates from other species of tobacco plant, i.e., *N. reparda* Willd., *N. nesphila* Johnst and *N. stocktonii* Brandog, were also found toxic to *M. sexta* (Parr and Thurston, 1968; Thurston, 1970).

583. *Nicotiana glauca* Linn. (Solacaneae)

Common name: Mastard tree

This is a perennial shrub/small tree found in temperate and sub-tropical zones. Its stem, branch and leaf extracts were reported to be toxic to aphids (Usher, 1973).

584. *Nicotiana gossei* Domin. (Solanaceae)

This is an annual herb found in tropical zones and commonly cultivated in Andhra Pradesh and Tamil Nadu in India. Its leaf extract in water was reported to be toxic to tobacco leaf eating caterpillar, *S. litura* (Joshi and Sitaramiah, 1979); the whitefly, *Bemisia tabaci* (Patel et al., 1976); the stem borer, *Scrobipalpa heliopa* (Patel et al., 1975); and the green peach aphid, *Myzus persicae* (Joshi et al., 1978).

The various fractions of its leaf extract were tested for their biological activity to the early instar larvae of *S. litura* and cytoplasmic polar lipid fraction was found to be the most toxic fraction. Its active components were identified as a mixture of saponins apart from the alkaloids present in the leaf extract (Prabhu et al., 1981).

585. *Nicotiana rustica* Linn. (Solanaceae)

Common name: East-Indian tobacco or Turkish tobacco

This is an annual herb found in tropical, semi-arid and alpine zones. It is a native to Central America and now available in Punjab, Bihar and West Bengal in India. Its root, branch and leaf extracts in water were reported to be toxic to the whitefly, *Bemisia tabaci* (Jacobson, 1975); the winter moth, *Cheimatobia bruma*; and the apple sucker psyllid, *Psylla mali* (McIndoo, 1983).

586. *Nicotiana sylvestris* Speggazini & Comes (Solanaceae)

Common name: Wild tobacco

This is a perennial shrub found in tropical and sub-tropical zones and available in Central and North America. Its leaf extract in water was reported to be toxic to the bean aphid, *Aphis fabae* (McIndoo, 1983). A number of pyridine containing alkaloids isolated from its leaf extract were reported to be toxic to the insects (Gordon, 1961; Schmeltz, 1971). Parr and Thurston (1972) showed that tobacco hormworms, despite their ability to efficiently excrete, ingested tobacco alkaloids. Self et al. (1964) reported that insects reared on diets containing 2% or higher concentrations of nicotine had significantly reduced survival of their larvae. The alkaloid content in undamaged tobacco leaves was reported to increase many-fold if the plant was damaged or top leaves were chopped off. The increase in alkaloid content is of great use for pesticidal commercialization. The induced alkaloidal response appeared to be triggered by a phloem-borne cue in its roots that allowed the plant to distinguish between different types of leaf damage (Baldwin, 1988, 1989; Baldwin et al., 1990).

587. *Nicotiana tabacum* Linn. (Solanaceae)

Common name: Tobacco

This is an annual herb found in tropical and semi-arid zones. It is native to Central America and now grown in Andhra Pradesh, Uttar Pradesh, Maharastra, West Bengal and Karnataka in India and several other countries in Europe and Australia. It is reported that in 1560 John Nicot sent seeds to the French King, describing them as germs of a medicinal plant of great value. Thus the value of tobacco, a plant introduced into Europe about 1560 as an insecticide, was known to Peter Collinson, who in 1764 wrote from England advising Bartram, an American botanist, to use water in which tobacco leaves were soaked against the plum curculionids (Waite, 1925). Tobacco leaf extract in water was used to protect crops from aphids and other soft bodied insects (Charles, 1929). Giles (1974) reported tobacco leaf powder to protect stored wheat from insect attack when admixed with the grains. These dried leaves were also reported to protect stored paddy against Angoumois grain moth, *Sitotroga cerealella* when admixed with the grains (Satpathy, 1983). Its whole plant, leaf, branch and root aqueous extracts were reported to show insecticidal and repellent activities against the jassid, *Amrasca* sp.; the cowpea aphid, *Aphis craccivora*; the green peach aphid, *Myzus persicae*; and the citrus leaf miner, *Phyllocnistic citrella* (Krishnamurthy Rao, 1982).

Active Component

The active component (an alkaloid called nicotine) of toxic property was traced in the tobacco plant as early as 1828 by Posselt and Reimann. In the market, both free alkaloid and as the sulphate were sold under the name of 'Black-leaf-40' containing 40% of the base or 40% of nicotine sulphate. Reddy and Venugopal Rao (1990) isolated nicotine sulphate from this plant and found it to protect the cotton crop from *Aphis gossypii*. Similarly, Prabhu et al. (1990) and Patil et al. (1990) reported nicotine sulphate isolated from waste tobacco leaves to show toxicity to *Heliothis armigera*, *Spodoptera litura*, *Myzus persicae* and *Bemisia tabaci*. Two formulations of tobacco, viz., nicotine 40% solution and 10% DP, have been registered in India for their commercial use and export (Parmar and Devakumar, 1993).

Nicotine (Figure 79), a colourless liquid, acts as a fumigant due to its volatility and also penetrates directly through insect integument. Nicotine preparation were reported to be toxic to lepidopterous pests of the turk crops (Swingle and Cooker, 1935). Nicotine dust was first used to control the orchard pests in California in 1917 using equal parts of ground tobacco and sulphur and was found very effective to kill the aphids, *Chromaphis juglandicola*. The spray application of nicotine with oils or soaps as carrier of nicotine were commonly used (De-Ong, 1948). Nicotine also acted as nonpersistent contact insecticide against the aphids, jassids, leaf-miners, codling moth and thrips on a wide range of crops (Cremlyn, 1978). However, its use rapidly declined and it was replaced by synthetic insecticides due to its high mammalian toxicity (LD50 [oral] to rats = 50 mg/kg) (De-Ong, 1971) and lack of effectiveness in cold weather.

Chemical structures of nicotine ($C_{10}H_{14}N_2$) and its isomer components were established as 3-(1-methyl-2-pyrrolidyl pyridine I). This component was first isolated in 1828 and its structure was elucidated in 1893 (Hartley and West, 1969; Martin, 1973). Other alkaloid components, which possessed insecticidal activity were nornicotine (Figure 79), neonicotine or anabasine (Figure 80), nicotyrine (Figure 81) and metanicotine (Figure 82). Nicotine, noricotine and neonicotine were the potent insecticides but nicotyrine and meta-nicotine were comparatively less toxic to *Aphis rumicis*.

Figure 79. Nicotine.

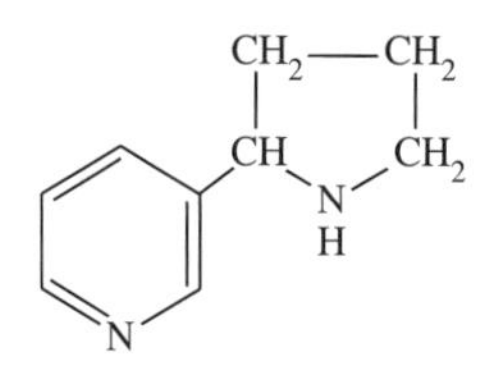

Figure 80. Nornicotine.

Figure 81. Nicotyrine.

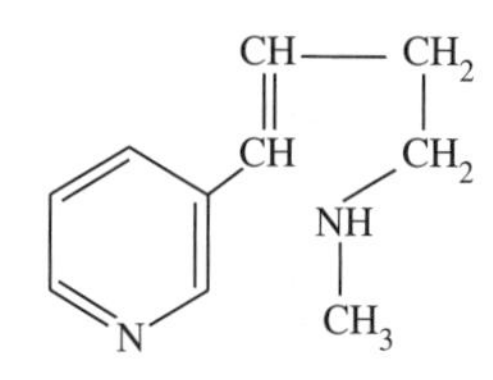

Figure 82. Metanicotine.

588. *Nigella sativa* Linn. (Ranunculaceae)

Common name: Black-cumin

This is an annual herb found in temperate and Mediterranean zones and available in South India and Sri Lanka. Its leaf extract in water was reported to show insecticidal and repellent activities against the Japanese beetle, *Popillia japonica* (McIndoo, 1983). Its seed extract in petroleum ether also showed toxicity to the defoliating beetle, *Epilachna vigintioctopunctata* when tested under laboratory conditions (Chandel et al., 1987). Despande et al. (1974) reported oleic and linoleic acid as insecticidal components from this plant, which were found to be toxic to the pulse beetle, *Callosobruchus chinensis*.

589. *Nyctanthes arbontristris* Linn. (Nyctanthaceae)

Common name: Tree of sorrow

This is a large perennial shrub or small ornamental tree grown in tropical and sub-tropical zones and commonly available in the Himalayan tract of India. Its 50 and 100% crude leaf extract in water were found to show significant antifeedancy to the 3rd instar larvae of the jute semi-looper, *Anomis sabulifera* and the 5th instar larvae of Bihar hairy caterpillar, *Spilosoma obliqua* (Sharma et al., 1992).

590. *Nymania capensis* K.Schum. (Euphorbiaceae)

This is a perennial herb widely distributed in tropical, sub-tropical and temperate regions. From its bark, stem and wood extracts the active component 'prieurianin' was isolated and reported to show strong antifeedant activity against the tobacco bud worm, *Heliothis virescens*; the southern armyworm, *Spodoptera eridania*; and Mexican bean beetle, *Epilachna varivestis* (Lidert et al., 1985).

591. *Nyssa sylvatica* Marsh. (Nyssaceae)

Common name: Sour or Cotton gum tree or Ogeechu lime

This is a perennial tree found in sub-tropical and temperate zones and available in North America. Its seed powder in ethyl ether showed attractancy to the female and the male of the melon fly, *Dacus cucurbitae* (Keiser et al., 1975).

592. *Ocimum americanum* Linn. (Lamiaceae)

(Syn.-*O. canum* Sims.)

Common name: Hoary basil or Black basil

This is an annual herb found in tropical and sub-tropical zones and commonly available in Rwanda and Burundi in Africa. Its whole plant and leaf extracts in water were found to protect the grains from storage insect infestation (Giles, 1964) and also showed toxicity to the hairy caterpillar, *Euproctis fraterna* (McIndoo, 1983). Whole leaves of this annual mint are usually added to stored foodstuffs to prevent insect damage within the traditional closed structures in Rwanda (Dunkel et al., 1986). Prakash and Rao (1995b) recommended dried leaves of *O. canum* to be admixed with stored rice @ 1% w/w to protect it from damage caused by coleopterans in rural storage of eastern India. Linalool (3,7-dimethyl-1,6-Octadien-3-ol), a major component of its essential oil, representing 60–90% of the total volatiles collected (Ntezurubanza, 1987), was reported to be biologically active and showed toxicity to *Sitophilus oryzae* and *Zabrotes subfasciatus* (Weaver et al., 1991).

593. *Ocimum basilicum* Linn. (Lamiaceae)

Common name: Sweet basil or Forest basil

This is an annual aromatic herb found in tropical and Mediterranean zones. This is available throughout India, Sri Lanka, Malaysia, Indonesia and Japan. Its seed extract was found to show toxicity to the 5th instar larvae of the potato tuber moth, *Gnorimoschema operculella* (Pandey et al., 1982). Its whole plant and leaf extracts and oil were reported to be toxic to cotton aphid, *Aphis gossypii* (Jacobson, 1975) and showed growth inhibiting activity to the cotton stainer, *Dysdercus cingulatus* (Balasubramanian, 1982) and repellency to Colorado beetle, *Leptinotarsa decemlineata* (McIndoo, 1983). Kareem (1984) found its stem and leaf extracts to inhibit the development and to reduce the longevity of the cotton stainer, *Dysdercus cingulatus* and *Spodoptera litura*. However, its leaf odour was reported to decline the hatchability of *Earias fabia* when exposed to moth held during mating in its environment. Mohiuddin et al. (1987) found its oil to show repellency to the red flour beetle, *T. castaneum* under laboratory test. Methanolic extract of this plant also showed toxicity to 3rd instar larvae of *Spodoptera litura* (Deva Prasad, 1990). Shaaya et al. (1991) found its essential oil to show fumigant toxicity to saw-toothed grain beetle, *Oryzaephilus surinamensis*.

Jacobson (1989) reported that its whole plant extract contained clerodanes known as juvocinine-I and juvocinine-II, which were found to show juvenile hormone analogue mimic activity against milkweed bug, *Oncopeltus fasciatus*. Methyl chavicol, a chemical constituent of sweet basil oil, also reduced oviposition of rice brown planthopper, *Nilaparvata lugens* (Subaharan and Nachiappan, 1994).

594. *Ocimum gratissimum* Linn. (Lamiaceae)

Common name: Shrubby basil (Ramtulsi)

This is an annual or perennial shrub found in tropical zones and commonly available in India. Bhattacharya and Bordoloi (1986) reported its essential oil to show growth retarding activity by inhibiting the pupal formation and adult emergence of *D. melanogaster* and *T. castaneum* when tested incorporating in their diets. Its seed extract in petroleum ether also reduced oviposition of brown planthopper, *Nilaparvata lugens* when sprayed on the rice plant as 5% solution (Reddy and Urs, 1988).

595. *Ocimum sanctum* Linn. (Lamiaceae)

Common name: Holy basil

This is an annual herb found in tropical zones and available throughout India and other South East Asian countries, South Africa and Australia. Oil extracted from its leaves was reported to be an attractant to the male of the mango fruitfly, *Dacus correctus* (Shah and Patel, 1976). Rajendran and Gopalan (1979) reported its leaf and whole plant extracts to show insecticidal activity to the 5th instar larvae of the cotton stainer, *Dysdercus cingulatus*; the 3rd instar larvae of the cotton leaf armyworm, *Spodoptera litura*; and a castor pest, *Pericalia ricini*. Volatiles from its leaf extract were found to inhibit oviposition of the cotton leafhopper, *Amrasca devastans* (Saxena and Basit, 1982). Its leaf and stem aqueous extracts were also reported to be toxic to the cotton leaf armyworm, *Spodoptera litura* (Balasubrahmanjan, 1982) and the cotton stainer, *Dysdercus cingulatus* (Kareem, 1984). Verma and Singh (1985) further observed its aqueous leaf extract to show antifeedant activity to the hairy caterpillar, *Amsacta moorei*. Aqueous leaf extract of holy basil when sprayed on the rice plant was found to reduce survival of the green leafhopper, *Nephotettix virescens* and rice tungro virus (Narsimhan and Mariappan, 1988). Mallick and Banerji (1989) reported 5 and 10% extracts of its leaf in methanol to show antifeedant activity to the final instar larvae of the jute semi-looper, *Anomis sabulifera* when tested under laboratory conditions on the treated leaves of the jute. From its oil, an active component, 'methyl eugenol', was isolated and found to attract the males of the mango fruit fly, *Dacus correctus* (Shaha and Patel, 1976).

596. *Ocimum suave* Linn. (Lamiaceae)

This is an annual/perennial shrub found in tropical Africa. Its leaf powder was reported to check the multiplication of the storage insects. An active component, 'eugenol', was found in its leaf extract, which was reported to show repellent activity against the corn weevil, *Sitophilus zeamais* (Prakash et al., 1978–1987).

597. *Oenanthe crocata* Linn. (Apiaceae)

This is a perennial herb found in temperate, tropical, Mediterranean and alpine zones. Its root extract was reported to show insecticidal activity to the cabbage white butterfly, *Pieris brassicae* (McIndoo, 1983).

598. *Oldenlandia umbellata* Linn. (Rubiaceae)

Common name: Indian madar

This is an annual/perennial herb found in tropical zones and distributed in eastern and southern India and Bangladesh. Its dried root powder was reported to inhibit the development and population build up of the flour beetle, *Tribolium castaneum* in stored rice (Prakash et al., 1978–1987).

599. *Onsomodium occidentale* Michx. (Boraginaceae)

This is a perennial herb found in temperate zones and commonly available in Mexico and the United States. Its leaf extract was reported to show insecticidal activity to the red-legged grasshopper, *Melanoplus femmurubrum* when tested under controlled conditions (Jacobson, 1975).

600. *Origanum majorana* Linn. (Lamiaceae)

(Syn.-*Majorana* hortensis Moench.)

Common name: Sweet majoram

This is an aromatic perennial herb found in tropical, Mediterranean and arid zones. This is a native to South Europe but available in Central America and Asia Minor. Its leaf extract in acetone and oil was reported to show insecticidal and repellent activities to the cotton leaf armyworm, *Spodoptera litura* (Krishnamurthy Rao, 1982).

601. *Orixa japonica* Thunb. (Rutaceae)

This is an annual/perennial shrub found in temperate and sub-tropical zones and commonly available in North America, Japan and North China. Its whole plant extract was reported to be toxic to the cotton leaf armyworm, *Spodoptera litura* (Waespe and Hans, 1979). Jacobson (1989) reported its root bark extract to show antifeedant activity against *Spodoptera litura* and *Callosobruclus phaseoli*. Yajima et al. (1977) isolated coumarins as insect antifeeding principles form this plant against lepidopteran larvae.

602. *Orophea katschalica* Kurtz. (Annonaceae)

Common name: Tonyoge plant
This is a perennial tree found in tropical and sub-tropical zones. Its leaf extract was reported to repel the honey bee, *Apis dorsata* (Bhargava, 1983).

603. *Orthodon grossesarratus* Benth. & Oliv. (Lamiaceae)

This is a perennial herb found in the temperate and sub-tropical zones and commonly available in the hills of the Himalayas in India. Its branch, leaf and flower aqueous extracts were reported to be toxic to *Drosophila hydei* (Jacobson, 1958).

604. *Oryza sativa* Linn. (Poaceae)

Common name: Rice
This is a biannual herb grown as an important cereal crop in tropical and sub-tropical zones and distributed in all South East Asian countries, China, Japan, the Philippines, Australia and Central America. Rice is cultivated throughout India and known to show biological activities against a wide range of insects as follows:

Biological Activity Against Storage Insects

Its bran extract in ether exposed to the larvae of Indian meal moth, *Plodia interpunctella* terminated their diapause (Tsuji, 1966). Rice bran oil was also found to show toxicity to the almond moth, *Ephestia cautella* (Golob and Ashman, 1974). Its bran oil, however, was reported to inhibit the multiplication of the storage insects (Anonymous, 1979). Doharey et al. (1983, 1984a,b, 1987) found rice bran oil to protect the greengram seed in storage against the pulse beetles, *Callosobruchus chinensis* and *C. maculatus* and reported it to show repellancy to these beetles by reducing their oviposition and developmental period in the stored seeds. Jain and Yadav (1990) found paddy husk ash to enhance the toxicity of deltamethrin by 3.2-fold as compared to other diluents like pyrophyllite against the pulse beetle, *Callosobruchus chinensis*.

Singh et al. (1993) found rice bran oil to protect gram seed from the attack of *C. chinensis* when treated at the rate of 3 ml/kg.

Biological Activity Against Field Insects

The extract of rice variety TKM-6 strongly inhibited the oviposition of the moths of the striped stem borer, *Chilo suppressalis* and caused juvenile hormonal effects on its 3rd instar larvae (Dhaliwal et al., 1990a). However, extracts of Taitung-16 and Chianan-2 showed high mortality to its 5th instar larvae and pupae. The pupae also developed to the deformed moths without chitinous scales on their abdomen. Further, Dhaliwal et al. (1990b) reported steam distillate extracts of 90-day-old Taitung-16 (a moderately resistant rice variety) fractioned in a solvent mixture of hexane and diethyl ether (9:1) to be highly toxic to *C. suppressalis* and *N. lugens*.

Active Components

Sterols and asparagin were reported to inhibit the development of the brown plant hopper, *Nilaparvata lugens*, a pest of rice plant (Yoshio et al., 1982). An active component, 'pentadecanol', isolated for a resistant rice variety TKM-6 was reported to cause absolute mortality to the rice stem borer larvae, which was one of the major factors for resistance in rice to the striped stem borer, *Chilo suppressalis* (Dale, 1990, 1992).

605. *Osmunda japonica* Thunb. (Osmundaceae)

Common name: Japanese flowering fern
This is an annual/perennial flowering fern found in Japan. Its whole plant extract in ether showed antifeedant activity when tested against the larvae of the yellow butterfly, *Eurema hecabe mandarina* under controlled conditions. Osmundalactone and 5-hydroxy-2-hexan-4-olide were found to be the active components in its extracts to show antifeedancy (Numata et al., 1983).

606. *Ostryoderris gabonica* Dunn. (Leguminaceae)

This is an annual vine found in tropical zones. Its root and branch extracts in water were reported to show toxicity to the bean aphid, *Aphis fabae* (McIndoo, 1983).

607. *Ougeinia dalbergioides* Benth. (Fabaceae)

(Syn.-*O. oojeinensis* [Roxb.] Hochr.)
Common name: Sandan
This is a perennial small to medium size tree found in tropical, sub-tropical and temperate zones and commonly available in the outer Himalayas and

Northern and Central States of India. Its leaf and bark extracts and also their dried powders were toxic to the pulse beetle, *Callosobruchus chinensis* (Puttarudriah and Bhatia, 1955). Its leaf extracts in water, ethanol and diethylether were also reported to be insecticidal to the cabbage leaf webber, *Crocidolomia binotalis*; the hairy caterpillar, *Euproctis fraterna*; the cotton leaf armyworm, *Spodoptera litura*; and the diamond-back moth, *Plutella xylostella* (McIndoo, 1983).

608. *Oxalis deppei* Lodd. (Oxalidaceae)

Common name: Lucky clover

This is an annual/perennial herb found in tropical zones and commonly available in South America and Mexico. Aqueous leaf extract of this herb was reported to show antifeedant and growth inhibiting activities to the diamond-back moth, *Plutella xylostella* (Jacobson, 1975).

609. *Pachyrhizus angulatus* Rich. ex. D.C. (Fabaceae)

(Syn.-*P. erosus* [Linn.] Urb.)

Common name: Chinese yam or Yam bean

This is an annual/perennial vine found in tropical zones. It is native to Central America but commonly available in West Bengal, Orissa, Andhra Pradesh, Assam and Bihar in India. Its whole plant, fruit and seed extracts in water, alcohol or ether and also their latex/sap were reported to show insecticidal, antifeedant and growth inhibiting activities to the green stink bug, *Acrosternum hilare*; the bean beetle, *Cerotoma ruficornis*; the melon worm, *Diaphania hyalinata*; the fall armyworm, *Spodoptera frugiperda* (Jacobson, 1975); the common cutworm, *Spodoptera litura* (Hameed, 1983); the turnip aphid, *Hyadaphis erysimi* (Yingchol, 1983); and also to the bean aphid, *Aphis fabae* and Mexican bean beetle, *Epilactna varivestis* (McIndoo, 1983).

Jacobson and Crosby (1971) reported a rotenoid, 'pachyrrhizone' ($C_{20}H_{14}O_7$) (Figure 83), in its fruit and seed extracts which showed toxicity to *E. varivestis*, *M. persicae*, *Aphis rumicis* and *Macrosiphum psi*. Another active component obtained by oxidizing and hydrolytic procedure from pachyrrhizone, called 'pachyrrhizin' ($C_{19}H_{12}O_6$) (Figure 84), also found toxic to *E. varivestis* and the imported cabbage worm, *Pieris rapae* (Jacobson and Crosby, 1971).

610. *Pachyrhizus palmatibolus* Spreng. (Fabaceae)

This is an annual/biennial herb found in tropical zones. Its aqueous seed extract showed toxicity to *D. hyalinata*; *Dysdercus* spp. and *S. frugiperda* (Jacobson, 1975).

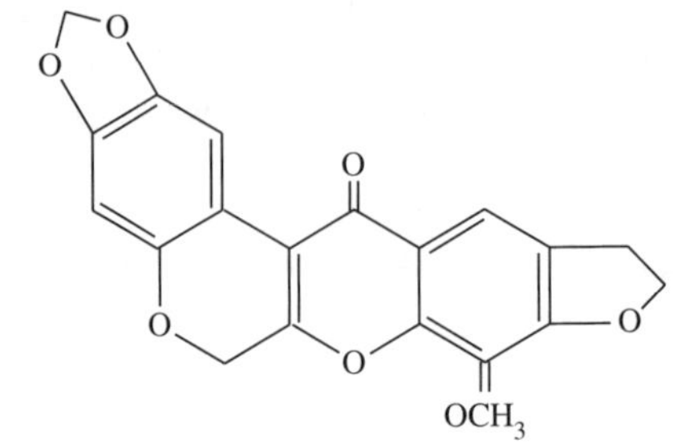

Figure 83. Pachyrrhizone.

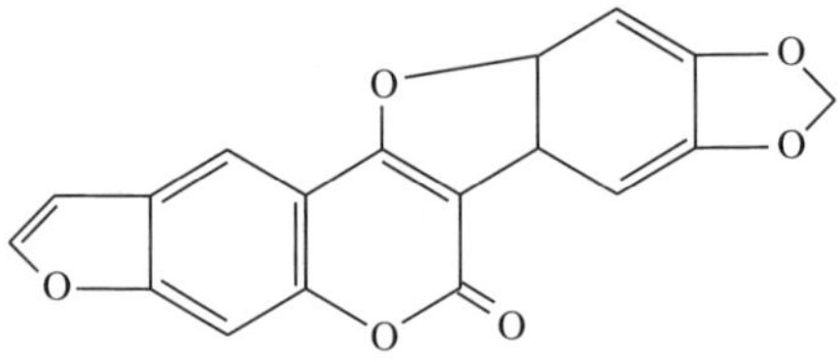

Figure 84. Pachyrrhizin.

611. *Pachyrhizus tuberosum* Spreng. (Fabaceae)

Common name: Yam bean

This is a perennial shrub found in tropical zones and native to South America. Dry powder of its seed was found to be toxic to the cabbage aphid, *Brevicoryne brassicae* and Mexican bean beetle, *Epilachna varivestis* (Jacobson, 1975). Pachyrrhizone, a toxic rotenoid isolated from its fruit and seed powders, was also reported to show insecticidal activity against the cabbage worm, *Pieris rapae* (Crosby, 1971).

612. *Padus racemosus* Mill. (Rosaceae)

Common name: European bird cherry

This is a perennial shrub found in temperate and sub-tropical zones and commonly available in North and South America and Europe. Its whole plant or bark extracts in water were reported to be toxic to the midge pests of agricultural importance (Jacobson, 1975).

613. *Paeonia suffruticosa* Andr. (Paeoniaceae)

This is a perennial rhizomatous herb found in temperate zones and commonly available in North America. Its stem and leaf extracts in ethyl ether were reported to attract both sexes of the Mediterranean fruit fly, *Ceratitis capitata* and the melon fly, *Dacus cucurbitae* (Keiser et al., 1975).

614. *Parabenzoin tribolum* Nakai (Lauraceae)

This is a perennial shrub found in temperate and alpine zones. Its whole plant or leaf extracts in water were reported to be antifeedant to the cotton leaf armyworm, *Spodoptera litura* (Waespe and Hans, 1979).

615. *Parthenocissus guinguafolia* (L.) Planch. (Vitidaceae)

Common name: Virginia creeper

This is a perennial woody climber of temperate and sub-tropical regions and commonly found in Virginia. Its leaf extracts in acetone and water were reported to show toxicity to the woolly aphid of sugarcane, *Ceratovacuna longigera* (McIndoo, 1983).

616. *Partthenium hysterosphorus* Linn. (Asteraceae)

Common name: Carrot grass

This is a biennial/annual herb found in tropical zones and commonly available in South Indian States. Tilak (1977) reported its whole plant aqueous extract to show toxicity to a castor pest, *Pericalia ricini* when tested under control conditions. Its whole plant and leaf extracts in petroleum ether were also reported to show juveno-mimetic activity to the 5th instar larvae of the cotton stainer, *Dysdercus cingulatus* and as a result showed morphogenetic changes in the larvae (Rajendran and Gopalan, 1978; Balsubramanian, 1982) and also showed antifeedant activity to the cotton leaf armyworm, *Spodoptera litura* (Balasubramanian, 1982) and brinjal spotted leaf beetle, *Henosepilachna vigintioctopunctata* (Dhandapani et al., 1985) and toxicity to the cabbage leaf webber, *Crocidolomia binotalis* and the migratory grasshopper, *Melanoplus sanguinipes* (Fagoonee, 1983). Kareem (1984) reported its whole plant extract to inhibit the growth and development of the cotton stainer, *Dysdercus cingulatus*. Bhaduri et al. (1985) found its leaf extracts in petroleum ether, benzene and alcohol to show significant grain protection against the pulse beetle, *Callosobruchus maculatus* infestating cowpea seeds.

617. *Passiflora mollissima* (Hb. & Kunth.) Bailey (Passifloraceae)

(Syn.-*Tacsonia mellissina* Hb. & Kunth.)

Common name: Banana passion fruit

This is a perennial small tree found in Tamil Nadu in India and available in tropical and sub-tropical regions. Tripathi et al. (1987) reported antifeedant activity in its whole plant extract against the jute hairy caterpillar, *Spilosoma obliqua*.

618. *Pastinaca sativa* Linn. (Apiaceae)

Common name: Parsnip
This is an annual herb found in temperate zones. Its leaf extract was reported to possess 'myristicin', an active component toxic to the fruit fly, *Dacus* sp. (Lavie et al., 1967).

619. *Paullinia pinnata* Linn. (Sapindaceae)

Common name: Nistmal
This is a perennial tree found in sub-tropical zones. Its alcoholic seed extract was reported to be toxic to the aphids (Worsley and Le, 1934).

620. *Pavonia zeylanica* Cav. (Malavaceae)

This is an annual/perennial shrub or small tree found in tropical and sub-tropical zones. Its whole plant extract in petroleum ether was reported to show juveno-mimetic activity and also showed morphogenetic changes in 5th instar larvae of the cotton stainer, *Dysdercus cingulatus* (Rajendran and Gopalan, 1978). Further, its flower extract in petroleum ether was found to be toxic to *D. cingulatus* (Balsubramanian, 1982). However, its leaf and stem extracts inhibited the growth and development of this bug (Kareem, 1984).

621. *Pelargonium sp.* (Geraniaceae)

Common name: Geranium
This is a perennial shrub found in sub-tropical zones. Its branch and flower extracts and oil from its extract distillation were reported to show antifeedant, repellent and attractant activities when tested against the diamond-back moth, *Plutella xylostella* (Jacobson, 1975); the screw worm, *Cochliomyia hominivorax*; and the Japanese beetle, *Popillia japonica* (McIndoo, 1983).

622. *Pelargonium gravcolens* (Linn.) Herit. (Geraniaceae)

Common name: Rose geranium
This is a perennial shrub found in temperate zones. It is native to Cape Province in South Africa and grown chiefly in France, Spain, Malagasy, Congo, USSR and Morocco as an oil crop. Acetone leaf extracts of rose geranium were found to show insecticidal activity to the cotton stainer, *Dysdercus cingulatus* (Balasubramannian, 1982). The leaf and stem extracts of this plant and its essential oil were also reported to inhibit the growth and development and to reduce the longevity of *D. cingulatus* and the cotton leaf armyworm, *Spodoptera litura* when tested under controlled conditions (Kareem, 1984).

623. *Pellionia pulchera* Gaudich. (Utricaceae)

This is an annual herb found in temperate zones. Its whole plant aqueous extract was reported to be toxic to the moths of general importance (Jacobson, 1975).

624. *Pellionia scabra* Wedd. (Utricaceae)

This is an annual herb found in temperate and sub-tropical regions. Its root, branch and leaf extracts in water were reported to be toxic to the vinegar fly, *Drosophila hydei* (Jacobson, 1975).

625. *Perezia nana* Lag. (Asteraceae)

This is an annual herb found in sub-tropical and semi-arid zones. Its meristamatic tissue extract and also its dried powder were reported to be toxic to the southern beat webworm, *Pachyzancla bipunctalis* and the southern armyworm, *Spodoptera eridania* (Jacobson, 1975).

626. *Petiveria allicaea* Linn. (Phytolaccaceae)

Common name: Guinea henweed

This is an annual herb found in tropical zones. It is native to America ?? but was later introduced to India. Its whole plant and root extracts were reported to show antifeedant activity when tested against the leaf cutting ants (Jacobson, 1975).

627. *Petunia axillaris* Lam. (Solanaceae)

This is an annual herb available in tropical and sub-tropical regions. Its leaf extract in ethanol showed toxicity to the tobacco hornworm, *Manduca sexta* (Thurston, 1970). The active components in its leaf extract were found to be steroidal ketones called petuniasterons (Elliger and Waiss, 1990).

628. *Petunia hybrida* Vilim. (Solanaceae)

Common name: Petunia

This is an annual herb found in tropical and temperate zones. Foliage of petunia plants is poisonous to various insects. Fraenkel et al. (1959) indicated that although tobacco hornworm, *Manduca sexta* larvae found this plant material to be highly acceptable (i.e., attractive or at least non-repellent) as food, growth of the caterpillars was not supported and premature death occurred. It was also stated that petunia was toxic to Colorado potato beetle, *Leptinotarasa decemlineata*. Parr and Thurston (1968) and Thurston (1970) confirmed that several species of petunia were extremely toxic to 1st and 2nd instar larvae of *M. sexta*. Its whole plant extract in water also showed toxicity to the tomato hornworm, *Manduca sexta* (Jacobson, 1975). Elliger and Waiss (1989, 1990) mentioned its leaf extract wash in ethanol to show antifeedant activity against *M. sexta*.

Petunia sp. contained an array of more than three dozen steroidal components formally related to ergostane, which were involved in the resistance of the plant toward the attack of lepidopteran larvae. Among these components are A-ring dienones and 1-acetoxy-4-en-3-ones. Epoxy functionalities could be found on the side chain in certain cases on the A and D rings. Hydroxy and acetoxy substituents may be located at various positions upon the steroid nucleus and the side chain. In a large number of examples, the side chain possesses a bicyclic orthoester moiety, which in certain cases could also have a thiolester functionality attached. Alteration of A from the usual steroid pattern yielded to a set of components, which had spirolactone at position-5. The biological activity of these substances was dependent upon various structural features. Most notably the orthoester was essential for toxicity of the components towards insects. These components were steroidal ketones called 'petuniasterones', which showed potent insect development inhibitor activity when tested against the corn earworm, *Heliothis zea*. Based on the ergostane skeleton, thirty seven petuniasterones were isolated from several species of *Petunia* (Elliger and Waiss, 1989). Petuniasterone A (Figure 85) showed many typical structural features. All these components differed in their inhibitory potential when fed to the larvae of the insect with the artificial diets but only those compounds having a bicyclic orthoester system in their side chain could be active to reduce the larval growth.

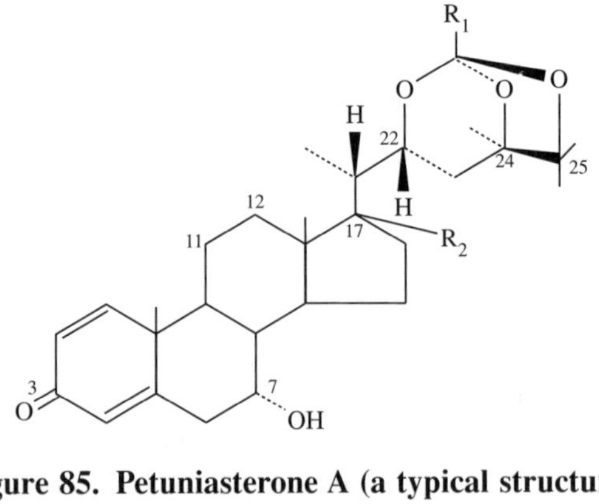

Figure 85. Petuniasterone A (a typical structure).

629. ***Petunia inflata*** **Fries (Solanaceae)**

Common name: White petunia

This is an annual herb found in temperate zones. Its leaf juice and exudate were reported to show insecticidal activity to the tomato hornworm, *Manduca sexta* (Thurston, 1970; Jacobson, 1975) and also to the cabbage white butterfly, *Pieris brassicae* (McIndoo, 1983).

630. *Petunia violacea* Lindl. (Solanaceae)

This is an annual herb found in tropical and temperate zones. Its leaf exudate and extract in alcohol were reported to be toxic to the tomato hornworm, *Manduca sexta* (Thurston, 1970; Jacobson, 1975; Elliger and Waiss, 1990). Petuniasterones have also been extracted from its leaf extract and were found to show insect growth inhibitory activity under controlled conditions (Elliger and Waiss, 1990).

631. *Peucedanum sp.* (Apiaceae)

Common name: Bonjira

This is an annual herb found in tropical zones and distributed in India, Sri Lanka, Pakistan, Malaysia and Indonesia. Its whole plant and seed extracts in water were reported to show antifeedant activity when tested against the tobacco whitefly, *Bemisia tabaci* (Islam, 1983).

632. *Phellodendron amurense* Rupr. (Rutaceae)

This is a perennial tree found in temperate zones and commonly available in North America and Europe. Its fruit extract in acetone was reported to be toxic to the codling moth, *Laspeyresia pomonella* (McIndoo, 1983). Crosby (1971) reported a toxic aporphine alkaloid, 'magnoflorine' (Figure 86), in its root, bark and seed extracts to possess insecticidal property.

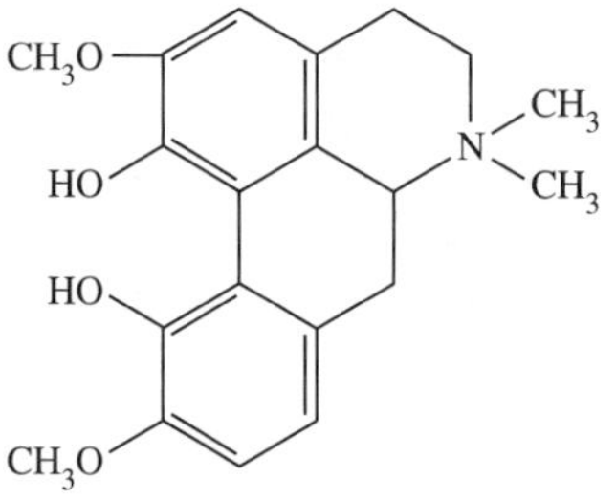

Figure 86. Magnoflorine.

633. *Phryma leptostachya* Linn. (Phrymaceae)

This is an annual herb found in temperate zones and available in Canada and North America. Its whole plant, fruits, bulb and corn extracts in petroleum ether were found to show antifeedant activity to *Drosophila melanogaster* (Jacobson, 1975).

634. *Phryma oblongifolia* Forsk. (Phrymaceae)

This is a perennial herb found in temperate and sub-tropical zones. Its root, leaf and flower extracts were toxic to *Drosophila melanogaster* (Jacobson, 1975).

635. *Phyllanthus acuminatus* Berth. (Euphorbiaceae)

Common name: Berry leaf flower
This is a perennial shrub/herb found in tropical and sub-tropical zones. Dried powder of its root was reported to be insecticidal when tested against the melon worm, *Diaphania hyalinata* and the diamond-back moth, *Plutella xylostella* (Jacobson, 1975).

636. *Physalis angulata* Linn. (Solanaceae)

Common name: Ground beetle
This is a perennial shrub found in temperate and sub-tropical zones. Its leaf extract in methanol was reported to increase oviposition of the tobacco budworm, *Heliothis subflexa* sprayed on a non-host tobacco plant (Mitchell and Heath, 1987).

637. *Physalis peruviana* Linn. (Solanaceae)

(Syn.-*P. edulis* Sims.)
Common name: Cape gooseberry
This is a perennial shrub found in tropical and sub-tropical zones and grown as edible fruit throughout India. Its leaf extract was reported to show antifeedant activity to *Spodoptera littoralis* (Michael et al., 1985). However, an active steroidal lactone, i.e., '4-β-hydroxy withanolide-E' (Figure 87), was isolated from its leaf extract and reported to possess antifeedant property when tested against *Spodoptera littoralis* and *Epilachna varivestis* (Ascher et al., 1987; Ascher, 1983) and was toxic to *Heliothis zea* (Elliger and Waiss, 1989). Further, another active steroidal lactone called '2-β-glucoside of perulactone' (Figure 88) was found effective to inhibit the growth of *Heliothis zea* (Elliger and Waiss, 1989).

638. *Physocarpus capitatus* (Pursh.) O. Kuntze. (Rosaceae)

This is a perennial shrub found in temperate and sub-tropical zones. Its stem, wood and bark extracts in ethyl ether showed attractancy to the female of Mediterranean fruit fly, *Ceratitis capitata* (Keiser et al., 1975).

639. *Physostigma venenosum* Balf. (Leguminosae)

Common name: Calabar bean
This is an annual vine/herb found in tropical zones. Dried powder of its whole

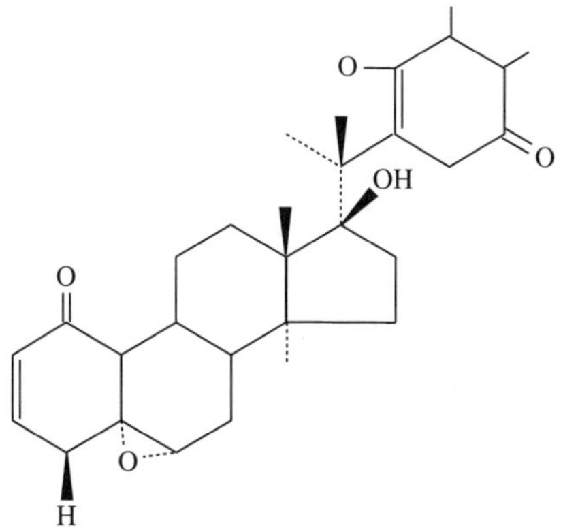

Figure 87. 4-β-hydroxy withanolide-E.

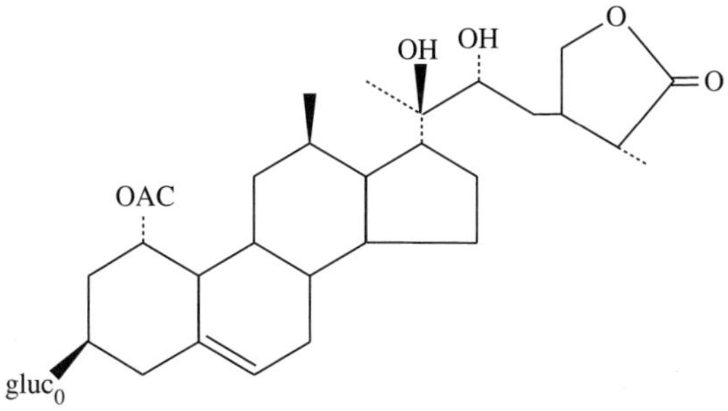

Figure 88. 2-β-Glucoside of perulactone.

plant was reported to show toxicity to Colorado potato beetle, *Leptinotarsa decemlineata* (Jacobson, 1975).

640. Phytolacca acinoca Roxb. (Phytolaccaceae)

Common name: Indian poke

This is a perennial herb found in tropical regions and commonly available in Bhutan, Jammu and Kashmir in India. Dried powder of its roots was toxic to Mexican bean beetle, *Epilachna varivestis* (Michael et al., 1985).

641. *Phytolacca americana* Linn. (Phytolaccaceae)

Common name: Poke berry

This is an annual herb found in tropical and sub-tropical zones but native to South America. Its root and bark extracts in alcohol were reported to be antifeedant to caseworms of general importance (McIndoo, 1983).

642. *Picea rubens* A. Dietr. (Pinaceae)

Common name: Red spruce
This is a perennial tree found in temperate zones and commonly available in North America. Dried powder of its bark was reported to show repellent activity to the white pine weevil, *Pissodes strobi* (Jacobson, 1975).

643. *Picrasma excelsa* (Swz.) Lindl. (Simaroubaceae)

Common name: Jamaica quassia
This is a perennial tree found in tropical and sub-tropical zones. Its active component, 'picrasmin', was reported to be toxic to the aphids (Martin, 1964), whereas powders of its dried leaf and branch were reported to show insecticidal activity to the sawfly, *Phymatocera aterrima* (Khan, 1982).

644. *Pimpinella ansium* Linn. (Apiaceae)

Common name: Anise
This is an annual herb found in semi-arid zones and native to the Mediterranean region but now cultivated in Uttar Pradesh, Assam, Orissa and Punjab. Its seed extract in water and also seed oil showed repellency and toxicity to the milkweed bug, *Oncopeltus fasciatus* (Hael et al., 1950). Shaaya et al. (1991) found fumigant toxicity in its essential oil to red flour beetle, *Tribolium castaneum.*

645. *Pimpenella heyneana* Wall. ex Kurz. (Apiaceae)

This is a slender erect and annual herb found in tropical and sub-tropical zones and commonly available in Chhotanagpur, Kalahandi (Orissa), Konkan and Western Ghats in India. Its whole plant crude extract showed antifeedant activity to *S. litura* and 'xanthotoxin', an active component, was reported to be the main feeding inhibitor in its plant extract (Banerji, 1988).

646. *Pinus sp.* (Pinaceae)

This is a perennial tree found in temperate regions. Its leaf extract in water was found to show insecticidal activity when evaluated against the Japanese beetle, *Popillia japonica* (McIndoo, 1983).

647. *Pinus densiflora* Linn. (Pinaceae)

This is a perennial tree found in tropical and sub-tropical zones and commonly available in Japan and North Korea. Its freshly collected needles extract was reported to show repellency to the citrus longhorns, *Monochamus alternatus* and ethane was found to be an active component in the extract (Sumimoto et al., 1975).

648. *Pinus insularis* Endl. (Pinaceae)

(Syn.-*P. khasya* Royle. ex Parl.)

Common name: Benguet pine or Khasi pine

This is a perennial tree found in sub-tropical and temperate zones and distributed in South East Asia, the Philippines and Japan. In India, it is commonly available in the Khashi and Jaintia hills of the Himalayas. Its leaf and branch extracts in water were found to show repellency to the insect pests in the rice fields (Litsinger et al., 1978).

649. *Pinus ponderosa* Linn. (Pinaceae)

This is a perennial small tree found in tropical and sub-tropical zones. Its secretions and vapours of its bark extract were found to show toxicity to the western pine beetle, *Dendroctonus brevicornis* (Menasa, 1961).

650. *Pinus rigida Mill.* (Pinaceae)

Common name: Pitch pine

This is a perennial small tree found in temperate zones. Its leaf, stem and fruit extracts showed juvenile hormone analogue activity against the milkweed bug, *Oncopeltus fasciatus* (Jacobson et al., 1975a).

651. *Piper arcuatum* Blume (Piperaceae)

(Syn.-*P. aurantiacum* Wall. ex. D.C.: *P. walcichii* Hand-Mazz., *P. auritum* Linn.)

This is a perennial stout climber found in tropical and sub-tropical regions and commonly available in Bihar. Four components isolated and characterized as prenylated phenols showed repellency activity to *Atta cephalotes* (Ampofo et al., 1987).

652. *Piper betle* Linn. (Piperaceae)

Common name: Betel pepper

This is a perennial climbing shrub commonly grown in tropical, sub-tropical and temperate zones. It is very common in coastal states of India. Its root and leaf aqueous extracts and also their powders were reported to show insecticidal activity tested against the bean leaf beetle, *Cerotoma trifurcata*; the melon worm, *Diaphania hyalinata*; and the cotton stainer, *Dysdercus flavidus* (Jacobson, 1975).

653. *Piper excelsum* Gn. (Piperaceae)

(Syn.-*Macropiper excelsum*)

This is a perennial climbing shrub found in sub-tropical and temperate zones. From its leaf extract an active component, 'juvadecene' [(E)-5-(3-decenyl)-

1,3-benzodioxole], was isolated and found to produce super numeralcy 6th instars when topically applied @ 30 μg/5th instar grub of the milkweed bug, *Oncopeltus fasciatus*, and also reported to show insecticidal activity to its adults (Nishida et al., 1983).

654. *Piper guineese* Schum. (Piperaceae)

Common name: West African pepper

This is a perennial shrub found in tropical zones and commonly available in West Africa. Its fruit dry powder was reported to be toxic to the armyworm, *Pseudoletia unipunctata* (Jacobson, 1958). Its oil was also reported to be toxic to the adults of *Dysdercus superstiosus* and larvae of *Acraea eponina*, *Ootheca mutabilis* and *Riptortus dentipes* (Olaifa et al., 1987). Cowpea seeds treated with its oil using 2 ml/kg seeds showed 100% mortality to the pulse beetle, *Callosobruchus maculatus* even after 24 h of treatment. Oji (1991) also demonstrated the use of its fruits' dry powder to protect stored maize against corn weevil, *Sitophilus oryzae*.

Wisanine and piperine, two active components isolated from its seed extract, were found to be toxic to the rice weevil, *Sitophilus oryaze* (Su, 1977; Su and Sondengam, 1980).

655. *Piper longum* Linn. (Piperaceae)

Common name: Chilli pepper or Long pepper

This is a perennial shrub found in tropical and alpine zones. It is native to India and commonly grown in Western Ghats, Tamil Nadu and Karnataka. Its seed oil was reported to show repellency to the insect pests of stored grains (Khan, 1982).

656. *Piper nigrum* Linn. (Piperaceae)

Common name: Black pepper or Round pepper

This is a perennial vine found in tropical regions and native to the Indo-Malaysian region. In India, it is commonly cultivated in Western Ghats, Karnataka, Kerala, Maharastra and Assam. Its dried seed powder was reported to be an effective repellent against the corn earworm, *Heliothis zea* (Freeborn and Wymore, 1929). Scott and McKibben (1978) reported black pepper extract as insecticidal to the adults of the boll weevil, *Anthonomus grandis*. Morallo-Rajessus (1984) reported its fruit/seed extract to show repellent activity to the cotton stainer, *Dysdercus cingulatus*. Osmani et al. (1987) found its seed extract in acetone to cause morphogenetic effects associated with juvenile hormone activity to the 5th instar nymphs of *Dysdercus cingulatus* and also to the final instar larvae of noctuid, *Achoea janata*.

Dried seed powder of black pepper and its extracts both were found to be effective grain protectants against the bean weevil, *Acanthoscelides obtectus*

(Lathrop and Keirstead, 1946) and also against the rice weevil, *Sitophilus oryzae* and the pulse beetle, *Callosobruchus maculatus* (Su, 1977, 1978; Prakash et al., 1978–1987; Mathur et al., 1985). Ponce De Leon (1983) and Javier and Morallo-Rajessus (1986) reported insecticidal properties in its fruit extract when tested against the storage insects, viz., *S. zeamis*, *R. dominica*, *T. castaneum* and *O. surinamensis*. Sighamony et al. (1986) found its oil to show toxicity to the adults of *S. oryzae* and *R. dominica* when tested using 25–100 ppm wheat grain treatment and it protected the stored wheat up to 60 days. Osmani et al. (1987) further reported its seed extract to cause morphogenetic changes during larval development of the rice moth, *Corcyra cephalonica* when tested under laboratory conditions. Black pepper seeds were also found to protect stored rice against infestation of *S. cerealella* (Prakash et al., 1989); *S. oryzae* (Prakash et al., 1990c); and *R. dominica* (Prakash et al., 1991) and minimized their population build up in the rice grains treated with its extract or admixed with its powder and stored.

The active component, 'piperine' (Figure 89), was isolated and reported to be toxic to *S. oryzae* (Su, 1977). Miyakado et al. (1979) reported a new pipercide (2E,4E,10E)-11-(1,3-benzodioxol-5-yl)-N-(2-methyl propyl)-2,4,10-undecatrienamide called 'piperaceal amide-I', which was more toxic to the pulse beetle, *Callosobruchus chinensis* than another component 'pelitorine' [(2E,4E)-N-(2-methyl propyl)-2-4-decadienamide] or 'piperine' [(2E,4E)-1-[5-(-1,3-benzodioxol-5-yl)-1-oxo-2,4-pentadienyl] piperidine]. Su and Robert (1981) also isolated 3 amides as the active components, which were characterized as I. (E,E)-N-(2-methyl propyl)-2,4-decadienamide, II. (E,E,E)-13-(1,3-benzodioxol-5-yl)-N-(2-methyl propyl)-2,4,10-4-N-decatrienamide. These components showed toxicity at higher rates when tested against the cowpea weevil, *Callosobruchus maculatus*. Myakado et al. (1989) reported from its fruit extract five unsaturated amides, i.e., pipercide (Figure 90), dehydropipercide (Figure 91), quineesine (Figure 92), pellitorine and piperine, called peperceal amids, that were also toxic to the pulse beetle, *C. chinensis*.

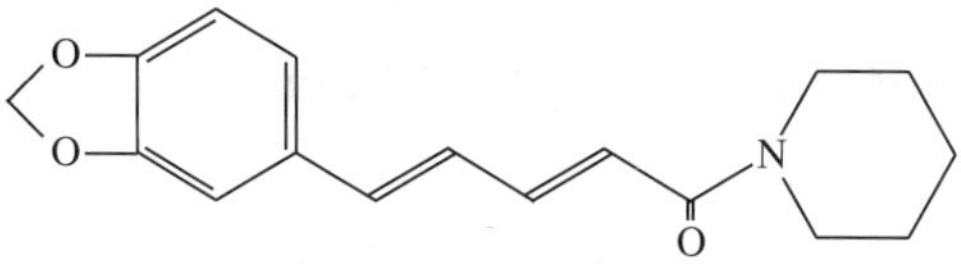

Figure 89. Piperine.

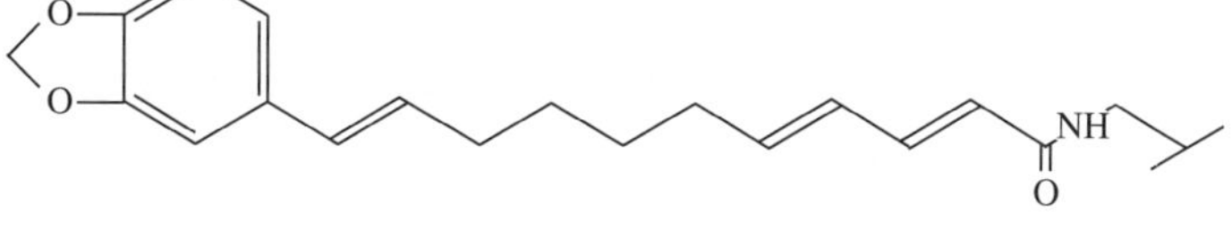

Figure 90. Pipercide.

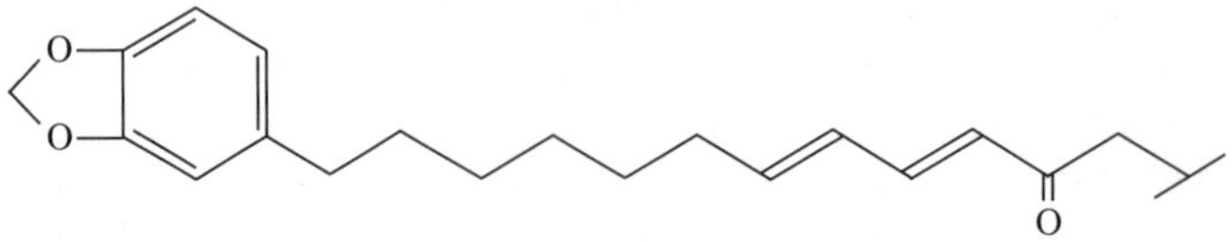

Figure 91. Dehydropipercide.

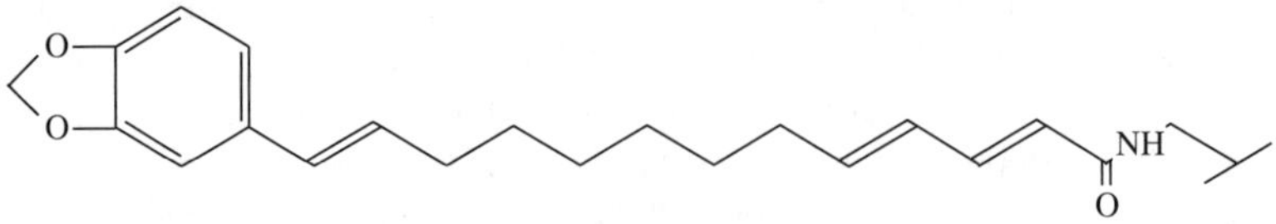

Figure 92. Quineesine.

657. *Piscidia acuminata* Linn. (Fabaceae)

This is a perennial tree found in tropical regions. Its root and leaf dried powders were reported to show insecticidal activity when tested against melon worm, *Diaphania hyalinata* and diamond-back moth, *Plutella xylostella* (Jacobson, 1975).

658. *Piscidia piscipula* (Linn.) Sarg. (Fabaceae)

(Syn.-*P. erythrina* Linn.)

Common name: Jamaica dogwood

This is a perennial tree found in tropical regions. Its root, bark and leaf extracts in water, alcohol and ether and also their dried powder were reported to show insecticidal and antifeedant activities when tested against melon worm, *Diaphania hyalinata*; diamond-back moth, *Plutella xylostella*; cotton stainer, *Dysdercus cingulatus*; southern beet worm, *Pachyzancla bipunctalis*; and southern armyworm, *Spodoptera eridania* (Jacobson, 1958).

659. *Pisum sativum* Linn. (Papillionaceae)

Common name: Garden pea

This is a seasonally grown vegetable as well as pulse crop in tropical zones and commonly available in North and Central States of India. Its leaf extract and fruit pericarp juice were reported to inhibit the oviposition of the bollworm, *Earias vitella* (Dongre and Rahalkar, 1984).

660. *Plumbago auriculata* Lamk. (Plumbaginaceae)

(Syn.-*P. carpensis* Thunb.)

Common name: Cape leadwort

This is an ornamental perennial bushy shrub found in tropical zones and commonly available in West Africa. An active component, 'plumbagin' [5-

hydroxy-2-methyl-1,4-naphthalendione] (Figure 93) isolated from its leaf extract, showed antifeedancy to the larvae of African armyworm, *Spodoptera exempta* and also inhibited ecdysis during development of the pink cotton bollworm, *Pectinophora gossypiella*; corn earworm, *Heliothis zea*; and tobacco budworm, *Heliothis virescens* (Kubo et al., 1983).

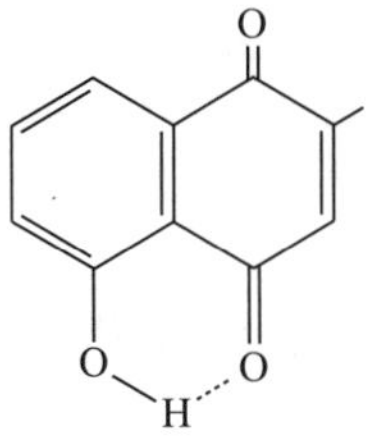

Figure 93. Plumbagin.

661. *Plumbago zeylanica* Linn. (Plumbaginaceae)

Common name: Leadwort

This is an annual/biennial herb found in temperate and sub-tropical zones. Its root and bark extracts in alcohol and ether were reported to show toxicity to the aphid, *Lipaphis erysimi* (Borle, 1982); cotton stainer, *Dysdercus koenigii*; Mexican bean beetle, *Epilachna varivestis*; and hairy caterpillar, *Euproctis fraterna* (McIndoo, 1983). Plumbagin, an active component of its leaf extracts, was also found to interfere with the synthesis of new cuticle during development of red cotton bug, *Dysdercus koenigii* by inhibiting the protein synthesis activity and finally resulted in deformed adults.

662. *Poa annua* Linn. (Poaceae)

Common name: Meadow grass

This is an annual tufted grass found in sub-tropical and alpine zones. It is a native to the North Mediterranean region and available in the Himalayas, Khasi hills and Nilgiris in India. Its aqueous leaf extract was reported to show repellent activity to Colorado potato beetle, *Leptinotarsa decemlineata* (Jacobson, 1958).

663. *Podophyllum peltatum* Linn. (Berberidaceae)

Common name: Mandrake

This is a perennial herb found in temperate zones and available in the Himalayan region. Its root and leaf extracts in water and ether showed antifeedant and attractant activities against Colorado potato beetle, *Leptinotarsa decemlineata* (Mathews, 1981).

664. *Pogostemon heynemus* Benth. (Lamiaceae)

(Syn.-*P. patchouli* Hook. f. non. Pellet, *P. perilloides* [Linn.] Mansf.)
Common name: Patchouli

This is a perennial herb found in tropical and sub-tropical zones and commonly available in India. Its root and branch and meristematic tissue and distilled oil showed toxicity and repellency to the cabbage leafhopper, *Crocidolamia binotalis* and cotton leaf worm, *Spodoptera litura* (McIndoo, 1983). Despande et al. (1974) isolated an active component, 'eugenol', from this plant extract, which showed toxicity to *Callosobruchus chinensis* and *Stegobium paniceum*.

665. *Polugonatum japonicum* Mill. (Polygonaceae)

This is a perennial herb found in temperate zones. Its aqueous leaf extract was reported to be toxic to the vinegar fly, *Drosophila hydei* (Jacobson, 1958).

666. *Polygonum auberti* Linn. (Polygonaceae)

Common name: Fleece vine

This is an annual/perennial vine found in temperate zones. Its leaf extract in water was toxic to the Japanese beetle, *Popillia japonica* (Jacobson, 1958).

667. *Polygonum hydropiper* Linn. (Polygonaceae)

(Syn.-*P. flaccidum* Meissn.)
Common name: Water pepper or Pepper wort

This is a perennial herb found in tropical and Mediterranean zones. This is available throughout India. Its whole plant, root and leaf extracts in water and petroleum ether showed repellency to *Sitophilus* spp. (Zaman, 1983) and *Spilosoma obliqua* (Islam, 1983). Zhang et al. (1993) isolated (-)-polygodial as an effective antifeedant against the aphids.

668. *Polypodium walkarae* Linn. (Polygonaceae)

This is a perennial fern found in tropical zones. Its whole plant ether extract was found to toxic to the cotton stainer, *Dysdercus cingulatus* (Balsubramanian, 1982) and also inhibited its growth and development (Kareem, 1984).

669. *Polyscias guilfoylii* Bailey (Araliaceae)

This is a perennial shrub found in tropical zones and distributed in South East Asia and West Africa. Its stem and leaf extracts in ether were reported to show juveno-mimetic activity to the cotton stainer, *Dysdercus cingulatus* and also to cause morphogenetic changes in its 5th instar larvae (Rajendran and Gopalan, 1978, 1985). However, its bark and leaf extracts in petroleum ether were reported to show toxicity to *D. cingulatus* (Balasubramanian, 1982).

Kareem (1984) also found its stem and leaf aqueous extracts to inhibit the growth and development of *D. cingulatus*.

670. *Pongamia glabra* Vent. (Fabaceae)

(Syn.-*P. pinnata* [Linn.] Pierre)

Common name: Puna oil tree or Pongamia or Kharanja

This is a perennial tree found in tropical and sub-tropical zones. Pongamia tree is commonly available in coastal India, Sri Lanka, Bangladesh and Malaysia. Its root, leaf, flower, seed and fruit extracts in alcohol, their dried powders and also its seed oil and oil-cake were reported to show insecticidal, antifeedant and repellent activities against a wide range of insects infesting both storage and field ecosystems.

Biological Activity Against Storage Insects

Its leaves admixed with paddy grains were reported to protect the grains from storage insects (Govindaswami and Patnaik, 1958). Its leaf extract and dried powder also showed antifeedant activity against the rice weevil, *Sitophilus oryzae* (Jacobson, 1975). Its seed oil was reported to show synergistic activity with pyrethrins against red flour beetle, *Tribolium* sp. (Parmer, 1977) and toxicity to the pulse beetle, *Callosobruchus chinensis* when admixed with red gram and it protected these grains in storage (Sangappa, 1977). Dried root powder of pongamia and its seed oil were also found to show repellent activity against the lesser grain borer, *Rhyzopertha dominica* and Angoumois grain moth, *Sitotroga cerealella* (Balasubramanian, 1982; Krishnamurthy Rao, 1982). Seed oil of kharanja was also reported to show antifeedant activity against red flour beetle, *Tribolium castaneum* and the saw toothed grain beetle, *Oryzaephilus surinamensis* (Prakash and Rao, 1986).

Kharanja oil @ 1% w/w protected cowpea grains in storage against *C. maculatus* (Naik and Dumbre, 1984, 1985; Rajsekharan and Kumarswami, 1985a; Singh and Satyavir, 1987; Ramesh Babu and Hussain, 1990a); the rice weevil, *Sitophilus oryzae*; and the lesser grain borer, *Rhyzopertha dominica* (Sighamonx et al., 1986). Babu et al. (1989) and Khaire et al. (1989) tested the ovipositional preferences of *C. chinensis* in an olfactometer and found pongamia oil to reduce oviposition of this weevil at par with neem oil.

Biological Activity Against Field Insects

In India, oil of pongamia was reported to show insecticidal activity as early as 1956 (Osmani and Naidu, 1956). Pongamia cake application in the soil effectively controlled the ground beetle, *Mesomorphus villiger* in tobacco nurseries (Joshi and Rao, 1968; Joshi et al., 1974) and in addition with rice bran application in the soil enhanced the effectiveness of the cake (Joshi et al., 1974). Soil application of its cake also reduced the incidences of white grub, *Holotrichia insularis* in chilies (Sachan and Pal, 1976) and *H. consan-*

guinea in groundnut crop (Nigam, 1970). Joshi et al. (1990) reported its cake aqueous extract also to protect the tobacco seedlings from the attack of tobacco caterpillar, *Spodoptera litura* in nursery.

Its leaf, root and seed extracts in alcohol were found to show insecticidal activity against aphid, *Lapaphis erysimi* and tobacco beetle, *Mesomorphus villiger* (Jacobson, 1975). Krishnamurthy Rao (1982) reported its seed extract to show antifeedant activity against cotton leaf armyworm, *Spodoptera litura*. Its seed oil showed toxicity to *S. litura* (Rajasekaran and Kumarswami, 1985a) but could not significantly control the gram pod borer, *Heliothis armigera* on chickpea (Sharma and Dahiya, 1986). Shelke et al. (1987) found its seed extract to show effective ovicidal activity against potato tuber moth, *Phthorimaea operculella* under laboratory tests using 5 and 10% concentrations of the extract. Koshiya and Ghelani (1990) also reported its 5% seed extract to show phago-deterrency on treated groundnut leaves against *S. litura*. Its leaf extract in water showed toxicity to larvae of Gujrat hairy caterpillar, *Amsacta moorei* (Patel et al., 1990a).

Its oil was reported to reduce the survival of green leafhopper, *Nephotettix virescens* and rice leaffolder, *Cnaphalocrosis medenalis* (Mariappan et al., 1988). Similarly, oily fraction of seed extract of pongamia showed antifeedant activity against *Amsacta moorei* (Verma and Singh, 1985) and *Diacrisia obilqua* (Mohanty et al., 1988). Ramaraju and Sundara Babu (1989) found its 1% oil and 2% seed extract to reduce the emergence of brown planthopper, *Nilaparvata lugens* and white backed planthopper, *Sogatella furcifera* in rice. Sardana and Kumar (1989) also reported its 2% oil to protect okra fruits from infestation of *Earias vittella* when sprayed on the crop. Its 1% oil was also reported to show 57 to 90% nymphal mortality to the grape mealy bug, *Maconellicoccus hirsutus* on treated grape twigs and also showed repellency to its nymphs and adults (Abraham and Tandon, 1990). Similarly, 3 sprays of its 2 and 4% oil on citrus plants at intervals of 15 days significantly reduced the infestation of citrus leaf miner, *Phyllocnistic citrella* (Dhara et al., 1990a) and populations of citrus aphid, *Toxoptera citricidus* (Dhara et al., 1990b). Sharma and Bhatnagar (1990) also reported its oil to reduce the survival of the larvae of *Chilo partellus* in maize whorl under a laboratory test. Mixture of pongamia and neem oils (1:1) safely protected summer paddy crop from gall midge and stem borer attack (Nanda et al., 1993). Bajpai and Sehgal (1994) also found its 2% oil to show insect growth regulator activity up to 24 days and prolonged larval development of *Helicoverpa armigera* when tested under controlled conditions.

Active Components

Parmar and Gulati (1969) reported ethanolic extract of de-oiled karanja cake, its defatted seed and pure karanjin to show insecticidal activity to mustard aphid, *Lipaphis erysimi*. Chakraborty et al. (1976) identified the active components of karanjin group from its oil extracted in water medium, which were toxic to *S. litura*. Rao and Niranjan (1982) reported an active component, 'karanjin' (Figure 94), isolated from its seed oil to show juveno-mimetic activity against the larvae of *T. castaneum*. Its final instar larvae provided with treated (0.5 to 2 mg karanjin/ml/g) flour checked the moulting of the pupae by the end of 4th week. Srinmannarayan and Rao (1985) and Geetanjali et al. (1987) also isolated a flavone called 'karanjin', an active component from its seed extract and found it to show antifeedant activity to *Spodoptera litura* when tested under controlled laboratory conditions.

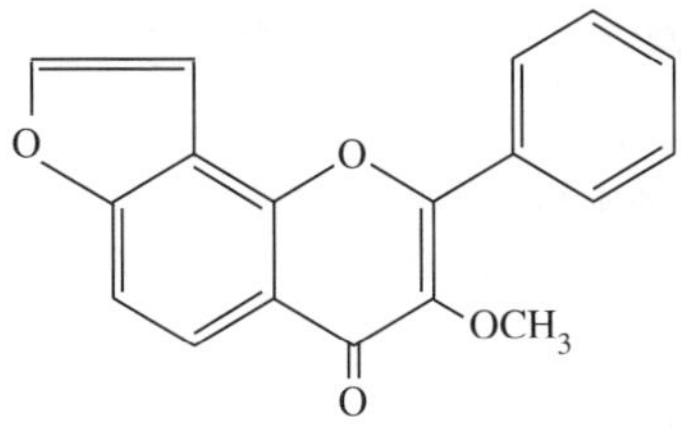

Figure 94. Karanjin.

671. *Populus heterophylla* Linn. (Salicaceae)

This is a perennial tree found in temperate zones and commonly available in North America and Europe. Its bark extract in ethyl ether was reported to show attractancy to Mediterranean fruit fly, *Ceratitis capitata* (Keiser et al., 1975).

672. *Portulaca oleraceae* Linn. (Portulacaceae)

Common name: Kitchen garden purslane or Common purslane
This is an annual herb found in tropical and arid zones. Its leaf extract in water was reported to be toxic to the long headed flour beetle, *Latheticus oryzae* (Prakash et al., 1978–1987).

673. *Proboscidea locisianica* Schmidl. (Martyiaceae)

Common name: Unicorn plant
This is an annual creeper found in tropical zones. Powder of its dried branch was reported to be toxic to the melon worm, *Diaphania hyalinata* (Jacobson, 1958).

674. *Purnus persica* (Linn.) Batch. (Rosaceae)

(Syn.-*Amygdalus persica* Linn., *Persica vulgaris* Mill.)
Common name: Peach
This is a perennial tree found in temperate zones. This is native to China but now grows in Himachal Pradesh, Jammu and Kashmir, Kumaun hills and Western Uttar Pradesh. Its leaf and flower extracts in water were reported to

show repellent activity against the silkworm, *Bombyx mori* and the screw worm, *Cochliomyia hominivorax* (McIndoo, 1983).

675. *Prunus burgeriana* Linn. (Rosaceae)

This is a perennial tree found in temperate zones. Its branch and leaf aqueous extracts were reported to be toxic to the vinegar fly, *Drosophila hydei* (Jacobson, 1958).

676. *Prunus grayana* Linn. (Rosaceae)

Common name: Gray's artichoke

This is a perennial tree found in sub-tropical and temperate zones. Its leaf and branch extracts in water were reported to be toxic to the vinegar fly, *Drosophila* sp. (Jacobson, 1958).

677. *Prunus japonica* Batsch. (Rosaceae)

Common name: Flowering almond

This is a perennial tree found in temperate zones. Its aqueous leaf extract was reported to be toxic to the vinegar fly, *Drosophila hydei* (Jacobson, 1958).

678. *Prunus padus* Linn. (Rosaceae)

(Syn.-*P. avium* Mill.)

Common name: European bird cherry or Sweet cherry

This is a perennial deciduous tree found in the temperate zones and commonly available in the temperate Himalayan region in India. Its whole plant and bark extracts in water and also their powder formulations were reported to show insecticidal and repellent activities against the midges and the vinegar fly, *Drosophila hydei* (Jacobson, 1958).

679. *Pseudoelephantopus apicatus* Rhor. (Asteraceae)

This is an annual herb found in tropical zones. Its flower and root extracts showed toxicity to the maize weevil, *Sitophilus zeamais* and the red flour beetle, *Tribolium castaneum* when applied topically (Carino, 1981). Its leaf juice/crude extract was also reported to show insecticidal activity to the cotton stainer, *Dysdercus cingulatus* (Carino, 1983).

680. *Psoralea corylifolia* Linn. (Fabaceae)

Common name: Babchi

This is an annual/biennial erect herb found in tropical zones and commonly available in various parts of India. Its leaf extract was reported to show toxicity to the pulse beetles, *Callosobruchus chinensis* and *C. maculatus* when tested under laboratory conditions (Anonymous, 1980–1981). Its seed extract in petroleum ether also showed juvenile-hormone activity against the red cotton bug, *Dysdercus koenigii* and an active component from this seed extract was isolated and identified as 'bakuchiol' (Figure 95) (Joshi et al., 1974).

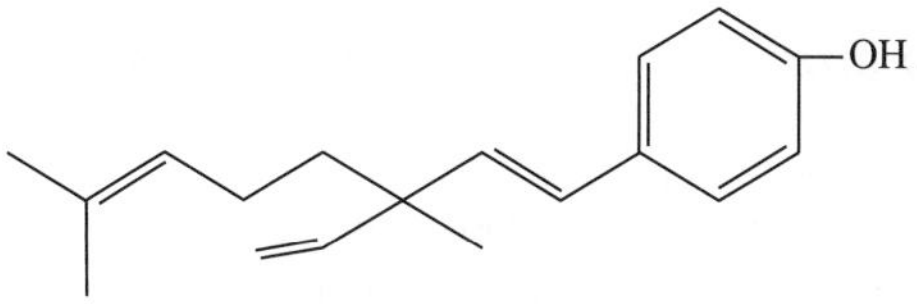

OH

Figure 95. Bakuchiol.

This component was also reported to show chemosterilant activity and reduced hatching of *D. koenigii* when its one-day-old females were treated with this component (Joshi et al., 1975). Oily fraction of its seed extract was reported to show antifeedancy to the larvae of the jute hairy caterpillar, *Diacrisia obliqua* when fed on the treated castor leaves (Mohanty et al., 1988).

681. *Pterocarya stenoptera* Kunth. (Juglandaceae)

This is a perennial tree found in tropical zones. Its bark and leaf extract in water and also its dried powders were tested to be toxic to Mexican bean beetle, *Epilachna varivestis* (Khan, 1982).

682. *Pueraria yunnanensis* Willd. (Fabaceae)

This is a perennial woody climber found in tropical zones. Its root and branch aqueous extracts were reported to be toxic to the aphids and Mexican bean beetle, *Epilachna varivestis* (Khan, 1982).

683. *Punica gronatum* Linn. (Punicaceae)

Common name: Pomegranate
This is a perennial shrub or small tree found in tropical, Mediterranean and alpine zones. It is native to Iran but now available throughout India. Its dried leaf powder showed toxicity to the diamond-back moth, *Plutella xylostella* (Jacobson, 1958).

684. *Purshia tridentata* D.C. (Rosaceae)

This is a perennial shrub found in tropical and sub-tropical zones. Its essential oil, polyphenols and coumarins were reported to be antifeedant to the Colorado potato beetle, *Leptinotarsa decemlineata* (Jermy et al., 1980).

685. *Pycanthemum rigidus* G. Don. (Lamiaceae)

This is a perennial herb found in tropical and temperate zones. Its dried leaf powder was reported to be toxic to the aphids and Mexican bean beetle, *Epilachna varivestis* (Jacobson, 1958).

686. *Quassia amara* Linn. (Simaroubaceae)

(Syn.-*Picraena excelsa* [Swz.] Lindl.)

Common name: Surinam quassia or Jamaica quassia

This is a perennial ornamental shrub/small tree found in tropical and sub-tropical zones. That is a native to Brazil and commonly available in New Guinea, Surinam, Jamaica, Panama, the West Indies and India. In fact, quassia is the wood of two trees, commercially called Jamaica quassia-wood of *Picraena excelsa* (Swz.) Lindl. and Surinam quassia-wood of *Quassia amara* Linn. The extract of quassia wood was first used as an insecticide against the hop aphid, *Phorodon humuli* during the years 1880–1884 (Ormerod, 1884). Later it was found to be toxic to several insect species. Jacobson (1954) reported its toxicity to melon worm, *Diaphania hyalinata* and also to the sawfly, *Phymatocera aterrima* (Poe, 1982).

The active components, viz., fenulin, helenalin, neo-quassin and quassin (Figure 96), were isolated from its wood extract and reported to be toxic to the 2nd instar of the bean beetle (McGovern et al., 1942) and also to the aphids of agricultural importance (Crosby, 1971).

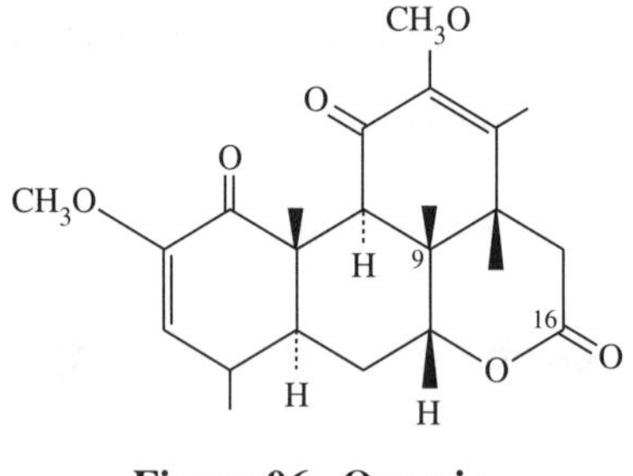

Figure 96. Quassin.

687. *Quassia indica* (Gaertn.) Nooteb. (Simaroubaceae)

(Syn.-*Samadera indica* Gaertn., *S. lucida* Wall.)

This is a perennial evergreen tree found in tropical zones. This is a native to Malagasy and found in Kerala in India. Its leaf extract in water was reported to be toxic to termites (Usher, 1973).

688. *Quercus alba* Linn. (Fagaceae)

Common name: White oak

This is a perennial shrub found in tropical, sub-tropical and semi-arid zones. Its leaf and bark extracts in alcohol were reported to show antifeedant activity to the Colorado potato beetle, *Leptinotarsa decemlineata* when tested under laboratory conditions (Drummond et al., 1985).

689. *Quercus elutina* Linn. (Fagaceae)

Common name: Black oak

This is a perennial tree found in temperate zones. Its aqueous leaf extract was reported to be repellent to the Japanese beetle, *Popillia japonica* (McIndoo, 1983).

690. *Randia nilotica* Linn. (Rubiaceae)

This is a perennial tree found in tropical zones. Dried powder of its root was reported to be toxic to the diamond-back moth, *Plutella xylostella* (Jacobson, 1958).

691. *Randia spinosa* Poir. (Rubiaceae)

(Syn.-*R. dumetorum* Poir.)

Common name: Emetic nut

This is a perennial deciduous thorny shrub/small tree found in India, Pakistan, Malaysia, South and East Africa. Its fruit and root aqueous extracts were reported to protect grains from insects in storage. Its fruit extract contained two saponins, i.e., randia and randia acid, as active components (Watt and Breyer-Bradwijk, 1962).

692. *Ranunculus flagelliformis* Linn. (Ranunculaceae)

Common name: Buttercup

This is an annual herb found in tropical, temperate and alpine regions. Its root, leaf and branch extracts in water were reported to be toxic to the vinegar fly, *Drosophila hydei* (Jacobson, 1958).

693. *Ranunculus mirissimus* Linn. (Ranunculaceae)

Common name: Buttercup

This is an annual herb found in tropical, temperate and alpine regions. Its root, leaf and branch extracts in water were also reported to be toxic to the vinegar fly, *Drosophila hydei* (Jacobson, 1958).

694. *Ranunculus scelerater* Linn. (Ranunculaceae)

Common name: Blister buttercup
This is an annual erect herb found in temperate zones. Its root and branch extracts were reported to be toxic to the vinegar fly, *Drosophila hydei* (Jacobson, 1958).

695. *Ranunculus vernyii* var. *glaber* Hook. (Ranunculaceae)

Common name: Buttercup
This is an annual herb found in tropical, temperate and alpine zones. Its leaf extract in water was reported to be toxic to the vinegar fly, *Drosophila hydei* (Jacobson, 1958).

696. *Ranunculus vernyii* var. *quelpaertensis* Linn. (Ranunculaceae)

Common name: Buttercup
This is an annual herb found in tropical, temperate and alpine zones. Its root, branch and leaf extracts in water were reported to show toxicity to the vinegar fly, *Drosophila hydei* (Jacobson, 1958).

697. *Ranunculus zuccarini* Linn. (Ranunculaceae)

Common name: Buttercup
This is an annual herb found in tropical, temperate and Mediterranean zones. Its root, leaf and flower extracts in water were reported to be toxic to the vinegar fly, *Drosophila hydei* (Jacobson, 1958).

698. *Ratibida columuifera* Rafin. (Asteraceae)

Common name: Prairie corn flower
This is an annual/perennial herb found in temperate zones. Its leaf and flower extracts in water were reported to be toxic to the grasshoppers of general importance (Jacobson, 1975).

699. *Rauvolfia serpentina* (Linn.) Benth. ex. Kurz. (Apocynaceae)

(Syn.-*Ophioxylon serpentium* Linn.)
Common name: Java devil pepper or Serpentine root or Rauvolfa root
This is a perennial small tree or shrub found in tropical zones. In India, it is available in Sikkim, Punjab, Nepal belt, Assam, Western Ghats and Andaman Islands. Its root extract in water and also its dried powder were reported to be toxic and inhibited the growth of the red flour beetle, *Tribolium castaneum* in a laboratory test (Jacobson, 1975).

700. *Rhamnus crenata* Linn. (Rhamnaceae)

Common name: Buckthorn genus

This is a perennial small tree and shrub found in sub-tropical and temperate zones. Its dried root powder was toxic to the bean aphid, *Aphis fabae* (Jacobson, 1958).

701. *Rhaponticum integrifolium* C. Winkle. (Asteraceae)

This is an annual/perennial herb found in temperate zones. Leaf and inflorescence extracts containing phyto-ecdysones showed a marked meta-toxic effect on the 3rd instar larvae of the gypsy moth, *Lymantria dispar* (Ganiev, 1987).

702. *Rhododendron hunnewellianum* Linn. (Ericaceae)

Common name: Nao-yang-wha

This is a perennial shrub/small tree available in temperate zones. Its dried flower powder was reported to show insecticidal activity when tested against the bean aphid, *Aphis fabae* (Roark, 1947).

703. *Rhododendron japonicum* Muell. (Ericaceae)

Common name: Japanese azalea

This is a perennial shrub of temperate zones. Its dried flower powder was reported to be toxic to the mulberry white caterpillar, *Rondotia nenciana* (McIndoo, 1983).

704. *Rhododendron molle* Linn. (Ericaceae)

This is a perennial shrub found in tropical zones. Its root, leaf and flower extracts in water, alcohol, acetone, and petroleum ether were reported to be toxic to the rice yellow stem borer, *Scirpophaga incertulas* (Chiu, 1982) and also to the cabbage leaf beetle, *Calophellus bowringi* (McIndoo, 1983).

Rhodojaponin III, a grayanoid diterpene, was isolated and identified as an active component from its dried flower extract and found to show antifeedant, growth inhibitory and insecticidal activities against the larvae of *Leptinotarsa decemlineata* and *Spodoptera frugiperda.*

705. *Rhus coriaria* Linn. (Ancistrocladaceae)

Common name: Sicilian sumac

This is a perennial shrub found in tropical and Mediterranean zones. Its dried leaf powder was reported to be toxic to the woolly apple aphid, *Eriosoma lonigerum* and it showed repellency to the grape phylloxae, *Phylloxera vitirolia* (McIndoo, 1983).

706. *Rhus continum* Linn. (Ancistrocladaceae)

(Syn.-*R. coggygria* Scop., *R. velutina* Wall.)

Common name: Ting or Tungla

This is a perennial deciduous shrub/small tree found in tropical regions and distributed in South China, Japan, the Philippines, Korea and Malaysia. In India, it is commonly available in the North West Himalayan tract. Its oil was reported to show antifeedant activity to the boll weevil, *Anthonomus grandis*. The active components in its oil were reported to be aeleostearic acid and erthro-9, 10-dihydroxy octadecyl acetate, which showed antifeedancy to *A. grandis*. This acid was found to be too unstable for its practical utility but its methyl ester was more stable and equally effective as an antifeedant to this weevil (Jacobson et al., 1981a).

707. *Rhus typhina* (D.C.) Wall. (Ancistrocladaceae)

This is a perennial shrub found in sub-tropical and temperate regions and available in Europe and North America. Six aphidicidal constituents, viz., phytol, linalool, tetradecanol, decosanol, tetradecanoic acid and hexadecanoic acid, were isolated by steam distillation of the freshly harvested leaves and reported to show toxicity to the aphids of agricultural importance (Bestmann et al., 1988).

708. *Ricinus communis* Linn. (Euphorbiaceae)

Common name: Castor

This is an annual/perennial shrub/small tree found in tropical and semi-arid zones. Its seed oil and cake both were reported to show insecticidal, antifeedant and repellent activities against a wide range of insects.

Castor oil was reported to inhibit the multiplication of the pulse beetles, *Callosobruchus* spp. and the storage weevils, *Sitophilus* spp. (Su et al., 1972; Talati, 1976; Mummigatti and Raghunathan, 1977; Sangapa, 1977; Naik and Dumbre, 1984) and showed repellent activity to the rice weevil, *Sitophilus oryzae* when its oil cake was admixed with maize grains (Bowrey et al., 1984). Ramesh Babu and Hussain (1990a) reported its oil to significantly reduce the oviposition of the pulse beetle, *C. chinensis* in the seeds treated with its oil @ 10 ml/kg. Khaire et al. (1989, 1993) found castor oil to significantly reduce the oviposition of *C. maculatus* and *C. chinensis*. Among the different oils of plant origin tested against infestation of *Callosobruchus rhodesianus* using cowpea seed treatments, castor oil was most effective in reducing infestation as well as fecundity and fertility of the beetle (Gija and Munetsi, 1990).

However, against field insect pests its seed oil was reported to be toxic to the leaf cutting ants (Jacobson, 1975). Saxena and Basit (1982) reported volatiles of castor oil to inhibit the oviposition of the leafhopper, *Amrasca devastans*. Rao et al. (1990) and Chitra et al. (1991a,b) found 1 and 5% aqueous and petroleum ether leaf extracts to protect the treated host leaves from 1st, 2nd and 3rd instar larvae of the spotted leaf beetle, *Henosepilachna vigintioctopunactata*. Puri et al. (1991) evaluated a formulation of 2% castor seed oil, i.e., Flytack-2, against the sweet potato white fly, *Bemisia tabaci* on the treated cotton leaves and found promising reduction in its adult population up to 7 days after treatment

The active component, 'ricinin', a toxic protein, and the alkaloid 'ricinin' both were isolated from its seed-oil extract and also from the other parts of this plant and found to show toxicity to the codling moth larvae (Haller and McIndoo, 1943; De-Ong, 1948). Frear (1948) also reported ricinin to show toxicity to the codling moth.

709. *Rosmarinus officinalis* Linn. (Lamiaceae)

Common name: Rosemary

This is a perennial, ornamental herb/shrub found in tropical and Mediterranean zones. Its leaf extract in petroleum ether and also its oil were reported to show antifeedant activity to the Japanese beetle, *Popillia japonica* (McIndoo, 1983).

710. *Rubia tenuifolia* Urv. (Rubiaceae)

Common name: Alizari or European madder

This is an annual/perennial shrub found in temperate and sub-tropical zones. Its whole plant extract in ether showed attractancy to both the male and female of the Mediterranean fruit fly, *Ceratitis capitata* (Keiser et al., 1975).

711. *Rubus japonica* Linn. (Rosaceae)

Common name: Bramble genus

This is an annual/perennial shrub found in temperate zones. Its whole plant extract in water was toxic to the vinegar fly, *Drosophila hydei* (Jacobson, 1958).

712. *Rubeckia hirta* Linn. (Asteraceae)

Common name: Black-eyed Susan

This is an annual/perennial herb found in sub-tropical and temperate zones. Its whole plant extract in water was reported to show repellent activity to the Japanese beetle, *Popillia japonica* (McIndoo, 1983).

713. *Ruta graveolens* Linn. var. *angustifolia* Hook.f. (Rutaceae)

(Syn.-*Ruta chalepensis* Linn., *R. bracteosa* D.C., *R. angustifolia* Pars.)

This is an aromatic shrub found in the Mediterranean zone and native to South Europe and North Africa. Its whole plant, roots, leaf and seed extracts in water were reported to show repellent activity when tested against the Japanese beetle, *Popillia japonica* (McIndoo, 1983).

714. *Ryania speciosa* Vahl. (Flacourtiaceae)

This is a perennial shrub/tree found in tropical zones and distributed in South India, Sri Lanka, Indonesia and the Philippines. Its dried root, leaf and stem powders passed through a 200 mesh screen commonly called 'ryania' contained about 0.25% alkaloid. This powder was recommended for use directly or mixed with an inert diluent such as talc, bentonite and pyrophyllite to provide a dust or wettable powder. It was also recommended as extracts of ryania in water, alcohol, acetone, chloroform etc. against a number of insects of agricultural importance but because of higher mammalian toxicity, its utility was restricted (Jacobson and Crosby, 1971). Ryania formulations were reported to show insecticidal and repellent activities against European corn borer, *Ostrinia* (*Pyraustica*) *nubilalis* (Rusbey, 1958); the codling moth, *Laspeyresia pomonella* (Hamilton and Cleveland, 1957); the melon worm, *Diaphania hyalinata* (Roark, 1947); New Guinea sugarcane weevil, *Rhabdoscelus obscurus*; Oriental fruit fly, *Dacus dorsalis* (Jacobson, 1958); the sugarcane shoot borer, *Argyria stricticrapsis* (Poe, 1982); the sweet potato weevil, *Cylas formicarius*; and the chilli thrip, *Scritothrips dorsalis* (Krishnamurthy Rao, 1982).

Crosby (1971) reported 'ryanodine' (Figure 97) and its derivatives in its dried root and leaf powders as the active components, which were found to show insecticidal activity against the corn earworm, *Heliothis zea*; the codling moth, *Carpocapsa pomonella*; the sugarcane borer, *Diatraea saccharalis*; and Oriental fruit moth, *Graphoitha molesta*. These toxic components were also reported in the root powder of other species of *Rhynia,* i.e., *R. pyrifera*, *R. dentata*, *R. acuminata*, *R. tomentosa*, *R. sagotiania* and *R. subuliflora*. Waterhouse et al. (1985) isolated a new ryanoid, i.e., 9,21-didehydroryanodine, from its powdered stem wood and reported it to be more toxic than ryanodine.

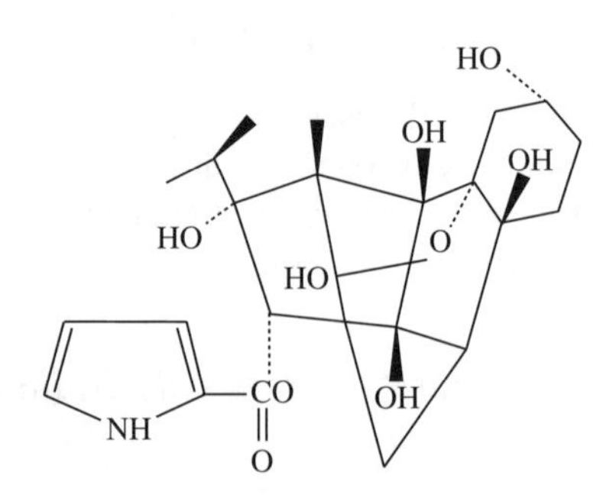

Figure 97. Ryanodine.

715. *Sabadilla* sp. (Liliaceae)

This is an annual herb found in tropical and temperate zones. From its methanolic seed extract an alkaloid, 'cevadine', was reported to show toxicity to the milkweed bug, *Oncopeltus fasciatus* (Allen et al., 1945).

716. *Salvia plebeia* Linn. (Lamiaceae)

This is an annual herb found in temperate and sub-tropical zones. Its seed extract and oil showed toxicity to the cotton aphid, *Aphis gossypii* (McIndoo, 1983).

717. *Salvia sclarea* Linn. (Lamiaceae)

This is a biennial/perennial shrub found in Mediterranean and semi-arid zones. Its oil and distillates of the oil were reported to be toxic to the cotton aphid, *Aphis gossypii* (McIndoo, 1983).

718. *Sambucus nigra* Linn. (Caprifoliaceae)

Common name: European elder

This is a perennial spreading shrub/small tree found in temperate zones. In India, it is available in the hilly areas of Uttar Pradesh. Its leaf extract in water was reported to be toxic to the sweet potato weevil, *Cylas formicarius* (McIndoo, 1983).

719. *Sanguinaria canadensis* Linn. (Papaveraceae)

Common name: Blood root

This is a perennial herb found in temperate zones. Its root and leaf extracts in water and ether were reported to show insecticidal and growth inhibiting activities when tested on the vinegar fruit fly, *Drosophila melanogaster* (Michael et al., 1985).

720. *Sapindus emarginatus* Vahl. (Sapindaceae)

(Syn.-*S. trifoliatus* Heirn.)

Common name: Florida soapberry or Soap nut

This is a perennial tree found in tropical, sub-tropical and semi-arid zones. In India, it is available in South Indian States, Bihar, Uttar Pradesh, Madhya Pradesh and West Bengal. Its seed extract and dried powder were reported to show antifeedant activity against the lesser grain borer, *Rhyzopertha dominica*; Angoumois grain moth, *Sitotroga cerealella*; and the rice weevil, *Sitophilus oryzae* (Balasubramanian, 1982). Its seed powder admixed @ 2% w/w with the cowpea seeds protected them from infestation of the pulse beetle, *Calloso-*

bruchus chinensis in storage and showed toxicity to the beetle (Yadav and Bhatnagar, 1987). Reddy and Urs (1988) also reported its seed extract in petroleum ether to reduce oviposition of the brown planthopper, *Nilaparvata lugens* when sprayed on the rice plant.

721. *Sapindus mukorossi* Benth. (Sapindaceae)

Common name: Chinese soap berry or Soap nut-tree
This is a perennial tree found in tropical and semi-arid zones and distributed in South China, Nepal, India, Bhutan and Burma. Its shell powder was reported to show insecticidal activity against the red flour beetle, *T. castaneum* and the lesser grain borer, *R. dominica* when admixed with their diets and it also showed synergistic action with DDT water dispersible powder (Dutta et al., 1978). Its fruit extract was also reported to be insecticidal when tested against the green peach aphid (McIndoo, 1983). Chandel et al. (1897) registered its fruit extract in acetone to show toxicity to *Epilachna vigintioctopunctata*, a coccinellid defoliator, when tested under controlled conditions.

722. *Sassafras albidum* (Nutt.) Nees. (Lauraceae)

This is a perennial tree found in tropical and temperate zones. Its leaf, stem and fruit extracts in water showed juvenile hormone mimic activity to the milkweed bug, *Oncopeltus fasciatus* (Jacobson et al., 1975).

723. *Saturija hortensis* Linn. (Lamiaceae)

(Syn.-*Calamintha hortensis* Linn.)
This is an annual herb found in the Mediterranean zone and commonly available in Jammu and Kashmir in India. Its oil and expressed juice were reported to show repellent activity to the cotton aphid, *Aphis gossypii* (McIndoo, 1983).

724. *Saussurea lappa* C. B. Clarke. (Asteraceae)

(Syn.-*Aucklandia costus* Falc.)
Common name: Costus or Kut root
This is an annual/perennial herb or shrub found in temperate zones. Its seed powder extract was reported to show toxicity to the pulse beetles, *Callosobruchus chinensis* and *C. maculatus* when tested under controlled conditions (Anonymous, 1980–1981). Its root and leaf dried powders were found to protect wheat grains from infestation of the small khapra beetle, *Trogoderma granarium* grubs and showed repellent activity against the red flour beetle, *T. castaneum* and antifeedancy against the lesser grain borer, *R. dominica* (Malik and Mujtaba, 1984).

However, its root extract in petroleum ether was reported to be toxic to the cotton stainer, *Dysdercus koenigii* (Chandel et al., 1984) and the surface grasshopper, *Chrotogonus trachypterus* (Hameed, 1983). Nigam et al. (1990) found its extract to significantly reduce the populations of the mustard aphid, *Lipaphis erysimi* and also to enhance the grain yield of mustard.

725. *Schima khasiania* Dyer. (Theaceae)

(Syn.-*S. wallichii* [D.C.] Dyer., *S. mollis* Dyer.)

Common name: Needle wood

This is a perennial large tree found in sub-tropical and temperate regions. In India, it is distributed in East Himalayas from Nepal to Assam and Khasi Hills and Manipur. Its leaf extract was reported to show moderate antifeedancy to the jute hairy caterpillar, *Spilosoma obliqua* (Tripathi et al., 1987).

726. *Schinus terebinthifolius* Raddi. (Anacardiaceae)

(Syn.-*S. aroeira*)

Common name: Christmas berry tree or Brazilian pepper tree

This is an ornamental perennial shrub found in temperate and sub-tropical zones and native to Brazil. From its dried leaf powder, a candidate alkaloid was isolated as an active component, which alone and also in combinations with synthetic insecticides was found to show toxicity to the larvae of Egyptian cotton leaf worm, *Spodoptera littoralis* (Abbassy, 1982).

727. *Schkuhria pinnata* (Lan.) O.Kurntz (Asteraceae)

This is an annual herb found in temperate and sub-tropical zones. Its leaf and whole plant extracts in water and petroleum ether were reported to show antifeedant activity against Mexican bean beetle, *Epilachna varivestis* and African armyworm, *Spodoptera exempta* (Waespe and Hans, 1979). From its whole plant extract, two germacranolids active components, i.e., schkurin-I (Figure 98) and schkurin-II (Figure 99), were isolated, identified and found to show antifeedant activity against African armyworm, *Spodoptera exempta* and Mexican bean beetle, *Epilachna varivestis* (Kubo and Nakanishi, 1978) and toxicity to Mexican bean beetle, *E. varivestis* but were also found to show carcino-static activity (Jacobson, 1979).

728. *Schoenocaulon officinale* A. Gray (Lilliaceae)

(Syn.-*Sabadila officinarum* Brant.)

Common name: Sabadilla

This is an annual herb found in tropical zones and commonly available in Mexico, Central and South America. Its bulb and seed extracts in water, alcohol and petroleum ether and their dried powders were reported to be toxic to the silkworm, *Bombyx mori* (McIndoo and Sievers, 1924); the lygaed bug, *Lygus elisum* (Feinstein, 1952); the melon worm, *Diaphania hyalinata*; the codling moth, *Laspey-*

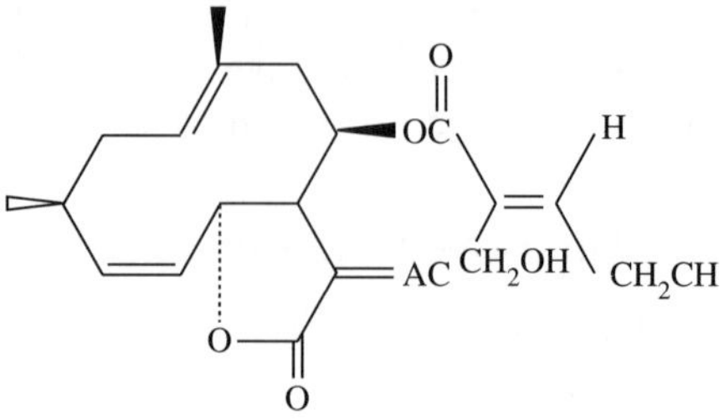

Figure 98. Schkurin-I.

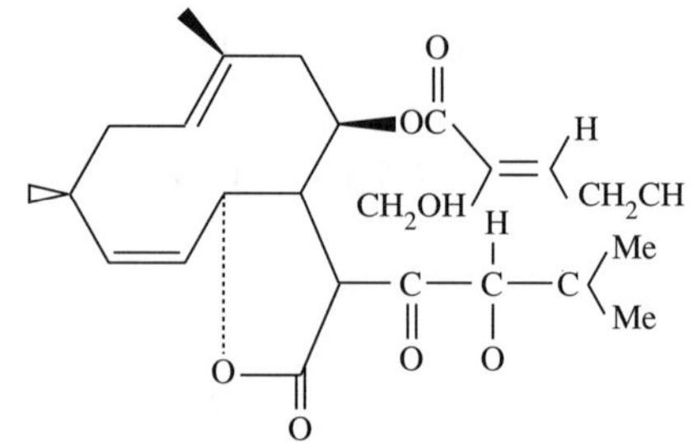

Figure 99. Schkurin-II.

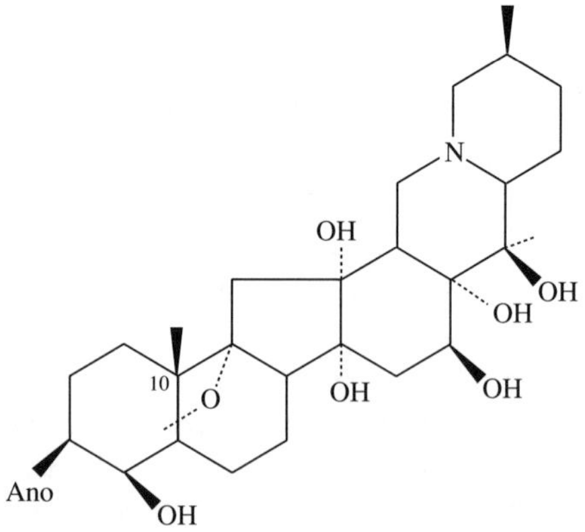

Figure 100. Cevadine.

resia pomonella; the red-legged grasshopper, *Melanoplus femmurubrum*; European corn borer, *Ostrinia nubilalis*; the armyworm, *Mythimna* (*Pseudoletia*) *unipunctata*; the southern armyworm, *Spodoptera eridania*; the cabbage looper, *Trichoplusia ni* (Jacobson, 1958); the potato leafhopper, *Empoasca fabae*; tobacco thrips, *Thrips tabci* (Jacobson, 1975); and the rice noctuid, *Spodoptera abyssina* (Litsinger et al., 1981). Allen et al. (1945) reported an active component, 'cevadine' (Figure 100) (C_{32} $H_{12}NO_9$), an alkaloid from its extract, and found it to be toxic to the red-legged grasshopper, *Melanoplus femmurubrum*.

729. *Semecarpus anacardium* Linn. (Ancistrocladaceae)

(Syn.-*Anacardium orientale* Linn.)
Common name: Varnish tree
This is a perennial tree found in tropical zones. Its fruit and seed expressed juice, oil and also their dried powders were reported to be toxic to the white ants and the caterpillars (Jacobson, 1958; Usher, 1973).

730. *Senecio canus* Linn. (Asteraceae)

This is a perennial herb found in arid zones. Its leaf and flower extract in water and their dried powder were reported to be toxic to the vinegar fly, *Drosophila melanogaster* (Jacobson, 1975).

731. *Senecio cineraria* Linn. (Asteraceae)

Common name: Dusty miller
This is a perennial herb found in the Mediterranean zone. Its whole plant extract in water was reported to show antifeedant activity to the diamond-back moth, *Plutella xylostella* (Jacobson, 1975).

732. *Serratula sogdiana* Bunge. (Asteraceae)

This is a perennial herb found in sub-tropical and temperate zones. Its leaf and inflorescence extracts containing phytoecdysones were reported to show meta-toxic effect on the 3rd instar larvae of the gypsy moth, *Lymantria dispar* (Ganiev, 1987).

733. *Seruvium portulacastrum* Linn. (Aizoaceae)

This is a palophytic plant found to grow abundantly along the seacoasts in India. Its whole plant was reported to be one of the best sources for 'ecdysterone' (Figure 101), a moulting hormone, and 'ecdysone' (Figure 102), a rare moulting hormone (Banerji and Chadha, 1970). The ecdysterone, another moulting hormone found in this plant, was used commercially in the synchronisation of spinning of the larvae of the silkworm, *Bombyx mori* and thus facilitated the collection of silk fibre.

734. *Sesamum indicum* Linn. (Pedaliaceae)

(Syn.-*S. orientale* Linn.)
Common name: Sesame or Gingilly
This is a seasonal herb grown in tropical zones. In India, it is commonly cultivated as an oil seed crop in Uttar Pradesh, Madhya Pradesh, Andhra Pradesh, Tamil Nadu, Rajasthan, Maharastra, Gujrat and Orissa. Sesame oil

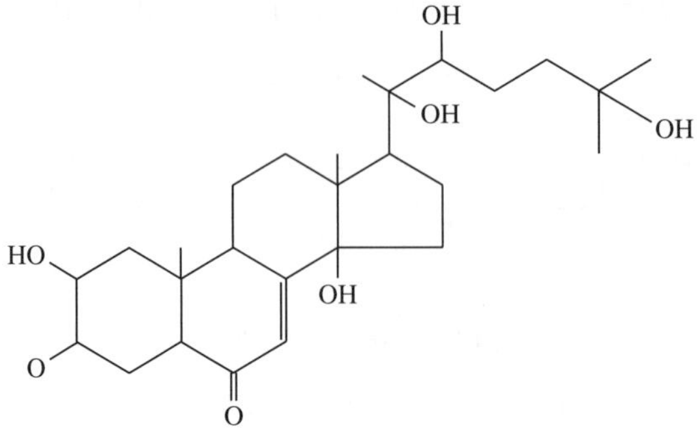

Figure 101. Ecdysterone.

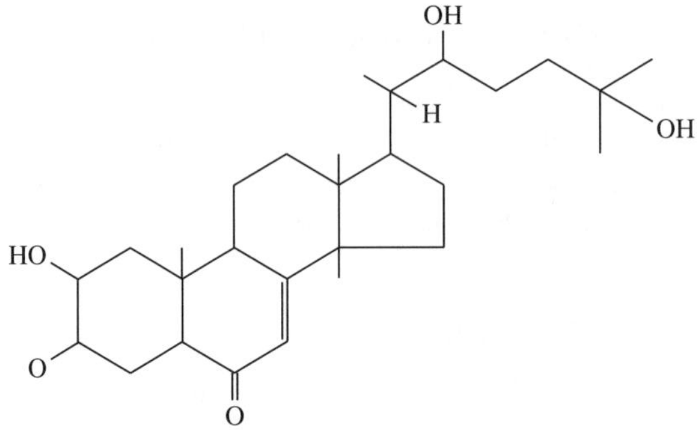

Figure 102. Ecdysone.

was reported to be a grain protectant when admixed with the wheat grain and it checked the infestation of the lesser grain borer, *R. dominica* and the rice weevil, *S. oryzae* (Fletcher, 1919). Su et al. (1972) reported its oil to inhibit the population build up of the insects in stored grains. However, its oil showed synergistic effect with synthetic insecticide to the red flour beetle, *T. castaneum* (Parmar, 1977). Majumder and Amla (1977) and Ali et al. (1983) reported toxicity to the pulse beetle, *Callosobruchus chinensis* and also checked its multiplication in stored greengram admixed with the sesame oil @ 0.3% w/w. Similarly, Doharey et al. (1983, 1984a,b, 1987) reported its oil to protect greengram seeds from infestation of the pulse beetles, *C. chinensis* and *C. maculatus* when admixed with the oil @ 1% w/w and found it to show toxicity to the eggs of these beetles and also to inhibit their multiplication. Sujatha and Punnaiah (1984) found 0.5% sesame oil to check 100% multiplication of *C. chinensis* in the treated blackgram. Trivedi (1987) reported sesame oil to enhance the effectiveness of permethrin dusts when admixed with its oil and tested against *T. cartaneum* and *R. dominica*. Further, Rao et al. (1990) reported its oil @ 0.3% w/w in the treated pigeonpea seeds to protect them against *C. chinensis*. Jacob and Sheila (1990) found sesame oil treatment of greengram seeds @ 1 ml/kg as a most effective grain protectant against *C. chinensis*.

Sesamin (Figure 103) ($C_{20}H_{18}O_{16}$) and sesamolin, two active components isolated from its oil, were reported to show synergistic activity by enhancing the toxicity of pyrethrins (Jacobson and Crosby, 1971).

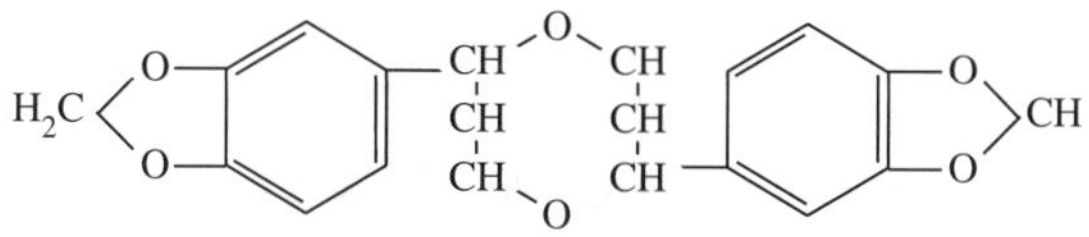

Figure 103. Sesamin.

735. *Shibataea kumasasa* Makino (Poaceae)

Common name: Bamboo

This is a perennial herb/shrub found in tropical and sub-tropical zones. Its branch and leaf extracts in water were reported to be toxic to the vinegar fly, *Drosophila hydei* (Jacobson, 1958).

736. *Shorea robusta* Roxb. (Dipteraceae)

Common name: Dammer

This is a perennial tree found in sub-tropical and temperate zones. It is commonly available in the Himalayan tract in India. Its seed oil was reported to protect stored greengram from the attack of the pulse beetles, *Callosobruchus chinensis* and *C. maculatus* (Pandey et al., 1981).

737. *Sidalcea oregana* (Nutt. ex. Torr. & Gray (Malvaceae))

This is a perennial shrub found in West and North America. Its root, stem, leaf and fruit extracts in ethyl ether attracted both the male and female of the melon fly, *Dacus cucurbitae* and Oriental fruit fly, *Dacus dorsalis* (Keiser et al., 1975).

738. *Simaba multiflora* Aubl. (Simaroubaceae)

This is a perennial evergreen tree found in tropical, sub-tropical and temperate zones. Some anti-leukomic and cytotoxic quassinoids isolated from this plant inhibited growth and development of fall armyworm, *Spodoptera frugiperda* and the tobacco bud worm, *Heliothis virescens* under laboratory tests (Klocke et al., 1989).

739. *Smilax stebold* Linn. (Liliaceae)

This is an annual/perennial vine found in tropical zones. Its seed extract in water was reported to be toxic to the vinegar fly, *Drosophila hydei* (Jacobson, 1958).

740. *Solanum aculeatissimum* Jacq. (Solanaceae)

This is a spiny perennial shrub found in tropical zones and commonly available in Assam and Kerala in India. Its aqueous leaf extract was reported to show insecticidal and antifeedant activities when tested against Colorado potato beetle, *Leptinotarsa decemlineata* (Jacobson, 1975).

741. *Solanum auriculatum* Linn. (Solanaceae)

Common name: Fuma bean

This is an annual herb found in tropical and temperate zones. Its leaf and fruit extracts in water were reported to show insecticidal and attractant activities when tested against Colorado potato beetle, *Leptinotarsa decemlineata* (Jacobson, 1975).

742. *Solanum brachystachys* Dun. (Solanaceae)

This is an annual herb found in temperate zones. Its twig, leaf, flower and fruit extracts in ethyl ether attracted the adults of *Ceratitis capitata* (Keiser et al., 1975).

743. *Solanum calvescens* Linn. (Solanaceae)

This is an annual herb found in tropical and temperate zones. Its whole plant extract in water was reported to show antifeedant activity against Colorado potato beetle, *Leptinotarsa decemlineata* (Jacobson, 1975).

744. *Solanum chacoense* (Linn.) Cav. (Solanaceae)

This is an annual herb found in temperate and sub-tropical zones. Its leaf extract in water was also reported to show antifeedant activity to Colorado potato beetle, *Leptinotarsa decemlineata* (Jacobson, 1975). However, earlier Dahlman and Hibbs (1967) isolated active components, viz., solanine, solanidine and demissidine, from the leaf extract of this herb and reported them to inhibit the larval development of the potato leafhopper, *Empoasca fabae*.

745. *Solanum demissum* Linn. (Solanaceae)

This is an annual herb found in tropical and temperate zones. Its leaf extract in water was reported to show insecticidal and antifeedant activities against Colorado potato beetle, *Leptinotarsa decemlineata* (Jacobson, 1975).

746. *Solanum dulcamara* Linn. (Solanaceae)

Common name: Woody night shade or Bitter-sweet

This is a perennial woody climbing shrub found in temperate zones. In India, it is available in the Himalayas from Jammu and Kashmir to Sikkim. Its aqueous seed extract in water was reported to be toxic to the cotton stainer, *Dysdercus koenigii* under laboratory test using dry film method (Chandel et al., 1984).

747. *Solanum indicum* Linn. (Solanaceae)

Common name: Indian night shade
This is an annual herb found in tropical regions and commonly available in North and Central States of India. Its seed extract in petroleum ether was reported to show toxicity to *Epilachna vigintioctopunctata*, a defoliating coccinellid pest, when tested under laboratory conditions (Chandel et al., 1987).

748. *Solanum jasminoides* Paxt. (Solanaceae)

Common name: Potato vine
This is an annual herb found in sub-tropical zones. Its leaf extract in water was reported to show toxicity and antifeedant activity when tested against Colorado potato beetle, *Leptinotarsa decemlineata* (McIndoo, 1983).

749. *Solanum luteum* Linn. (Solanaceae)

This is an annual herb found in tropical zones. Its leaf extract in water was reported to exhibit antifeedant activity against Colorado potato beetle, *Leptinotarsa decemlineata* (Jacobson, 1975).

750. *Solanum melongene* Linn. (Solanacae)

Common name: Eggplant or Brinjal
This is an annual herb found in temperate and sub-tropical zones. Its leaf and fruit extracts and juices were reported to inhibit the oviposition of the cotton bollworm, *Earias vittella* (Dongre and Rahalkar, 1984).

751. *Solanum nigrum* Linn. (Solanaceae)

Common name: Black night shade
This is an annual herb found in tropical zones. Its whole plant and fruit extracts in water were reported to show insecticidal and antifeedant activities against the sugarcane woolly aphid, *Ceratovacuna longigera* (McIndoo, 1983).

752. *Solanum polyadenium* Carr. (Solanaceae)

This is an annual herb found in temperate zones. Its leaf extract in water was reported to show antifeedant activity against Colorado potato beetle, *Leptinotarsa decemlineata* (Jacobson, 1975).

753. *Solanum tuberosum* Linn. (Solanaceae)

Common name: Potato

This is an annual bushy herb found in temperate zones. This is a native to South America but available throughout India. Its leaf extract in water was reported to show growth inhibiting activity when tested against the peach aphid, *Myzus persicae* (Agustin, 1983).

754. *Solanum xanthocarpum* Schrad. & Wendl. (Solanaceae)

(Syn.-*S. surattense* Burm f.)

Common name: Yellow berried night shade

This is a perennial spiny diffused herb found in tropical zones and available throughout India and in neighbouring countries. Its whole plant and leaf extracts in water were reported to show insecticidal and antifeedant activities when tested against the cotton stainer, *Dysdercus cingulatus*; the cotton leaf armyworm, *Spodoptera litura* (Balasubramanian, 1982); and brinjal spotted beetle, *Henosepilachla vigintioctopunctata* (Dhandapani et al., 1985). Its leaf crude extract was also reported to be toxic to the sugarcane top-shoot borer, *Tryporyza nivella* (Pandey, 1988).

755. *Solidago microglossa* Linn. (Asteraceae)

Common name: Goldenrod genus

This is a perennial herb found in temperate zones. Its whole plant extract in ether was reported to show antifeedant and insecticidal activities when tested against the red-legged grasshopper, *Melanoplus femmurubrum* (Hael et al., 1950).

756. *Solidago missouriensis* Mill. (Asteraceae)

Common name: Goldenrod genus

This is a perennial herb found in temperate zones. Its whole plant and leaf extract in water were reported to show insecticidal property to the tulip tree aphid, *Macrosiphum liriodendri* (Jacobson, 1975).

757. *Sophora flavescens* Linn. (Fabaceae)

Common name: Matrine

This is a perennial tree found in tropical and sub-tropical zones. Its root, branch and leaf extracts in water showed toxicity to the bean aphid, *Aphis fabae* and the spotted cucumber beetle, *Diabrotica undecimpunctata howardi* (Jacobson, 1958).

758. *Sophora pachycarpa* Linn. (Fabaceae)

This is a perennial tree found in sub-tropical zones. Its leaf and root extracts in water were reported to show insecticidal activity to the aphids (McIndoo, 1983).

759. *Sophora secundiflora* D.C. (Fabaceae)

Common name: Coral bean or Mescal bean
This is a perennial ornamental shrub/small tree found in sub-tropical and semi-arid zones. This is native to USA but also available in India. Dried powders of its branch and seed were reported to show insecticidal activity to the cotton leaf army-worm, *Spodoptera litura* (Jacobson, 1958).

760. *Sorghum biclor* (Linn.) Moerch. (Poaceae)

Common name: Sorghum
This is an annual/seasonal stout grass found in tropical, sub-tropical and semi-arid zones. Its whole plant and root extracts in water were reported to be toxic to the western corn root worm, *Diabrotica vergifera* (Jacobson, 1975).

761. *Soulamea soulameoides* Lam. (Simaroubaceae)

This is a perennial shrub/small tree found in New Guinea and Fiji. Several quassinoids isolated from its leaf extracts were reported to show insect growth regulating activity when tested against the fall armyworm, *Spodoptera frugiperda* and the tobacco budworm, *Heliothis virescens* (Klocke et al., 1989).

762. *Spatholobus roxburghii* Benth. (Papilionaceae)

(Syn.-*Butea parviflora* Roxb.)
This is a perennial gigantic woody climber found in tropical zones and available throughout India and Burma. Its dried root powder was reported to be toxic to Mediterranean fruit fly, *Ceratitis capitata* and the fruit fly, *Dacus sp.* (Loke-Wai-Hong, 1983). The active component from its root extract was also reported to be 'rotenone' (Jacobson and Crosby, 1971).

763. *Sphaernthus indicus* Linn. (Asteraceae)

This is an aromatic herb found in damp habitats in the plains of India. Its whole plant extract in petroleum ether was reported to show juveno-mimetic activity to the 5th instar larvae of the cotton stainer, *Dysdercus cingulatus* and also reported to cause morphogenetic changes in this bug (Rajendran and Gopalan, 1978).

764. ***Spilanythes oleraceae*** **Jacq. (Asteraceae)**

Common name: Brazilian cress or Para cress

This is an ornamental perennial and succulent herb found in Brazil and other parts of South America and now cultivated in India. From its flower head extract, an active component, 'spilanthol' (Figure 104), an unsaturated isobutylamide having a chemical formula $C_{14}H_{27}NO$ (N-isobutyl-2,6,8-decatrienamide), was isolated, which showed insecticidal activity against a wide range of insects of agricultural importance (Jacobson and Crosby, 1971).

$$CH_3CH{=}CHCH{=}CH(CH_2)_2CH{=}CHCO\ NH\ CH_2\ (CH_3)_2$$

Figure 104. Spilanthol.

765. ***Spilanthes mauritiana*** **Linn. (Asteraceae)**

This is an annual herb found in tropical zones and commonly available in Kenya. Hassanali and Lwande (1989) reported insecticidal isobutylamides from insecticidal isobutylamides from its whole plant aqueous extract, which were toxic to the insect pests of legume crops.

766. ***Spinaea nipponica*** **Rafin. (Rosaceae)**

Common name: Tosa spirea

This is a perennial/biennial shrub found in temperate and tropical zones. Its meristematic tissue extract in water was reported to be toxic to the vinegar fly, *Drosophila hydei* (Jacobson, 1958).

767. ***Spingelia anthelmia*** **Linn. (Spigeliaceae)**

Common name: India pink or Pink root

This is a perennial shrub found in tropical and sub-tropical zones and available in South America. Its leaf extract in water was reported to be toxic to the leaf eating caterpillars (Jacobson, 1958).

768. ***Stachys riederi*** **Linn. (Lamiaceae)**

This is an annual herb found in tropical zones. Its whole plant extract in water was reported to be toxic to the vinegar fly, *Drosophila hydei* (Khan, 1982).

769. ***Stachytarpheta mutabilis*** **(Jacq.) Vahl. (Verbenaceae)**

This is an annual shrub found in tropical zones and native to tropical America and India. An active component as a novel iridoid glucoside named 'ipolamide' isolated from its leaf extract showed antifeedancy activity when tested against the locust, *Schistocerca gregaria* and the noctuid, *Spodoptera littoralis* (Bernays and Luca, 1981).

770. *Stachytarpheta urticaefolia* (Salisb.) Sims. (Verbenaceae)

(Syn.-*S. indica* Auct. non Vahl.)

This is a perennial herb found in tropical zones and available in South India. It is commonly available/grown in gardens as a hedge plant with blue flowers. Its leaf extract was found to be toxic to *Spodoptera litura* when sprayed on the castor leaves (Subhadrabai and Kandaswamy, 1985).

771. *Stemmadenia glabra* Benth. (Apocynaceae)

This is an annual/perennial shrub found in tropical zones. Its dried root powder was reported to be toxic to the bean aphid, *Aphis fabae* (Jacobson, 1958).

772. *Stemona tuberosa* Lour. (Roxburgiaceae)

Common name: Nohn-tai-yahk

This is an annual climbing herb found in tropical zones. In India, it is available in Assam, West Bengal and southward to peninsular India. Its root extract in water and dried powder were reported to be toxic to the crickets, weevils and caterpillars of general importance (McIndoo, 1983).

773. *Strychnos nux-vomica* Linn. (Loganiaceae)

(Syn.-*S. lucida* Wall.)

Common name: Nux vomica or Struchnine tree

This is a perennial tree found in tropical and sub-tropical regions. It is commonly available in Tamil Nadu and North India up to the height of 1000 m and in the Philippines. Chandel et al. (1987) reported toxicity in acetone extract of its dried stem powder when tested against *Epilachna vigintioctopunctata*, a defoliating coccinellid. Quisumbing (1951) found alkaloids like strychnine and brucine from its leaf extract to possess insecticidal activity against termites.

774. *Swartzia madagascariensis* Schreb. (Leguminosae)

This is a perennial shrub/small tree found in tropical zones. Its seed and fruit dried powders were reported to show repellent activity to the stored grain insect pests and toxicity to the termites of agricultural importance (Jacobson, 1975).

775. *Swertia chirata* Buch. Han. ex. CB. Clarke (Gentianaceae)

(Syn.-*S. chirayita* Roxb. ex. Flem. Karst.)
Common name: Chirata or Chiretta
This is an annual and perennial herb found in alpine and tropical zones and available in North India. Its leaf extract sprayed on the jute plant inhibited the development and growth of the sunn-hemp pest, *Utethesia pulchella* (Singh and Pandey, 1979) and was found to be toxic to the Japanese beetle, *Popillia japonica* (McIndoo, 1983). Sharma et al. (1992) reported its whole plant extract in water to show antifeedant activity against 3rd instar larvae of the jute semi-looper, *Anomis sabulifera* and 4th instar larvae of Bihar hairy caterpillar, *Spilosoma obliqua*.

776. *Swientenia mahagoni* (Linn.) Jacq. (Meliaceae)

Common name: West Indian mahogany
This is a perennial tree found in tropical and sub-tropical zones but native to Central America. It is now cultivated as a timber tree in South India. Its seed extract in water was reported to be highly toxic to the common cutworm, *Spodoptera litura* (Rajesekaran and Kumaraswami, 1985b). Jacobson (1989) reported tetra-nor-triterponoid, i.e., swintenine, an active component, to show biological activity against the insect pests of agricultural importance.

777. *Syzygium aromaticum* (Linn.) Merr. & Ferry (Myrtaceae)

(Syn.-*Egenia aromatica* O. Kuntze., *E. caryophyllata* Thunb., *Caryophyllus aromaticus* Linn.)
Common name: Clove
This is a perennial evergreen tree found in tropical zones. It is commonly grown in Tamil Nadu and Kerala in India. Its seed oil was reported to show toxicity to the pulse beetles, *Callosobruchus chinensis* and *C. maculatus* (Anonymous, 1980–1981). Sighmony et al. (1986) reported toxicity in its oil when tested against the rice weevil, *Sitophilus oryzae* and the lesser grain borer, *Rhyzopertha dominica*.

778. *Tabebuia flavescens* Gomes (Begoniaceae)

Common name: Trumpet tree genus
This is a perennial tree found in tropical and sub-tropical zones. Its leaf extract in water was toxic to the termites (Jacobson, 1975).

779. *Tabebuia ipe* (Gomes) ex. D.C. (Begoniaceae)

This is a perennial tree found in tropical zones. Its leaf extract in water was reported to be toxic to the termites (Jacobson, 1975).

780. *Tabebuia rosea* (Bertol.) D.C. (Begoniaceae)

(Syn.-*Tecoma rosea* Bertol.)

Common name: Pink poni or Mauve tabebuia

This is an ornamental perennial small tree found in tropical zones. Its flower extract in petroleum ether was toxic to *Sitophilus oryzae* (Jacobson, 1975).

781. *Tacca pinnatifida* Forst. (Taccaceae)

(Syn.-*Tacca bontopetaloides* [Linn.] O. Kuntze.)

Common name: Indian arrowroot

This is an annual ornamental herb found in tropical zones and commonly cultivated in West Bengal, Central and Western Peninsula in India. Its root powder admixed with stored grains protected them from insect infestation in storage (Giles, 1964).

782. *Tagetes erecta* Linn. (Asteraceae)

Common name: African marigold or Big marigold or Aztee marigold

This is an annual ornamental herb found in tropical and sub-tropical zones and native to Mexico. African marigold is distributed in West Africa, Mexico and Japan. It is now commonly available in India. Its root extract in water was reported to show toxicity to the rice green leafhopper, *Nephotettix* spp.; rice brown planthopper, *Nilaparvata lugens*; and the black bean aphid, *Aphis rumicis* when applied topically to the test insects (Morallo-Rajessus and Eroles, 1978; Morallo-Rajessus and Decena, 1982). Rajendran and Gopalan (1978) found its whole plant extract to show juveno-mimetic activity to 5th instar nymphs of the cotton stainer, *Dysdercus cingulatus*. Its whole plant, root, leaf and flower extracts were also reported to show insecticidal activity to the red cotton bug, *Dysdercus cingulatus*; the cowpea aphid, *Aphis craccivora*; and Asian corn borer, *Ostrinia furnacalis* (Fagoonee, 1983). Mohiuddin et al. (1987) observed repellency in its oil to the red flour beetle, *T. castaneum*.

Morallo-Rajessus and Eroles (1978) isolated two purified active components, PA and PB, from this plant extract, which were toxic to the diamondback moth, *Plutella xylostella* when applied topically. These components were identified as 5-(3-buten-1-ynyl)-2-2' bithienyl (PA) and α-terthienyl (PB) and both the components were also reported to be toxic to the green leafhopper, *Nephotettix virescens* (Morallo-Rajessus and Decena, 1982).

783. *Tagetes minuta* Linn. (Asteraceae)

(Syn.-*T. glandulifera* Schrank)

Common name: Stinking Roger

This is an annual aromatic herb found in tropical zones but native to South America and now available in the North West Himalayas in India. Oil extracted from its leaf was reported to show juvenile hormone mimetic activity when tested on the cotton stainer, *Dysdercus cingulatus* (Saxena and Srivastava, 1973). Further, its leaf oil was also reported to be toxic to the maggots (Jacobson, 1975). Keiser et al. (1975) found its stem, leaf and flower extracts in ethyl ether as attractants to both male and female of the Mediterranean fruit fly, *Ceratitis capitata*.

784. *Tagetes patula* Linn. (Asteraceae)

Common name: French marigold

This is an annual herb/shrub found in tropical, sub-tropical and temperate zones. This is native to Mexico but now available in North India, France, Germany, USSR and North America. Morallo-Rajessus and Silva (1979) reported its root extract to show toxicity to the lesser grain borer, *Rhyzopertha dominica* and the red flour beetle, *Tribolium castaneum*. Its root extract in water was found to be toxic and also showed repellent activity against the cowpea aphid, *Aphis craccivora*; the rice green leafhopper, *Nephotettix virescens*; and Asian corn borer, *Ostrinia furnacalis* (Morallo-Rajessus and Eroles, 1980). Morallo-Rajessus (1982) reported its root extract to show toxicity to the rice brown planthopper, *Nilaparvata lugens*.

785. *Tamarindus indica* Linn. (Caesalpiniaceae)

This is a perennial tree found in tropical zones. It is commonly available in India and tropical Africa. Its planting was suitably used to control outbreak of the red locust, *Nomadacris septemfaciata* (Roberston, 1958).

786. *Tanacetum vulgare* Linn. (Asteraceae)

(Syn.-*Chrysanthemum vulgare* [Linn.] Bernmh.)

Common name: Tansy

This is a perennial aromatic herb or woody climber found in temperate zones and available in North America and Europe. Extract and oil prepared from its flower were reported to show toxicity and repellent activities when tested against the imported cabbage worm, *Pieris rapae* and the aphids and ants of agricultural importance (McIndoo, 1983). Similarly, steam distillates of its fresh leaf and flower also showed repellency to Colorado potato beetle, *Leptionotarsa decemlineata* (Schearer, 1984).

787. *Taraxacum officinalis* Weber ex. Wiggers (Asteraceae)

Common name: Dandelion

This is a perennial herb found in temperate zones. In India, it is commonly available in the Himalayas and in Khasi and Mishni hills, Gujarat and hills of

South India. Its leaf extract in water was reported to show repellent activity against Colorado potato beetle, *Leptinotarsa decemlineata* (Jacobson, 1975).

788. *Techtona grandi* Linn. (Verbenaceae)

Common name: Teak

This is a perennial tree found in tropical zones. In India, it is available in Western Peninsula, Central India and Bihar. Its wood extract in water was reported to show repellent activity to the termites (McIndoo, 1983).

789. *Teclea grandifolia* Erign. (Rutaceae)

This is a perennial small tree found in sub-tropical zones. An active component, 'tecleanin', a tetranortriterpenoid isolated from its root bark, was reported to show antifeedant activity against *Spodoptera frugiperda* (Jacobson, 1989).

790. *Teclea trichocarpa* Engn. (Rutaceae)

This is a perennial evergreen tree found in tropical Africa, the Comoros and Madagascar. Two active components isolated from its seed extract, viz., 'meli-copicene' and 'tecleanthine', were reported to show antifeedant activity against *Spodoptera exempta* at 1000 and 5000 ppm respectively in a leaf-disc method of laboratory test (Lwande et al., 1983).

791. *Tecoma indica* Juss. (Begoniaceae)

Common name: Yellow bell genus

This is a perennial tree found in tropical and sub-tropical zones. Its flower extract in water was reported to show insecticidal activity against the rice weevil, *Sitophilus oryzae* (Jacobson, 1975).

792. *Tephrosia apollinea* Pers. (Fabaceae)

This is a perennial wild shrub found in tropical and sub-tropical zones and commonly available in Egypt. Rotenoids extracted from its seeds at 0.1% concentration caused 100% mortality to *Aphis craccivora* in 48 h; however, its leaf rotenoids were less toxic to the aphids (Salem and Abdel-Hafex, 1990).

793. *Tephrosia candida* (Roxb.) D.C. (Fabaceae)

(Syn.-*Lonchocarpus fruiticosus* Span.)

Common name: White tephrosia

This is a perennial gregarious shrub found in tropical zones and distributed in India and other South East Asian countries and Japan. Krishnamurthi and Rao

(1950) reported its seed powder to protect wheat grains from the rice weevil, *Sitophilus oryzae* when admixed with the grains, whereas its seed extract was found to be toxic to *Spodoptera litura*, *Crocidolomia binotalis*, *Plutella maculipennis*, *Idiocerus* sp. and *Callosobruchus chinensis* (Puttarudriah and Bhatia, 1955). Its root bark, branch and leaf extracts were also reported to be toxic to the bean aphid, *Aphis favae* (McIndoo, 1983). Jha and Roychoudhury (1988) found its seed powder to protect stored grain from *Sitophilus oryzae*. Vir (1990) reported its seed extracts foliar spray on the moth bean to protect crop significantly from infestation of the white fly, *Bemisia tabaci*. The active components in its seed extract were reported to be 'rotenone' and 'tephrosin', which were found to show toxicity to the insects tested under controlled laboratory tests (De-Ong, 1948).

794. *Tephrosia elata* Deff. (Fabaceae)

This is an annual herb found in tropical and sub-tropical zones. Two active components, i.e., tephrosin and isopongaflavone, isolated from its methanol seed extract showed antifeedancy to southern armyworm, *Spodoptera exempta*; bean pod borer, *Maruca testulalis*; and African sugarcane borer, *Eldana saccharina*. Tephrosin was reported to be most effective to *S. exempta*, whereas isopongaflavone showed foremost antifeedancy to *M. testulalis* and *E. saccharina* (Bentley et al., 1987).

795. *Tephrosia hildebrandti* (Linn.) Pers. (Fabaceae)

This is an annual/perennial herb found in tropical zones. From its leaf and shoot extracts, a pterocarpan known as 'hildecarpin' (Figure 105) was found as an active component to show antifeedant activity against the legume pod borer, *Maruca testulalis* (Hassnali and Lwande, 1989).

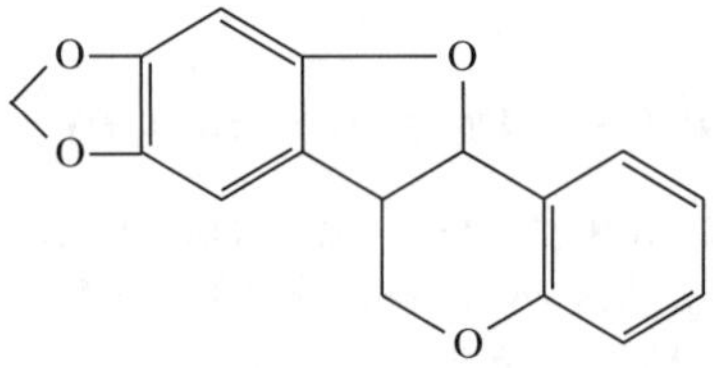

Figure 105. Hildecarpin.

796. *Tephrosia macropoda* Pers. (Fabaceae)

This is an annual herb found in tropical and sub-tropical zones. Its aqueous leaf extract was reported to be toxic to the winter moth, *Cheimatobia bruma* (Michael et al., 1985).

797. *Tephrosia noctiflora* Pers. (Fabaceae)

This is an annual herb found in tropical and sub-tropical zones. Its leaf extract in water was toxic to the aphids (Worsley and Le, 1934).

798. *Tephrosia nyikensis* (Linn.) Pers. (Fabaceae)

This is an annual herb found in tropical and sub-tropical zones. Its aqueous leaf extract was reported to be toxic to the aphids (Worsley and Le, 1934).

799. *Tephrosia purpurea* (Linn.) Pers. (Fabaceae)

Common name: Barbasco or Purple tephrosia

This is a perennial herb/small shrub found in tropical and sub-tropical zones and commonly available in India. Its root and leaf extracts in water and ether were reported to be toxic and repellent to the hairy caterpillar, *Euproctis fraterna* (Hameed, 1982) and also to the cotton leaf armyworm, *Spodoptera litura* (Krishnamurthy Rao, 1982). Mathew and Chauhan (1994) also found its leaf extract in methanol and dichloromethane (1:1) to show antifeedant and insecticidal activities when tested against 3rd and 4th instar larvae of *S. litura*. An active component, 'isoflavone', isolated from its root extract was reported to show antifeedant activity to *S. litura* (Srimmannarayan and Rao, 1985).

800. *Tephrosia rosea* Linn. (Fabaceae)

(Syn.-*T. cracca*)

This is a perennial tree found in tropical zones. It is commonly available in Central Asia. Its root and leaf extracts were reported to possess insecticidal property to the bean aphid, *Aphis fabae* and two active components, i.e., rotenone and tephrosin, isolated from its root extract also showed insecticidal activity to the bean aphid, *Aphis fabae* (De-Ong, 1948).

801. *Tephrosia toxicara* Pers. (Fabaceae)

Common name: Fish death tephrosia

This is an annual or perennial shrub found in tropical and sub-tropical zones. Its root extracts in water, alcohol and petroleum ether were reported to show insecticidal and antifeedant activities to the winter moth, *Cheimatobia bruma* (Jacobson, 1958).

802. *Tephrosia villosa* Pers. (Fabaceae)

This is an annual herb found in tropical and sub-tropical zones and available in Central India. Its root and seed extracts in ether were reported to show repellent and insecticidal activities to the mustard webworm, *Crocidolomia binotalis* (Puttarudriah and Bhatia, 1955); the cotton stainer, *Dysdercus koenigii*; and the mustard aphid, *Lipaphis erysimi* (Borle, 1982) and they were also toxic to the cotton leaf armyworm, *Spodoptera litura* (Krishnamurthy Rao, 1982).

803. *Tephrosia vogelii* Hook. f. (Fabaceae)

Common name: Vogel tephrosia
This is a perennial shrub found in tropical zones. It is a native to East Africa but now available in Assam and South India. Its leaf and seed extracts in water were reported to show insecticidal activity. The rotenoids present in these extracts also showed toxicity to the citrus aphid, *Aphis citri* and the bean aphid, *Aphis fabae* (McIndoo, 1983; Crooker, 1983). Its leaf and seed extracts were found to show antifeedant and repellent activities to the cabbage leaf webber, *Crocidolomia binotalis*; the spotted lady bird beetle, *Henosepilachna sparsa*; and the diamond-back moth, *Plutella xylostella* (Crooker, 1983). Delobel and Malonga (1987) reported its seed powder to be toxic to the groundnut beetle, *Caryedon serratus* and it reduced survival of the beetle by 90–98% when admixed @ 2.5% w/w with the groundnut.

Chiu (1989) reported its 10% leaf extract in acetone to remarkably inhibit the growth of 4th instar larvae of the diamond-back moth, *Plutella xylostella*, which were malformed by 86% and died after feeding on the treated host leaves for 48 h. Similarly, topical application @ 5% w/w of its extract to the larvae of the cabbage worm, *Pieris rapae* reduced 89% of its pupation because the leaves possessed nearly 5% of rotenone as a toxic active component.

804. *Terminalia catappa* Linn. (Combretaceae)

Common name: Tropical almond or Indian almond
This is a perennial tree found in tropical zones. Its aqueous leaf extract showed repellency to the Japanese beetle, *Popillia japonica* (McIndoo, 1983).

805. *Tetrapterys acutifolia* (Malpighiaceae)

This is an annual/perennial vine found in tropical and temperate zones. Its leaf extract in water was toxic to the leaf eating caterpillar (Jacobson, 1958).

806. *Thamnosma montana* Gamble (Rutaceae)

Common name: Turpentine broom
This is a perennial shrub or small tree found in sub-tropical zones and commonly available in India, South America and China. Hexane extract of this plant in artificial diet fed to 1st instar larvae of the tobacco budworm, *Heliothis virescens* caused growth inhibition and at a dosage of 2500 ppm caused toxicity. A number of phyto-chemicals as alkaloids and coumarins in this plant

were reported as active components. Psoralen (Figure 106), a furnocoumarin, was found to be highly phytotoxic, as were xanthotoxin (Figure 107) and bergapten and all these components inhibited the growth of 4th instar larvae of *H. virescens* (Klocke et al., 1989).

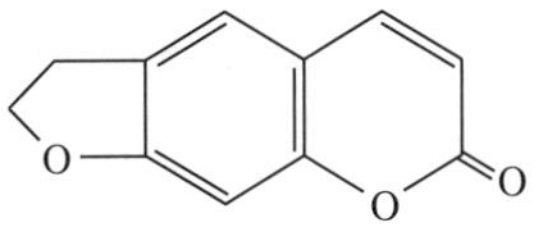

Figure 106. Psoralen.

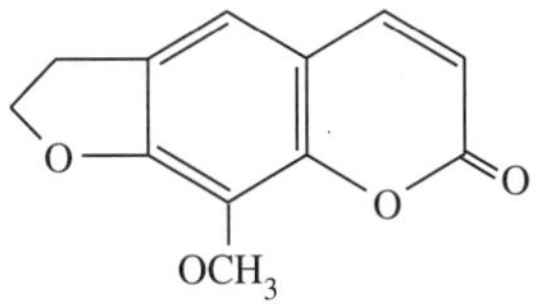

Figure 107. Xanthotoxin.

807. *Thevetia peruviana* (Pers.) K. Schum. (Apocynaceae)

(Syn.-*T. nerifolia* Juss. et. Steud., *Cerbera peruviana* Pers.)
Common name: Yellow oleander

This is a perennial small tree or shrub found in tropical zones. This is native to Central America but commonly available in India. Its leaf and fruit extracts and oil were found to be toxic to the aphids (Jacobson, 1958) and the pulse beetles, *Callosobruchus chinensis* and *C. maculatus* (Pandey and Singh, 1976). Pandey et al. (1986) reported its leaf, flower and bud extracts to show repellency to these beetles and reduced their oviposition and multiplication and finally protected stored greengram against the attack of *C. chinensis*. Dried leaf powder and also its extract, however, could not check multiplication of Angoumois grain moth, *Sitotroga cerealella* when admixed with stored paddy grains in the gunny bags for a period of 3 months (Tewari et al., 1989). Its leaf and fruit extracts in water were also reported to show toxicity to the cowpea aphid, *Aphis craccivora* (Morallo-Rajessus, 1982); jute hairy caterpillar, *Diacrisia obliqua* (Islam, 1983); *Aphis gossypii* (Saradamma, 1988); and also to *Sitophilus zeamais* and *Tribolium castaneum* (Morallo-Rajessus, 1984).

808. *Thevetia thevitioides* (H.B.K.) K. Schum. (Apocynaceae)

Common name: Mexican yellow oleander

This is a perennial small tree found in tropical zones and distributed in North India and Central America. Its seed extract in water was reported to possess two insecticidal active components, i.e., nerifolin and 2-acetylnerifolin, which were toxic to European corn borer, *Ostrinia nubilalis* (McLaughlin et al., 1980) and antifeedant and insecticidal activities against the two striped cucumber beetle, *Acalymma vittatum* and the codling moth, *Laspeyresia pomonella* (Reed et al., 1982).

809. *Thuja plicata D. Don.* (Cupressaceae)

(Syn.-*T. gigantica* Nutt.)

Common name: Western red cedar or Giant arbor-vitae

This is a perennial tree found in tropical and sub-tropical zones. This is a native of North America now available in India and Nepal. Its leaf and wood extracts in water were reported to be toxic to the mealworm, *Tenebrio molitor.* An active component named 'thujic acid' isolated from its wood extract was reported to show juveno-mimetic activity against this insect (Barton et al., 1972).

810. *Thymus vulgairs* Linn. (Lamiaceae)

Common name: Thyme or White clover

This is an annual/perennial aromatic herb native to Mediterranean regions and available in Nilgiri hills in India. Its seed and leaf extracts showed repellency to *P. xylostella* and antifeedancy to the larvae of *P. brassiae* (Dover, 1985).

811. *Tinospora rumphi* Boerl. (Menispermaceae)

(Syn.-*T. crispa* [Linn.] Miers ex. Hook. f. Thoms.)

Common name: Makabuhai

This is a perennial climbing shrub found in Assam in India, Sri Lanka, the Philippines and South Korea. Its root and stem incorporated with the soil were reported to show insecticidal property to the rice green leafhopper, *Nephotettix virescens* (Del Fierro and Morallo-Rajessus, 1976). Its submerged chopped stem incorporated with the soil of potted rice plant also showed toxicity to the rice green leafhoppers *Nephotettix* spp. (Morallo-Rajessus and Silva, 1979).

812. *Tithonia diversifolia* A. gray (Asteraceae)

Common name: Wild sunflower

This is a perennial small shrub/tree found in tropical zones and distributed in India, Japan and the Philippines. Its leaf extract were found to be toxic to the rice weevil, *Sitophilus oryzae* and the maize weevil, *Sitophilus zeamais* and the red flour beetle, *Tribolium castaneum* (Carino and Morallo-Rajessus, 1982; Carion, 1983). Dutta et al. (1986) observed feeding deterrency in leaf extract of this shrub against *Philosamia sicini*. Prakash et al. (1989) reported its leaf extract to reduce the population build-up of Angoumois grain moth, *Sitotroga cerealella* in treated paddy stored in the farm godown under natural conditions of insect infestation.

813. *Toddalia asiatica* (Linn.) Lam. (Rutaceae)

(Syn.-*T. aculeata* Pers.)

Common name: Wild orange tree

This is a perennial shrub found in tropical zones. Its leaf extract in ethyl ether attracted the Mediterranean fruit fly, *Ceratitis capitata* (Keiser et al., 1975).

814. *Toona ciliata* M.J. Roem. (Meliaceae)

(Syn.-*Cedrela toona* Roxb. ex. Rottl. & Willd.)

Common name: Red cedar or Cedrela tree

This is a perennial tree found in tropical zones and commonly available in the hill tracts of Central and South India, Malaysia, Indonesia, Sri Lanka and Japan. Its leaf extract was reported to show antifeedant activity against the attack of the mahogany shoot borer, *Hypsiphyla grandella* (Grijpma, 1970). Wheat seedlings soaked in its aqueous leaf extract or sprayed showed antifeedant activity to *Locusta* sp. (Omollo, 1983) and also to Mexican bean beetle, *Epilachna varivestis* (Waespe and Hans, 1979). Its leaf and root extracts in ethanol also showed antifeedant activity to the striped cucumber beetle, *Acalymma vittatum* (Jacobson, 1989). Sighamonx et al. (1986) reported its oil to show grain protection to the stored wheat grains against *S. oryzae* and *R. dominica*. Prakash et al. (1989) also found its leaf extract to protect the treated paddy in storage against *S. cerealella* and *R. dominica*.

From its leaf extract to β-secotetrano triterpenoids, viz., 'toonacilin' and '6-acetoxy toonacilin', showed strong antifeedant activity to the larvae of the moth, *Hipsiphyla grandella* (Kraus et al., 1978, 1980). A limonoid called 'cedrelone' (Figure 108) was reported to inhibit the moulting of the milkweed bug, *O. fascitus* and the variegated cutworm, *P. saucia* (Champagne et al., 1989). Another alkaloid known as 'anthothecol' (Figure 109) isolated from its seed extract was also found to inhibit the larval growth of *P. soucia* (Champagne et al., 1989).

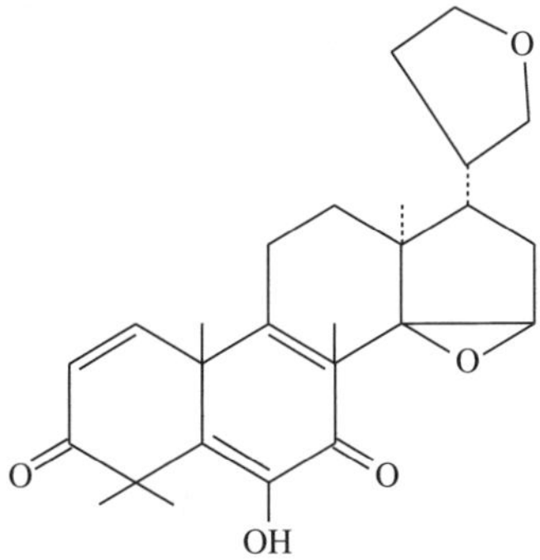

O O OH O O

Figure 108. Cedrelone.

O ACO O O O OH

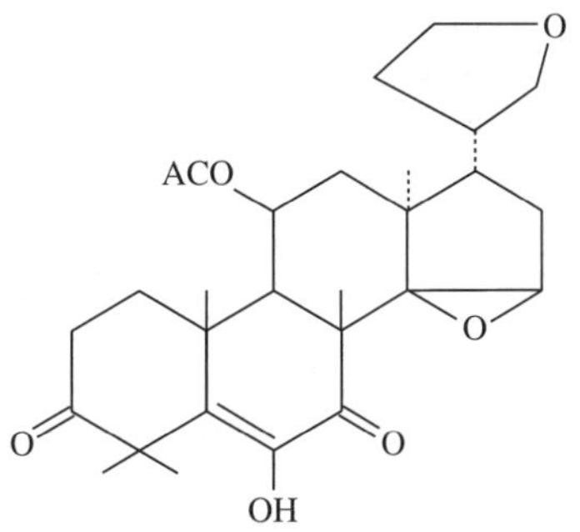

Figure 109. Anthothecol.

815. *Trewia nudiflora* Linn. (Euphorbiaceae)

Common name: False white teak

This is a perennial large tree found in tropical zones and commonly available in moist and coastal States of India and Bangladesh. Its seed extract in ethanol was reported to show antifeedant activity to the spotted cucumber beetle, *Diabrotica undecimpunctata howardi* and European corn borer, *Ostrinia nubilalis* and it showed morphogenetic changes in the codling moth, *Cydia pomonella*, disrupted the normal life cycle and reduced the progeny of the red banded leaf roller moth, *Argyrostaenia velutinana* and the plum curculio, *Conotrachelus nenuphar* and showed insecticidal activity to the striped cucumber beetle, *Acalymma vittatum* and also gave absolute control of *Menacanthus stramineus* for 28 days (Freedman et al., 1982).

Trewiasine, a metasteroid, was isolated as an active component from its seed extract and found to show toxicity to the adults of *Acalymma vittatum* and the larvae of *Cydia pomonella* (Reed et al., 1983).

816. *Tribulus terrestris* Linn. (Tetracentraceae)

Common name: Puncture vine or Land caltrops

This is an annual/perennial vine or prostrate herb available in tropical and sub-tropical zones and distributed in Indian desert, Afghanistan and Saudi Arabia. Its flower extract in ether was reported to show insecticidal and antifeedant activities when tested against the cotton stainer, *Dysdercus cingulatus* (Balasubramanian, 1982); the fall armyworm, *Spodoptera litura* (Balasubramanian, 1982); and the gram pod borer, *Heliothis armigera* and *Spodoptera litura* (Gunasekaran and Chelliah, 1985a,b). Narisimhan and Mariappan (1988) found the survival of the green leafhopper, *Nephotettix virescens* to be reduced when its flower and leaf extracts were sprayed on the rice seedlings.

Its leaf extract showed synergistic activity with synthetic pyrethroid when tested against *Antherigona soccata* (Kandalker and Narkhede, 1989). Deva Prasad et al. (1990) also reported synergistic activity in its methanolic extract along with nuclear polyhedrosis virus when tested against 3rd instar larvae of *Spodoptera litura*.

817. *Trichilia hispida* Penning (Meliaceae)

This is a perennial tree found in sub-tropical and tropical zones. Its branch extract in ethanol was reported to show antifeedant activity against the striped cucumber beetle, *Acalymma vittatum* and limonoids as hispindins A-C were found to be the active components to show biological activity (Jacobson, 1989).

818. *Trichilia roka* (Linn.) P.Br. (Meliaceae)

This is a perennial tree found in tropical zones and available in tropical America. Its aqueous seed extract showed toxicity to the spruce budworm, *Chroistoneura fumiferana*; Mexican bean beetle, *Epilachna varivestis*; and the southern armyworm, *Spodoptera eridania* (Synder, 1983).

Limonoids like trichilin A-F isolated from its root-bark were reported as antifeedants to the larvae of the southern armyworm, *Spodoptera eridania* and Mexican bean beetle, *Epilachna varivestis*. Another limonoid, 'sendanin', isolated from its fruit was also found as a potent growth inhibitor when tested against the tobacco budworm, *Heliothis virescens*; the fall armyworm, *Spodoptera frugiperda*; and the corn earworm, *Heliothis zea* (Jacobson, 1989). Kubo and Klocke (1981) extracted sendanin and limonin, two limonoids, from its fruit extract and found them to possess antifeedant activity against the larvae of *S. eridania* on the treated host plant.

819. *Trichilia trifoliata* P.Br. (Meliaceae)

Common name: Ukani-seed

This is a perennial tree found in tropical zones and available in the coastal states in India. Its seed extract in water was reported to be toxic to the lesser grain borer, *Rhyzopertha dominica* (Prakash et al., 1978–1987).

820. *Tridax procubens* Linn. (Asteraceae)

Common name: Coat buttons or Wild dairy

This is an annual/perennial herb found in tropical zones and distributed in India, the Philippines, Malaysia, Japan and South Korea. Its bark, flower and root extracts in water and petroleum ether were reported to be toxic to the corn weevil, *Sitophilus zeamais*; the red flour beetle, *Tribolium castaneum*; the rice weevil, *Sitophilus oryzae*; and the red cotton bug, *Dysdercus cingulatus* (Carino, 1981, 1983). Its flower extracts in petroleum ether, benzene and alcohol were also found to be most effective cowpea seeds protectants when tested against the pulse beetle, *Callosobruchus chinensis* using its 4% solutions for seed treatments (Bhaduri et al., 1985).

821. *Trifolium pratense* Linn. (Papilionaceae)

Common name: Red clover plant or Tripatra

This is a perennial tree found in temperate zones but commonly grown in Niligiris and Western Himalayas in India and also found in North America and Europe. The volatile active components isolated from its leaf, flower and seed pod extracts were reported to attract the bugs, *Lygus* spp. when tested under controlled conditions. These components were (Z)-3-hexenyl acetate, (Z)-3-hexen-1-ol and (E)- and (Z)-β-ocinenes (Buttery et al., 1984).

822. *Triglochin maritima* Linn. (Juneaginaceae)

Common name: Shore pod grass
This is an annual herb found in temperate zones. Its branch and leaf extracts were reported to show growth inhibiting activity when tested against the vinegar fly, *Drosophila melanogaster* (Michael et al., 1985).

823. *Trigonella foenum-gracecum* Linn. (Leguminosae)

Common name: Fenugreek
This is an annual herb found in tropical, Mediterranean, temperate and semi-arid zones. Its seed oil was reported to show repellency and toxicity to the rice weevil, *Sitophilus oryzae* (Jilani, 1980). Seed extract of fenugreek in petroleum ether showed repellency to the grain weevil, *Sitophilus granarius* (Jilani and Su, 1983).

824. *Trillium erectum* Lindl. (Liliaceae)

Common name: Stinking Benjamin
This is a perennial herb found in temperate zones. Its root extracts in water, alcohol and ether were reported to show antifeedant activity when tested against the Japanese beetle, *Popillia japonica* (McIndoo, 1983).

825. *Tripterygium forfestii* Hook.f. (Celastraceae)

Common name: Three winged nut
This is a perennial shrub found in temperate zones and distributed in North America and Europe. Its root extract in alcohol was reported to show insecticidal activity when tested against the bean aphid, *Aphis fabae*; the ugly nest caterpillar, *Archpos cerasivorana*; sugarcane woolly aphid, *Certovacuna lonigera*; the cabbage leaf beetle, *Calophellus bowringi*; Mexican bean beetle, *Epilachna varivestis*; Hawaiian beet webworm, *Hymenia recurvalis*; the codling moth, *Laspeyresia pomonella*; the cabbage beetle, *Phaedon brassicae*; the armyworm, *Pseudoletia unipunctata*; and the cotton leaf armyworm, *Spodoptera litura* (Jacobson, 1958, 1975).

826. *Tripterygium wilifodrii* Hook. f. (Celastraceae)

Common name: Thunder God-vine
This is an annual shrub/woody climber found in temperate zones. Its root and bark dried powders were reported to show insecticidal and antifeedant activities against the cross striped cabbage worm, *Evergestis rimosalis*; the codling moth, *Laspeyresia pomonella*; European corn borer, *Ostrinia nubilalis*; the imported cabbage worm, *Pieris rapae*; the diamond-back moth, *Plutella xylostella* (Jacobson, 1958); the pepper weevil, *Anthonomus eugenii*;

the armyworm, *Pseudoletia unipunctata* (Jacobson, 1975); the cabbage beetle, *Phaedon brassicae* (McIndoo, 1983); and also against the rice yellow stem borer, *Scirpophaga incertulas* (Chiu, 1982). Zhang and Zhao (1983) found its root powder @ 0.5% w/w to inhibit the development and subsequent multiplication of the rice weevils, *Sitophilus oryzae* and *S. zeamais*. Chiu (1989) reported its root bark 0.5% ethanolic extract +300 ppm toosendanin mixture to show synergistic activity in reducing the percent pupation of 4th instar larvae of the cabbage worm, *Pieris rapae* when tested under controlled conditions.

A number of alkaloids, i.e., wilforine ($C_{43}H_{49}NO_{18}$), wilfordine ($C_{43}H_{49}NO_{20}$), wilforgine ($C_{43}H_{49}NO_{18}$), wilfortrine ($C_{43}H_{49}NO_{19}$), and wilforzine ($C_{41}H_{49}NO_{18}$), were isolated from its root extract and reported to show insecticidal activity (Crosby, 1971). Sesquiterpene ester isolated from the root of *T. wilifordii* var. *regelii* was identified as triptofordine A-C-1 and reported to possess insecticidal activity (Takaishi et al., 1987).

827. *Tropaeolum majus* Linn. (Tropaeolaceae)

Common name: Garden nasturtium

This is an annual ornamental herb found in tropical zones and native to South America. Its leaf extract was reported to show repellent activity against the woolly apple aphid, *Eriosoma lonigerum* (McIndoo, 1983).

828. *Tsuga canadensis* (Linn.) Corr. (Pinaceae)

Common name: Hemlock spruce

This is a perennial tree found in temperate zones. It is commonly available in the Himalayan region in India and also in other East Asian countries and North America. Its leaf extract showed juvenile hormone analogue activity to the milkweed bug, *Oncopeltus fasciatus* (Jacobson et al., 1975a).

829. *Tylophora asthematica* (W & A) Wright. (Asclepiadaceae)

(Syn.-*T. pubescens* Wall, *T. vomitoria* Voigt., *Asclepias asthmatica* Wild., *A. tunicata* Hort. Calc., *A. voimtoria* Kasn., *Cynanchum vomitorium* Lam.)

Common name: Antamool

This is a perennial tropical and sub-tropical herb commonly found in South India. Chander and Ahmed (1986) reported its powdered leaf admixed with wheat grains to protect against larvae of the rice moth, *Corcyra cephalonica* for two months. Total alkaloid extracts isolated from its leaf showed antifeedancy to *S. litura* larvae when incorporated with a semi-synthetic diet (Chadha, 1986). From its alkaloid extract, an active component, 'tylophorine' (Figure 110), showed antifeedancy to *S. litura*, which was comparable with that of azardirachtin (Verma et al., 1986). Benerji (1988) isolated alkaloids, viz., tylophorinine, tylophorinidine and pergularine, which were also found to show antifeedant activity against *S. litura*.

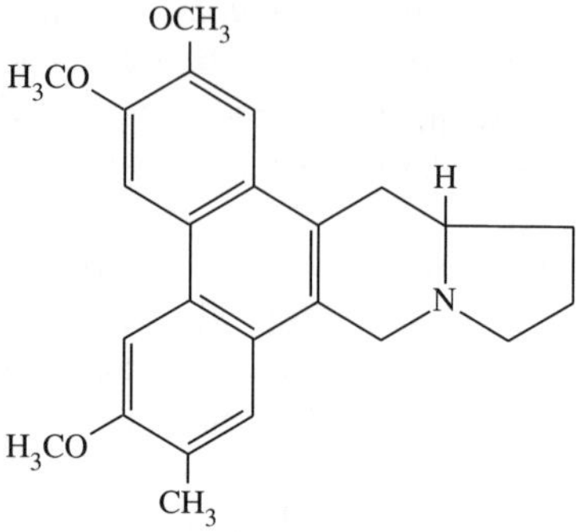

Figure 110. Tylophorine.

830. *Ulmus americana* Linn. (Ulmaceae)

Common name: American elm

This is a perennial woody tree found in tropical and sub-tropical zones. Bark extract of this tree showed feeding stimulant activity to the adults of European elm bark beetle, *Scolytus multistriatus*. An active component, 'pentacyclic triterpene', was isolated and identified as an antifeedant to this beetle (Baker and Norris, 1967).

831. *Urginea maritima* (Linn.) Baker (Liliaceae)

(Syn.-*U. silla* Steinn.)

Common name: Sea onion or Squill

This is a perennial herb found in sub-tropical and Mediterranean zones. It is a native to South Italy and North Africa. Its rhizome extract in water showed repellency to the lepidopterans, viz., *Cochylis* sp. and *Eudemis* sp. (McIndoo, 1983).

832. *Uvaria hookeri* King. (Annonaceae)

This is a perennial woody climber found in tropical and sub-tropical zones and commonly available in South India. Its 1% root bark extract in ethyl acetate showed 60% mortality to sweet potato weevil, *Cylas formicarius* and acetogenins isolated from this extract showed 10 and 50% mortality to this weevil at 0.0005 and 0.002% concentrations respectively (Padmaja et al., 1995).

833. *Uvaria narum* Wall. (Annonaceae)

This is a perennial woody climber found in tropical and sub-tropical zones and commonly available in South India, Sri Lanka and Malaysia. Its 1% root bark extract in ethyl acetate showed 60% mortality to sweet potato weevil, *Cylas formicarius* and acetogenins isolated from this extract showed 10 and 50% mortality to this weevil at 0.0005 and 0.002% concentrations respectively (Padmaja et al., 1995).

834. *Valeriana officinalis* Linn. (Valerianaceae)

Common name: Mushkhala
This is a perennial herb found in semi-arid zones. Its root powder was reported to show repellent and insecticidal activities when tested against the khapra beetle, *Trogoderma granarium* in the treated and stored wheat grains when tested under controlled conditions of insect infestation (Jilani, 1980).

835. *Vateria indica* Linn. (Dipterocarpaceae)

Common name: Indian copal tree or Malabar fallow tree or White damar
This is a perennial tree found in tropical zones and commonly available in Western Ghats and South Indian States and also distributed in Sri Lanka, Malaysia and Indonesia. Its bark and meristematic tissue extracts were reported to be toxic to the ants of agricultural importance (McIndoo, 1983). Parmar and Dutta (1982) found its bark extract to be toxic to the red flour beetle, *Tribolium castaneum* and also showed synergistic activity when combined with malathion and tested under controlled laboratory conditions.

836. *Veratrum album* Linn. (Liliaceae)

Common name: European white hellebore
This is a perennial herb found in temperate and sub-tropical zones. Its root and whole plant extracts in alcohol, petroleum ether and also their dried powders showed insecticidal and antifeedant activities against the melon worm, *Diaphania hyalinata* (Jacobson, 1958) and caterpillars (Jacobson, 1975). Quassin ($C_{22}H_{28}O_6$) and neo-quassin ($C_{21}H_{26}O_5$), two alkaloids, were isolated from its tuber extract and reported to show insecticidal activity (Crosby, 1971).

837. *Veratrum grandifolium* Linn. (Liliaceae)

Common name: False hellebore genus
This is a perennial herb found in temperate zones. Its whole plant extract in water was toxic to the vinegar fly, *Drosophila hydei* (Jacobson, 1958).

838. *Veratrum nigrum* (Linn.) Ait. (Liliaceae)

Common name: Black hellebore

This is a perennial herb found in tropical and temperate zones. Its root/bulb extract in water also dried powder were reported to show toxicity to the vinegar fly, *Drosophila hydei* and mulberry white caterpillar, *Rondotia nenciana* (McIndoo, 1983).

839. *Veratrum viride Ait.* (Liliaceae)

(Syn.-*V. irride*)

Common name: Green hellebore or Swamp hellebore or America false hellebore

This is a perennial herb found in temperate zones. Green hellebore is native to North America and also available in Jammu and Kashmir in India. Its root, leaf and bulb extracts in water and the dried powders were toxic to the larvae of the codling moth, *Laspeyresia pomonella* (Feinstein, 1952); the melon worm, *Diaphnania hyalinata*; Hawaiian beet webworm, *Hymenia recurvalis*; and European corn borer, *Ostrinia nubilalis* (Jacobson, 1958). Seiferle et al. (1942) reported two alkaloids as active components, i.e., 'jervine' and 'pseudo-jervine', from its plant extract and found them to control the tobacco thrips, *Thrips tabaci*. Crosby (1971) reported two more insecticidal alkaloids, i.e., quassin and isoquassin, in its tuber extract.

840. *Vernonia gigantea* Schreb. (Asteraceae)

This is an annual herb found in sub-tropical and tropical zones. Its whole plant aqueous extract was toxic to *S. eridania* and *S. frugiperda* (Carino, 1983). Sesquiterpene lactones were isolated as insect feeding deterrents in *Vernonia* sp. (Burnett et al., 1974).

841. *Vernonia gluaca* Schreb. (Asteraceae)

This is an annual herb found in tropical and sub-tropical zones. Its whole plant extract in water was toxic to *S. eridania* and *S. frugiperda* (Carino, 1983). Sarin (1995) isolated from the floral heads of *Vernonia* sp. toxic esters like pyrethrins I, II, jasmolin I, II and cinerin I, II.

842. *Vetiveria zizanioides* (Linn.) Nash. (Poaceae)

(Syn.-*Phalaris zizanoides* Linn., *Andropogon muricatus* Retz.)

Common name: Vetiver khas-khas

This is a perennial grass found in tropical zones and commonly available in Uttar Pradesh, Rajasthan, Punjab and West Bengal in India and distributed throughout South East Asian countries, the Philippines, West Africa and Australia. Its root oil and extracts in water and petroleum ether of its meristematic tissues were reported to show antifeedant activity when tested against the long headed flour beetle, *Latheticus oryzae* (Balsubramanian, 1982). However, its rhizome/root extracts showed inhibition in the growth and development of the cotton stainer, *Dysdercus cingulatus* and also reduced the longevity of the cotton leaf armyworm, *Spodoptera litura* (Kareem, 1984).

843. *Vicia cracca* Linn. (Leguminosae)

Common name: Bird vetch

This is a perennial herb found in temperate and sub-tropical zones and commonly available in North America and European countries. Its leaf extract in water was reported to show insecticidal activity to Mexican bean beetle, *Epilachna varivestis* (Jacobson, 1958).

844. *Vinca rosea* Linn. (Apocynaceae)

(Syn.-*Catharanthus roseus* [Linn.] G. Dong.)

Common name: Old maid

This is a perennial shrub found in tropical and sub-tropical zones and commonly available in India, Pakistan, Burma and Nepal. Its whole plant extract was reported to show mortality to the cotton stainer, *Dysdercus cingulatus* (Kareem, 1984). Dhandapani et al. (1985) found its plant extract to show antifeedant activity against the brinjal spotted leaf beetle, *Henosepilachna vigintioctopunctata* when sprayed on eggplant.

845. *Viola phalacrocarpoides* (Linn.) Muerr. (Violaceae)

This is an annual herb found in temperate zones. Its root, leaf and branch aqueous extracts were reported to be toxic to the vinegar fly, *Drosophila hydei* (Jacobson, 1985).

846. *Viola tekedana* var. *variegata* Lam. (Violaceae)

Common name: Violet genus

This is an annual herb found in temperate and alpine zones. Its root, leaf and branch extracts in water were reported to be toxic to the vinegar fly, *Drosophila hydei* (Jacobson, 1958).

847. *Vitex negundo* Linn. (Verbenaceae)

Common name: Indian privet or Chinese chaste tree

This is a perennial shrub or small tree found in tropical and sub-tropical zones and available throughout Coastal States, Bihar and North-Eastern regions in

India. This shrub is commonly available in Sri Lanka, China, the Philippines and Malaysia. Its branch, leaf and seed extracts in water and alcohol and also its seed oil were reported to show insecticidal, repellent, juvenile hormone mimetic and antifeedant activities against the wide range of insects in storage and the lepidopteran pests in field ecosystems.

Biological Activity Against Storage Insects

Chopra et al. (1928) reported its leaf extracts to show antifeedant property against storage insects in general. Its dried leaves protected stored wheat from *S. oryzae* when admixed with the grains (Krishnamurthi and Rao, 1950). Krishnaiah and Ganesalingam (1981) reported stem volatile constituents of *V. negundo* to exhibit significant repellent activity to *S. cerealella*. Rahman and Bhattacharya (1982) also found its leaves to protect stored pulses from infestation of *Latheticus sativus*. However, infestation of *S. cerealella* was found to be reduced significantly when its dried leaf powder was admixed with the stored paddy grains (Abraham et al., 1972, 1979; Pradash et al., 1981b, 1983, 1989; Sukumaria et al., 1987; Prakash and Rao, 1995a,b). Similarly, other coleopteran pests, i.e., *Callosobruchus chinensis*, *R. dominica*, *S. oryzae*, *S. zeamais* and *Latheticus oryzae*, were reported to be repelled by the leaf powder and its aqueous and alcoholic extracts when tested under natural and controlled conditions (Anonymous, 1980–1981; Prakash et al., 1981b, 1982, 1989, 1990c; Balasubramanian, 1982; Islam, 1983; Satpathy, 1983; Rout, 1986; David et al., 1988). Essential oil of Indian privet isolated from its leaf combined with citronella oil showed repellency to *S. cerealella* (Krishanrajah et al., 1985).

Biological Activity Against Field Insects

Leaf and branch extracts of this shrub were reported to show repellency to the field insect pests of paddy (Litsinger et al., 1978) and found to show antifeedant activity to the hairy caterpillar, *Euproctis fraterna* and cotton leaf armyworm, *Spodoptera litura* on the castor leaves (Subadrabai and Kandaswamy, 1985). Sukumaran et al. (1987) reported its petroleum ether leaf extract to produce malformed pupae of the rice leaffolder, *Cnaphalocrosis medinalis* at 200 to 1000 ppm concentrations when evaluated against the leaffolder larvae under controlled conditions.

Active Components

Viticosterone-E, iridoides and ecdysones isolated from *Vitex* sp. were reported to show juveno-mimetic activity against the insects (Rimpler and Schulz, 1967; Rimpler, 1969, 1972). Prakash (1986) and Prakash et al. (1990a) isolated and identified '2-heptatriacontanone' ($C_{35}H_{74}O$) (Figure 111) as an active component from leaf extract of Indian privet and reported it to inhibit the oviposition of stored grain boring insects like *S. cerealella*, *R. dominica* and *S. oryzae*.

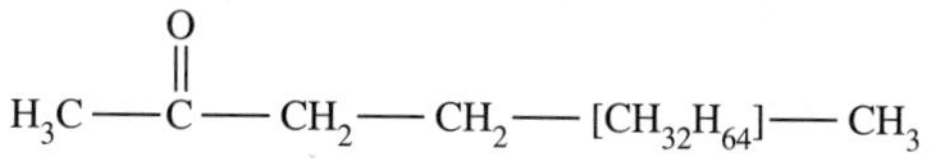

$$H_3C-\overset{\displaystyle O}{\overset{\|}{C}}-CH_2-CH_2-[CH_{32}H_{64}]-CH_3$$

Figure 111. 2-heptatriacontanone.

848. *Vitis girdiana* Munson. (Vitaceae)

This is an annual/perennial vine found in temperate zones and available in North America. Its stem and leaf extracts in ethyl ether showed attractancy to both the male and the female of Mediterranean fruit fly, *Ceratitis capitata* (Keiser et al., 1975).

849. *Warburgia ugandensis* Engl. (Canellaceae)

This is a perennial tree found in tropical and sub-tropical zones and commonly found in Syria, Palestine and South Africa. Its aqueous leaf extract was reported to show antifeedant activity to African armyworm, *Spodoptera exempta* and Mexican bean beetle, *Epilachna varivestis* when tested under controlled laboratory conditions (Jacobson, 1958).

An active component, 'warburganol', isolated from its methanolic leaf extract was reported to show strong antifeedant activity to *S. exempta* and *E. varivestis* (Kubo and Nakanishi, 1978). Kubo and Ganjian (1981) also isolated and identified a dialdehyde sesquiterpenoid called 'polygodial' (Figure 112) and reported it to show antifeedant activity against the larvae of *S. exempta* and *S. littoralis*. Jacobson (1989) also referred to a series of sesquiterpenoid dialdehydes like warburganol, polygodial and muzigodial. All these components were found to show antifeedant activity against the bollworms, *Heliothis armigera* and *H. virescens* when evaluated in a laboratory bioassay.

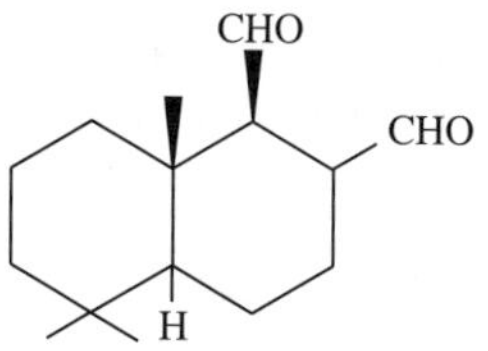

CHO
CHO
H

Figure 112. Polygodial.

850. *Wedelia prostrata* Jacq. (Asteraceae)

This is an annual herb found in tropical and sub-tropical zones and available in Korea, the Philippines and Japan. Its flowers' extract as expressed juice was reported to show toxicity to the cotton stainer, *Dysdercus cingulatus* (Carino, 1983).

851. *Willarida mexicana* Rosli. (Leguninosae)

Common name: Nesco

This is a perennial tree found in semi-arid zones and commonly available in Mexico and the West Indies. Its root, bark, leaf and branch extracts in water, alcohol and also their dried powders were reported to show antifeedant and insecticidal activities when tested against the melon worm, *Diaphania hyalinata*; the cross striped cabbage worm, *Evergestis rimosalis*; Hawaiian beet worm, *Hymenia recurvalis*; and the southern armyworm, *Spodoptera eridania* (Jacobson, 1958).

852. *Withania somnifera* (Linn.) Dunal. (Solanaceae)

Common name: Aswagandha

This is a perennial herb found in tropical zones and commonly available in North India and South China. The active components, i.e., 8-withanolids, and other related components isolated from this herb were reported to show antifeedant activity when tested against *Spodoptera littoralis* and another active component, 'nicalbin-A', was found to be the most effective antifeedant to *Epilachna varivestis* (Ascher et al., 1987; Ascher, 1983). Ascher et al. (1987) reported several new withanolides, a group of naturally occurring steroids in this plant, built on an ergostane type skeleton in which C_{22} and C_{26} were appropriately oxidised for the formation of 6-membraned ring acetone and were found to show antifeedant activity against *S. littoralis*, *E. varivestis* and *T. castaneum.*

853. *Xanthium candense* Linn. (Asteraceae)

This is an annual herb found in sub-tropical and temperate zones and commonly available in Japan and South Korea. An active component, 'trideca-1-ene-3,5,7,9,11-pentanyl', isolated from its methanolic root extract was reported to show ovicidal activity when tested against eggs of the vinegar fruit fly, *Drosophila melanogaster* (Nakajima and Kawazu, 1977).

854. *Xanthmiun strumarium* Linn. (Asteraceae)

Common name: Cocklebur or Burweed

This is an annual herb found in tropical zones and distributed throughout India and Sri Lanka. Its aqueous leaf extract was reported to inhibit the oviposition of the cotton bollworm, *Earias vittella* (Dongre and Rahalkar, 1984).

855. *Xeromphis spinosa* (Thunb.) Keoy. (Rubiaceae)

This is a perennial shrub/small tree found in tropical zones and distributed in all South East Asian countries and Australia. Its dried root and fruit powders and extracts in water and alcohol were reported to show insecticidal and antifeedant activities against the green scale, *Coccus viridis* and the stored grain insect pests (Jacobson, 1958) and also against the grasshopper, *Epacromia tumulus* and the hairy caterpillar, *Euproctis fraterna* (McIndoo, 1983). Chandel et al. (1987) reported acetone extract of its dried stem to show toxicity to *Epilachna vigintioctopunctata* when tested under controlled conditions.

856. *Xylocarpus moluccensis* Roem. (Meliaceae)

(Syn.-*Carpa obovata* Blume; *C. moluccenes* Lamk.)

Common name: Red wood

This is a perennial small tree found in tropical, sub-tropical and temperate zones and distributed in Japan, South Korea, tropical Africa, Australia, and India (Sudarban and Andamans). From its raw fruit an active component, 'xylomolin' (Figure 113), a non-glucosidic recoiridoid, was isolated and reported to show antifeedant and repellent activities when tested against African armyworm, *Spodoptera exempta* (Kubo and Nakanishi, 1978).

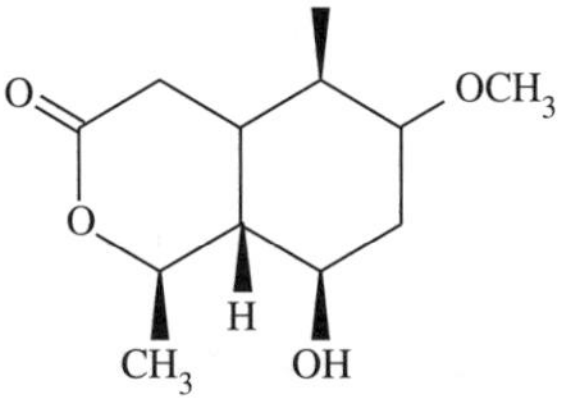

Figure 113. Xylomolin.

857. *Yucca schidigera* Linn. (Liliaceae)

Common name: Spanish dagger

This is a perennial shrub found in sub-tropical and arid zones and distributed in Hawaii, Japan and South India. Its dried leaf powder was reported to show insecticidal and repellent activities against the melon worm, *Diaphania hyalinata*; Hawaiian beet webworm, *Hymenia recurvalis*; the codling moth, *Laspeyresia pomonella*; and the bean leafroller, *Urbanus proteus* (Jacobson, 1958).

858. *Zanthoxylum alatum* Roxb. (Rutaceae)

(Syn.-*Z. arantum* D.C.)

Common name: Wing-leaf prickly ash

This is a perennial small aromatic tree found in temperate and sub-tropical zones. In India, it is commonly available in Punjab, Kumaun and Khasi hills. Its dried bark and fruit powders and also water extracts were reported to show antifeedant activity to the migratory locust, *Locusta migratoria* (Jacobson, 1975).

859. *Zanthoxylum americanum* Linn. (Rutaceae)

Common name: Toothache tree
This is a perennial tree found in temperate zones. Its branch and fruit extracts in acetone and also their dried powders were reported to be toxic to the Japanese beetle, *Popillia japonica* (McIndoo, 1983).

860. *Zanthoxylum clava-herculis* Linn. (Rutaceae)

Common name: Southern prickly ash
This is a perennial shrub/small tree found in sub-tropical zones. Its leaf and bark extracts in ether and their dried powders were reported to show insecticidal and repellent activities when tested against the melon worm, *Diaphania hyalinata*; the armyworm, *Psdudoletia unipunctata*; and the bean leafroller, *Urbanus proteus* (Jacobson, 1958). Jacobson (1971) reported 'herculin' ($C_{16}H_{25}NO$) and 'neo-herculin' ($C_{16}H_{25}NO$), two unsaturated isobutylamides, as active components from its root extract that showed insecticidal activity against a wide range of the insects of agricultural importance and mealworm, *Tenebrio molitor* (Jacobson and Crosby, 1971). Jacobson (1989) also isolated herculin (Figure 114), a purgent toxic component from its bark.

$$CH_3(CH_2)_2CH{=}CH\,(CH_2)_4CH{=}\,CHCO\,NH\,CH_2\,CH\,(CH_3)_2$$

Figure 114. Herculin.

861. *Zanthoxylum piperitum* D.C. (Rutaceae)

Common name: Japanese pepper
This is a perennial small tree found in sub-tropical and temperate zones and commonly distributed in Central America, Japan and China. From its root and bark extracts two isobutylamides were isolated as active components called 'sanshool-I' (Figure 115) and 'sanshool-II' (Figure 116), which showed toxicity to the soft bodied insects like the aphids and the jassids of agricultural importance (Jacobson and Crosby, 1971; Miyakado et al., 1989). Sanshool-I (α-sanshool) has a molecular formula $C_{16}H_{25}NO$ and called as N-isobutyl-2,4,8-didecatrienamide, whereas sanshool-II or β-sanshool ($C_{16}H_{25}NO$) is β all-trans of α-sanshool (Miyakado et al., 1989).

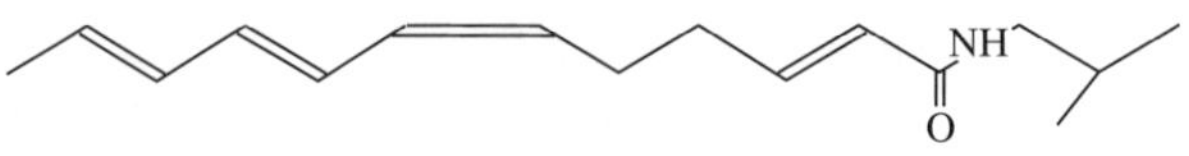

NH
O

Figure 115. Sanshool-I.

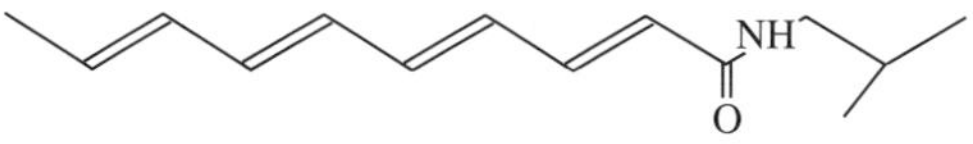

Figure 116. Sanshool-II.

862. *Zea mays* Linn. (Poaceae)

Common name: Maize or Corn

This is a seasonal cereal/tall grass found in tropical zones. It is native to South America but now commonly grows in Uttar Pradesh, Madhya Pradesh, Punjab, Andhra Pradesh, Bihar and Rajasthan in India as a cereal and fodder crop. This crop is widely distributed in West Africa, Japan, South China, Australia, America, Iran, etc. Oil extracted from its grains was reported to reduce the survival and progeny of the rice weevil, *Sitophilus oryzae* and corn weevil, *Sitophilus zeamais* when tested by treating stored sorghum and wheat grains with the maize oil (Oca et al., 1978). However, earlier corn oil was also found to show insecticidal or growth inhibiting activities against Mexican mealy bug, *Phenacoccus gossypii* and European corn borer, *Ostrinia nubilalis* (Jacobson, 1958; Beck, 1960). El-Sayed et al. (1991) found corn oil treatment with stored cowpea @ 2.5 ml/kg seeds to show 98.5% mortality to the adults of pulse beetle, *Callosobruchus chinensis* and to protect the grains.

Argandona et al. (1983) isolated a main hydroxamic acid, i.e., DIMBOA [2,4-dihydroxy-7-methoxy-2H-1, 4-benzoxazin-3(4H)-one] from its plant extract and reported it to show toxicity and antifeedant activities to the aphid, *Schizaphis graminum* when fed to the aphids through artificial diets. Hariprasad and Kanaujia (1992) also found DIMBOA in maize crop to show antibiosis to the jute hairy caterpillars, *Spilosoma obliqua*.

863. *Zephyranthes grandiflora* Lindl. (Amaryllidaceae)

Common name: Large flower lily

This is an annual/perennial herb native to South America and now found in India, North China and South East Asia. Mid-rib sap of its leaf was reported to show antifeedant activity against the desert locust, *Schistocerca gregaria* (Singh and Pant, 1980). Patel et al. (1990b) reported its bulb extract to reduce larval populations of Gujrat hair caterpillar, *Amsacta moorei* up to one week of the extract spray on the host crop.

864. *Zingiber hypogeon* Bochm. (Zingiberaceae)

Common name: Ginger

This is an annual/perennial aromatic herb found in tropical zones and commonly available in Uttar Pradesh, Himachal Pradesh, Andhra Pradesh, Kerala and Tamil Nadu in India. Its rhizome oil was reported to show toxicity to the pulse beetles, *Callosobruchus chinensis* and *C. maculatus* when tested under laboratory conditions (Mummigatti and Raghunathan, 1977; Anonymous, 1980–1981). Doharey et al. (1983, 1984a,b, 1987) reported its rhizome oil to protect the greengram seeds against the pulse beetles, *C. chinensis* and *C. maculatus* in storage when admixed @ 1% w/w and it also showed toxicity to their eggs and finally checked their population build up.

865. *Zingiber officinale* Rosc. (Zingiberaceae)

Common name: Ginger

This is an annual/seasonal aromatic herb found in tropical zones and available in Kerala, Uttar Pradesh, West Bengal, Maharastra, Himachal Pradesh and Andhra Pradesh in India. Its dried fruit powder was reported to show toxicity to the pulse beetle, *Callosobruchus chinensis* and it protected stored blackgram (Mathur et al., 1985). Ahamod and Ahmed (1991) found its rhizome powder admixed with stored grain @ 3% w/w to show 100% mortality to *C. cephalonica* and *S. oryzae*.

866. *Zingiber squarresum* Roxb. (Zingiberaceae)

This is an annual/biennual herb found in tropical zones and distributed in East Indian States and Bangladesh. Its whole plant sap and leaf expressed juice showed tranquilizing activity to the giant rock bee, *Apis dorsata* and helped the honey collectors (Dutta et al., 1985).

B BOTANICAL PESTICIDES AGAINST NEMATODES

1. *Acacia auriculiformis* A. Cunn. (Mimosaceae)

Common name: Australian wattle or Australian phyllode acacia
This is an evergreen perennial tree cultivated for its graceful pendant phyllodes and beautiful light-yellow flowers and available in tropical and sub-tropical zones. This tree is commonly found in Australia and East Indian States. Its alcoholic extracts and also air dried powdered funicles showed larval mortality to the root-knot nematode, *Meliodogyne incognita* when tested in the *in vivo* and *in vitro* conditions. Two triterpenoid saponins were found to be the biologically active components from this extract (Sinhababu et al., 1992). Its funicles crude alcoholic extract killed 55% juveniles of *M. incognita* in 80 minutes at 4 mg/ml concentration in a bioassay test.

Nematicidal principle isolated from the funicles consisted of two triterpenoid saponins known as 'acaciaside A and B', which killed 62.5% juveniles of *M. incognita* at the abovementioned concentration. Both the components when applied by soil drench as well as foliar spray at 10 mg/ml and 4 mg/ml concentrations respectively reduced the root-galling and nematode population in the roots of cowpea (Roy et al., 1993).

2. *Adhatoda vasica* Nees. (Acanthaceae)

Common name: Malabar nut tree or Ajuba
This is a perennial shrub found in tropical and sub-tropical zones. Pandey (1995) reported its powdered leaves @ 100 g/kg soil to significantly reduce the multiplication of root-knot nematode, *Meloidogyne incognita.*

3. *Ageratum conzoides* Linn. (Asteraceae)

Common name: Goat weed
This is an annual/perennial weed found in tropical and sub-tropical zones and distributed throughout India. Its whole plant extracts in ether and water were

reported to be toxic to the root-knot nematodes, *Meloidogyne javanica* and *M. incognita* (Mukherjee, 1983).

4. *Allium cepa* Linn. (Amaryllidaceae)

Common name: Onion

This is an annual/biennial herb found in tropical and sub-tropical zones. Its aqueous extract was reported to show nematicidal activity against the root-knot nematodes, *Meloidogyne incognita* and *Radopholus similis* (Sarra-Guzman, 1984). Hasan (1992) found its root exudates to minimize the gall formation by *M. incognita* and it showed toxicity to the 2nd, 4th and 5th stage juveniles of this nematode.

5. *Allium sativum* Linn. (Amaryllidacea)

Common name: Garlic

This is an annual/biennial herb found in tropical and sub-tropical zones. It is native to Central Asia but commonly grown in India. Crude extracts of its leaf and bulb in water, methanol and also steam distillates of their extracts showed toxicity to the sugarcane nematode, *Aphelenchoides sacchari* and the citrus root nematode, *Tylenchorynchus semipenetrans* (Nath et al., 1982). Sukul et al. (1974) and Mukherjee (1983) found its aqueous extract to show nematicidal activity to the root-knot nematodes, *Meloidogyne incognita* and *M. javanica*. Sarra-Guzman (1984) also reported its bulb extract as nematicidal to *Radopholus similis*. Its bulb extract applied in the soil showed toxicity to the root-knot nematode, *M. incognita* in the vegetable roots (Gupta et al., 1985). Its bulb crude extract was found to reduce fertility when eggs of *M. incognita* were dipped in the extract for 10 days and also showed larvicidal activity in laboratory tests (Gupta and Sharma, 1990a,b) and killed the larvae of *M. javanica* in tobacco nurseries (Krishnamurthy and Murthy, 1990). Gupta and Sharma (1985, 1990b) reported ovicidal activity of the garlic oil and extract when tested against the eggs of *M. incognita* and reduced hatchability of the eggs by 53% at 10 ppm concentrations. Nath et al. (1982) isolated and identified an active component, 'diallyl disulphide', from its bulb extract and found it to show nematicidal activity.

6. *Aloe barbadensis* Mill. (Liliaceae)

(Syn.-*A. vera* [L.] Webb. & Berth. [non-Mill.]; *A. perfoliata* var. *vera* Linn.)

Common name: Barbados aloe

This is a perennial stoloniferous succulent shrub native to the West Indies and now naturalized in India. Its whole plant extract in water was reported to inhibit the larval hatching of *M. incognita* in vegetable crops (Hussain and Masood, 1975). Mahmood et al. (1979) also reported its leaf extract to show toxicity to

M. incognita in potted vegetable plants. Pandey and Haseeb (1988) found its root and shoot extracts in distilled water to reduce the hatching of the eggs of *M. incognita* when tested under laboratory conditions and similar activity was recorded in the extracts of other species of this plant, viz., *A. penyli*.

7. *Amaranthus gracilis* Linn. (Amaranthaceae)

This is an annual herb found in tropical zones. Its whole plant extract in water was reported to show nematicidal activity against the root-knot nematode, *Meloidogyne javanica* (Mandal and Bhatti, 1983).

8. *Anagallis arvensis* Linn. (Primulaceae)

Common name: Poor man's weather-glass

This is an annual/perennial herb found in tropical zones. Extract of its leaf in water was reported to show toxicity to the root nematode, *Rotylenchulus reniformis* (Mahmood et al., 1982).

9. *Andrographis paniculata* (Burm. f.) Wall. ex. Nees (Acanthaceae)

(Syn.-*Justicia paniculata* Burm.f.)

Common name: King of bitters

This is an annual/perennial herb found in tropical and sub-tropical zones and commonly available in Coastal States of India. Its leaf and whole plant extracts were reported to show nematicidal activity against the root-knot nematode, *Meloidogyne incognita* (Mukherjee and Sukul, 1978). Goswami and Vijayalakshmi (1985) reported its dried plant amended in the soil to reduce incidences of the root-knot nematode, *M. incognita* in vegetables. Poornima and Vadivelu (1993) tested its plant extract 1.5 dilution applied at the base of the brinjal plants by forming holes around the plant in the soil and found it to significantly reduce the population of *M. incognita*, *R. reniformis* and *Pratylenchus delattrei*.

10. *Angelica pubscens* Linn. (Umbelliferae)

This is a perennial herb found in temperate zones. Its whole plant extract was reported to be toxic to the white-tip nematode, *Aphelenchoides besseyi* (Katsura, 1981).

11. *Annona squamosa* Linn. (Annonaceae)

Common name: Sugar apple or Sweet-sop

This is a perennial shrub or small tree found in tropical regions. It is native to South America and the West Indies but now available throughout India. Its leaf extracts in water and ethanol were reported to inhibit larval hatching and also showed toxicity to the root-knot nematode, *M. incognita* when tested under controlled conditions in vegetable roots (Hussain and Masood, 1975; Mahmood et al., 1979).

12. *Anthocephalus chinensis* (Lamb.) Rich. ex. Walp. (Naucleaceae)

(Syn.-*A. indicus* A. Rich, *Nauclea cadamba* Roxb.)

Common name: Kadam

This is a moderate-sized graceful deciduous and ornamental tree grown in tropical zones and commonly available in Assam, West Bengal and Andaman Islands in India. Aqueous root extract and its purified fractions were reported to be toxic to the root-knot nematodes, *Meloidogyne incognita* and *Radopholus similis* (Sarra-Guzman, 1984). Chatterjee and Sukul (1979) found its leaf extract to check the incidences of *M. incognita* in okra roots.

13. *Arachis hypogea* Linn. (Fabaceae)

Common name: Groundnut

This is an annual herb found in tropical and sub-tropical zones and commonly available in India. Its seed kernel oil and oil cake were reported to show nematicidal activity when tested against a wide range of nematodes by incorporating them in the soil. Results are summarized in Table 15.

14. *Araucaria cookii* Don. (Araucariaceae)

This is a perennial tree found in tropical zones and commonly available in northern India. Its aqueous leaf extract was reported to be toxic to the root-knot nematode, *M. incognita* in the vegetable crops (Jain et al., 1986).

15. *Argemone mexicana* Linn. (Papaveraceae)

Common name: Mexican prickly poppy

This is an annual bushy herb/undershrub found in tropical and sub-tropical zones and commonly available in India. Its root and the leaf aqueous extracts were reported to show nematicidal activity when tested against the root-knot nematodes, *Meloidogyne incognita* and *M. javanica* (Ramnath et al., 1982; Mukherjee, 1983). Reddy et al. (1990) found its chopped leaves to reduce the number of root-galls caused by *M. incognita* in the papaya when admixed with the infested soil. Its root exudates showed highest reduction in the adult female populations and enhanced the life cycle span twice to reduce the incidences of *M. incognita* (Hasan, 1992). Sharma and Prasad (1995) reported

Table 15. Nematicidal activity of groundnut oil cake

Nematode	Result of the test	Reference
Meloidogyne javanica	Oil cake amended in the soil @ 0.2% w/w reduced egg laying, larval population and root galls in okra.	Singh et al., 1967
M. incognita	Oil cake water extract inhibited larval development.	Khan et al., 1967
M. javanica	Oil cake amendment in the soil reduced the incidences of the root-knot nematode in tomato.	Hameed, 1968
M. incognita	Its oil cake amendment in the soil before transplanting reduced gall of the root-knot nematode in okra and tomato.	Singh and Sitaramaiah, 1969
Helicotylenchus sp., *Hoplolaimus* sp., *Tylenchorhynchus* sp., *Rotylenchulus reniformis*, *M. incognita*	Oil cake amendment in the soil reduced the nematodes' populations around the guava and the citrus roots.	Mobin and Khan, 1969
M. incognita	Its oil cake amended in the soil reduced the root-knot incidences in tomato and brinjal.	Srivastava, 1971
M. javanica	Oil cake amendment in soil reduced the root-knot incidences in the tomato.	Gowda, 1972
M. incognita	Oil cake amendment in soil reduced the nematode population in tomato and other vegetable roots.	Desai et al., 1973; Khan et al., 1973
Rotylenchulus reniformis, *Tylenchorhynchus brassicae*, *M. incognita*, *Hoplolaimus indicus*	Its oil cake extracts inhibited the larval hatching and finally killed the adults of the test nematodes.	Mishra and Prasad, 1973

Table 15. Nematicidal activity of groundnut oil cake (Continued)

Nematode	Result of the test	Reference
H. indicus, T. brassicae, M. incognita, Aphelenchus avenae	Its oil cake amendment in the soil reduced the populations of these nematodes in the vegetable crops.	Alam and Khan, 1974; Khan, 1974
H. indicus, T. brassicae, M. incognita, Aphelenchus avenae	Liberation of ammonia from the oil cake was reported to show nematicidal activity.	Alam et al., 1976
M. incognita	Its oil cake amendment in the soil @ 1% w/w promisingly controlled its population.	Varma, 1976; Singh and Sitaramaiah, 1976
M. incognita	Its oil cake amendment in the soil @ 100 N/kg soil reduced the nematode development in tomato roots.	Azam and Khan, 1976
M. incognita	Active components in oil cake extract that showed toxicity were formaldehyde (0.478%) and acetone (0.18%).	Alam et al., 1976
T. brassicae, H. indicus, M. incognita, Helicotylenchus erythrine	Its oil cake amendment in the soil @ 1% w/w significantly reduced the population of nematodes.	Khan et al., 1976
M. incognita	Its oil cake amendment in the soil effectively reduced the nematode populations in the roots of tomato.	Singh et al., 1976; Siddiqui, 1976
Meloidogyne exigua	Its oil cake amendment in the soil @ 1% w/w reduced the populations of the nematode in the coffee seedlings.	Moraes, 1977
Tylenchorhynchus brassicae, H. indicus	Its oil cake amendments in the soil suppressed the nematode populations in the plots sown tomato.	Alam et al., 1977

Table 15. Nematicidal activity of groundnut oil cake (Continued)

Nematode	Result of the test	Reference
M. incognita	Water soluble fractions of its oil cake applied in the soil checked the nematode incidences in the okra.	Bhatnagar et al., 1978
M. incognita	Its oil cake released phenol 0.245 mg/100 mg cake, which was toxic to the nematode.	Aham et al., 1979
M. incognita, M. javanica	Neem cake amended in the soil reduced the incidences of this nematode in the tobacco nursery.	Desai et al., 1979
M. incognita	Its oil cake extract in water showed toxicity to this nematode.	Singh, 1980
M. incognita	Leaf powder admixed with soil @ 40 gm/pot reduced the root knot incidence in the okra.	Bhatnagar et al., 1980
A. avenae, Ditylenchus cypei, Hoplolaimus indicus, T. brassicae	Its oil cake reduced the populations of the test nematodes.	Mookherjee, 1983
M. incognita	Oil cake amendment in the soil reduced the incidences of this nematode in the tomato.	Goswami and Vijayalakshmi, 1983; Bhattacharya and Goswami, 1987a,b
M. incognita	Its oil cake amendment in the soil reduced the incidences of the nematode in the mulberry.	Sikdar et al., 1986
M. incognita	Groundnut testas extract was found to show toxicity to this nematode.	Prasad et al., 1987
M. incognita	Its oil cake amendment in the soil reduced the root-gall index in the eggplants.	Singh and Singh, 1988
Tylenchorynchus spp., *M. incognita*	Castor cake @ 1.7 t/h incorporated in the soil 10 days prior to sowing mung seeds reduced the nematodes' populations.	Mishra and Gaur, 1989

Table 15. Nematicidal activity of groundnut oil cake (Continued)

Nematode	Result of the test	Reference
M. incognita	Oil cake application in the soil at the rate of 1 t/ha minimized root galling in the okra.	Reddy and Khan, 1990
Aphelenchoides sacchari	Oil cake @ 2% w/w amended in the soil in mushroom compost reduced the nematode population.	Rao et al., 1991
M. arenaria	Castor oil cake @ 2% t/h amended in the soil reduced the incidences of root-knot nematode.	Vaishnav et al., 1993
M. incognita, R. reniformis, P. delattrei	Oil cake @ 100 kg N/ha decomposed in the soil reduced the nematodes' populations in the brinjal crop.	Poornima and Vadivelu, 1994

its aqueous shoots extract to reduce penetration of *M. incognita* in the brinjal and the soyabean crops when tested *in vivo*.

16. *Artabotrys odoratissimus* R.Br. (Annonaceae)

(Syn.-*Artabotrys uncinatus* [Lam.] Merr.; *A. hexapetalus* [L.F.] Bhandari)
Common name: Climbing ylang-ylang

This is a perennial ornamental shrub with greenish yellow fragrant flowers often grown in the gardens in tropical and sub-tropical zones. Chattopadhyaya and Mukhopadhyay (1989b) found its aqueous leaves extract to significantly reduce the larval hatching of the root-knot nematode, *Meloidogyne incognita* in 48 and 72 hrs of exposures in a laboratory test.

17. *Artimisia vulgaris* Auct. non. Linn. (Asteraceae)

(Syn.-*A. nilagirica* [Clarke] Pamp. *A. vulgaris nilagirica* O.B. Clarke)
Common name: Indian wormwood fleabane

This is a perennial aromatic herb found in arid zones and commonly available in Sikkim, Khasi, Hills, Darjeeling, Manipur and Western Ghats in India. Its root extract was reported to show nematicidal activity to *Meliodogyne incognita* and *Radopholus similis* (Sarra-Guzman, 1984; Sherif et al., 1987).

18. *Asparagus officinalis* Linn. (Liliaceae)

Common name: Garden asparagus

This is a perennial/biennial herb found in sub-tropical zones and native to Europe and West Asia. Its aqueous root extract was reported to show nematicidal activity when tested against the reniform nematode, *Rotylenchulus reniformis* and the root-knot nematode, *Meloidogyne incognita* (Das, 1983). 'Asparaguric acid' (Figure 117), an active component isolated from its root extract, was also found to show toxicity to *Heterodera rostochinensis*, *Meloidogyne halpa*, *Pratylenchus penetrans* and *P. curvitatus*.

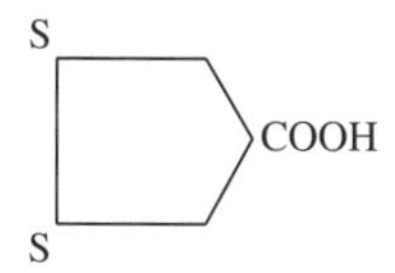

Figure 117. Asparaguric acid.

19. *Azardirachta indica* A. Juss. (Meliaceae)

Common name: Neem tree

This is a perennial tree found in sub-tropical and tropical zones. It is commonly available in India and other tropical countries. Its leaf and kernel extracts and oil cake were reported to show nematicidal activity against a wide range of the nematodes of agricultural importance.

Neem Leaf as Nematicide

Applications of 5–10% w/w leaves of neem in the infested soil reduced the incidences of the root-knot nematode in tomato and okra (Singh and Sitaramaiah, 1967). Similarly, chopped leaves of neem incorporated with the soil were found to reduce the populations of the root-knot nematode, *Meloidogyne incognita* in okra and eggplant (Lall and Hameed, 1969; Hussain and Masood, 1975; Kuriyan, 1988). Egunjobi and Afolami (1976) found neem leaf extract to reduce the population of *Pratylenchus brachyurus* in maize roots. Alam (1976) reported neem seedlings planted in the periphery of tomato, eggplant and chillies in naturally infested soil in a mixed cropping pattern to reduce populations of the nematodes. Neem leaf powder incorporated in the soil @ 40 gm/pot was reported to reduce the incidences of the root-knot nematode, *M. incognita* in okra seedlings (Bhatnagar et al., 1980). Kaliram and Gupta (1980) also reported neem leaf extract in water to reduce root-galls caused by *M. javanica* in infested chickpea. Khan and Hussain (1988a,b) found reduced incidences of *M. incognita* when cowpea seeds were treated with its leaf extract and sown. Its leaves amended in the soil also reduced the incidences of the root-knot nematode, *M. javanica* in chickpea (Singh and Sitaramaiah, 1967; Kaliram and Gupta, 1980, 1982; Gupta and Kaliram, 1981), *M. incognita* in papaya (Reddy et al., 1990) and also reduced the nematode populations and enhanced an average bunch weight in plantain (Jacob et al., 1990). Saxena et al. (1993) found neem leaves aqueous extract to show mortality to *Criconemoides annulatum* and *Xiphinema basiri* when tested under controlled conditions. Neem leaves soil amendment @ 20 g/kg soil showed 85% increase in the yield of wheat over control when tested against cyst nematode, *Heterodera avenae* (Singh et al., 1995). Similarly, chopped neem leaves admixed with the soil in the plant pit @ 1500 gm/m^2 area of the pointed gourd under field conditions against *M. incognita*, *H. indicus*, *T. vulgaris* and *H. dihystera* showed significant reduction in the populations of these nematodes in the roots (Verma and Anwar, 1995).

Neem Kernel as Nematicide

Root exudates of one-month-old neem seedling were toxic to the adults of *Helicotylenchus indicus*, *Rotylenchulus reniformis*, *Tylenchorhynchus brassicae*, *Tylenchusfiliformis* and the larvae of *Meloidogyne incognita* (Alam et al., 1975). Proct and Kornprobst (1986) found that pretreatment of tomato seedlings with the neem seed extract inhibited the penetration of the nematode, *M. javanica* juveniles to its roots. Neem oil fraction in hexane was also found to show nematicidal activity against the root-knot nematode, *Meloidogyne incognita* (Devakumar et al., 1985). Dried neem kernel powder applied to wheat seeds with the help of gum @ 800 mg/100 g seed successfully inhibited the penetration and development of the wheat gall nematode, *Anguina tritici* in the seedlings (Gokte and Swarup, 1985). Mishra and Mojumder (1993) found decomposed neem seed/kernel to reduce the root-knot nematode, *M. incognita* when amended in the soil @ 10% w/w. Mojumder and Mishra (1995) reported 20% w/w neem kernel powder coated on chickpea seed (cv-Pusa 240) to show very effective reduction in the population of *M. incognita* and *R. reniformis* and enhance grain yield.

Neem Oil Cake as Nematicide

Water extract of neem oil cake was reported to inhibit the larval emergence of *Meloidogyne incognita* (Khan et al., 1967). Singh and Sitaramaiah (1969) found neem oil cake incorporated in the infested soil in field plots 3 weeks before planting of okra and tomato to reduce disease incidences and also nematodes' populations in the root system. Similarly, Mobie et al. (1969) found suppressed populations of *Tylenchorhynchus* sp., *Hoplolaimus* sp., *Helicotylenchus* sp., *Rotylenchulus reniformis* and *Meloidogyne incognita* around roots of the eggplant, guava and citrus plants when neem oil cake was amended in the infested soil. One month before sowing of wheat, the amendment of neem cake in the soil also reduced parasitic nematode populations (Gaur and Prasad, 1970). Further, amendment of neem oil cake was reported to reduce the populations of root-knot nematodes, *M. incognita* and *M. javanica* and root galls in okra, tomato and other vegetables (Singh and Sitaramaiah, 1971; Sharma et al., 1971; Gowda, 1972; Desai et al., 1973; Khan et al., 1973; Mishra and Prasad, 1973). However, neem oil cake influenced and modified the relationship between plant and the nematode and the extract of neem cake was found to be toxic to *M. incognita* (Mishra and Prasad, 1973).

Water soluble fractions of neem oil cake were reported to show toxicity to the nematodes, viz., *Hoplolaimus indicus*, *Rotylenchulus reniformis*, *Tylenchorhynchus brassicae* and *M. incognita* (Khan et al., 1973). Alam and Khan (1974) reported neem seed cake applied in the soil of nursery bed @ 45 kg N per acre and in urea treated beds showed highest reduction in the populations of the stylet bearing nematodes, *H. indicus*, *T. brassicae*, *M. incognita*, *Ditylenchus cypei* and *Aphelenchus avenae*. Khan et al. (1974) found neem oil cake toxic to the soil inhabiting nematodes fauna when amended in the soil @ 1 g/49 g soil and ammonia liberated through decompositions of the cake was detrimental to these nematodes.

Azam et al. (1976) reported neem oil cake @ 100 N/kg soil to promote plant growth and to reduce the root-knot populations in the soil individually and also in combination with nemagon. Khan et al. (1976) and Siddiqui (1976) further reported the suppression of the nematode populations, viz., *T. brassicae*, *H. indicus*, *Helicotylenchus erythrinae* and *M. incognita*, in rhizosphere of the vegetable plants when neem oil cake was amended in the soil. Singh and Sitaramaiah (1976) reported neem oil cake amendment in the soil as an organic manure and also found it to reduce the root-knot nematode populations in the tomato.

Aqueous extract of neem cake amended in the soil was reported to inhibit the oviposition, hatching and caused mortality to *M. javanica*, *A. avenae* and *H. indicus* (Sitaramaiah and Singh, 1977). Khan (1979) and Vijayalakshmi and Goswami (1983) found an effective inhibition in hatching of the larvae from the freshly laid eggs of *M. incognita* when aqueous extract of neem oil cake was evaluated against the eggs of the nematode. Neem oil cake in the soil also reduced the incidences of *M. incognita* in the mungbean (Vijayalakshmi and Prasad, 1979). Neem oil cake amended in the soil was reported to decrease the population of *Tylenchorhynchus elegans* @ 1% w/w in the soil (Sitaramiah and Singh, 1977). Desai et al. (1979) found neem cake to reduce the incidences of *M. incognita* in tobacco when incorporated in the soil. Similarly, Acharya and Padhi (1988) and Jagadale et al. (1985) reported neem cake to reduce infestation of the root-knot nematode, *M. incognita* in betelvine when incorporated in the soil. Its oil cake extracts also showed toxicity to *M. incognita* (Singh, 1980).

Sharma et al. (1985) reported neem oil cake extracted in hot water and methyl alcohol to suppress the population of *M. incognita* in a laboratory test.

Neem cake amendment in the soil also reduced the incidences of *M. incognita* and *M. javanica* in the mulberry (Sikdar et al., 1986), in betelvine (Jagadale et al., 1985a,b; Acharya, 1985) and in the tomato and okra (Srivastava, 1971; Routray and Sahoo, 1985; Shukla and Bankar, 1985; Vijayalakshmi and Goswami, 1985; Bhattacharya and Goswami, 1987a,b, 1988; Gupta and Bhattacharya, 1995). Neem oil cake amendment in the soil also controlled the nematode, *Radopholus similis* in yellow leaf disease affecting arechanut palms. Seed treatment of the cowpea with its oil cake extracts was reported to reduce the incidences of *R. reniformis* and *M. incognita* after sowing (Khan and Hussain, 1988a,b). Neem oil cake amended in the sterilized soil @ 1 gm/kg was reported to significantly reduce the root galling caused by *M. incognita* in the potted plants of *Ocimum basilicum* (Haseeb et al., 1988). Neem cake at the rate of 17 quintal/ha significantly controlled the nematodes, *Helicotylenchus* sp. *Hoplolaimus indicus*, *Rotylenchulus reniformis* and *Hemecriconemoides* sp. when applied in the soil (Gaur and Mishra, 1989). Reddy and Khan (1990) found neem oil cake incorporated in the soil @ 1 t/ha to reduce the incidences of root-galls in okra caused by *M. incognita*. Prasad and Khan (1990) found neem cake application in the soil alone or in combination with aldicarb to eliminate the populations of *H. indicus*, *Heliocotylenchus elegans*, *Pratylenchus zeae* and *Tylenchorhynchus vulgaris* in the soybean. Post-pruning top dressing of neem cake (1–2 Kg/vine) to the crop, *Vitis* sp. was effective in suppressing the root-knot nematode, *M. incognita* (Jayaraj, 1991). Hasan (1992) reported its oil cake and also root exudates to reduce the gall formation by *M. incognita*. Haseeb and Butool (1991) found neem cake amendment in the soil @ 1 gm N/kg soil to significantly suppress the population of *M. incognita* in the roots of davana, *Artemisia pallens*.

Hussaini et al. (1993) found powdered neem cake amended in the soil @ 400 gm/m^2 to check root-knot nematode in the tobacco nurseries. Neem cake amended in the soil also showed nematicidal activity when tested in the mungbean crop against *M. incognita* and *R. reniformis* (Tiyagi and Alam, 1993). Haseeb and Butool (1993) also reported neem cake amended in the potted soil @ 1 gm/kg to reduce the root-knot index of *M. incognita* on *Trachyspermum ammi*. Neem cake @ 100 kg N/ha decomposed in the soil mixture in the pots for 1 week significantly reduced the populations of *M. incognita*, *R. reniformis* and *P. delattrai* in the brinjal crop (Poornima and Vadivelu, 1993). In Japanese mint cultivation the *M. incognita* population was reduced effectively when neem cake was admixed with soil @ 20% w/w alone (Pandey, 1995) and also in combination with carbofuran (Singh and Kumar, 1995).

Active Components

Two active components, viz., nimbidin and thionimone, isolated from neem extracts were reported to kill the populations of *Hoplolaimus indicus*,

Rolylenchulus reniformis, *T. brassicae* and *M. incognita* and also inhibited the growth of their larvae (Khan et al., 1973). Khan et al. (1974) found ammonia liberated through decomposition of neem oil cake in the soil as a toxic chemical to the soil inhabiting nematodes. Alam et al. (1976) reported formaldehyde in the oil cake manure as a chemical responsible for killing the nematodes in the soil. However, a considerable amount of phenol was detected from neem oil cake in the soil as a toxicant to the nematodes (Alam et al., 1979). Limonoids extracted from neem kernels were also reported to inhibit the hatching and showed a high larval mortality to *M. incognita* (Devakumar et al., 1985). Rao and Reddy (1993) reported repelin and welgro at 2.5 and 5% concentrations to significantly reduce the population of *M. incognita* when tomato nursery beds were drenched with their suspensions. Neem allelochemicals like azardirachtin, nimbidic acid and quercetin were also found to show nematicidal activity when tested against *M. incognita*, *R. reniformis*, *H. indicus* and *T. brassicae* under controlled conditions (Siddiqi and Alam, 1993).

20. *Azolla pinnata* R. Br. (Azollaceae)

Common name: Azolla

This is an annual/seasonal aquatic blue-green algae commonly found in lowland rice fields and ponds in India. Thakar (1988) found its leaf extract to show toxicity to the eggs of the root-knot nematodes, *M. incognita* and *M. javanica*. Thakar et al. (1987) found an economic and effective use of dry azolla as a biofertilizer, which could significantly reduce the infections of *M. incognita* in the roots of the okra plants. Patel et al. (1990b) reported its dried leaf powder to reduce the incidences of *M. incognita* and *M. javanica* when applied in the soil below the crop row 1–2 days prior to the okra seedling.

21. *Bocconia cordata* Linn. (Papaveraceae)

This is an annual/perennial herb found in tropical and sub-tropical zones. From the roots of this herb the nematicidal substances were isolated and identified as three alkaloids, viz., sanguinarine (Figure 118), chelerythrine (Figure 119) and bocconine (Figure 120). The alkaloids were reported to show toxicity to *Rhabditis* sp. and *Panagrolaimus* sp. (Rhode, 1972).

22. *Boehmerea nivea* (Linn.) Gauel. (Urticaceae)

Common name: Ramie

This is a perennial small shrub found in South East Asia and now commonly grown in Assam and West Bengal in India and Bangladesh. Its oil cake and leaves amended in the soil reduced population of the root-knot nematode, *Meloidogyne arenaria* (Mian and Rodriguez, 1982).

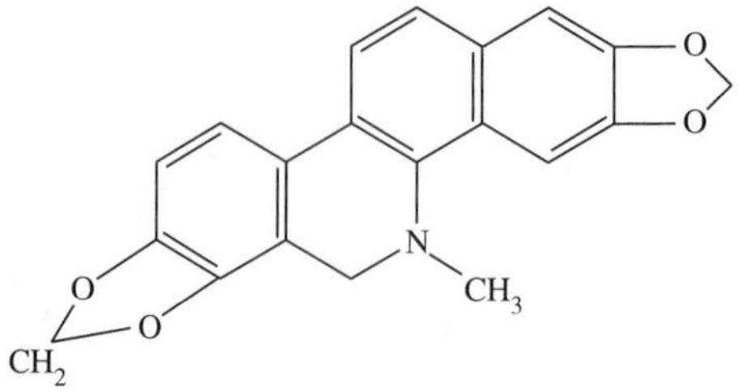

Figure 118. Sanguinarine.

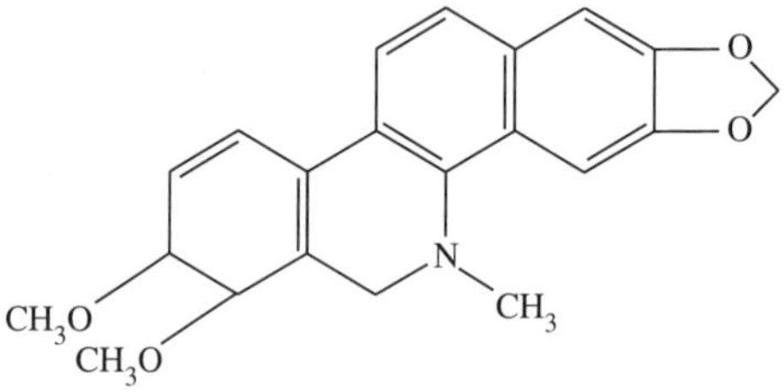

Figure 119. Chelerythrine.

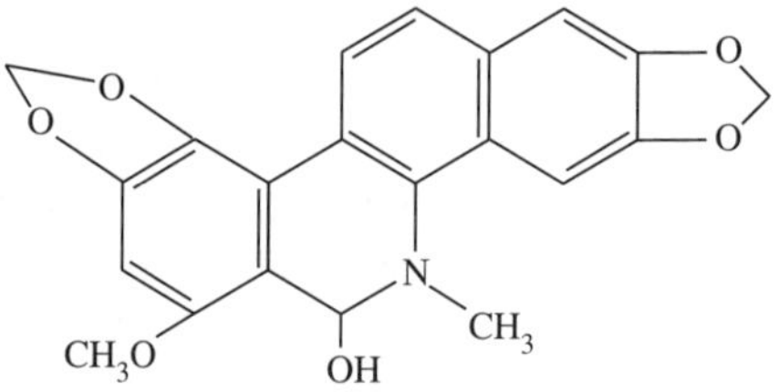

Figure 120. Bocconine.

23. *Brassica campestris* Linn. var. *sarson* Prain (Brassicaceae)

(Syn.-*B. campestris* Linn. var. *glauca* Duth. and Full., *B. campestris* Linn. var. *dichotoma* Watt.)

Common name: Yellow mustard or Black mustard or Indian colza

This is an annual herb cultivated for its oil in tropical and sub-tropical zones. Its oil cake was reported to show nematicidal activity to a wide range of the nematodes of agricultural importance. Hameed (1968) and Singh and Sitaramaiah (1971) found its oil cake amended in the soil @ 2.5 t/ha to reduce the intensity of root-galls formation due to *Meloidogyne javanica* in tomato and in okra. However, in a laboratory test water extract of its oil cake proved to be toxic to *M. incognita* (Mishra and Prasad, 1973). Further, oil cake amendment in the soil was also reported to suppress the populations of the nematodes, viz., *H. indicus*, *M. incognita*, *A. avenae* and *T. brassicae* (Khan et al., 1974). Azam et al. (1976) also found amendment of its oil cake in the soil at the rate of 100 kg N/ha to promote the growth of the potato and also reduced development of the root-knot nematode. Similarly, Singh and Sitaramaiah (1976) reported its oil cake application in the soil to suppress the populations of the root-knot nematodes in the tomato. Alam et al. (1977) reported mustard oil-cake amended in the soil to decrease the populations of *H. indicus*, *T. brassicae*, *Tylenchus filiformis* and *M. incognita* up to 6 months in the maize and the sunnhemp crops. Sharma et al. (1981) reported mustard cake amendment in the soil to reduce the cyst populations and hatching of *Hetrodera avenae*. Water soluble fractions of its oil cake applied in the soil also reduced the incidences of *M. incognita* and *M. javanica* in okra and tomato (Srivastav, 1971; Bhatnaga et al., 1978; Patel, 1985a,b).

Active Components

Khan et al. (1974) reported 'ammonia' liberated from its oil cake amended in the soil as an active component, which was detrimental to the soil nematodes like *T. brassicae*, *H. indicus*, *M. incognita* and *A. avenae*. Further, the phenolic contents found in its oil cake were as high as 0.5 mg/100 mg and showed toxicity to the soil inhabiting nematodes (Alam et al., 1977, 1979).

24. ***Brassica hirta*** **Moench. (Brassicaceae)**

(Syn.-*B. alba* [Linn.] Rabenha., *Sinapis alba* Linn.)

Common name: White mustard

This is a seasonal herb found in temperate and Mediterranean zones. Its oil cake aqueous extract showed nematicidal activity to the golden nematode, *Heterodera rostochinensis* (Ellenby, 1945a,b).

25. ***Brassica integrefolia*** **Linn. (Brassicaceae)**

Common name: Mustard

This is a seasonal tropical herb commonly cultivated in India. Its oil cake application in the soil was reported to reduce the incidences of the root-knot nematode, *M. incognita* and also enhanced the plant vigour (Verma, 1976).

26. ***Brassica nigra*** **(Linn.) Koeh. (Brassicaceae)**

Common name: Black rye or Black mustard

This is a seasonal herb found in tropical regions and commonly cultivated in India. Khan et al. (1967) found water extract of its oil cake to inhibit the larval emergence of the root-knot nematode, *M. incognita*. Singh and Sitaramaiah

(1969, 1971) and Siddiqui (1976) reported its oil cake amendment in the soil 3 weeks before transplanting okra and tomato to reduce the root-knot galls caused by *M. javanica*. However, soil applications of mustard oil cake @ 40 kg N/acre reduced the populations of the stylet bearing nematodes (Alam et al., 1976). Mukherjee (1983) reported nematicidal activity in the water extract of its oil cake against the oat nematode *Aphelenchus avenae*; the stem and bulb nematode, *Ditylenchus cypei*; the golden nematode, *Heterodera rostochinensis*; the root-knot nematode, *M. incognita*; and stunt nematode, *T. brassicae*. Its plant extract (1:10 dilution) applied in the soil of a tobacco nursery reduced the incidences of root galls caused by *M. incognita* (Krishnamurthy and Murthy, 1990).

Allyl isothiocynates as main components of its seed oil were reported to control the potato root eelworm, *Heterodera rostochinensis* by inhabiting its larval emergence (Ellenby, 1945a,b).

27. *Caesalpinia crista* Linn. (Caesalpiniaceae)

(Syn.-*C. bonducella* Flem., *C. bonduc* Roxb.)

Common name: Molucea bean

This is a perennial straggling prickly shrub distributed throughout the Coastal States of India. Its leaf extract (1:10 dilution) applied in the soil of the tobacco nurseries reduced the incidences of the root-knot nematode, *M. incognita* (Krishnamurthy and Murthy, 1990).

28. *Calendula officinalis* Linn. (Asteraceae)

Common name: Pot marigold

This is an ornamental seasonal/annual herb grown for its bright orange yellow colored heads in tropical India. Its dried plant amended in the soil was found to reduce the incidences of the root-knot nematode, *M. incognita* in the vegetables (Goswami and Vijayalakshmi, 1985).

29. *Callophyllum inophyllum* Linn. (Clusiaceae)

Common name: Undi or Indian laurel or Laurel wood

This is a perennial evergreen tree found in tropical zones and commonly available in Orissa, Karnataka, Maharastra and Andamans in India. Its oil cake amended in the soil gave significant control of the root-knot nematode, *M. javanica* in tobacco (Pillai and Desai, 1976). Similarly, its oil cake extract was also reported to show toxicity to the root-knot nematodes, *M. incognita* and *M. javanica* (Mukherjee, 1983; Goswami and Vijayalakshmi, 1987). Undi cake amended in the soil of the potted rice plants not only suppressed the populations of root-knot nematode, *M. graminicola* (*oryzaecola*) but also promoted the rice plant growth (Prakash et al., 1990d).

30. *Calotropis gigantea* Ait. (Asclepidaceae)

Common name: Crown plant or Madar

This is a perennial undershrub found in tropical and arid zones and available throughout India. Its leaves amended in the soil were reported to reduce the incidences of the root-knot nematode, *M. incognita* and also enhanced the growth of the okra plant (Kumar and Nair, 1976). Madar leaf powder incorporated in the soil @ 40 gm/pot also reduced the incidences of the root-knot nematode, *M. incognita* (Bhatnagar et al., 1980). Vijayalakshmi and Goswami (1985) reported its 100 gm fresh flower or plant incorporated with per kg of the soil to significantly reduce the larval penetration of the nematode, *Meloidogyne incognita* in the tomato roots when tested in the potted plants. Dried leaves of madar amended in the soil @ 2500 kg/ha also effectively reduced the population of *M. incognita*, *H. indicus* and *R. reniformis* in betelvine (Sivakumar and Marimuthu, 1986).

31. *Calotropis procera* Willd. (Asclepidaceae)

Common name: Akund or Swallow-wart

This is an evergreen perennial herb distributed in Western and Southern India. Its aqueous leaf extract was reported to show nematicidal activity against the root-knot nematode, *M. incognita* when tested in potted plants of the vegetables (Jain et al., 1986). Reddy et al. (1990) found its chopped leaves admixed with the infested soil to reduce the formation of galls caused by *M. incognita* in the papaya. Its chopped leaves admixed with the soil in the potted tomato effectively reduced the population of *M. incognita* and also root-gall index (Sundarabubu et al., 1993). Sharma and Trivedi (1995) found its leaf extract to show toxicity to the eggs of the cyst nematode, *Heterodera avenae* on barley. Similarly, Verma and Anwar (1995) recorded significant reduction in the populations of *M. incognita*, *H. indicus*, *T. vulgaris* and *H. dihystera* in the pointed gourd fields when chopped castor leaves were admixed with the soil in the plant pit @ 1500 gm/m^2.

32. *Camellia sinensis* (Linn.) O. Kuntz. (Theaceae)

(Syn.-*Camellia thea* Link., *C. theifera* Griff., *Thea sinensis* Linn.)

Common name: Tea

This is a perennial evergreen shrub cultivated in tropical and sub-tropical zones. In India, it is commonly grown in Assam, Darjeeling, Travancore, Nilagiris, Malabar, West Bengal, Deheradun and Kumaun hills. Its decaffeinated tea waste was reported to inhibit the oviposition in the root-knot nematode, *M. graminicola* but larval penetration into rice roots could not be reduced (Roy, 1976).

33. *Cannabis sativa* Linn. (Moraceae)

Common name: Hemp or Marijuana
This is an annual/perennial tall aromatic herb found in tropical and sub-tropical zones. It is native to Central Asia but available throughout India. Its root and shoot aqueous extracts were reported to show nematicidal activity against the nematodes, viz., *T. brassicae*, *H. indicus* and *R. reniformis* (Haseeb et al., 1982). Goswami and Vijayalakshmi (1986) reported soil amendment of its whole plant to reduce the incidences of the root-knot nematode, *M. incognita*.

34. *Capsicum annum* Linn. (Solanaceae)

Common name: Red pepper or Chillies
This is a seasonal/annual herb native to the West Indies and tropical America and is now cultivated throughout India. Its whole plant extract (1:10 dilution) applied in the soil of the tobacco nurseries was reported to reduce the root-galling caused by *M. incognita* (Krishnamurthy and Murthy, 1990).

35. *Carica papaya* Linn. (Caricaceae)

Common name: Papaya
This is a perennial small tree found in tropical and semi-arid zones. It is native to the West Indies and Central America but available throughout India. Its unripe fruit extract in water was reported to show nematicidal activity when tested against the root-knot nematode, *Meloidogyne incognita* (Hoan and Davide, 1979).

36. *Carthamus oxycantha* M.Bieb. (Asteraceae)

Common name: Wild safflower or Kara
This is an annual spiny pernicious herb found in arid zones. Its whole plant and leaf extracts in water showed nematicidal activity against the nematodes of agricultural importance (Mukherjee, 1983).

37. *Carthamus tinctorius* Linn. (Asteraceae)

Common name: Safflower or Kusum
This is an annual herb found in tropical and Mediterranean zones but native to India. Its flower and root aqueous extracts were reported to be toxic to the white-tip nematode, *Aphelenchoides besseyi* (Katsura, 1981) and also to other nematodes of agricultural importance (Mukherjee, 1983). Oil of safflower admixed with the soil was found to be toxic to *Pratylenchus penetrans* (Miller, 1979). Its oil-cake amendment in the soil was reported to reduce the incidences of the root-knot nematodes, *M. incognita* and *M. javanica* in the tomato (Hameed, 1968; Goswami and Vajayalakshmi, 1986).

The active components isolated from its root and flower extracts were found to be two isomeric polyacetylene compounds, viz., 3-trans-trans and 3 cis-trans, 11 trans trideca, 1,3,11 triene 5,7,9 triyene (Figures 121 and 122), and it showed toxicity to *A. besseyi* (Kosigo et al., 1976). Munakata (1978) described the chemical structures of these two nematicidal components.

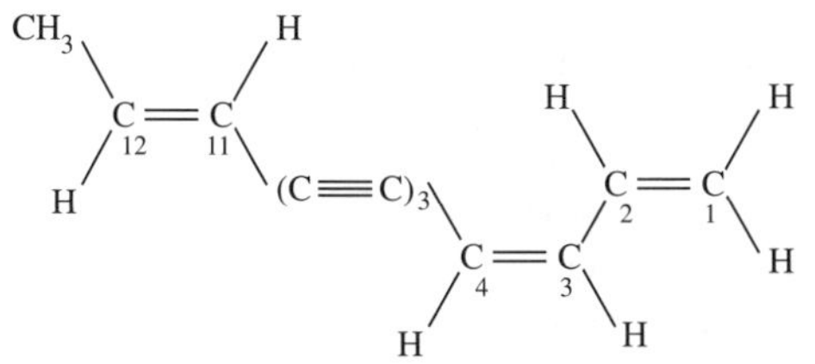

Figure 121. 3-cis, 11-trans-trideca-1, 3, 11-triene-5, 7, 9-triyene.

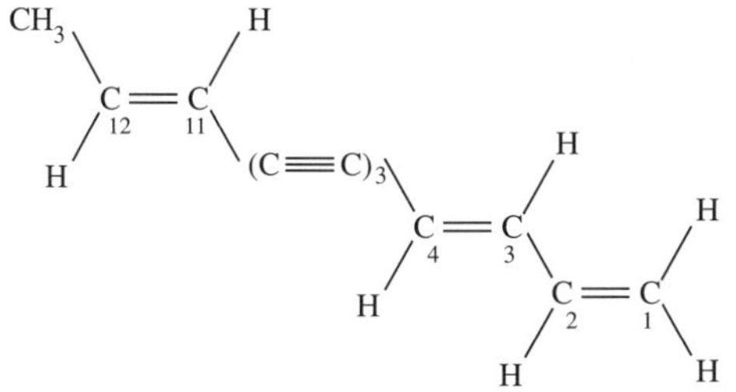

Figure 122. 3-trans, 11-trans-trideca-1, 3, 11-triene-5, 7, 9-triyene.

38. *Cassia carandas* Linn. (Caesalpiniaceae)

This is a perennial shrub found in tropical and sub-tropical zones. This shrub is commonly available in Uttar Pradesh, Bihar, Madhya Pradesh, Orissa, West Bengal and Kerala in India. Its leaf extract was reported to show strong nematicidal activity when tested against the larvae of root-knot nematode, *M. incognita* and the adults of reniform nematode, *R. reniformis* (Haseeb et al., 1982).

39. *Cassia fistula* Linn. (Caesalpiniaceae)

This is an ornamental perennial tree found in tropical zones. This tree is commonly available in Uttar Pradesh, Bihar, Orissa, West Bengal and Madhya Pradesh in India and also in Nepal and Sri Lanka. Soil amendment of its leaves @ 5–10% (w/w) in potted plants of okra and tomato was reported to reduce incidences of the root-knot nematode, *M. javanica* (Singh and Sitaramaiah, 1967). Haseeb et al. (1982) also reported its leaf extract toxic to the adults of reniform nematode, *R. reniformis* in a laboratory test.

40. *Cassia occidentalis* Linn. (Caesalpiniaceae)

This is an ornamental perennial shrub/small tree found in tropical zones and commonly available in North Indian States. Its 5–10% (w/w) leaves' applications in the soil of the potted tomato and okra plants reduced incidences of the root-knot nematode, *M. javanica* (Singh and Sitaramaiah, 1967).

41. *Catharanthus roseus* (Linn.) G.Don. (Apocynaceae)

Common name: Madagascar periwinkle or White flower winkle

This is a perennial small ornamental shrub native to the West Indies but available in India. Amendment of its leaves and whole plant in the soil were reported to inhibit the larval development of *M. incognita* and *M. javanica* when applied in the soil in rows 1–2 days before seeding of the okra and also inhibited the development and populations of the reniform nematode, *Rotylenchulus reniformis* and the stunt nematode *Tylenchorynchus vulgaris* in the inoculated and infested crop fields with these nematodes (Patel et al., 1990a,b). Krishnamurthy and Murthy (1990), however, reported its plant extract (1:10 dilution) to reduce root-galling in the tobacco nursery caused by root-knot nematode, *M. incognita*.

42. *Centrosema pubescens* Benth. (Papailionaceae)

This is a perennial herb/vine grown as cover crop for coffee in tropical zones and also available in Tamil Nadu and Karnataka in India. Its aqueous leaf extract was reported to be nematicidal to the root-knot nematode, *Meloidogyne incognita* (Hoan and Davide, 1979).

43. *Chenopodium album* Linn. (Chenopodiaceae)

Common name: Pigweed or Lambsquarters

This is an annual herb found in tropical zones and commonly available in India as a weed in the wheat crop fields. Its whole plant extract in water showed nematicidal activity to the root-knot nematode, *M. javanica* under laboratory conditions (Nandal and Bhatti, 1983).

44. *Chenopodium ambrosioides* Linn. (Chenopodiaceae)

(Syn.-*C. anthelminticum* [Linn.] A. Gray)

Common name: Wormwood seed or American worm seed

This is an annual/perennial strongly scented herb found in tropical and temperate zones. It is a native to the West Indies and South America but commonly available in West Bengal, Jammu, Kashmir and Maharastra in India. Hussian and Masood (1975) reported its leaf extract toxic to *M. incognita* larvae in the vegetable crops when tested in the pot experiment. Its root and shoot extracts in water were reported to show nematicidal activity to *T. brassicae*, *H. indicus* and *R. reniformis* (Haseeb et al., 1978). This herb was used as a potential nematicide in combating with phyto-parasitic nematodes (Garciaespinosa, 1980).

45. *Chromolaena odorata* D.C. (Asteraceae)

Common name: Hagonoy

This is an annual herb found in tropical and temperate zones. Its leaf extract in water was reported to show nematicidal activity to the root-knot nematode, *Meloidogyne incognita* under laboratory tests (Hoan and Davide, 1979).

46. *Chrysanthemum indicum* Linn. (Asteraceae)

(Syn.-*C. japonicum* Thunb., *Pyrethrum indicum* D.C.)

Common name: Japanese chrysanthemum

This is a biennial/perennial ornamental herb native to China and Japan and found in tropical zones. It is available throughout India. Its whole plant extract (1:10 dilution) applied in the soil of tobacco nurseries was reported to reduce the root galls of *M. incognita* (Krishnamurthy and Murthy, 1990).

47. *Cirsium arvense* Mill. (Asteraceae)

Common name: Canadian thistle

This is an annual herb found in tropical zones. Its whole plant aqueous extract showed nematicidal activity to the nematodes of agricultural importance (Mukherjee, 1983).

48. *Citrus reticulata* Blanco. (Rutaceae)

(Syn.-*C. nobilis* Andrews. non. Lour.)

Common name: Mandarin or Tangerine or Orange

This is a perennial small tree found in tropical and sub-tropical zones and commonly available in Maharastra, Assam, Sikkim, Karnataka, Punjab, West Bengal, Jammu and Kashmir in India. Its fruit peel and leaf extracts were reported to show nematicidal activity to *M. incognita* and *M. javanica* (Mukherjee, 1983).

49. *Clerodendrum inerme* (L.) Garetin. (Verbenaceae)

Common name: Vilayati mehndi

This is a perennial straggling and trailing shrub commonly grown in India as a hedge plant. Patel et al. (1985a,b) reported its whole plant amendment in the soil to reduce incidences of the root-knot nematodes, *M. incognita* and *M.*

javanica in the okra roots. Further, its leaf powder also reduced incidences of both these nematodes when applied in the soil before seeding of the okra seeds (Patel et al., 1990a). Application of its leaf extract in the soil was reported to inhibit root galling of *M. incognita* in the tobacco nursery (Krishnamurthy and Murthy, 1990) and its leaves amendment in the soil using 5 kg/plant in plantain cultivation reduced the nematode population and enhanced an average bunch weight (Jacob et al., 1990).

50. *Cloasia antiqurum* Remy. (Asteraceae)

This is an annual/perennial herb found in tropical zones. Its leaf aqueous extract was reported to be nematicidal to *R. reniformis* when tested under controlled laboratory conditions (Haseeb et al., 1982).

51. *Coffea robusta* Linden (Rubiaceae)

(Syn.-*C. canephora* Pierre ex. Frochner)

Common name: Congo coffee or Robusta coffee

This is a robust evergreen shrub or small tree native to Congo and commonly grown in Tamil Nadu and Karnataka in India. The coffee powder was reported to show nematicidal activity against the root-knot nematode, *Meloidogyne arenaria* (Mian and Rodriguez, 1982).

52. *Cosmos bipinnatus* Cav. (Asteraceae)

This is an annual ornamental herb found in sub-tropical and arid zones. Its flower extract in water was toxic to the juveniles and adults of the root-knot nematode, *M. incognita* and inhibited oviposition of the female nematode (Bano et al., 1986).

53. *Crodia myxa* Linn. (Borganinaceae)

This is a perennial tree found in tropical zones and commonly cultivated in Egypt and other parts of Africa. Its leaf extract in water was toxic to the adults of reniform nematode, *Rolylenchulus reniformis* (Haseeb et al., 1982).

54. *Crotalaria juncea* Linn. (Fabaceae)

Common name: Sunn-hemp or Junjunia

This is an annual/seasonal tall shrub grown in tropical and sub-tropical zones and commonly cultivated as a fibre crop in India. Its leaves amendment in the soil at the rate of 5–10% w/w in potted tomato and okra reduced the population of the root-knot nematode, *M. javanica* (Singh and Sitaramaiah, 1967). Its whole plant aqueous extract was also reported to be toxic to the root-knot nematode, *M. arenaria* (Mian and Rodriguez, 1982) and to *M. incognita* (Dar, 1983). Dahlya et al. (1986) found chopped leaves of the sunn-hemp to increase the shoot length of the wheat crop and also to reduce the population of the cyst nematode, *Heterodera avenae*.

55. *Crotalaria spectabilis* Medik. (Fabaceae)

Common name: Rattle box

This is an annual herb distributed in tropical and sub-tropical zones. Its aqueous leaf and seed extracts were toxic to *M. incognita* (Hoan and Davide, 1979).

56. *Croton sparsiflorum* Marong. (Euphorbiaceae)

Common name: Croton

This is an annual ornamental herb found in tropical and sub-tropical zones. Its leaf aqueous extract was toxic to *M. incognita* and *R. reniformis* (Islam, 1983).

57. *Cymbopogon citratus* (D.C.) Stapf. (Poaceae)

(Syn.-*Andropogon citratus* D.C., *A. citrioclorum* Desf., *A. roxburghi* Nees.)

Common name: Lemon grass

This is a perennial tufted grass found in tropical and sub-tropical zones commonly available in India. Its leaf extract in water showed nematicidal activity to the root-knot nematode, *M. incognita* (Kumar and Nair, 1976; Hoan and Davide, 1979).

58. *Cynodon dactylon* (Linn.) Pers. (Poaceae)

(Syn.-*Panicum dactylon* Linn., *Digitaria dactylon* [Linn.] Cop.)

Common name: Bermuda grass or Bahama grass

This is a perennial grass found in tropical and sub-tropical zones. Its leaf and whole plant extracts showed toxicity to *M. incognita* (Hoan and Davide, 1979).

59. *Cyperus rotundus* Linn. (Cyperaceae)

(Syn.-*C. hexastachyos* Roxb.)

Common name: Nut grass

This is a perennial sedge found in tropical zones and commonly available in India. Its whole plant extract showed toxicity to *M. incognita* (Hoan and Davide, 1979).

60. *Dalbergia sisso* Roxb. (Fabaceae)

Common name: Sisso

This is a perennial tree found in tropical and sub-tropical zones. Its sawdust amended in the soil alone or in combination with neem kernel powder reduced the population of *M. javanica* and admixing of its sawdust alone increased the phenolic contents in the plant as well as in the soil (Sitaramaiah and Singh, 1978). Aqueous leaf extract of sisso was also toxic to *R. reniformis* (Haseeb et al., 1982).

61. *Daphne odora* Linn. (Thymelacaceae)

Common name: Sweet daphne

This is a perennial shrub found in sub-tropical and temperate zones and also available in India. Its root extract in water was reported to be toxic to the white-tip nematode, *Aphelanchoides besseyi* (Munakata, 1978; Katsura, 1981). Two nematicidal active components as diterpenes having an ortho-easter group named 'odoracin' (Figure 123) and 'odoratrin' (Figure 124) were isolated from its roots' extract (Munakata, 1978).

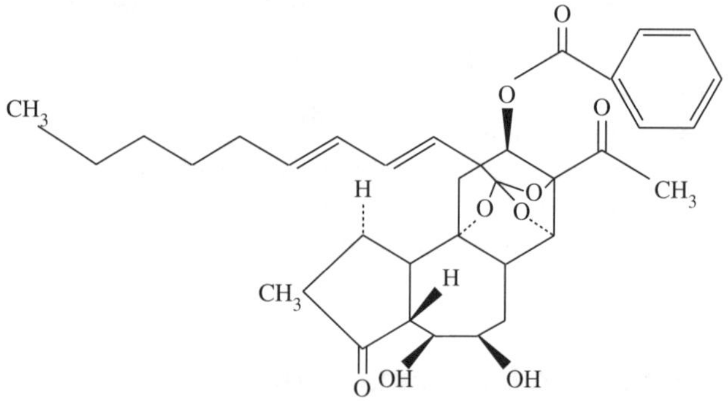

Figure 123. Odoracin.

62. *Datura metel* Linn. (Solanaceae)

(Syn.-*D. alba* Nees., *D. fastuosa* Linn.)

Common name: Angel trumpet

This is an annual/perennial herb or small shrub found in tropical zones and commonly available in India. Datura leaves amended in the soil were reported to reduce the incidences of root galls in the chickpea caused by *M. javanica* (Kaliram and Gupta, 1980, 1982; Gupta and Kaliram, 1981). Its aqueous leaf extract showed nematicidal activity to *M. incognita* and *M. javanica* (Mukherjee, 1983).

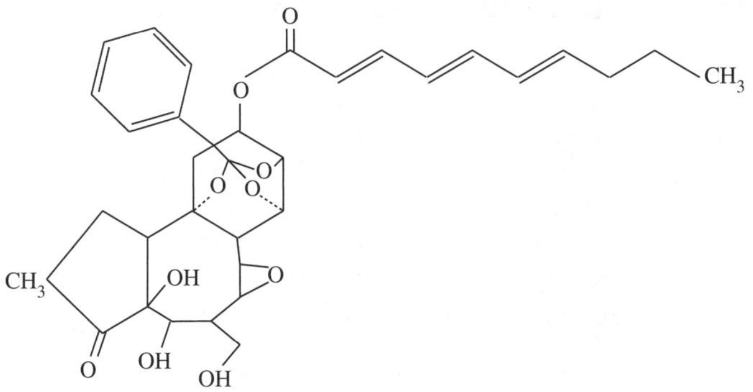

Figure 124. Odoratrin.

63. *Datura stramonium* Linn. (Solanaceae)

Common name: Jimson weed or Thorn apple or Stramonium
This is a perennial shrub found in tropical and temperate zones. Its seed and leaf aqueous extracts were reported to kill the root-knot nematodes, *Meloidogyne incognita* and *M. javanica* (Mukherjee, 1983). Water and methanolic extracts of its leaf, bulb and stem showed 75–100% mortality to the 2nd stage larvae of *Tylenchulus semipenetrans* and *Anguina tritici*. Further oil extracted from its seed also showed 65–90% mortality to the larvae of the test nematodes (Kumar et al., 1986).

The alkaloids, viz., atropine, nicotine and scopalamine, isolated from its leaf, stem and bulb extracts were found to show toxicity to *T. semipenetrans* and *A. tritici* (Uhlenbrock and Bijloo, 1959a; Winoto, 1959; Kumari et al., 1986).

64. *Derris elliptica* (Wall.) Benth. (Fabaceae)

Common name: Tuba-root or Bakal bip or Derris
This is a perennial shrub/woody climber found in tropical and sub-tropical zones and also available in India. Its root and seed extracts in water were reported to be nematicidal in nature when tested against the root-knot nematode, *Meloidogyne incognita* (Hoan and Davide, 1979).

65. *Desmodium zangelicum* (Linn.) D.C. (Leguminosae)

This is a perennial herb found in sub-tropical and temperate regions. Its root extract in water was reported to be toxic to the root-knot nematodes, *M. incognita* and *Radopholus similis* (Sarra-Guzman, 1984).

66. *Digitaria decumbens* (Linn.) Nees. (Poaceae)

Common name: Pangola digit grass

This is an annual grass found in tropical zones. Its whole plant aqueous extract was reported to kill the eggs of *M. incognita* prior to hatching and those that hatched failed to survive (Haroon et al., 1982). Further, population density of this nematode was also reduced in the soil planted with pangola digit grass more than in the soil left fallow or planted with tomato (Haroon and Smart, 1983a). Haroon and Smart (1983b) also reported its root extract to reduce hatching of the eggs and to cause mortality to the larvae of *M. incognita*.

67. *Eclipta alba* Hussk. (Asteraceae)

Common name: Morchand

This is an annual herb found in tropical zones and commonly available in Coastal States in India. Its root, shoot and whole plant extracts in water showed nematicidal activity to the rice root-knot nematode, *Meloidogyne graminicola* (Prasad and Rao, 1979) and to other species of *Meloidogyne* and *Rolylenchulus* (Islam, 1983). Bano et al. (1986) reported its whole plant extract to kill the root-knot nematode, *M. incognita* and also to inhibit hatching of this nematode's eggs. Goswami and Vijayalakshmi (1986) reported its leaves amended in soil to reduce incidences of the root-knots caused by *M. incognita* in the tomato.

Prasad and Rao (1979) reported a phenolic component (α-tetra thienyl and dithiophene derivative) called 'wadelectone', as nematicidal active component from its whole plant extract.

68. *Eichornia crassipes* (Mart.) Salms. (Pontedariaceae)

Common name: Water hyacinth

This is an annual/perennial aquatic weed found in tropical and sub-tropical zones. It is commonly available in India and used as compost fertilizer. Its compost amendment in the soil reduced hatching of the eggs of rice root-knot nematode, *Meloidogyne graminicola* but did not check the larval penetration into rice roots (Roy, 1976). Similarly, its root extract was toxic to *Meloidogyne incognita* and *Radopholus simillis* and its purified fractions were observed to be more toxic to these nematodes (Sarra-Guzman, 1984). Jain et al. (1986) reported its aqueous leaf extract toxic to the root-knot nematode, *M. incognita* in the vegetable crops.

69. *Emblica officinalis* Gaertn. (Euphorbiaceae)

(Syn.-*Phyllanthus emblica* Linn.)

Common name: Indian gooseberry or Emblic

This is a perennial deciduous tree found in tropical zones but native to South East Asia and now available throughout India. Its leaf extract was reported to be toxic to the root-knot nematode, *M. incognita* larvae in a laboratory test (Haseeb et al., 1982).

70. *Eragrostis amabillis* Jacq. (Poaceae)

This is an annual/perennial grass found in tropical and temperate zones. Its leaf extract in water was reported to be toxic to the root-knot nematode, *Meloidogyne incognita* (Hoan and Davide, 1979).

71. *Eragrostis curvula* (Schrad.) Nees. (Poaceae)

(Syn.-*Poa curvula* Schrad.)
Common name: Weeping lovegrass
This is an annual/perennial tufted grass found in tropical and sub-tropical zones and native to Africa and Rhodesia. Its cultivation in the nematode infested soil reduced the population build up of *Pratylenchus loosi* in the roots of tea plants (Gnanapragaram, 1981).

72. *Eruca vesicaria* (L.) Cav. Subsp. *sativa* (Mill.) Thell. (Cruciferae)

Common name: Roquette, Rocket salad or Taramira
This is a seasonal herb native to South Europe but commonly grown in Jammu and Kashmir, Himachal Pradesh, Kumum region of Uttar Pradesh, Arunachal Pradesh and North East Indian States for its oil used for burning of the lamps. Bhatnagar et al. (1978) reported water soluble fractions of its oil cake extract to show nematicidal activity against the root-knot nematode, *M. incognita* in the roots of okra plant when admixed with the soil.

73. *Erythrina indica* Lamk. var. *parcellii* Hort. (Papilionaceae)

(Syn.-*E. parcellii* Bull.)
Common name: Indian ceral tree
This is a perennial tree found in tropical and sub-tropical zones. In India, it is grown as an ornamental tree for its bright cinnamon-red flowers. Its leaf extract was found to be toxic to the root-knot nematode, *M. incognita* in the potted vegetable plants (Mohanty and Das, 1987, 1988).

74. *Eupatorium odoratum* Linn. (Asteraceae)

This is a perennial/annual herb found in tropical and sub-tropical zones. Its leaves amendment in the soil reduced the populations of the root-knot galls caused by *M. incognita* in okra and other vegetable crops (Kumar et al., 1976; Subrahmaniyam, 1985; Kuriyan, 1988).

75. *Euphrobia cyparissias* Lindl. (Euphorbiaceae)

Common name: Cypress spurge

This is a perennial herb found in temperate zones. Its whole plant extract in ether was reported to show nematicidal activity when tested against the nematodes of general importance (McIndoo, 1983).

76. *Ficus elastica* Roxb. (Moraceae)

Common name: India rubber tree

This is a small to medium-sized perennial evergreen tree native to India and Malaysia. Its chopped shoots of seedlings and early plants incorporated with the soil were found to reduce root-knot development and also reduced populations of the plant parasitic nematodes, *M. intognita*, *R. reniformis* and *T. brassicae* in tomato and eggplant (Siddiqui et al., 1987, 1992).

77. *Fleurya interrupta* Gandich. (Urticaceae)

This is an annual/perennial herb or shrub found in tropical zones. Its leaf extract in water was reported to be toxic to the root-knot nematodes, *M. incognita* and *M. javanica* (Mukharjee and Sukul, 1978; Mukherjee, 1983).

78. *Gaillardia picta* Foug. (Asteraceae)

This is an annual herb found in tropical and semi-arid zones. Its flower, stem and leaf extracts in water were reported to show toxicity to the juveniles of the root-knot nematode, *M. incognita* (Tiyagi et al., 1985).

79. *Giliricidia sepium* (Jacq.) Walp. (Papilionaceae)

(Syn.-*G. maculata* [H.B. and Kunth.] Steud., *Lonchocarpus maculatus* D.C.)

Common name: Madre tree or Spotted giliricidia

This is a perennial small tree found in tropical and sub-tropical regions. This is a native to tropical America but commonly grown as an ornamental tree and also for green manure in Tamil Nadu, Karnataka, Maharastra and Kerala in India. Kuriyan (1988) reported its leaves amended in the soil to reduce incidences of the root-knot nematode, *M. inognita* in the okra crop. Jasy and Koshy (1992) found its leaf extract lethal to *Radopholus similis*. Further, its chopped leaves @ 10 g/kg soil as green manure were found to reduce the population of this nematode and enhanced growth of black pepper grown under pot conditions.

80. *Gloriosa superba* Linn. (Liliaceae)

Common name: Glory lily or Tiger's claws

This is an annual/biennial ornamental climbing herb found in tropical India. Its root and shoot extracts in water were reported to show toxicity to the eggs of *M. incognita* when tested in laboratory conditions (Pandey and Haseeb, 1988).

81. *Gossypium hirsutum* Linn. (Malvaceae)

Common name: Cotton

This is an annual/seasonal shrub cultivated in tropical zones and commonly grown as cash crop in Uttar Pradesh, Punjab, Andhra Pradesh, Tamil Nadu and Maharashtra in India. Its seed oil admixed with the soil was reported to be toxic to the nematode, *Hetrodera tabacum* (Miller, 1957) and also to the root nematode, *Pratylenchus penetrans* (Miller, 1979). Lanjeswar and Shukla (1986) observed its oil cake to reduce incidences of *M. incognita* in the vegetable crops when amended in the soil.

82. *Guizotia obyssinica* Cass. (Asteraceae)

This is an annual herb found in tropical regions. It is native to tropical Africa and commonly cultivated in Madhya Pradesh, Andhra Pradesh, Karnataka and Orissa in India. Its seed oil cake amended in the soil was reported to reduce incidences of the root-knot nematode, *M. incognita* in the tomato (Gowda, 1972).

83. *Hannoa undulata* Planch. (Simaroubaceae)

This is a perennial shrub found in tropical, sub-tropical and temperate zones. Extract of its seeds admixed with the soil before sowing inhibited the penetration of *M. javanica* (Prot and Komprobst, 1983). Its root extract was also known to possess quassinoids as the active components and reported to inhibit penetration and reproduction of *M. javanica* in the tomato (Proct and Komprobst, 1985).

84. *Hannoa klaineana* Planch. (Simaroubaceae)

This is a perennial shrub found in tropical and sub-tropical zones. Treatment of the soil with crude aqueous extract of its seed was reported to inhibit the penetration of *M. javanica* juveniles (Proct and Komprobst, 1983).

85. *Helenium* (hybrid) Linn. (Asteraceae)

Common name: Moreheim beauty

This is an annual herb found in sub-tropical and temperate zones and commonly found in Western American states. Two nematicidal active components, i.e., 1-tridecanene-3,5,7,9,11-pentayne and 2,3-dihydro-2-hydroxy-3-methylene-6-methyl benzofuran, were isolated and identified from its methanolic extracts. These components were reported to show nematicidal activity when tested against *Pratylenchus penetrans* (Gommers, 1971).

86. *Helianthus annuus* Linn. (Asteraceae)

Common name: Sunflower

This is an ornamental annual/seasonal shrub found in tropical and temperate zones. Its root, flower and seed extract in water were reported to be toxic to the root-knot nematode, *Meloidogyne incognita* (Dar, 1983). Singh and Singh (1988) reported its oil cakes amendment in the soil to significantly reduce the root-knot index in the eggplant caused by *M. incognita.*

87. *Holarrhena antidysenterica* R.Br. (Apocynaceae)

Common name: Tellicherry bark

This is a perennial small tree found in tropical zones and commonly available in India, the Philippines and Malaysia. Its leaf, root and bark extracts in water were reported to be toxic to the root-knot nematodes, *Meloidogyne javanica* and *M. incognita* (Mukherjee, 1983).

88. *Hydrocarpus laurifolia* (Donnstr.) Sleumer (Flacourtiaceae)

(Syn.-*H. wightyiana* Blume)

Common name: Marotti

This is a perennial tree found in tropical zones and available in Western Ghats in India. Desai et al. (1973) reported *in vitro* the nematicidal property of the marotti cake. Its flower, fruit and seed aqueous extracts were also found to be toxic to the root-knot nematodes, *Meloidogyne incognita* and *M. javanica* (Mukharjee, 1983).

89. *Imperata cylindrica* (Linn.) P. Beauv. var. *major* (Nees.) C.E. Hubb. ex. Hubb. and Vaughan (Poaceae)

(Syn.-*I. arundinace* Cyr., *Lagurus cylindricus* Linn., *Saccharum cylindricum* [Link.] Lank.)

Common name: Sword grass or Congo grass

This is a perennial herb found in tropical and sub-tropical zones and available throughout India. Its leaf extract in water was toxic to the root-knot nematode, *M. incognita* (Hoan and Davide, 1979).

90. *Ipomea palmata* Forsk. (Convolvulaceae)

Common name: Railway creeper

This is a perennial climbing shrub found in tropical and sub-tropical zones and grown as an ornamental plant throughout India. Kumar et al. (1986) reported water and methanolic extracts of its leaf and stem to show mortality to the 2nd stage larvae of *Tylenchulus semipenetrans* and *Anguina tritici.*

91. *Iris japonica* Linn. (Iridaceae)

Common name: Japanese iris

This is a perennial herb found in temperate zones. Its aqueous leaf extract was reported to be toxic to the white-tip nematode, *Aphelenchoides besseyi* (Katsura, 1981).

92. *Jasminum arborescens* Roxb. (Oleaceae)

(Syn.-*J. roxburghianum* Wall.)

Common name: Tree jasmine

This is a perennial climbing herb found in tropical and sub-tropical zones and grown in India for its scented flowers. Its aqueous leaf extract inhibited larval hatching of *M. incognita* in roots of the vegetable crops (Hussain and Masood, 1975).

93. *Jatropha gossypifolia* Linn. (Euphorbiaceae)

This is a perennial ornamental shrub found in tropical zones and distributed in South East Asia. Sharma and Trivedi (1995) reported its leaves' extract in water to show toxicity to the egg of the cyst nematode, *Heterodera avenae* in a laboratory test.

94. *Lantana camara* Linn. (Verbenaceae)

Common name: Lantana

This is a perennial climbing aromatic herb found in tropical zones native to tropical America and now completely naturalized and available throughout India. Its chopped leaves admixed with the infested soil were found to reduce root galling in the papaya caused by *M. incognita* (Reddy et al., 1990).

95. *Lawsonia inermis* Linn. (Lathyraceae)

(Syn.-*L. alba* Lamk.)

Common name: Henna or Egyptian privet

This is a perennial branched shrub or small tree native to Arabia and Persia but now being cultivated mainly in West Indian States. Kumar et al. (1986) reported water and methanolic extracts of its leaves to show significant larval mortality to the 2nd stage larvae of *Tylenchulus semipenetrans* and *Anguina tritici*. Palmitic, linoleic and oleic acids were found to be the active components in this plant to show nematicidal activity (Badami and Patil, 1975).

96. *Leucaena leucocephala* (Lamk.) de Wit. (Mimosaceae)

(Syn.-*L. glauca* Benth., *Acacia gluca* Willd., *Mimosa leucocephala* Lamk.)

Common name: White popinac or Horse tamarind

This is a perennial shrub or small tree found in tropical zones and native to tropical America but now available throughout India. Its aqueous root extract was reported to show toxicity to *M. incognita* (Hoan and Davide, 1979) and *Radopholus similis* (Sarra-Guzman, 1984). Its chopped leaves amendment in the soil was also found to effectively reduce the population of cyst nematode, *Heterodera avenae* in wheat crop (Dahlya, 1986). Leaf extract of the plant was reported to be toxic to *Aphelenchoides composticola* (Grewal and Sohi, 1988). Sudarababu et al. (1993) found its chopped leaves admixed with the soil in the pot experiment @ 5 g/kg to reduce the gall index and population of *M. incognita* in the tomato crop.

97. *Linum usitatissimum* Linn. (Linaceae)

Common name: Linseed or Flax

This is a seasonal shrub grown in India. Its oil cake reduced the incidences of the root-knot galls in their roots caused by *M. incognita* and *M. javanica* when amended in the soil 3 weeks before transplanting of the okra and tomato seedlings (Hameed, 1968; Singh and Sitaramaiah, 1969, 1971, 1976; Srivastava, 1971; Lanjeswar and Shukla, 1986). Extract of its leaf in water was reported to be toxic to *Rotylenchus reniformis* and *M. incognita* (Mahmood et al., 1982). Its oil admixed with the soil was also toxic to *Pratylenchus penetrans* (Miller, 1979) and *M. incognita* (Sen and Das Gupta, 1985). Singh and Singh (1988) reported its oil cake to reduce root-knot index and also population of *M. incognita* infesting eggplant and this could be due to a significant increase in the content of phenols and amino acid in the roots.

98. *Madhuca indica* J.F. Gmel. (Sapotaceae)

(Syn.-*M. latifolia* [Roxb.] Macb., *Bassia latifolia* Roxb.)

Common name: Mahua

This is a perennial tree found in tropical and sub-tropical zones and commonly available in India. Water extract from its oil cake was reported to inhibit the larval emergence of the root-knot nematode, *M. javanica* (Khan et al., 1967). Singh and Sitaramaiah (1969) reported its oil cake amendment in the soil 3 weeks before transplanting to reduce the galls in okra and tomato roots caused by the root-knot nematode, *M. incognita*. Similarly, population of stylet bearing nematodes, *Tylenchorhynchus* sp., *Hoplolaimus* sp., *Helicotylenchus* sp., *Rotylenchulus reniformis* and *M. incognita* were reported to be suppressed in/around the roots of eggplant (brinjal), guava and citrus in the soil amended with oil cake of the mahua (Mobin et al., 1969). Srivastava (1971) and Singh and Sitaramaiah (1971, 1976) found a reduced intensity of the root galls in okra, tomato and eggplant caused by *M. javanica*. Amendments of its oil cake in the soil @ 100 kg N/ha reduced the population of *M. incognita* in the tomato and other vegetable crops (Khan et al., 1973; Azam and Khan, 1976; Siddiqui, 1976; Lanjeswar and Shukla, 1986). Water extract from its oil cake was also reported to show toxicity to *M. incognita* (Mishra and Prasad, 1973; Khan et al., 1973; Khan, 1974) to *H. indicus*, *R. reniformis*, *T. brassicae* (Khan, 1973) and also to *Aphelenchus avenae* and *Ditylenchus cypei* (Mukherjee, 1983). Haseeb et al. (1988a,b) reported mahua cake amendment in the soil @ 1 gm/kg soil to reduce the root galling effectively in *Ocimum basilicum* caused by *M. incognita*. Poornima and Vadivelu (1993) found mahua cake to reduce the population of *M. incognita*, *R. reniformis* and *P. delattrei* in the brinjal crop when amended in the soil @ 100 kg N/ha and decomposed for 1 week.

Alam and Khan (1974) reported ammonia liberated from mahua cake amended in the soil to kill the populations of *H. indicus*, *M. incognita*, *T. brassicae* and *A. avenae*. Further, Alam et al. (1976) analyzed its oil cake chemically and found it to possess 0.966% formaldehyde, which showed toxicity to the abovementioned test nematodes. Later, it was reported to be phenol, which was detrimental to the nematodes and phenolic contents were estimated to be 0.22 mg/100 mg of the cake (Alam et al., 1979).

99. *Mangifera indica* Linn. (Anacardiaceae)

Common name: Mango

This is a perennial tree available in tropical and sub-tropical zones and extensively grown in Uttar Pradesh, Andhra Pradesh, Punjab, Maharastra and Tamil Nadu in India. Its leaf extract (1:10 dilution) amended in the soil of tobacco nurseries was reported to reduce root galling caused by *M. incognita* (Krishnamurthy and Murthy, 1990). Green leaves of the mango admixed with the soil @ 5000 kg/ha were found to control *M. incognita* in okra (Kumar and Nair, 1976). Vijayalakshmi et al. (1979) observed its leaf extract to show toxicity to 2nd stage juveniles of *M. incognita*. Jain and Saxena (1993) reported solution of dried mango leaves soaked in distilled water to show 88% mortality to *Criconemoides annulatum*.

100. *Melia azedarach* Linn. (Meliaceae)

Common name: China berry or Persian lilac or Dharek

This is a perennial tree found in tropical and sub-tropical zones and commonly available in India. Singh and Sitaramaiah (1967) reported its leaves to reduce incidences of the root-knots caused by *M. javanica* in okra and tomato roots when amended in the soil. Its root and shoot extracts showed nematicidal activity against *T. brassicae*, *H. indicus* and *R. reniformis* (Haseeb et al., 1978).

Siddiqui and Alam (1987) and Siddiqui and Saxena (1987) reported interculture of China berry with the tomato and okra plants to reduce incidences of the root-knot nematode, *M. incognita* and reniform nematode, *R. reniformis*. Its oil cake soil application @ 5000 kg/ha significantly reduced the root-knot index of *M. incognita* (Verma, 1993).

101. *Mentha spicata* (Linn.) Huds. (Lamiaceae)

(Syn.-*M. viridis* Linn.)

Common name: Spearmint

This is an annual herb found in temperate and tropical zones. Its leaf extract in water was toxic to the larvae of *M. incognita* in laboratory tests (Haseeb et al., 1982).

102. *Mimosa pudica* Linn. (Mimosaceae)

Common name: Sensitive plant or Touch me not

This is a perennial spreading undershrub found in tropical and sub-tropical zones and commonly available in India. Its leaf extract in water was reported to be toxic to the root-knot nematode, *M. incognita* (Hoan and Davide, 1979).

103. *Momordica charantia* Linn. (Cucurbitaceae)

Common name: Bitter gourd

This is an annual vine found in tropical and sub-tropical zones. Its whole plant extracts in alcohol and water were reported to be toxic to the adults and juveniles of the root-knot nematodes, *M. incognita* and *M. javanica* (Hussain and Masood, 1975; Mahmood et al., 1979; Mukherjee, 1983).

104. *Morinda pterygosperma* Gaertn. (Rubiaceae)

Common name: Horse tree

This is a perennial small tree found in temperate and sub-tropical zones. Its leaf extract was reported to be toxic to *M. incognita* (Hoan and Davide, 1979; Sarra-Guzman, 1984) and to *Radopholus similis* (Sarra-Guzman, 1984).

105. *Moringa oleifera* Lamk. (Moringaceae)

(Syn.-*M. peterygosperma* Gaertn., *Guilandina moringa* Linn.)

Common name: Drumstick tree or Horseradish tree

This is a perennial small tree found in tropical regions and native to India. Its leaf extract (1:10 dilution) incorporated in the soil of tobacco nurseries reduced the root galling of *M. incognita* (Krishnamurthy and Murthy, 1990).

106. *Nasturtium officinalis* R. Br. (Brassicaceae)

(Syn.-*N. fontanum* Aschers., *Rorippa nasturtium aquaticum* [Linn.] Hayek.)
Common name: Watercress

This is an annual small herb found in tropical zones and commonly grown as a vegetable in West Bengal, Orissa and Punjab in India. Its aqueous leaf extract was reported to be toxic to *Heterodera rostochinensis* (Ellenby, 1945a).

107. *Nicotiana tobacum* Linn. (Solanaceae)

Common name: Tobacco

This is an annual herb found in tropical and semi-arid zones but a native to Central America. This is commonly grown as a cash crop in Andhra Pradesh, West Bengal, Maharastra and Karnataka. Its root and shoot aqueous extracts showed nematicidal activity to *T. brassicae*, *H. indicus* and *R. reniformis* (Haseeb et al., 1978).

108. *Ocimum sanctum* Linn. (Lamiaceae)

Common name: Holy basil

This is an annual/perennial shrub found in tropical zones and commonly available throughout India. The crude leaf extract of holy basil obtained by squashing 100 g leaves in 100 ml of water and also its nematicidal principle 'eugenol' at 4 mg/ml concentrations killed 100% of the juveniles of *M. incognita* in 120 and 80 minutes respectively in a bioassay test (Chatterjee et al., 1982). Its leaf extracts in ethanol and acetone were reported to be toxic to the root-knot nematodes, *M. incognita* and *M. javanica* (Mukherjee, 1983). Haseeb and Butool (1990) found its root and shoot aqueous extracts to reduce hatching of the eggs of *M. incognita* when tested under laboratory conditions. Extracts of other *Ocimum* spp. like *O. kilimondscharicum*, *O. basilicum* and *O. canum* showed similar activity against this nematode. Sharma and Trivedi (1995) also reported its aqueous leaf extract to show toxicity to the eggs of the cyst nematode, *Heterodera avenue* in a laboratory test.

109. *Oryza sativa* Linn.

Common name: Rice

This is a seasonal herb grown as an important cereal crop in tropical and sub-tropical zones and cultivated throughout India. Its leaf extract in water was reported to be toxic to the root-knot nematode, *M. incognita* (Hoan and Davide, 1979).

110. *Papaver rhoeas* Linn. (Papaveraceae)

Common name: Corn poppy

This is an annual/perennial herb found in sub-tropical and temperate zones but a native to Europe and Asia. Its root and shoot extracts in water were reported to be nematicidal to *T. brassicae*, *H. indicus* and *R. reniformis* (Haseeb et al., 1978).

111. *Parthenium hysterophorus* Linn. (Asteraceae)

This is an annual weed found in tropical zones and commonly available in South India. Whole plant extract of this weed was reported to show toxicity to the root-knot nematodes, *M. incognita* and *H. dihystera* in the vegetable crops (Hasan and Jain, 1984). Its chopped leaves admixed with the infested soil were found to reduce root-galling in papaya caused by *M. incognita* (Reddy et al., 1990).

112. *Peristropha bicalyculata* (Retz.) Nees. (Acanthaceae)

This is a perennial shrub found in tropical and sub-tropical regions and available throughout India. Chatterjee and Sukul (1979) reported amendments of its leaves in the soil to reduce incidences of the root-knot nematode, *M. incognita* in okra.

113. *Phaseolus lunatus* Linn. (Papilionaceae)

Common name: Sieva bean or Lima bean

This is a seasonal/annual herb native to tropical America but now grown in North India. Its seed extract was reported to inhibit the larval hatching and it showed toxicity to the root-knot nematode, *M. incognita* in the vegetable crops (Hussain and Masood, 1975; Mahmood et al., 1979).

114. *Piper nigrum* Linn. (Piperaceae)

Common name: Black pepper

This is a biennial/perennial climbing shrub native to the Indo-Malaysian region and cultivated in Western Ghats, Kerala, Karnataka and Maharastra in India. Its whole plant extract (1:10 dilution) was found to inhibit root-galling of *M. incognita* when amended in the soil of tobacco nurseries (Krishnamurthy and Murthy, 1990).

115. *Polyantha longifolia* (Sonn.) Thw. (Annonaceae)

This is a perennial tree having edible fruits found in tropical zones and commonly grown as an avenue tree on roadsides in Central India. Its aqueous leaf extract was toxic to the eggs of the cyst nematode, *Heterodera avenae* (Sharma and Trivedi, 1995).

116. *Pongamia glabra* Vent. (Fabaceae)

(Syn.-*P. pinnata* [L.] Pierre)

Common name: Puna oil tree or Pongamia or Karanja

This is a perennial tree found in tropical zones and commonly available in Maharastra, Gujarat and Madhya Pradesh in India. Water extract of its oil cake was reported to be toxic to *M. incognita* (Desai et al., 1973; Mishra and Prasad, 1973; Mishra et al., 1973), whereas its oil cake amended in the soil reduced the populations of the root-knot nematode, *M. javanica* in tomato (Hameed, 1968; Singh and Sitaramaiah, 1976) and *M. incognita* in tobacco (Desai et al., 1979) and in betelvine (Jagadale et al., 1985a,b). Rao and Pandey (1982) reported 5% karanja leaf extract to significantly reduce the population of *Aphenchoides composticola* at par with even carbofuran treatment. Govindaiah et al. (1989) found mulching of green leaves of pongamia in the soil to effectively reduce incidences of *M. incognita* in the mulberry and increased 28 to 45% leaf yield as compared to 16 to 39% in neem leaves.

117. *Populus deltoides* Marsh. (Salicaceae)

Common name: Caroline poplar

This is a perennial deciduous tree found in temperate zones and commonly available in North America and in Jammu and Kashmir in India. Its leaf aqueous extract was toxic to *Aphelenchoides composticola* (Grewal and Sohi, 1988).

118. *Portulaca oleraceae* Linn. (Portulacaceae)

Common name: Kitchen garden parslane or Common parslane

This is an annual herb found in temperate and arid zones. Its aqueous leaf and stem extracts were reported to show nematicidal activity to the root-knot nematode, *M. incognita* (Hoan and Davide, 1979).

119. *Prosopis juliflora* (sw.) D.C. (Mimosaceae)

(Syn.-*Prosopis chilensis* [Molino] D.C.; *Mimosa juliflora* Sw.)

Common name: Algaroba or Mesquite

This is a perennial small to medium-sized evergreen tree indigenous to North and South America and abundantly found in Buenos Aires, Chile, Peru, Central America, Mexico and West and South India. Sundarababu et al. (1993) reported chopped leaves of this tree to reduce the gall index and population of *M. incognita* in leaf-bits and tomato when admixed with the soil @ 5 g/kg in the pot experiment.

120. *Puerasia lobata* (Willd.) Ohwi. (Fabaceae)

(Syn.-*P. thunbergiana* Benth.)

Common name: Thumberry or Kudzu vine

This is a perennial shrub found in sub-tropical and tropical zones but native to China and Japan. Its leaf and root extracts were reported to show toxicity to the root-knot nematode, *M arenaria* (Mian and Rodriguez, 1982).

121. *Raphanus sativus* Linn. (Brassicaceae)

Common name: Radish

This seasonal herb found in tropical zones and commonly grown in India as a vegetable crop. Its root extract in water was reported to be toxic to the root-knot nematode, *M. incognita* (Das, 1983).

122. *Ricinus communis* Linn. (Euphorbiaceae)

Common name: Castor or Castor bean

This is an annual/perennial shrub or small tree found in tropical, sub-tropical and semi-arid zones. It is commonly grown in Andhra Pradesh, Maharastra and Karnataka as an oil seed crop. Its oil cake amended in the soil reduced intensity of root galls in the okra and tomato caused by *M. javanica* (Singh and Sitaramaiah, 1969, 1976), whereas water extract from its oil cake was reported to be toxic to the larvae of *M. incognita* (Khan et al., 1976; Sharma et al., 1985); to the adults of *H. indicus*, *R. reniformis*, and *T. brassicae* (Khan et al., 1973); and to *M. javanica* (Nadal and Bhatti, 1983). Mobin and Khan (1969), Alam and Khan (1974) and Alam et al. (1976) found its oil cake amended in the soil to reduce the populations of *Hoplolaimus indicus*, *Helicotylenchus* sp., *Rotylenchulus reniformis*, *Tylenchorynchus brassicae*, *M. incognita* and *Aphelenchus avenae* around the citrus, brinjal, wheat and guava roots. Similarly, its oil cake amendment in the soil reduced the incidences of *M. javanica* and *M. incognita* in the tomato, okra and brinjal (Hameed, 1968; Singh and Sitaramaiah, 1971; Srivastava, 1971; Gowda, 1972; Azam and Khan, 1976; Siddiqui, 1976; Reddy and Khan, 1990) and in the mulberry (Sikdar et al., 1986).

Oil cake amendment in the soil was reported to suppress the populations of *H. indicus*, *T. brassicae*, *Tylenchus filiformis* and *M. incognita* in the potato and radish roots (Alam et al., 1977). Mukherjee (1983) reported distilled water extract from its oil cake to show toxicity to *A. avenae*, *H. indicus*, *T. brassicae* and *Ditylenchus cypei*.

Leaf powder of castor when incorporated in the soil @ 40 gm/pot was found to reduce incidences of the root-knot nematodes in the okra roots (Bhatnagar et al., 1980). Its chopped leaves admixed with the infested soil were found to reduce the number of root-galls formed by *M. incognita* and *M. javanica* in the tomato (Goswami and Vijayalakshmi, 1986a,b; Ravidutt and Bhatti, 1986b; Zaki and Bhatti, 1989) and by *M. incognita* in the papaya (Reddy et al., 1990). Dutta and Bhatti (1986) also found its aqueous leaf extract application @ 44 g/kg soil to show maximum reduction in the penetration, development and larval hatching of *M. incognita* in the tomato crop. Castor cropping in between the vegetables was found to reduce the number of root-knots caused by *M. incognita* (Prasad et al., 1987). Hasan (1992) found its root exudates to reduce gall formation by *M. incognita*. Verma and Anwar (1995) reported chopped leaves of castor mixed with the soil in the plant pit @ 1500 g/m^2 pit area showed significant reduction in the populations of plant parasitic nematodes, *M. incognita*, *H. indicus*, *Helicotylenchus dihystera* and *Tylenchorhynchus vulgaris* in the pointed gourd plants.

The active components isolated from its oil cake, formaldehyde and phenol, showed toxicity to these nematodes (Alam et al., 1976, 1979).

123. *Scilla indica* Roxb. (Liliaceae)

(Syn.-*S. maculata* Baker, *Lederbouria hyacinthia* Roth., *L. maculata* Dall., *Barnaria indica* Wight)

This is an annual herb found in tropical and sub-tropical regions and commonly available in coastal South India in Deccan Peninsula from Konkan and Nagpur southwards. Pandey and Haseeb (1988) reported its root and shoot extracts to inhibit the larval hatching of root-knot nematode, *Meloidogyne incognita*.

124. *Sesamum indicum* Linn. (Pedaliaceae)

(Syn.-*S. orientale* Linn.)

Common name: Sesame or Gingilly

This is an annual/seasonal herb found in tropical zones. It is commonly grown in Uttar Pradesh, Madhya Pradesh, Andhra Pradesh, Tamil Nadu, Rajsthan, Maharastra, Gujrat and Orissa as an oil seed crop. Sesamum oil cake amended in the soil showed nematicidal activity to the plant parasitic nematodes in general (Midha and Chabra, 1974) and to *M. incognita* (Sikdar et al., 1986). Bhatnagar et al. (1978) found soil application of the water soluble fractions of its oil to reduce incidences of *M. incognita* in the okra roots. Reddy et al. (1990) reported its chopped leaves admixed with infested soil to reduce incidences of the root-knot nematode, *M. incognita* in the roots of the vegetable crops.

125. *Sesbania aculeata* (Willd.) Poir. (Papilionaceae)

(Syn.-*S. bispinosa* [Jacq.] W.f. Weight.)

Common name: Dhaincha

This is an annual shrub found in tropical zones and available in Peninsular India. Its leaves amended in the soil @ 5–10% w/w in the pot experiments reduced population of the root-knot nematode, *M. javanica* in the tomato and okra roots (Singh and Sitaramaiah, 1967).

126. *Shorea robusta* Gaertn. f. (Dipterocarpaceae)

Common name: Sal

This is a perennial tree found in tropical zones and commonly available in the Indo-Gangetic delta in India. Its sawdust from the wood amended in the soil alone or in combination with neem kernel powder reduced populations of the root-knot nematode, *M. javanica* in the tomato roots (Sitaramaiah and Singh, 1978). Sharma et al. (1985) reported its seed oil cake amended in the soil to reduce larval populations of the root-knot nematode, *M. incognita* but also reduced the wheat yield when the doses were high.

127. *Sida cordifolia* Linn. (Malvaceae)

Common name: Country-mallow

This is a perennial shrub found in tropical zones and available throughout India. Its leaf extract in water was reported to show toxicity to the root-knot nematodes, *M. incognita* and *R. reniformis* (Mahmood et al., 1982).

128. *Solanum hispidium* Linn. (Solanaceae)

This is a biennial herb found in tropical zones. Its leaf and root extracts in water were toxic to *T. brassicae* and *H. indicus* (Haseeb et al., 1978).

129. *Solanum pampasense* Linn. (Solanaceae)

This is an annual herb found in temperate and tropical zones. Its aqueous root extract was reported to be toxic to the golden nematode, *Heterodera rostochinensis* (Jacobson, 1958).

130. *Solanum sucrense* (Linn.) Mill. (Solanaceae)

This is an annual herb found in temperate and sub-tropical zones. Its root extract in water was reported to be toxic to the golden nematode, *Heterodera rostochinensis* (Jacobson, 1958).

131. *Solanum xanthocarpum.* Schrad. and Wendlo. (Solanaceae)

Common name: Yellow berried night shade

This is an annual spiny herb found in tropical regions and distributed throughout India, Sri Lanka, Bangladesh, Pakistan, Indonesia and Malaysia. Its root exudates significantly reduced the gall formation by the root-knot nematode, *M. incognita* (Hasan, 1992).

132. *Sonchus oleraceus* Linn. (Asteraceae)

This is an annual herb found in tropical Africa and Mediterranean zones. Its leaf and root extracts in water were reported to be highly toxic to the root-knot nematode, *M. incognita* and they also inhibited hatching of the eggs (Bano et al., 1986).

133. *Sophora flavescens* Linn. (Fabaceae)

This is a perennial tree found in tropical and sub-tropical zones. Two alkaloids, i.e., cystine and (-)-N-methyl cystine, were isolated from its root aqueous extract and found to show nematicidal activity against pine wood nematode, *Bursaphelenchus zylophilus* (Matsuda et al., 1991).

134. *Sorghum vulgare* Pers. (Graminae)

(Syn.-*Andropogon sorghum* [L.] Brot., *Holcus sorghum* Linn.)
Common name: Sorghum

This is a seasonal cereal/grass extensively grown in Uttar Pradesh, Punjab, Madhya Pradesh, Andhra Pradesh, Karnataka, Rajasthan and Tamil Nadu in India. Its whole plant extract (1:10 dilution) amended in the soil of tobacco nurseries inhibited the root-galls formed by the root-knot nematode, *M. incognita* (Krishnamurthy and Murthy, 1990).

135. *Sphenocloea zeylanica* Gaertn. (Sphenocleaceae)

This is a succulent herb found in tropical and sub-tropical zones. It is commonly available as a weed in the rice fields in coastal regions in India. Its green manuring in rice fields was also reported to reduce populations of the root nematode, *Hirschmanniella oryzae* under field conditions. Further, its leaf crude extract (50 g/100 ml water) in water was found to be toxic to *H. oryzae* when tested under controlled conditions (Mohandas et al., 1982).

136. *Tagetes erecta* Linn. (Asteraceae)

Common name: Big marigold or Aztee marigold

This is an annual herb found in tropical and sub-tropical zones and commonly grows as an ornamental herb throughout the country. Intercropping of marigold with the main crops of tomato, brinjal and wheat showed more significant reduction in populations of the root-knot nematode, *M. incognita* and the root nematodes, *Heterodera rostochinensis* and *Pratylenchus penetrans* than without intercropping of the marigold (Ruelo, 1976; Ruelo and Davide, 1979). Hussain and Masood (1975) found its leaf extract to inhibit the larval hatching of *M. incognita* in the vegetable crops. Alam et al. (1976) reported that seedlings of the tomato, brinjal and chillies surrounded by seedlings of the marigold reduced population of the plant parasitic nematodes. Root extract of its seedlings showed mortality to *H. indicus*, *R. reniformis*, *T. brassicae* and *M. incognita* and also inhibited their larval hatching (Alam et al., 1975). Its root extracts in water and ether were reported to be toxic to the root-knot nematode, *M. incognita* in a laboratory test (Hoan and Davide, 1979).

Similarly, its root exudate showed nematicidal activity to the root nematodes under field conditions (Rao et al., 1985). Leaf extract of this marigold was also found to be toxic to *M. incognita* in the vegetable crops (Mahmood et al., 1979). Kumar et al. (1986) reported aqueous and methonolic extracts of its leaf and stem to show 75–100% mortalities to 2nd instar larvae of *Tylenchulus semipenetrans* and *Anguina tritici*. However, its chopped leaves admixed with the infested soil reduced the root galling of *M. incognita* in the papaya (Reddy et al., 1990).

Govindaiah et al. (1990) studied marigold intercropping in the mulberry against *M. incognita* under field conditions and found that intercropping of marigold at 15 cm spacing reduced the root-galls and also enhanced leaf yield. Verma and Anwar (1995) found its chopped leaves admixed with the soil @ 500 g/m^2 in plant pit of pointed gourd to show significant reduction in the populations of *M. incognita*, *H. indicus*, *T. vulgaris* and *H. dihystera*. Similarly, mixed cropping of the aztee marigold along with tomato cultivation in the mini-plot experiment was found to effectively reduce the soil population of the root-knot nematode, *M. incognita* in the tomato roots (Siyanand et al., 1995).

Uhlenbrock and Bijloo (1959a) reported two active components from its root extract, which were found to show nematicidal activity to *Ditylenchus dipsaci*, *Anguina tritici*, *Heterodera rostochinensis* and *Pratylenchus penetrans*. These components were identified as (i) α-terthienyl (Figure 125) and (ii) 5-(3-butens-1-ynyl)-2,2′-bithienyl (Figure 126).

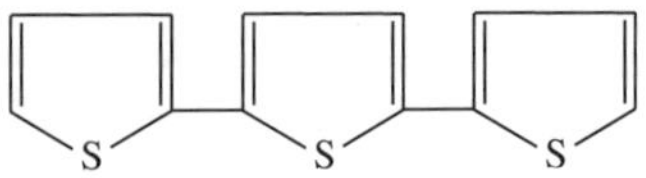

Figure 125. α-terthienyl.

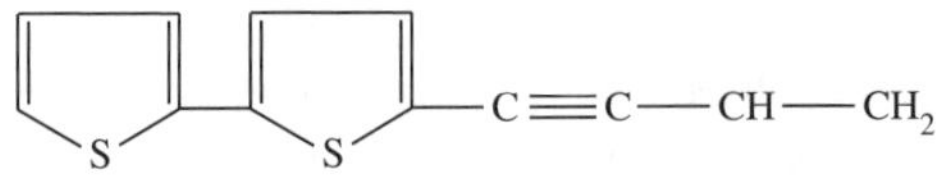

Figure 126. 5-(e-butens-1-ynyl)-2,2′-bithienyl.

137. *Tagetes lucida* Cav. (Asteraceae)

Common name: Sweet-scented marigold
This is an annual tall herb found in tropical and sub-tropical zones but native to Mexico. It is commonly grown as an ornamental herb for its orange yellow flowers throughout India and South East Asia. Its root extract was reported to show nematicidal activity to the nematodes of agricultural importance (Mukherjee, 1983). Siddiqui et al. (1988a,b) found its flower aqueous extract to show highest mortality and inhibition in hatching of the juveniles of *M. incognita*, *R. reniformis*, *T. brassicae*, *H. indicus* and *T. filiformis*.

138. *Tagetes minuta* Linn. (Asteraceae)

(Syn.-*T. glandilifera* Schrank.)
Common name: Stinking Roger
This is an annual aromatic herb found in tropical and sub-tropical zones and also available in the North West Himalayas. Oil extracted from its leaf was reported to be toxic to the nematodes of agricultural importance (Mukherjee, 1983). Siddiqui et al. (1988b) found nematotoxic principle to a varying degree in its root exudates and *Helicotylenchus indicus* was more highly susceptible to the exudates than *Hoplolaimus indicus*.

139. *Tagetes patula* Linn. (Asteraceae)

Common name: French marigold
This is an annual herb/shrub found in tropical, sub-tropical and temperate zones. Its alcoholic root extract was reported to show nematicidal activity when tested against the root-knot nematode, *M. incognita* (Morallo-Rajessus and Eroles, 1978). Siddiqui et al. (1988b) found nematotoxic principle to a varying degree in its root exudates and *Helicotylenchus indicus* was more highly susceptible to the exudates than *Hoplolaimus indicus*. Subramaniyam and Selvaraj (1988) also reported nematotoxic effect of its aqueous leaf extract when tested in a laboratory experiment against *Radopholus similis*.

Chemicals like theophene and α-tetrathienyl derivatives isolated from its leaf and stem extracts were found to show nematicidal activity against *Tylnchulus semipenetrans* and *Anguina tritici* (Uhlenbrock and Bijloo, 1959b; Winoto, 1969; Kumari et al., 1986).

140. *Tagetes tenuifolia* Cav. (Asteraceae)

(Syn.-*T. signata* Bartl.)
Common name: Striped marigold
This is an annual herb found in tropical zones and native to Mexico. It is commonly cultivated as an ornamental herb in India. Its aqueous root extract

was reported to show toxicity to the plant parasitic nematodes (Mukherjee, 1983). Siddiqui et al. (1988b) found nematotoxic principle to a varying degree in its root exudates and *Helicotylenchus indicus* was more highly susceptible to the exudates than *hoplolaimus indicus*.

141. *Tamarindus indica* Linn. (Caesalpiniaceae)

Common name: Tamarind or Tamarindo

This is a perennial tree indigenous to tropical Africa and now distributed throughout the plains and sub-Himalayan tracts of India. Its leaf extract was reported to inhibit the larval hatching of the root-knot nematode, *M. incognita* in the vegetable crops when tested in the pot experiments (Hussain and Masood, 1975). Krishnamurthy and Murthy (1990) also found its leaf extract to inhibit root-galling of *M. incognita* when 1:10 dilution of the extracts was applied in the soil of a tobacco nursery.

142. *Thuja orientalis* Linn. (Cupressaceae)

(Syn.-*Biota orientalis* Endl.)

Common name: Oriental arbore vitae

This is a perennial ornamental shrub or small tree found in sub-tropical and temperate zones. It is a native to China but is now being cultivated throughout North India. Aqueous leaf extract of this shrub was found to show toxicity to the root-knot nematode, *M. incognita* in the vegetable crops (Jain et al., 1986).

143. *Tithonia diversifolia* A. Gray. (Asteraceae)

Common name: Wild sunflower

This is a perennial shrub/small tree found in tropical zones. It is also available in Rajasthan and Madhya Pradesh in India. Its flower extract in water was reported to be toxic to the juvenile stages of the root-knot nematode, *M. incognita* (Tiyagi et al., 1985).

144. *Trichosanthes anguina* Linn. (Cucurbitaceae)

Common name: Snake gourd

This is an annual or seasonal climbing herb found in tropical zones. Snake gourd is cultivated throughout India, particularly in South India. Seed extract of this herb was reported to inhibit larval hatching of the root-knot nematode, *M. incognita* in the vegetable crops (Hussain and Masood, 1975; Mahmood et al., 1979).

145. *Typhonium trilobatum* (Linn.) Schott. (Araceae)

This is an annual tuberous herb found in Peninsular India from Jamuna eastwards in North India and also grown in South India as an edible tuber. Mukhopadhyaya and Chattopadhayaya (1979) reported its green plant extract to reduce incidences of the root-knot nematode, *M. incognita* in eggplants. Its corn powder admixed with the soil in the pot experiment reduced population of *M. incognita* at a higher rate than the amendment of seed dust of *Melia azedarach* (Chattopadhyay and Mukhopadhyay, 1989a).

146. *Vernonia anthelmintica* Willd. (Asteraceae)

(Syn.-*Centratherum anthelminticum* [L.] O. Kuntze.)

This is an annual/seasonal herb found in tropical zones and commonly available in India. Its leaf and root extracts in water were reported to show nematicidal activity to the root-knot nematodes, *M. incognita* and *M. javanica* (Mukherjee, 1983).

147. *Vinca rosea* Linn. (Apocynaceae)

(Syn.-*Catharanthus roseus* [Linn.] G. Dong.)

Common name: Old maid

This is a perennial shrub found in tropical and sub-tropical zones and also available in India. Its whole plant and leaf extracts were reported to be toxic to the root-knot nematodes, *M. incognita* and *M. javanica* (Mukherjee, 1983). Reddy et al. (1990) found its chopped leaves admixed with the infested soil to reduce the number of root-galls caused by *M. incognita* in the papaya.

148. *Withania somnifera* (Linn.) Dunal. (Solanaceae)

This is an annual herb found in sub-tropical and temperate zones. Its root and shoot aqueous extracts were reported to show nematicidal activity to *T. brassicae*, *H. indicus* and *R. reniformis* (Haseeb et al., 1978).

149. *Xanthium strumarium* Linn. (Asteraceae)

Common name: Cockbur or Burweed

This is an annual/seasonal herb found throughout India and used as an organic manure. Its aqueous leaf extracts showed nematicidal effect when tested against the root-knot nematode, *M. incognita* (Nandal and Bhatti, 1983; Bano et al., 1986). Ghosh and Sukal (1992) reported essential oil extracted from its leaves to show toxicity to 3rd instar juveniles of *M. incognita*.

150. *Zinnia elegans* Jacq. (Asteraceae)

Common name: Youth and old-age

This is a biennial ornamental herb grown in tropical zones but native to Mexico. It is commonly available in India. Its whole plant extract in water was reported to be toxic to the juveniles and adults of the root-knot nematode, *M. incognita* and it also inhibited hatching of the eggs of test nematode (Bano et al., 1986).

C BOTANICAL PESTICIDES AGAINST MITES, RODENTS AND MOLLUSCAN PESTS

BOTANICAL PESTICIDES AGAINST MITES

A number of plant products have been reported to possess miticidal property when tested against phytophagous mites. Details about the habit and habitat of these plants have already been mentioned in Sections A and B. Biological activities of these products and their formulations are presented in Table 16.

Active Components

The 6-methoxy analogue of pisiferic acid was extracted from the root and leaf of *Chamaecyparis pisifera* and reported to inhibit the feeding of the adults and hatching of the eggs of spotted spider mite, *Tetranychus urticae* (Ahn et al., 1984). Three flavones isolated from the root and bark of *Daphne odora*, i.e., daphnodorous A, B and C, showed miticidal activity when tested against *Tetranychus urticae* (Inamori et al., 1987). From ethanolic extracts of its seed and bark acetogenin an active component called 'asimicin' was reported to show toxicity to the two spotted spider mite, *T. urticae* (Alkofahi et al., 1989). Jacobson (1989) referred isomers of arjugarins as clerodane diterpenes from *Ajuga remota*, which were toxic to *T. urticae*. Reda et al. (1990) found alkaloids isolated from *Abrus precatorius* as deterrents against the adult females of *T. urticae* and reduced life span and oviposition of this mite. Mansour (1993) reported 'repelin', a neem based formulation (having azardirachtin as an active component), a highly toxic formulation to the phytophagous mite, *Tetranychus cinnabarinus*. Neemark (0.5%) also showed toxicity to the spider mite, *Tetranychus macfarlanei* (Rae et al., 1993). Further, Patel et al. (1993) evaluated neem based products against spider mites, *Tetranychus macfarlanei* and *T. cinnabasinus* and found that repelin 1%, margocide CK 0.1%, margocide OK

Table 16. Plant products that show biological activity against mites

Generic name of plant	Plant parts and its formulations	Biological activity	Reference
Ajuga remota Linn. (Lamiaceae)	Juice and crude extract of the leaf	Toxicity to the spotted spider mite, *Tetranychus urticae*.	Schauer and Schmutterer, 1981
Aloysia triphylla Ort. and Palaue. (Verbenaceae)	Oils from its plant, leaf and seed	Toxicity to the red spider mite, *Tetranychus telarius*.	Jacobson, 1975
Allium sativum Linn. (Liliaceae)	Acetone extract of its bulb with deltamethrin	Enhanced the miticidal activity to *T. urticae*.	Barkat et al., 1986b
Artemisia sp.	Worm wood powder (60 gm dose)	Reduced attack of *Acarapis woodi* to bee colonies.	Abu-Zaid and Salem, 1991
Artemisia saissanica (Asteraceae)	Chloroform extracts of its buds, leaves and twigs	Repellency to *T. urticae*.	Adekenov et al., 1990
Anabasis aphylla Linn. (Chenopodiaceae)	Branch and leaf extracts in water	Toxicity to the mites.	Poe, 1980
Azardirachta indica A. Juss. (Meliaceae)	Kernel extract	Showed repellency to the carmine spider mite, *Tetranychus cinnabarinus*.	Mansour and Ascher, 1983
Azardirachta indica A. Juss. (Meliaceae)	Neem seed kernel extracts in methanol, ethanol and acetone	Showed toxicity to carmine spider mite, *Tetranychus cinnabarinus*.	Mansour et al., 1987
A. incida	Leaf extracts (1.0%) and seed kernel extract (5%)	Controlled *T. urticae* on brinjal.	Devraj Urs, 1990

Table 16. Plant products that show biological activity against mites (Continued)

Generic name of plant	Plant parts and its formulations	Biological activity	Reference
Boswellia dalzielli Roxb. (Burseraceae)	Bark extract in water	Toxicity to the mites in general.	Jacobson, 1975
Brassica oleracea var. rapifera Metz. (Brassicaceae)	Root extract in water	Toxicity to the spider mite, *Tetranychus atlanticus*.	Jacobson, 1975
Cardaria draba Desv. (Brassicaceae)	Leaf and root extracts in water	Toxicity to the mites in general.	McIndoo, 1983
Chamaecyparis pisifera (Sieb and Zucc.) Edl. (Cupressaceae)	Leaf and root extracts in water	Regulated the growth of *Tetranychus pisifera*.	Ahn et al., 1984
Chrysanthemum spp. (Asteraceae)	Pyrethrum extract and piperonyl-butoxide aerosols	Toxicity to the red spider mite, *Tetranychus telarius*.	Tiitanen, 1964
Citrullus colocynthis (Linn.) Kuntze. (Cucurbitaceae)	Leaf and fruit extracts in water	Toxicity to the red spider mite, *T. telarius*.	McIndoo, 1983
Consolida regalis Gilib. (Rananculaceae)	Root, stem and leaf extracts in water	Toxicity to the red spider mite, *T. telarius*.	McIndoo, 1983
Coriandrum sativum Linn. (Umbelliferae)	Oil emulsion (2%) spray	Killed the spider mites.	Feinstein, 1952
Croton klotzsclianus Roxb. (Euphorbiaceae)	Leaf extract in water	Toxicity to the red spider mite, *T. telarius*.	McIndoo, 1983
Crucuma longa Koening non-Linn. (Zingiberaceae)	Tumeric powder extract in water	Toxicity to the red spider mite, *T. telarius*.	McIndoo, 1983
Cymbopogan citratus (D.C.) Stapf. (Poaceae)	Leaf extract in water	Toxicity to the red spider mite, *T. telarius*.	McIndoo, 1983

Table 16. Plant products that show biological activity against mites (Continued)

Generic name of plant	Plant parts and its formulations	Biological activity	Reference
Cynodon dactylon (Linn.) Pers. (Poaceae)	Leaf extract in water	Toxicity to the red spider mite, *T. telarius*.	McIndoo, 1983
Daphne odora (Thymelaceae)	Root and bark extracts	Toxicity to *T. utricae*.	Inamori et al., 1987
Delphinium staphysagria Linn. (Ranunculaceae)	Branch, seed and leaf extracts in water	Toxicity to the red spider mite, *T. telarius*.	McIndoo, 1983
Hyptis savelolens (L.) Poit. (Labiatae)	Leaf extract in water	Toxicity to *Tetranychus necoldonicus*.	Roy and Pande, 1991
Lepidium ruderale Linn. (Brassicaceae)	Leaf extract in water	Toxicity to the spider mites.	McIndoo, 1983
Leucas cephalotis (Roth.) Spreng. (Lamiaceae)	Leaf extract in water	Toxicity to the spider mites.	McIndoo, 1983
Melia azedarach Linn. (Meliaceae)	Oil and seed extract in water	Toxicity to the citrus red mite, *Panonychus citri*.	Chiu, 1982
Melia toosandan Linn. (Meliaceae)	Seed extracts and its oil	Toxicity to the mites in general.	Chiu, 1982, 1989
Nicotiana tabacum Linn. (Solanaceae)	Leaf and whole plant extracts	Toxicity to the mites.	Nagarajah, 1983
Ocimum basilicum Linn. (Lamiaceae)	Leaf extract in water	Toxicity to *T. telarius*.	McIndoo, 1983
Piper nigrum Linn. (Piperaceae)	Seed extracts in acetone and diethyl-ether	Toxicity to adults of *Tetranychus urticae*.	Barakat et al., 1986a
Pragos pabularia Hybrid. (Celasteraceae)	Whole plant extract in water	Toxicity to the mites.	Jacobson, 1958
Trichilia cuneata P.Br. (Meliaceae)	Leaf and fruit aqueous extracts	Toxicity to the mites.	McIndoo, 1983

Table 16. Plant products that show biological activity against mites (Continued)

Generic name of plant	Plant parts and its formulations	Biological activity	Reference
Trichilia trifolia P.Br.	Ethonolic seed extract and its fractions	Toxicity to *Caloglyphus berlesei.*	Rao and Prakash, 1995
Trigonella foenumgraecum (Papilionaceae)	Diethyl-ether extract of its seed with deltamethrin	Enhanced the miticidal activity to *T. urticae.*	Barkat et al., 1986b

0.8% and neemark 0.5% were effective against *T. cinnabarinus* infesting the brinjal crop and Indian bean but failed to kill *T. macfarlanei* infestation in okra. However, their effectiveness varied with different plant hosts.

BOTANICAL PESTICIDES AGAINST RODENTS

A number of plant products have also been reported to show rodenticidal activity when tested against a wide range of rodent pests of agricultural importance. These plant products generally possessed repellency and also showed toxicity to the rats and squirrels and only in a few cases were reported to exhibit sterilant activity. Details about habit and habitat of these plants have already been dealt with in Sections A and B. Biological activities of the products are presented in Table 17.

Active Components

The active components isolated from the dried and ripened seed extract of *Strychnos nux-vomica* was reported to be an alkaloid called 'strychnine' ($C_{21}H_{22}N_2O_2$), which showed rodenticidal activity against ground squirrels, *Citellus beecheyi* and *C. richardsoni* when tested in the laboratory (De-Ong, 1948). Similarly, from the bulb extract of *Urginea burkei*, a red-glucoside isolated as an active component was also reported to show toxicity to the rats when fed through their baits (De-Ong, 1948).

BOTANICAL PESTICIDES AGAINST MOLLUSCAN PESTS

Plant products known to possess molluscicidal activity against the snails of agricultural importance are presented in Table 18 along with details of their generic names, plant parts tested to possess biological activity and their formulations. Kubo and Nakanishi (1978) found 'warburganol', an active component isolated from the leaf extract of *Warburgia ugandensis*, to show toxicity

Table 17. Plant products that show biological activity against rodents

Generic name of plant	Plant part and its formulations	Biological activity	Reference
Anthemis ervensis Linn. (Asteraceae)	Aqueous leaf extract	Repellent	McIndoo, 1983
Antioris toxicaria (Pers.) Lesch. (Moraceae)	Leaf and flower extracts	Antifertility	Michael et al., 1985
Arachis hypogea Linn. (Papilionaceae)	Groundnut oil	Attractant with rodenticidal bait.	Chopra and Sood, 1980
Azadirachta indica A. Juss. (Meliaceae)	Neem oil	Antifertility	Tewari et al., 1986
A. indica	Neem oil	Impaired the structure and functional integrity in epididymis.	Sampathraj et al., 1993
A. indica	Neem oil (Chronic administration of neem oil to adult albino rats for 8 days)	Showed histopathological changes and toxicity.	Narayan et al., 1993
Brassica compestris Linn. (Brassicaceae)	Mustard oil	Attractant with rodenticidal bait.	Chopra and Sood, 1980
Cocos nucifera Linn. (Palmaceae)	Coconut oil	Attractant with rodenticidal baits.	Chopra and Sood, 1980
Depcadi cowanii Medik. (Liliaceae)	Bulb extract	Toxicity	Usher, 1973
Dianella ensifolia Lam. (Liliaceae)	Leaf extract	Toxicity	Usher, 1973
Dichapetalum toxicaria (Dichapetalaceae)	Seed extract	Toxicity	Usher, 1973
Gliricidia sepium (Jacq.) Walp. (Fabaceae)	Bark and seed extracts in water	Toxicity	Ines-Farias, 1983

Table 17. Plant products that show biological activity against rodents (Continued)

Generic name of plant	Plant part and its formulations	Biological activity	Reference
Madhuca latifolia Roxb. Chevai. (Sapotaceae)	Aqueous seed extract	Toxicity	McIndoo, 1983
Nerium oleander Linn. (Apocynaceae)	Bark and root extracts in water	Toxicity	Usher, 1973
Schoenocanlon officinale A. Gray. (Liliaceae)	Aqueous seed extract	Toxicity	Usher, 1973
Spondianthus ugandensi Engl. (Euphorbiaceae)	Powder and extract of bark	Toxicity	Usher, 1973
Strychnos nux-vomica Linn. (Lamiaceae)	Aqueous seed extract	Toxicity	De-Ong, 1948
Thecostele porlana Reichb. (Orchidaceae)	Leaf extract in water	Toxicity	Usher, 1973
Tylophora fasciaculata Buch. Ham, ex. Wight (Asclepidaceae)	Root and leaf extracts in water	Toxicity	McIndoo, 1983
Urginea burkei Steinh. (Liliaceae)	Aqueous bulb extract	Toxicity	De-Ong, 1948
Urginea martima Linn.	Aqueous bulb extract	Toxicity	De-Ong, 1948
Verbascum eychnitis Linn. (Scrophulariaceae)	Flower extract in water	Toxicity	Usher, 1973

to the snails like *Biumphalaria glabrata*, *B. pfeiferi* and *Lymnaea natalensis*. Kishor and Sati (1990) reported 'spirostanol glycoside' from an ornamental plant, *Yacca aloifolia* to be 100% toxic at 10 ppm when tested against *B. glabrata*.

Azardirachtin, an active component isolated from neem kernel extract, was also reported to show molluscicidal activity against *Lymnea luteda* (Ramesh Babu, 1983).

Table 18. Molluscicidal properties of plant products

Generic name	Part of plant and its formulations	Biological activity	Reference
Azardirachta indica A. Juss. (Meliaceae)	Kernel extract of neem and neem cake extract	Toxicity to *Lymnaea natalensis* and *L. auricularea.*	Reynaud, 1986; Sasmal, 1991
Balantis aegyptica (Linn.) Delite (Simaroubaceae)	Nut and leaf powders	Toxicity to snails.	Belen, 1982
Camellis sinessis (Linn.) (Theaceae)	Tea cake in standing water	Toxicity to *Lymnaea* sp.	Anonymous, 1974b; Liu, 1979
Citrus mitis (Linn.) Macf. (Rutaceae)	Dried fruit powder and its aqueous extract	Toxicity to the snails.	Belen, 1982
Coryzadenia balsamifera Griff. (Hernandiaceae)	Fruit and leaf aqueous extracts	Toxicity to the snails.	Belen, 1982
Croton figluim Linn. (Euphorbiaceae)	Leaf extract in water	Toxicity to the snails.	Belen, 1982
Detaricum heudelotianum	Bark powder	Toxicity to *Lymnaea natalensis.*	Reynaud, 1986
Entada gigas (Linn.) Fawc. and Rendle (Mimosaceae)	Leaf seed and bark extracts in water	Toxicity to the snails.	Belen, 1982
Fagara microphylla Desf. (Rutaceae)	Bark extract in water	Toxicity to *Biumphalaria glabrata.*	Kubo et al., 1984b
Jatropha curcas Linn. (Euphorbiaceae)	Fruit and seed aqueous extracts	Toxicity to the snails.	Belen, 1982
Menispermum coculus Linn. (Menisperma-ceae)	Leaf, branch, fruit and seed extracts in water	Toxicity to the snails.	Belen, 1982
Prago pabularia Hybrid (Cactaceae)	Aqueous leaf extract	Toxicity to the snails.	McIndoo, 1983

Table 18. Molluscicidal properties of plant products (Continued)

Generic name	Part of plant and its formulations	Biological activity	Reference
Thevetia nerifolia Juss. ex Steud. (Apocynaceae)	Alcoholic fruit extract	Toxicity to the garden snail, *Helix asparsa*	Johri et al., 1993
Warburgia ugandensis Eig. (Rubiaceae)	Leaf extract in water	Toxicity to the snails, viz., *Biumphalaria glabrata*, *B. pfeiferi* and *Lymnaea natalensis*	Kubo and Nakanishi, 1978

D BIOLOGICALLY ACTIVE CHEMICAL COMPONENTS

For ready reference and chemical indexing, this section contains an alphabetical listing of all the biologically active chemical components, their chemical grouping, biological activity like insecticidal, repellent, antifeedant, insect growth, regular/juvenile hormone mimic activity, nematicidal and/or activity of any other kind and also the name of the plant from which the component was isolated. The chemical components, which are dealt in sections A, B, C, are presented in Table 19. However, detailed chemical structures and chemical formulae of the components are mentioned in the text of the respective plant.

Table 19. Biologically active chemical components

Chemical component	Chemical group	Biological activity	Plant isolated from
Acaciaside A and B	triterpenoid saponins	nematicidal	*Acacia auriculiformis*
Acetogenins	tetrahydrofuran	insecticidal	*Uvaria hookeri, U. narum*
6-acetoxy toonacilin	terpenoid	antifeedant	*Toona ciliata*
Acetophenone	acetone derivative	attractant	*Trifolium pratense*
Acoragermacrone	cyclodecadinoid	attractant	*Acorus calamus*
Affinin	unsaturated isobutylamide	insecticidal	*Erigeron affinis*
Ageratochromene (Precocene)	chrominoid	insecticidal	*Ageratum conyzoides*
Ageratochromene	chrominoid	insecticidal	*Ageratum conyzoides*
Ajugarin-I, II, III	terpenoids	insecticidal, nematicidal	*Ajuga remota*

Table 19. Biologically active chemical components (Continued)

Chemical component	Chemical group	Biological activity	Plant isolated from
Ammonia	gas	nematicidal	*Azardirachta indica*
Anabasine (Neo-nicotine)	alkaloid	insecticidal	*Anabasis aphylla, Duboisia hopwoodi*
Anacyclin	unsaturated isobutylamide	insecticidal	*Anacyclus pyrethrum*
(Z)-2-angeloyl oxyracethyl-2-betenoic acid	—	insecticidal	*Anthriscus silvestris*
Anonanine	alkaloid	insecticidal	*Anona muricata*
Anthothecol	alkaloid	antifeedant	*Toona cilaita*
Anthricin	deoxy-podophyllo toxoid	insecticidal	*Anthriscus silvestris*
Anthriscinol methyl ether	ether derivative	insecticidal	*Anthriscus silvestris*
ar-turmerone	phenyl-lepyen derivative	repellent	*Curcuma longa*
β-Asarone	benzene derivative	attractant	*Acorus calamus*
Asaryldehyde	benzaldehyde derivative	attractant	*Acorus calamus*
Asimicin	acetogenin	miticidal	*Daphne odora*
Asparagin	protein derivative	growth inhibitor	*Oryza sativa*
Asparaguric acid	—	nematicidal	*Asparagus officinalis*
Aurimillone	isoflavonoid	antifeedant	*Millettea auriculata*
Azardirachtin	terpenoid	antifeedant, growth inhibitor	*Azardirachta indica*
Bakuchiol	terpenoid	growth inhibitor, chemo sterilant	*Psoralea corylifolia*
Benzaldehyde	aldehyde	nematicidal	*Laurus nobilus*

Table 19. Biologically active chemical components (Continued)

Chemical component	Chemical group	Biological activity	Plant isolated from
(E,E,E)-13-(1,3-benzodioxol-5-yl)-N-(2-methyl propyl)-2,4,12-tridecatrien-amide	benzodioxol-amide derivative	insecticidal	*Piper nigrum*
(E,E,E)-11-(1,3-benzodixol-5-yl)-N-(2 methyl propyl)-2,4,10-undecatrien-amide	benzodioxol-amide derivative	insecticidal	*Piper nigrum*
Bergaptan	coumarin	growth inhibitor	*Chloroxylon swietenia*
Bocconine	alkaloid	nematicidal	*Bocconia cordata*
Bornyl acetate	volatile oil	insecticidal	*Coniferous lumber, Artemisia capillaris*
(5-buten-1-ynyl)-2-2 bithienyl	—	nematicidal, insecticidal	*Tagetes erecta*
Brucine	alkaloid	insecticidal	*Strychnos nux-vomica*
Capillarin	alkaloid	antifeedant	*A. capillaris*
Capillin	alkaloid	antifeedant	*Artemisia capillaris*
Carvone	unsaturated isobutylamide	repellent	*Anethum gravelous, Mentha spicata*
Cavadine	alkaloid	insecticidal	*Schoenocaulon officinale*
Cederelone	limonoid	antifeedant	*Toona ciliata*
Celagalin	terpenoid	antifeedant	*Celastrus angulata*
Chaconine	steroidal glyco alkaloid	nematicidal	*Solanum melongena*
Chelerythrin	alkaloid	nematicidal	*Bocconia cordata*
Cinerin, I.II	esters	insecticidal	*Chrysanthemum cineraraefolium*
Cis-spiro-enol ether	—	insecticidal	*Chrysanthemum coronarium*

Table 19. Biologically active chemical components (Continued)

Chemical component	Chemical group	Biological activity	Plant isolated from
3-Cis, 11-trans-trideca-1,3,11-triene-5,7,9-triyene	polyacetylene	nematicidal	*Carthamus tinctorius*
Citronellol	limonoid	insecticidal	*Citrus aurantum*
Cleistanthin B	—	antifeedant	*Cleistanthus collinus*
Clerodane diterpenes	terpenoid	miticidal	*Ajuga remota*
Clerodendrin	trans decalin unit	antifeedant	*Clerodendrum infortunatum*
Cocculalidine	alkaloid	repellent	*Cocculus tribolus*
α-copaene	volatile oil	attractant	*Coniferous lumber*
ar-curcumine	curcuminoid (ketone derivative)	antifeedant	*Artemisea capillaris, Curcuma longa*
Daphnodorous A,B,C	flavonoids	miticidal	*Daphne odora*
Decosanol	alkaloid	insecticidal	*Rhus typhina*
Dedrelone	limonoid	antifeedant	*Toona ciliata*
Deguelin	isoflavonoid	insecticidal	*Derris malaccensis*
Demissidine	ecdysone	growth inhibitor	*Solanum chacoense*
4-desoxy-8-epi-evangustin	terpenoid	repellent	*Eupatorium quadrangularea*
5-[(3,7-dimethyl-2,6-octadienyl) oxy]-7-methoxy-2H-1-benzopyran-2-one	limonoid	insecticidal	*Citrus aurantifolia*
4,4-di-O-methyl lonchocarpic acid	coumarin	antifeedant	*Derris scandens*
1,3-diphynyl thio urea	—	insecticidal	*Allium cepa, Allium sativum*
Diplophyllalide	terpenoid	repellent	*Eupatorium quadrangularae*
Diallyl-di-sulphide	sulphide derivative	insecticidal	*Allium sativum*
Di-alyl-tri-sulphide	sulphide derivative	insecticidal	*Allium sativum*

Table 19. Biologically active chemical components (Continued)

Chemical component	Chemical group	Biological activity	Plant isolated from
6a,12a-dihydro-α-toxicanol	—	insecticidal, repellent	*Amorpha fruiticosa*
2,3-dihydro-2-hydroxy-3-methylone-6-methyl-benzofuran	benzofuran derivative	nematicidal	*Helenium hybrid*
2,4-dihydroxy-7-methoxy-2H-1,4-benzocazin-3(4H)-one (DIMBOA)	hydroxamic acid	antifeedant	*Zea mays*
Dill-apiol	—	antifeedant	*Anethum graveolus*
5,7-dimethoxy-2-H-1-benzo-pyran-2-one	limonoid	insecticidal	*Citrus aurantifolia*
9-[(3,7-dimethoxy-2,6-octadienyl)-oxyl]-7-H-furo[3,2-g] [1] benzopyran-7-one	limonoid	insecticidal	*Citrus aurantifolia*
4-[(3,7-dimethyl-2,6-octadienyl)-oxyl]-7-H-furo[3,2-g] [1]-benzopyran-7-one	limonoid	insecticidal	*Citrus aurantifolia*
Eburnamine	alkaloid	insecticidal	*Haplophyton cimicidum*
Ecdysone	steroid	growth inhibitor	*Seruvium portulacastrum*
Echinacein	isobutylamide	insecticidal	*Echinacea angustifolia*
Echinolone	steroid	growth inhibitor	*Echinaceae angustifolia*
Ecydysterone	steroid	growth inhibitor	*Seruvium portulacastrum*
α-Eleostearic acid	fatty acid	antifeedant	*Aleurites fordii, Rhus cotinum*
Elliptone	isoflavonoid	insecticidal	*Derris malaccensis*

Table 19. Biologically active chemical components (Continued)

Chemical component	Chemical group	Biological activity	Plant isolated from
Encecalin	acetyl chromene	insecticidal	*Encelia farinosa*
Ent-Kaur-15-3en-3b	kauramoid	insecticidal	*Croton aromaticus*
Epicatechin	proantho-cyanidin monomer	antifeedant	*Acacia catechu*
Epinimbin	terpenoid	antifeedant	*Azardirachta indica*
Eriancorin	—	antifeedant	*Eriangea cordifolia*
Erythro-9,10-dihydroxy octadecyl acetate	—	antifeedant	*Aleurites fordii, Rhus cotinum*
Eucalyptol	1,8-cineole (oil)	growth inhibitor	*Eucalyptus camaldulensis*
Eugenol	propanoid	insecticidal, nematicidal	*Pogostemon heynemus, Ocimum sanctum*
Evanine	terpenoid	antifeedant	*Celastrus angulata*
Formaldehyde	aldehyde	nematicidal	*Azardirachta indica, Brassica compestris, Madhuca indica, Ricinus communis*
Friedelin	terpenoid	antifeedant	*Acokanthera spectibilis*
Geranion	—	repellent	*Laurus nobilis*
2-β-glucoside of perulatone	steroidal lactone	growth inhibitor	*Physalis peruviana*
Gramine	protein derivative	insecticidal	*Hordeum vulgare*
Grayanotoxins	alkaloid	antifeedant	*Kalmia latifolia*
Haplophytine	alkaloid	insecticidal	*Haplophyton amicidinum*
Haplostachine	alkaloid	insecticidal	*Lolium pernae*
Hardwickiic acid	—	insecticidal	*Croton aromaticus*
Harrisonin	chromine	antifeedant	*Harrisonia abyssinica*

Table 19. Biologically active chemical components (Continued)

Chemical component	Chemical group	Biological activity	Plant isolated from
Heldecarpin	pterocarpan	antifeedant	*Tephrosia heldebrantii*
Helenalin	alkaloid	antifeedant	*Helenium aromaticus*
Helicocide H	terpenoid	antifeedant	*Gossypium hirsutum*
Helietin	—	antifeedant	*Chloroxylon swietenia*
Heliopsin	isobutylamide	insecticidal	*Heliopsis scabra*
2-heptatria-contanone	acetyl derivative	repellent	*Vitex negundo*
Herculin	unsaturated isobutylamide	insecticidal	*Zanthoxylum clavaherculis*
Hexa-decanoic acid	—	insecticidal	*Rhus typhina*
Hispidins A-C	limonoids	antifeedant	*Trichilia hispida*
Hordenine	—	insecticidal	*Hordeum vulgare*
5-hydroxy-2-hexane-4-olide	—	antifeedant	*Osmunda japanica*
4-β-hydroxy withanolide	withanolide derivative	antifeedant	*Physalis peruniana*
Iso-alantolactone	terpenoid	insecticidal	*Eupatorium quadrangularae*
Iso-butylamides	—	insecticidal	*Spilanthes mauritiana*
Iso-flavone	flavonoid	antifeedant	*Tephrosia purpurea*
Isopimpinellin	coumarin	antifeedant	*Chloroxylon swietenia*
Iso-quassin	alkaloid	insecticidal	*Veratrum viride*
Jasmolin, I, II	esters	insecticidal	*Chysanthemum cineraraefolium*
Jervine	alkaloid	antifeedant	*Veratrum viride*
Juvabione	esteroid	growth inhibitor	*Abies balsama*
Juvadocene	benzodioxoid	insecticidal	*Piper excelsum*
Juvocinine I, II	clerodanes	insecticidal	*Ocimum basilicum*
Kalmintoxin-I	alkaloid	antifeedant	*Kalmia latifolia*
Karanzin	flavonoid	antifeedant	*Pongamia glabra*
Linalool	alkaloid	insecticidal	*Rhus typhina, Ocimum canum*

Table 19. Biologically active chemical components (Continued)

Chemical component	Chemical group	Biological activity	Plant isolated from
Linoleic acid	fatty acid	insecticidal	*Nigella sativum*
Liriodenine	alkaloid	insecticidal	*Anona glabra*
Lonchocarpic acid	coumarin	insecticidal	*Derris scandens*
Luvangetin	coumarin	antifeedant	*Atlantia racemosa*
Lynanol	oil-distillate	insecticidal	*Mentha ciliata*
Lyoniol-A	alkaloid	antifeedant	*Kalmia latifolia*
Magnoflorine	alkaloid	insecticidal	*Phellodendron amurense*
Malaccol	isoflavonoid	insecticidal	*Derris malaccensis*
Mammein	alkaloid	insecticidal	*Mammea americana*
Maxima-substance-C	isoflavonoid	antifeedant	*Tephrosia purpuria*
Melianthin	limonoid	antifeedant	*Melia azedarach*
Meliantriol-I	oil-fraction	antifeedant	*A. indica, M. azedarach*
Melicopicene	terpenoid	antifeedant	*Teclea trichocarpa*
Metanicotine	alkaloid	insecticidal	*Nicotiana tabacum*
(-)-N-methyl-cystisine	alkaloid	nematicidal	*Sophora flavescens*
Methyl-euginol	—	antifeedant	*Laurus nobilis*
7-methyl juglone	naphthalenedione	insecticidal	*Diospyros virginiana*
(E,E)-N-(2-methylpropyl)-2,4-decadienamide	—	insecticidal	*Piper nigrum*
4-methyl thio-1,3-propanedithiol	alkaloid	insecticidal	*Chara globularis*
Momordicine II	terpenoid	antifeedant	*Momordica charantica*
Mundoserone	rotenoid	insecticidal	*Derris elliptica*
Mundulone	isoflavonoid	antifeedant	*Mundulea suberosa*
Muzigodial	sesquiterpenoid	antifeedant	*Warburgia ugandensis*
Neo-herculin	unsaturated isobutylamide	insecticidal	*Zonthoxylum piperitum*

Table 19. Biologically active chemical components (Continued)

Chemical component	Chemical group	Biological activity	Plant isolated from
Neo-nicotine	alkaloid	insecticidal	*N. tabacum*
Neo-quassin	alkaloid	insecticidal	*Veratrum album*
Nicablin-A	withanolids (steroid)	antifeedant	*N. tabacum, Nicandra physaloides, Withania somnifera*
Nicotine	alkaloid	insecticidal	*N. tabacum*
Nicotyrine	alkaloid	insecticidal	*Nicotiana tabacum*
Nimbicidin	triterpenoid	nematicidal	*Azardirachta indica*
Nimbidin	triterpenoid	insecticidal, antifeedant	*A. indica*
Nimbin	terpenoid	antifeedant	*A. indica*
Nomilin	limonoid	antifeedant	*Citrus aurantium*
Nor-nicotine	alkaloid	insecticidal	*N. tabacum*
Nuciferine	alkaloid	insecticidal	*Annona reticulata*
Obacumone	limonoid	antifeedant	*Citrus aurantium*
(E)-β-Ocinene	alkaloid	attractant	*Trifolium pratense*
(Z)-β-Ocinene	alkaloid	attractant	*Trifolium pratense*
Odoracin	ester	nematicidal	*Daphne odora*
Odoratrin	ester	nematicidal	*Daphne odora*
Oleic acid	fatty acid	insecticidal	*Nigella sativa*
Osmundalactone	ketone derivative	antifeedant	*Osmunda japanica*
Pachyrrhizone	rotenoid	insecticidal	*Neorantaneria pseudo-pachyrrhiza*
Pedonin	triterpenoid	insecticidal	*Harrisonia abyssinica*
Pellitorine	unsaturated isobutylamide	antifeedant	*Anabasis aphylla*
Penta cyclic triterpene	terpenoid	antifeedant	*Ulmus americana*
Peraisin (derivative)	terpenoid	antifeedant	*Melia azedarach*
Pergularine	alkaloid	antifeedant	*Tylophora asthematica*
Petuniasterone (A)	steroidal ketone	growth inhibitor	*Petunia hybrid, P. vialacca*

Table 19. Biologically active chemical components (Continued)

Chemical component	Chemical group	Biological activity	Plant isolated from
Phenol	phenol	nematicidal	*Azardirachta indica, Brassica compestris, Melia indica*
2-phenyl-ethyl isothiocyanate	—	insecticidal	*Brassica oleracea*
Phenyl propanoids	propanoid	antifeedant	*Acorus calamus*
1-pheylthanol	phenol derivative	attractant	*Trifolium pratense*
Phytol	alkaloid	insecticidal	*Rhus typhina*
α- and β-pinenes	volatile oils	repellent	*Coniferous lumber, Laurus nobilis*
Pipecolic acid	nonprotein imino acid	antifeedant	*Calliandra calothyrsus*
Piperaceal amide-I	amide	insecticidal	*Piper nigrum*
Piperine	amide	insecticidal	*Piper guineense*
Piperside	amide	insecticidal	*Piper nigrum*
Plumbagin	naphthoquinone	growth inhibitor	*Plumbago zeylanica*
Polygodial	sesquiterpenoid	antifeedant	*Warburgia ugandensis, Polygonum hydropiper*
Precocene I, II	benzopyranoids	insecticidal	*Ageratum conyzoides*
Procyanidin-B	procyandin dimer	antifeedant	*Annona squamosa*
Pseudo-gervine	alkaloid	insecticidal	*Veratrum virde*
Psoralen	furno-coumarin	growth inhibitor	*Petunia hybrid, P. violacea*
Pulegone 1,2-epioxide	terpenoid	insecticidal	*Lippia geminata*
Purpuranin A&B	isoflavonoid	antifeedant	*Tephrosia purpurea var. maxima*
Pyrethrins I, II	esters	insecticidal	*Chrysanthemum cineraraefolium*
Quadrangolide	terpenoid	repellent	*Eupatoruim quadregu-larae*
Quassin	alkaloid	insecticidal	*Veratrum album, V. oride, Simba multiflora*
Randia acid	saponin	insecticidal	*Randia spinosa*

Table 19. Biologically active chemical components (Continued)

Chemical component	Chemical group	Biological activity	Plant isolated from
Red-glucoside	glucoside	rodenticidal	*Urginia burkei*
Rhodojaponin III	grayanoid diterpene	antifeedant, insecticidal, growth inhibitor	*Rhododendron molle*
Ricin	protein	insecticidal	*Ricinus communis*
Ricinin	alkaloid	insecticidal	*R. communis*
Robustic acid	coumarin	antifeedant	*Derris robusta*
Rotenone	rotenoid	insecticidal	*Derris elliptica*
Ryanodine	alkaloid	insecticidal	*Ryania speciosa*
9,21-didehydro-ryanodine	alkaloid	insecticidal	*Ryania speciosa*
Salannin	terpenoid	antifeedant	*Azardirachta indica*
Sanquinasure	alkaloid	nematicidal	*Bocconia cordata*
Sansool I, II	isobutylamide	insecticidal	*Zanthoxylum piperitum*
Saponin	steroid	antifeedant	*Balanites aegyptica*
Scabrin	isobutylamide	antifeedant	*Heliopsis scabra*
Schkurin I, II	germacranolids	antifeedant	*Schkuhria pinnata*
Sendanin	limonoid	growth inhibitor	*Trichilia roka*
Sesamin	lignan	synergist with pyrethrins	*Sesamum indicum*
Sesamolin	lignan	synergist	*Sesamum indicum*
Solanidine	alkaloid	growth inhibitor	*Spilanthus oleraceae*
Solanine	alkaloid	antifeedant	*Solanum chacoense*
Specionin	iridoid	antifeedant	*Catalpa speciosa*
Spilanthol	unsaturated isobutylamide	insecticidal	*Spilanthus oleraceae*
Spirostanol glycoside	—	molluscicidal	*Yucca aloifolia*
Sterol	isobutylamide	antifeedant	*S. oleraceae*
Strychnine	alkaloid	rodenticidal, insecticidal	*Strychnos nux-vomica*
Sumatrol	isoflavonoid	insecticidal	*Derris malaccenses*

Table 19. Biologically active chemical components (Continued)

Chemical component	Chemical group	Biological activity	Plant isolated from
Tephrosin	isoflavonoid	antifeedant	*Tephrosia candiata, T. rosea, T. elata*
a-terthienyl	thienyl-sulphur derivative	insecticidal, nematicidal	*Tagetes erecta*
Tetra-decanoic acid	coumarin	insecticidal	*Rhus typhina*
Tetra-decanol	alkaloid	insecticidal	*R. typhina*
Thionimon	terpenoid	antifeedant	*Azardirachta indica*
Thioninone	limonoid	nematicidal	*Azardirachta indica*
Thujic acid	—	juveno-mimetic activity	*Thuja plicata*
Tilinosule	—	antifeedant	*Eremocarpus setigerus*
Trans-aconitic acid	tricarboxylic acid derivative	antifeedant	*Echinochloa crus-galli*
3-trans, 11-trans-tridecca-3,5,7,9,11-triene-5,7,9-triyene	polyacetylene	nematicidal	*Carthamus tinctorius*
Trewiasin	metasteroid	insecticidal	*Trewia nandi*
Trichilins A-C	limonoids	antifeedant	*Trachilia roka*
1-tridecanene 3,5,7,9,11-pentayne	—	nematicidal	*Helenium hybrid*
Trideca-1-ene-3,5,7,9,11-pentanyl	—	ovicidal	*Xanthium canadens*
2-tridecanone	isoflavonoid	antifeedant	*Lycopersicon esculentum*
Triptofordiene A-C-1	sesquiterpenoid	insecticidal	*Trypterygium wilfordii*
α-tomatine	steroidal glyco-alkaloid	nematicidal	*L. esculentum*
Toonacilin	terpenoid	antifeedant	*Melia toosandan*
Toxicarol	isoflavonoid	insecticidal	*Derris malaccensis*

Table 19. Biologically active chemical components (Continued)

Chemical component	Chemical group	Biological activity	Plant isolated from
Trilobolide	sesquiterpenoid lactone	antifeedant	*Laser trifolium, Laserpitum siler, L. archanaelica*
Tubotoxin	rotenoid	insecticidal	*Derris elliptica*
Turmerone, Ar-turmerone	curcuminoid	repellent	*Curcuma longa*
Tylophoridine	alkaloid	insecticidal	*Tylophora asthematica*
Tylophorine	alkaloid	insecticidal	*Tylophora asthematica*
Tylophorinine	alkaloid	insecticidal	*Tylophora asthematica*
Unedoside	—	antifeedant	*Canthium euriodes*
Viticosteron-E	iridoid	juvenile hormone mimic activity	*Vitex negundo*
Vulpinic acid	—	antifeedant	*Letharea vulpina*
Waldelectone	phenol	nematicidal	*Eclipta alba*
Warburganol	sesquiterpenoid	antifeedant	*Warburgia ugandensis*
Wilfordine	alkaloid	insecticidal	*Tripterygium wilfordii*
Wilforgine	alkaloid	insecticidal	*Tripterygium wilfordii*
Wiforine	alkaloid	insecticidal	*Tripterygium wilfordii*
Wilforzine	alkaloid	insecticidal	*T. wilfordii*
Wisanine	amide	insecticidal	*Piper guineense*
Xanthotoxin	furno-coumarin	growth inhibitor, antifeedant	*Thamnosma montaria*
Xanthylatin	pyrano-cumarin	antifeedant	*Atlantia racemosa*
Xylomolin	reco-iridoid	antifeedant	*Xylocarpus moluccensis*
Xylotenin	furano-coumarin	insecticidal	*Chloroxylon swietenia*

REFERENCES

Aassen, A.J., Culvenor, C.J.J., Finnie, E.P.F., Kellock, A.W. and Smith, L.W. 1969. Alkaloids as possible cause of ryegrass staggers in grazing livestock. *Aust. J. Agric. Res.* **20**: 71–86.

Abbassy, M.A. 1982. Naturally occurring chemicals for pest control III. Insecticidal and synergistic alkaloids isolated from *Schinus terebinthifolius* Raddi. *Mededingen vancle Facultit Landbouwwetens chappen Rihkshunivesisteit Gent.* **47** (2): 695–699.

Abbassy, M.A., El-Gayar, F. and El-Shazli, A. 1977. Phagostimulants from *Acokanthera spectabilis* Hook tested on *Spodoptera littoralis* Boisd. (Lepidoptera: Noctuidae). *Zeitschrift fur Angewandte Entomologie* **82** (4): 441–444.

Abbassy, M.A., El-Shazli, A. and Gayar, E. A new antifeedant to *Spodoptera littoralis* Boisd. (Lepidoptera Noctuidae) from *Acokanthera spectabilis* Hook. (Apocynaceae). *Zeitschrift fur Angewandte Entomologie* **83** (3): 317–322.

Abraham, C.C. and Ambika, B. 1979. Effect of leaf and kernel extract of neem on moulting and vitellogensis in *Dysdercus cingulatus* Fabr. (Heteroptera: Pyrhocoridae). *Curr. Sci.* **48** (12): 554–556.

Abraham, C.C., Thomas, B., Karunakaran, K. and Goplakrishnan, R. 1972. Relative efficacy of some plant products in controlling infestation by Angoumois grain moth, *Sitotroga cerealella* Oliv. Infesting stored paddy in Kerala. *Agric. Res. J. Kerala* **10** (1): 56–59.

Abraham, V. and Tandon, P.L. 1990. Efficacy of certain oils on the first instar nymphs of the grape mealybug, *Maconelliccocus hirsutus* Green. *National symposium on Problems and Prospects of Botanical Pesticides in Integrated Pest Management at C.T.R.I., Rajahmundry, India,* on 21–22 Jan., 1990 (Abstract): 13.

Abu-Zaid, M.I. and Salem, M.M. 1991. Evaluation of certain doses of wormwood as bio-active agent against the acarine mite, *Acarapis woodi* Rennie. *Bull. Entomol. Soc.* Egypt Econ. Series (1988–1989 Publ. 1991) No. 17: 121–125.

Acharya, A. 1985. Control of *Meloidogyne incognita* on betelvine with neem oil cake and saw dust under field conditions. *Proc. IV Nematology Symp. at Udaipur (India),* on 17 May, 1985: 13.

Acharya, A. and Padhi, N.N. 1988. Effect of neem oil cake and saw dust against root-knot nematode, *M. incognita* on betelvine. *Indian J. Nematol.* **18** (1): 105–107.

Adeknov, S.M., Kupriyanov, A.N., Gafurov, N.M. and Kurmanova, R. Sh. 1990. Sesquiterpene lactones of *Artemisia saissanica. Chem. Natural Compounds* **26** (6): 716–717.

Agrawal, I.L. 1993. Ovicidal activity of four plant extracts on *Dysdercus koenigii* Fabr. (Heteroptera: Pyrrhocoridae). *Indian J. Ent.* **55**: 440–443.

Agrawal, I.L. and Mall, S.B. 1988. Studies on the insecticidal and antifeedant activities of some plant extracts on Bihar hairy caterpillar, *Diacrisia obliqua* Walk. (Lepidoptera: Arctidae). *J. Appl. Entomol.* **105** (5): 529–532.

Agustin, J. 1983. Mindanao State Univ. Philippines. Personal commn. in Michael et al., 1985. Plant species reportedly found to possess pest control properties. *An EWC/UH DATA BASE, Univ. of Hawaii* p 249.

Ahamed, P.Y.A. and Ahmed, S.M. 1991. Potential of some rhizomes of Zingiberaceae family as grain protectants against storage insect pests. *J. Food Sci. Technol.* **28** (6): 375–377.

Ahmed, S. 1984. Use of neem materials by Indo-Pakistan farmers. *Presented at the Research Planning Workshop of the Botanical Pest Control Project at I.R.R.I. Los Banos, Philippines,* 6–10 Aug., 1984 p 9.

Ahmed, S.M., Gupta, M.R. and Bhavanagary, H.M. 1976. Stabilisation of pyrethrins for prolonged residual toxicity part II. Development of new formulations. *Pyrethmum Post.* **13**: 113–121.

Ahn, J.W., Wada, K., Maruio, S., Tanaka, H. and Osaka, Y. 1984. A mite growth regulator from *Chamaecyparis picifera*, which inhibits hatching and larval feeding of the two spotted spider mite, *Tetranychus urticae. Agric. & Biol. Chem.* **48**: 2167–2169.

Alam, M.M. 1976. Effect of mix cropping of marigold and margosa with some vegetables on plant growth and parasitic nematodes. *All India Symposium on Modern Concepts in Plant Protection and Udaipur,* 6–9 March, 1976 p 97.

Alam, M.M. and Khan, A.M. 1974. Control of phyto-nematodes with oil cake amendments in spinach field. *Indian J. Nematol.* **4**: 239–240.

Alam, M.M., Masood, A. and Hussain, S.I. 1975. Effect of margosa and marigold roots-exudates on mortality and larval hatch of certain nematodes. *Indian J. Expt. Biol.* **13** (4): 412–414.

Alam, M.M., Khan, M. and Saxena, S.K. 1976. Mechanism of control of plant parasitic-nematodes as a result of the application of oil-cake to soil: Role of aldehyde and ketones. *All India Symposium on Modern Concepts in Plant Protection at Udaipur,* 6–9 March, 1976: 104–105.

Alam, M.M., Khan, A.M. and Saxena, S.K. 1977. Persistent action of oil-cake on the population of nematode in field. *Botyu-Kagitu.* **42**: 119–124.

Alam, M.M., Khan, A.M. and Saxena, S.K. 1979. Mechanism of control of plant parasitic nematodes as a result of the application of organic amendments to the soil. V. Role of phenolic compounds. *Indian J. Nematol.* **9** (2): 136–142.

Alford, A.R., Cullen, J.A., Storach, R.H. and Bentley, M.D. 1987. Antifeedant activity of limonin against Colorado potato beetle, *Leptinotarsa decemlineata* Say (Coloptera: Chrysomelidoe). *J. Econ. Ent.* **80** (3): 575–578.

Ali, S.I., Rawat, R.R. and Singh, O.P. 1981. Effectiveness of certain forms of custard apple seed powder as seed protectant against the pulse beetle, *Callosobruchus cinensis.* Linn. *Bull. Grain Technol.* **19** (1): 53–56.

Ali, S.I., Singh, O.P. and Misra, U.S. 1983. Effectiveness of plant oils against pulse beetle, *Callosobruchus chinensis* Linn. *Indian J. Ent.* **45** (1): 6–9.

Alkofahi A., Rupprecht, J.K., Anderson, J.E., Mclanghlin, J.L., Mikolajczak, J. and Scott, A.B. 1989. Search for new pesticides from higher plants. In *Insecticides of plant origin. ACS Symp. Series 387 of III Chemical Congress of North America 195th National Meeting of American Chemical Society held at Toronto, Ontario, Canada,* 5–11 Jan., 1988: 23–43.

Allen, T.C., Link, K.P., Ikawa, M. and Brunn, L.K. 1945. The relative effectiveness of the principal alkaloids of *Sabadilla* seeds. *J. Econ. Ent.* **38**: 293–296.

Altieri, M.A., Lippmann, M., Schmidtt, L.L. and Kubo, I. 1984. Antifeedant effect of nomilin on *Spodoptera frugiperda* Smith. (Lepidoptera: Noctuidae) and *Trichoplusia ni* Hubn. (Lepidoptera: Noctuidae) under laboratory and net-house conditions. *Prot. Ecol.* **6** (1): 91–94.

Amonkar, S.V. and Banerji, A. 1971. Isolation and characterisation of larvicidal principle of garlic. *Science* (1971): 1343–1344.

Ampofo, S.A., Roussis, V. and Wiener, D.F. 1987. New pronyloated phenolics from *Piper auritum. Phytochem.* **26** (8):2367–2370.

Anonymous, 1961–1962. Use of neem cake as repellent against ground beetle. *Ann. Report, Central Tobacco Research Institute, Rajamundry, India* p 68.

Anonymous, 1972. Neem kernel suspension as an antifeedent against tobacco caterpillar, *Spodoptera litura* Fabr. *Ann. Report, Central Tobacco Research Institute, Rajahmundry, India* p. 85.

Anonymous, 1973–1974. Neem seed products a study of the insecticidal properties: Trial-1. Effect of neem seed products on *Callosobruchus chinensis* Linn. A stored grain pest: Trial-2. Effect of neem oil on *Triboluim confurum* Hrbst. A stored grain pest: Trial-3. Effect of neem products and 73 some conventional insecticides on sorghum shootfly, *Atheritgona varia scooata* Rond. *VII. Report of Nimbkar Agric. Res. Institute, Phaltan, Dist. Satara (M.S.) India* p 82.

Anonymous, 1974a. Pesticidal Manual. *Eds. Martin, H. and Warthing, C.R., 4th edn. British Crop Protection Council* p 263.

Anonymous, 1974b. Checking Prounica soil Fertilizer Institute. Cultivation, Propagation and utilisation of Azolla, *Agric. Publisher Peking* p 267.

Anonymous, 1975. *Annual Technical Report. All India co-ordinated scheme on Harvest and Post-harvest Technol. (ICAR), Central Rice Research Institute, Cuttack, India:* 2–10.

Anonymous, 1978a. Guidelines for the control of insects and mites of foods, fibers, feeds, ornamentals, livestock, forests and forest products. *USDA, Bull. No.* AH-554.

Anonymous, 1978b. Approved products for farmers and growers. *Ministry of Agriculture, Fishers and Food. London, Report for the Year* 1978.

Anonymous, 1979. Improvement of storage structures and storage practices. *Proc. IV. Ann. Works. IDRC/ICAR, Central Rice Research Institute, Cuttack, India:* 5–7.

Anonymous, 1980–1981. Annual Report of Central Food and Technology Research Institute, Mysore, India for the years 1980–1981: 11.

Anonymous, 1985. The Wealth of India, Raw Materials. *Vol. 1A. Publications and Information Directorate, CSIR, New Delhi.*

Anonymous, 1986. Studies on antifeedant and insecticidal activity of indigenous plant extracts against fruit borer, *Earias vitella* in okra. *Annual Report of Association Development Foundation, New Delhi*: 123–125.

Anonymous, 1987. Significant Research Achievements of Division of Entomology. IARI, New Delhi. *Report on 26th Convocation of Indian Agric. Res. Institute, New Delhi,* 4 Feb., 1987 p 63.

Anonymous, 1989a. *Annual Report of Central Tobacco Research Institute, Rajahmundry, India*: 35.

Anonymous, 1989b. Research on botanical pest control on rice based cropping systems (TA No. 5208). *Project Completion Report, IRRI, Los Banos, Philippines* p 47.

Anonymous, 1991. International Rice Research Institute —Researches on Botanical Pest Control. *Proc. Botanical Pest Control Project. Phase-II, Mid Term Project Meeting at Dhaka, Bangladesh,* 28–31 July, 1991.

Argandona, V.H., Corcuera, L.J., Niemeyer, H.M. and Campbell, B.C. Toxicity and feeding deterrency of hydroxamic acid from Gramineae in synthetic diets against the green bug, *Schizaphis graminum. Entomologia Experimentalis et Applicata* **34** (2): 134–138.

Arnason, J.T., Philogene, B.J.R., Donskov, N., Hudson, N., McDougall, C., Forster, G., Mornd, P., Gardener, D., Lambert, J., Morris, C. and Nozzolilla, C. 1985. Antifeedant and insecticidal properties of azardirachtin to European corn borer, *Ostrinia nubilalis. Entomologia Experimentalis et Applicata* **38** (1): 29–34.

Arun K., Tewari, G.C. and Pandey, N.D. 1979. Antifeeding and insecticidal properties of betel gaurd, *Momordica charantia* Linn. against *Athalia proxima* Klug. *Indian J. Ent.* **41** (2): 103–106.

Ascher, K.R.S. 1983. Withanolids and related steroids from Solanaceae as antifeedant. *Proc. X Intn. Congr. Pl. Prot. Vol-I held at Brighton, England* on 20–25 Nov., 1983.

Ascher, K.R.S. and Sell, G.R. 1981. The effect of neem seed kernel extract on *Epilachna varivestis* Muls. larvae. *Zeitschrift fur Planzankeran Kheiton and Pflanzenschutz* **88** (12): 764–767.

Ascher, K.R.S., Ahu, M.E., Glotfa, E., Glottu, A., Krison, I., Abraham, A., Jacobson, M. and Schmutterer, H. 1987. The antifeedant effect of some new withanolides on three insect species, *Spodoptera littoralis, Epilachna varivestis* and *Tribolium castaneum. Phytoparasitca* **15** (1): 15–29.

Attari, B.S. and Singh, R.P. 1977a. A note on the biological activity of the oil of *Lantana camara* Linn. *Indian J. Ent.* **39** (4): 384–385.

Attari, B. and Singh, R.P. 1977b. Effect of time and temperature on the antifeedant activity of neem seed to desert locust, *Schistocerca gregaria. Indian J. Ent.* **39** (4): 383–384.

Atwal, A.S. and Pajni, H.R. 1964. Preliminary studies on the insecticidal properties of drups of *Melia azardirach* against caterpillars of *Pieris brassicae* Linn. (Lipidoptera: Pieridae). *Indian J. Ent.* **26**: 221–227.

Atwal, A.S. and Sandhu, G.S. 1970. Preliminary studies on the efficacy of some vegetable oils and inert dusts as grain protectants. *J. Res. (PAU)* **7**: 52–54.

Ayyangar, G.S.G. and Rao, P.J. 1989. Neem (*Azardirachta indica* A. Juss.) extracts as larval repellent and ovipositional deterrent to *Spodoptera litura* Fabr. *Indian J. Ent.* **51** (2): 121–124.

Ayyangar, G.S.G. and Rao, P.J. 1990. Azardirachtin effects on the consumption and utilisation of food and mid-gut enzymes of *Spodoptera litura* Fabr. *Indian J. Ent.* **51** (4): 373–376.

Azam, M.F. and Khan, A.M. 1976. Combined effect of organic additives and nematicides in controlling root-knot nematode on tomatoes. *All Indian Symposium on Modern Concepts in Plant Protection. Udaipur, India,* 6–9 March, 1976 (Abstract): 98.

Babu, D.C.S. and Rajasekaran, B. 1984. Evaluation of certain synthetic pyrethroids and vegetable products for the control of Bengalgram borer, *Heliothis armigera* Hubn. *Pesticides* **18** (10): 58–59.

Babu, T.H. and Beri, Y.P. 1969. Efficacy of neem (*Azardirachta indica* A. Juss.) seed extracts in different solvents as deterrent to larvae of *Euproctis lunata. Andhra Agric. J.* **16** (4): 107–111.

Babu, T.R., Reddy, V.S. and Hussaini, S.H. 1989. Effect of edible and non-edible oils in the development of pulse beetle, *Callosobruchus chinensis* Linn. and on viability and yield of mungbean (*Vigna radiata* Linn.). *Tropical Sci.* **29**: 215–220.

Badami, R.C. and Patil, K.P. 1975. Minor seed oils, physico-chemical characteristics and components of acids of four seed oils. *J. Oil Tech. Assoc.* **7** (3): 79–81.

Bajaj, R., Chang, C.J. and McLaughlin, J.L. 1986. Tiliroside from the seeds of *Eremocarpus setigerus. J. Natural Products* **49** (6): 1174–1175.

Bajpai, N.K. and Sharma, V.K. 1993. Effect of neem, *Azardirachta indica* leaf extract on oviposition of *Chilo partellus* Swinhoe. *J. Appl. Zool. Res.* **4** (2): 147–148.

Bajpai, N.K. and Sehgal, V.K. 1994. Effect of neem products, nicotine and karanj on survival and biology of pod borer, *Helicoverpa armichira* Hubn. on chickpea. *Proc. II National Conf. of AZRA on 'Recent trends in plant, animal and human pest management: Impact on environment' at Madras Christian College, Madras,* 27–29 Dec., 1994 (Abstract): 48.

Baker, J.E. and Norris, D.M. 1967. A feeding stimulant for *Scolytus multistriatus* (Coleoptera: Scolytidae) isolated from the bark of *Ulmus americana. Ann. Ent. Soc. Am.* **60** (6): 1213–1215.

Balasubrahmanian, M. 1977. Efficacy of neem extractive in preventing the stored pest infestation with pulses like green gram and red gram. *Neem Newslett.* **11** & **12**: 16.

Balasubrahmanian, M. 1979. Pest management studies for rice brown plant hopper in Tamil Nadu, Agric. Univ. In *recent trends in brown plant hopper control. Colloquium on rice brown plant hopper, Coimbatore, India,* 24 June 1979: 9–10.

Balasubrahmanian, M. 1982. (TNAU, India). Pers. Commn. in Michael et al., 1985. Plant species reportedly possessing pest control properties. *An EWC/UH-DATA BASE, Univ. of Hawaii* p 249.

Balde, A.M., Vanhaelem, M. and Ottingeri, R.O. 1987. A chromone from root-bark of *Harrisonia abyssinica. Phytochem.* **26** (8): 2415–2416.

Baldwin, I.T. 1988. Damage induced alkaloids in tobacco pot-bound plants are not inducible. *J. Chem. Ecol.* **14** (4): 1113–1120.

Baldwin, I.T. 1989. Mechanism of damage induced alkaloid production in wild tobacco. *J. Chem. Ecol.* **15** (5): 1661–1680.

Baldwin, I.T., Cynthio, L.S. and Kean, S.E. 1990. The reproductive consequences associated with inducible alkaloidal responses in wild tobacco. *Ecology* **71** (1): 252–262.

Bandara, K.A.N.P., Kumar, V., Ranasinghe, M.A.S.K. and Kasunaratane, V. 1987. Laboratory evaluation of plant extracts in black bean aphid, *Aphis fabae* Scopli (Homoptera: Aphidae). *Quauantely Nawsl. Asia and Pacific Plant Prot. Comm. FAO, Thailand* **30** (2): 26–27.

Bandara, B.M.R., Wimalasiri, W.R. and Bandana, K.A.N.P. 1987. Isolation and insecticidal activity of (-) hardwickiic acid from *Croton aromaticus. Planta Medica* **53** (6): 575–577.

Bandara, B.M.R., Wimalasiri, W.R. and Macleod, J.K. 1988. Ent-Kauranes and oleanenes from *Croton lacciferus. Phytochem.* **27** (3): 869–871.

Bandhopadhyay, G. 1988. Effect of crude extracts of some plants and insects on the mushroom mite. *Rhyzoglyphus echinopus* Fabr. *In Integrated Pest Control: Progress and Perspectives. Ed. by Mohandas, N. and Koshy, G. Trivendrum Assoc. of Advancement of Entomology*: 392–396.

Banerji, A. 1988. Alternate strategies of plant control natural products as potential biocides. *Bhabha Atomic Research Centre, Bombay, India Publ. No.* **1416**: 62.

Banerji, A. and Chadha, M.S. 1970. Insect moulting hormone from *Achyranthes aspera. Phytochem.* **9**: 1671.

Banerji, A., Luthsia, D.L. and Prabhu, B.R. 1987. Chemical components of *Akalantia racemosa*: isolation, characterisation and synthesis of two new pyranaflavones. *National Symposium on Phytochemical Studies on Indian Medical Plants. Bharthidarsan Univ., Tiruchirapalli, India,* (Abstract): 3.

Bano, M., Anver, S., Tiyagi, S.A. and Alam, M.M. 1986. Evaluation of nematicidal properties of some members of composite. *Intn. Nematol. Network. Newsl.* **3** (1): 10.

Barkat, A., Shereef, G.M., Abdallah; S.S. and Amer, S.A.A. 1986a. Toxic action of some plant extracts against *Tetranychus urticae* Koch. *Bull. Ent. Soc. Egypt Econ. Series.* **14**: 233–242.

Barkat, A., Shareff, G.M., Abdallah, S.S. and Amer, S.A. 1986b. Joint action of some pesticides and plant extracts against *Tetranychus urticae* Koch. *Bull. Ent. Soc. Egypt Econ. Series.* **14**: 243–249.

Barton, G.M., McDonald, B.F. and Sahota, T.S. 1972. Juvenile hormone-like activity of thujic acid from wood extract of western cedar. *Bimonthly Res. Notes* **24** (4): 22–23.

Batra, R.C. and Sandhu, G.S. 1981. Comparison of different insecticides for the control of citrus leaf miner in the nursery. *Pesticides* **15** (2): 5–6.

Beck, S.D. 1960. The European corn borer, *Ostrinia nubilalis* Hubn. and its principal host plant. VII. Larval feeding behaviour and host plant resistance. *Ann. Ent. Soc. Am.* **53**: 206–212.

Begum, S. and Quiniones, A.C. 1990. Protection of stored mungbean seeds from bean weevil, *Callosobruchus chinensis* by vegetable oil application. *Bangladesh J. Zool.* **18** (2): 203–210.

Behera, Jasobanta 1992. Bio-ecology of Angoumois grain moth, *Sitotroga cerealella* and its management in stored paddy. *Ph.D. Thesis Utkal Univ., Bhubaneswar, India* p 182.

Belen, E.M. 1982. Plant extracts possessing rodenticidal and molluscicidal activities. *Monitor* **10** (1): 6–7.

Bentley, M., Hassanali, A. and Lwande, W. 1983. Studies on limonoid insect antifeedant *Proc. X. Intn. Congr. Plant. Prot. 1983, Vol. I Brighton, England,* 20–25 Nov., 1983.

Bentley, M., Hassenali, A., Lwande, W.N., Joroge, P.E.W., Ole Sitayo, E.N. and Yatagai, M. 1987. Insect antifeedants from *Tephrosia elata* Deflers. *Insect Science and Its Application* **8** (1): 55–58.

Bergot, B.J., Schooley, D.A. and deKort, C.D.A. 1981. Identification of juvenile hormone III as the principal juvenile hormone in *Locusta migratoria*. *Expermentia* **37**: 909–912.

Bernays, E. and Luca, C. De. 1981. Insect antifeedant properties of iridoid glucoside ipolamide. *Experimentia* **37**: 1289–1290.

Bestmann, H.J., Classen, B., Kobold, U., Vostrowsky, O. and Klingauf, F. 1987. Botanical insecticides. IV. The insecticidal effect of the essential oil from costmary chrysanthemum, *Chrysmathemum balsamita* Linn. *Angeiges fui Schadlingkunde Pflangenschutz Umweltschuts* **60** (2): 31–34.

Bestmann, H.J., Classen, B., Kobold, U., Vostrowsky, O., Klingauf, F. and Stein, U. 1988. Steam volatile constituents from leaves of *Rhus typhina*. *Phytochem.* **27** (1): 85–90.

Bhaduri, N., Ram, S. and Patil, B.D. 1985. Evaluation of some plant extracts as protectants against the pulse beetle, *Callosobruchus maculatus* Fabr. Infesting cowpea seed. *J. Entomol. Res.* **9** (2): 183–187.

Bhanotar, R.K. and Srivastava, Y.N. 1984. Effect of neem kernel suspension on the development of eggs of desert locust, *Schistocerca gregaria* Forsk. *Neem Newlett.* **1** (3): 30.

Bhargava, N. 1983. Ethnobotanical studies of the tribes of Andaman and Nicobar Islands, India. *I. Onge Econ. Bot.* **37**: 110–119.

Bhargawa, M.C. 1994. Effectiveness of four essential oils against *Corcyra cephalonica* Staint. (Pyralidae: Lepidoptera). *Proc. II National Conf. of AZRA on 'Recent trends in plant, animal and human pest management: Impact on environment' at Madras Christian College, Madras,* 27–29 Dec., 1994 p. 51.

Bhat, N.S. 1988. Chemical control of pod-borer complex of green gram, *Vignaradiata*. *Indian J. Res. Assam Agric. Science* **58** (11): 864–866.

Bhathal, S.S. and Singh, D. 1993. Effect of AZT-VR-K and some commercial neem products on development, reproduction and mortality of mustard aphid, *Lipaphis erysimi* Kalt. *World Neem Conf. Bangalore, India,* 24–28 Feb. 1993. (Abstract): 48.

Bhathal, S.S., Singh, S. and Dhillon, R.S. 1993. Efficacy of alantolides and coumarin as feeding deterrents against the hair caterpillar, *Spilosoma obliqua* Walk. *Pest Management & Economic Zool.* **1**: 20–26.

Bhatia, D.R. 1940. Observations on the biology of desert locusts. In Utilisation of neem and its byproducts. Edited by Ketkar, C.M. 1969–1970. *Report of Modified Neem Cake Manurial Project, Directorate of Oils and Soap Industries, K.V.I.C., Bombay*: 113–176.

Bhatia, A., Alam, N., Hameed, S.F. and Singh, R. 1994. Effectiveness of neem products (*Azardirachta indica* A. Juss.) against whorl maggot, *Hydrellia philippina* Ferino in rice field. *Oryza* **31**: 69–70.

Bhatnagar, A., Mukharjee, T.K. and Tiagi, B. 1978. Control of root-knot nematode on okra by application of water soluble fraction of oil cakes to soil. *Acta Botanica Indica* **6**: 188–189.

Bhatnagar, A., Mukherjee, T.K. and Taigi, B. 1980. Effect of leaves of some plants on *Meloidogyne incognita*. *The Andhra Agric. J.* **27** (5 & 6): 312–314.

Bhatnagar-Thomas, P.L. and Pal, P.K. 1974a. Studies on insecticidal activities of garlic oil. I-Differential toxicity of the oil to *Musca domestica nebulo* Fabr. and *Trogoderma granarium* Everts. *J. Food Sci. Tech.* **11** (3): 110–113.

Bhatnagar-Thomas, P.L. and Pal, P.K. 1974b. Studies on insecticidal activity in garlic oil. II-Mode of action of the oil as a pesticide in *Musca domestica nebulo* Fabr. and *Trogoderma granarium* Everts. *J. Food Sci. Tech.* **11** (3):153–158.

Bhattacharya, D. and Goswami, B.K. 1987a. Comparative efficacy of neem and groundnut oil cakes with aldicarb against *Meloidogyne incognita* in tomato. *Revue de Nematologie* **10** (4): 467–470.

Bhattacharya, D. and Goswami, B.K. 1987b. A study on the comparative efficacy of neem and groundnut oil cakes against root-knot nematode, *Meloidogyne incognita* as influenced by micro-organism on sterilized and unsterilized soil. *Indian J. Nematol.* **17** (1): 81–83.

Bhattacharya, D. and Goswami, B.K. 1988. Effect of different doses of neem and groundnut oil cakes on plant growth characters and populations of root knot nematode, *Meloidogyne incognita* in tomato. *Indian J. Nematol.* **18** (1): 125–127.

Bhattacharya, P.R. and Bordoloi, D.N. 1986. Insect growth retardant activity in some essential oil bearing plants. *Indian perfumer* **30** (2 and 3): 361–371.

Bidman, H.J. 1986. Ultrastructural changes in the prothoracic glands in *Manduca sexta* larvae (Lepidoptera) untreated and treated with azardirachtin. *Entomol. Gener.* **12**: 1–3.

Bilton, J.N., Broughton, H.B., Jones, P.S., Ley, S.V., Lidert, Z., Morgan, E.D., Rzepa, H.S., Sheppard, R.N., Slawin, A.M.Z. and Williams, D.J. 1987. An X-ray crystallo-graphic mass spectroscopic and *nmr* study of the limonoid insect antifeedant 'azardirachtin' and related derivatives. *Tetrahedron* **43**: 2805–2811.

Bloszyk, E., Nawrot, J., Harmatha, J., Drozadk, B. and Chemielewicz, Z. 1990. Effectiveness of antifeedants of plant origin in protection of packaging materials against storage insects. *J. Appl. Entomol.* **110** (1): 96–100.

Borah, D. and Saharia, D. 1982. Field evaluation of certain insecticides for control of *Henosepilachna viginctioctopuncta* Fabr. (Coleoptera Coccinelidae). *J. Res. Assam Agric. Univ.* **3** (2): 224–226.

Borle, M.N. 1982. Department of Entomology P.K.V Akola (Maharastra) India, Pers. Commn. in Michael et al., 1985. Plant species reportedly possessing pest control properties. *An EWC/UH DATA BASE, Univ. of Hawaii* p 249.

Bowry, S.K., Pandery, N.D. and Tripathy, R.A. 1984. Evaluation of certain oil seed cake powders as grain protectant against *Sitophilus oryzae* Linn. *Indian J. Ent.* **48** (2): 196–200.

Brajendra, Singh. 1987. Effect of neem kernel extract mixed with insecticides on the oviposition and mortality of the desert locust, *Schistocerca gregaria* Forsk. *Indian J. Plant Prot.* **15** (2): 184–187.

Brajendra, Singh. and Singh, A.P. 1987a. Neem kernel suspension as an oviposition deterrent to the desert locust, *Schistocerca gregaria* Forsk. *Plant. Prot. Bull. (India)* **39** (1–2): 9–10.

Brajendra, Singh. and Singh, A.P. 1987b. Effect of neem kernel suspension on the hatchability of eggs of the desert locust, *Schistocerca gregaria* Forsk. *J. Advanced Zool.* **8** (1): 52–54.

Brajendra, Singh. and Singh, A.P. 1988a. Effect of neem, *Azardirachta indica* cake on the oviposition of the desert locust, *Schistocerca gregaria* Forsk. *Plant Prot. Bull. (India)* **40** (1):33–34.

Brajendra, Singh. and Singh, A.P. 1988b. Effect on neem kernel suspension on the hatching of the desert locust, *Schistocerca gregaria* Forsk. *Plant Prot. Bull. (India)* **40** (1): 14–17.

Britski, VA.V. 1982. Antifeedant against the Colorado potato beetle. *Zashchita Rastenii* No. 2: 38–39.

Brown, A.W.A. 1951. Insect control by chemicals. *Wiley New York, USA* p 246.

Broza, M., Butler, G.D. Jr. and Heneberry, T.J. 1988. Cotton seed oil for control of *Bemisia tabaci* on cotton. *Beltwide Cotton Production Res. Conf. New Orleans, Louisiana*: 301.

Burnett, W.C., Jones, S.B., Mabry, T.J. and Padolina, W.G. 1974. Sesquiterpene lactones-insect feeding deterrents in *Vernonia. Biochem. Systematic Ecology* **2**: 25–29.

Butani, P.G. and Mittal, V.P. 1990. Comparative efficacy of botanical insecticide (Neem seed kernel suspension) and other insecticides against gram pod borer, *Heliothis armigera* Hubn. *National. Symp. Problems and Prospects of Botanical Pesticides in Integrated Pest Management at C.T.C.R.I., Rajahmundry, India*, 21–22 Jan., 1990 (Abstract): 12.

Butler, G.D. Jr., Coudrict, D.L. and Heneberry, T.J. 1988. Toxicity and repellency of soybean and cotton seed oils to the sweet potato whitefly. *Southwest Entomol.* **13**: 81–86.

Butterworth, J.H. and Morgan, E.D. 1971. Investigation of the locust feeding inhibition of the seeds of neem tree, *Azardirachta indica. J. Insect Physiol.* **17**: 969–977.

Butterworth, J.H. and Morgan, E.D. 1986. Isolation of the substance that suppressed feeding in locusts. *Chem. Commun.*: 23.

Butterworth, J.H., Morgan, E.D. and Percy, G.R. 1972. The structure of azardirachtin and functional group. *J. Chem. Soc. Perkin* (1972): 2445–2450.

Buttery, R.G., Kamm, J.A. and Ling, L.C. 1984. Volatile compounds of red clover leaves, flower and seed pods: possible insect attractants. *J. Agric. Food Chem.* **32** (2): 254–256.

Caasi, M.T. 1983. Morphogenetic effects and antifeedant properties of *Aristolochia tagala* Cham. and *A. elegans* Motsch. on several lepidopterous insects. *B.S. Thesis, College of Laguna, Univ. of Philippines, Los Banos* p 79.

Carino, M.F.A. 1981. Insecticidal screening of crude extracts from nine compositae species and their characterisation of insecticidal fractions from *Tithonia diversifolia* A. Gray. *M.S. Thesis, Los Banos College Laguna, Philippines* p 121.

Carino, M.F.A. 1983. Visayas State College of Agriculture, Baybay, Philippines. Personal Commn. in Michael et al., 1985. Plant species reportedly possessing pest control properties. *An EWC/UH DATA BASE, Univ. of Hawaii* p 249.

Carino, M.F.A. and Morallo-Rajessus, B. 1982. Isolation and characterisation of the insecticides from *Tithonia diversifolia* leaves. *Ann. Trop. Agric.* **4**: 1–11.

Carter, F.L., Garlo, A.M. and Stanley, J.B. 1975. Termiticidal components of wood extracts: 7-methyljuglone from *Diospyros virginiana. J. Agric. Food Chem.* **26** (4): 896–899.

Chachoria, H.S., Chandrate, M.T. and Ketkar, C.M. 1971. Insecticidal trial against stored grain peson maize seed with neem kernel powder carried out at Agric. School. Manjari Mahatashtra. In *Ketkar, C.M. (1969–1976) Utilisation of Neem and its Byproducts. A report of the modified neem cake manurial project, Directorate of Non-edible Oils and Soap Industries, K.V.I.C., Bombay*: 116–175.

Chadha, M.S. 1986. Trends in application of natural products in plant protection. *Proc. Indian National Sci. Acad.* (B) **52** (1): 25–34.

Chadha, M.S., Lathika, M.S., Joshi, N.K. and Banerji, A. 1986. Biological effect of plumbagin on the red cotton bug. *VI Intn. Cong. Pesticide Chem. Ottawa,* 1986 (Abstract): 38.

Chakraborthy, M.K., Prabhu, S.R. and Joshi, B.G. 1976. Isolating toxicants from pongamia oil to evaluate their insecticidal properties on tobacco caterpillar, *Spodoptera litura. Tobacco Res.* **2** (1): 38–44.

Champagne, D.E., Isman, M.B. and Towers, G.H.N. 1989. Insecticidal activity of phytochemicals and extracts of Meliaceae. *In insecticides of plant origin. ACS Symposium Series 387 of III Chemical Congress of North America, 195th National Meeting of American Chemical Society at Toronto, Ontario, Canada,* 5–11 June, 1988: 95–109.

Chandel, B.S., Pandey, U.K. and Singh, A.K. 1984. Insecticidal evaluation of some plant products against red cotton bug, *Dysderus koenigii* Fabr. *Indian J. Ent.* **46** (2): 187–191.

Chandel, B.S., Pandey, U.K. and Kumar, A. 1987. Insecticidal evaluation of some plant extracts against *Epilachna vigintioctopunctata* Fabr. (Coleoptera: Coccinellidae). *Indian J. Ent.* **49** (2): 294–296.

Chander, H. and Ahmed, S.M. 1982. Extractives of medicinal plants as pulse protectants against *Callosobruchus chinensis* Linn. infestation. *J. Food Sci. Tech.* **19** (2): 50–52.

Chander, H. and Ahmed, S.M. 1985. Efficacy of natural embelin against the red flour beetle, *Tribolium castaneum* (Herbst.) *Insect Sci. Applic.* **6**: 217.

Chander, H. and Ahmed, S.M. 1986. Effect of some plant materials on the development of rice moth, *Corcyra cephalonica* Staint. *Entomol.* **11** (4): 273–276.

Chander, H. and Ahmed, S.M. 1987a. Insecticidal activity of *Embelia ribes* Burm. *J. Food Sci. Tech.* **24** (4): 198–199.

Chander, H. and Ahmed, S.M. 1987b. Laboratory evaluation of natural embelin as a grain protectant against some insect pests of wheat in storage. *J. Stored Prod. Res.* **25** (1): 41–46.

Chander, R. 1990. Possible use of mahuwa cake for control of snow bound earth worms. *National Symp. on Problems and Prospects of Botanical Pesticides in Integrated Pest Management at C.T.R.I., Rajahmundry, India,* 21–22 Jan., 1990 (Abstract): 25.

Chandrakantha, J. 1988. Effect of plant chemicals on food utilisation in *Callosobruchus maculatus. III National Symp. on Nutritional Ecology of Insects and Environment at S.D. (PG) College, Muzaffarnagar (UP),* 2–4 Oct., 1988 (Abstract): 25.

Chandramohan, N. and Sivasubramanian, P. 1987. Evaluation of neem products against leaf miner, *Aproacrema modicella* D. *Neem Newslett.* **4** (4): 44.

Chandravadana, M.V. 1987. Identification of triterpenoid feeding deterrent of red pumpkin beetle, *Aulcophora foveicollis* from *Momordica charantia. J. Chem. Ecol.* **13** (7): 1689–1694.

Chandravadana, M.V. and Pal, A.B. 1983. Terpenoid feeding deterrent of *Raphidopalpa foveicollis* Linn. (red pumpkin beetle) from *Momordia charantia* Linn. *Curr. Sci.* **52**: 87–88.

Chang, C.C. and Nakanishi, K. 1983. Specionin and iridoid insect antifeedant from *Catalpa speciosa. J. Chem. Soc. Comm.* **11**: 605–606.

Chari, M.S. and Muralidharan, C.M. 1985. Neem (*Azazrdirachta indica* A. Juss.) as feeding deterrent of castor semi-looper, *Achoea janata* Linn. *J. Entomol. Res.* **9** (2): 243–245.

Chari, M.S., Bharpoda, T.M. and Patel, S.N. 1985. Studies on integrated management of *Spodoptera litura* Fabr. in tobacco nursery. *Tobacco Res.* **11** (2): 93–98.

Chari, M.S., RamPrasad, G., Sitaramaiah, S. and Murthy, P.S.N. 1990. Bioefficacy of neem formulations against *Spodoptera litura* Fabr. in tobacco nursery. *National Symp. on Problems and Prospects of Botanical Pesticides in Integrated Pest Management at C.T.R.I., Rajahmundry, India,* 21–22 Jan., 1990 (Abstract): 8.

Chari, M.S., RamPrasad, G., Sitaramaiah, S. and Murthy, P.S.N. 1993. Bioefficacy of repelin and some new formulations against tobacco caterpillar, *Spodoptera litura* Fabr. *World Neem Conf. Bangalore, India,* 24–28 Feb., 1993 (Abstract): 4.

Charles, K.B. 1929. Insect pests and their control in South Africa. *Die Nasonale Pers. Beperk, Kerrom Stree Cape Town* p 468.

Chatterjee, A. and Sukul, N.C. 1979. Total protein content of galled roots an index of root-knot nematode infestation of lady's finger plants. *Phytopath.* **71** (4): 372–374.

Chatterjee, A., Biswanath, D., Adityachandraya, N. and Debkritaniya, S. 1980a. Studies on aryl and diaryl thio-ureas and their insecticidal activity. *Indian J. Chem.* **19** (B): 163–164.

Chatterjee, A., Biswanath, D., Adityachandraya, N. and Debkritaniya, S. 1980b. Note on the insecticidal properties of the seeds of *Jatropha gossypifolia* Linn. *Indian J. Agric. Sci.* **50** (8): 637–638.

Chatterjee, A., Sukul, N.C., Laskar, S. and Shoshmajumdar, S. 1982. Nematicidal principles from two species of Lamiaceae. *J. Nematol.* **14**: 118–120.

Chattopadhyay, P.R. and Mukhopadhyay, M.C. 1989a. Comparative studies on the nematicidal properties of *Typhonium trilobatum* and *Melia azedarach. India J. Nematol.* **19** (1): 5–9.

Chattopadhyay, P.R. and Makhopadhyay, M.C. 1989b. Effect of baf extracts of *Artabotrys odoratissimus* R. Br. (Annonaceae) on the hatching of eggs of *Meloidogyne incognita. Indian J. Nemotol.* **19** (1): 28–31.

Chauhan, S.P.S., Ashok, Kr., Chandrlekha, Singh and Pandey, U.K. 1987. Toxicity of some plant extracts against rice moth, *Corcyra cephalonica. Indian J. Ent.* **49** (4): 532–534.

Chauvin, R.C. 1946. Sur la substance guin dans less fauilles de *Melia azardirach* rosponse less criquez. *Acad. Sci. Paris* **222**: 412–414.

Chellappa, K. and Chelliah, S. 1976. Studies on the efficacy of malathion and certain plant products on control of *Sitotroga cerealella* Oliv. and *Rhyzopertha dominica* Fabr. infesting rice. *Proc. All India Symp on Modern Concepts of Plant Prot. held at Udaipur, India,* 26–28 March, 1976: 80–86.

Cherian, M.C. and Gopal, M.E.R. 1944. Preliminary trials with neem oil emulsion for control of insect pests. *Madras Agric. J.* **1**: 10–11.

Chitra, S. and Kandswamy, C. 1988. Efficacy of certain new neem constituents against insect pests. *In Integrated pest control progress and perspective. Ed. by Mohandas, N. and Koshy, G. Trivandrum, published by Assoc. Adv. Entomology*: 388–391.

Chitra, S., Kandaswamy, C. and Srimmannarayan, G. 1987. Laboratory evaluation of four fractions of neem (nembidin, vermidin, nemol and nemicidin) against *Henospepilachna vigintioctopunctata*. Fabr. (Coleoptra: Cocceinellidae) *Proc. Symp. Alternatives to Synthetic Insecticides*: 203–207.

Chitra, K.C., Reddy, P.V.R. and Rao, Kameswara, P. 1991a. Effect of certain plant extracts in the control of brinjal spotted leaf beetle, *Henosepilachna vigintioctopunctata* Fabr. *J. Appl. Zool. Res.* **2** (1): 37–38.

Chitra, K.C., Reddy, P.V.R. and Rao, Kamesware, P. 1991b. Studies on the efficacy of petroleum ether extracts of certain plants in the control of brinjal spotted leaf beetle, *Henosepilachna vigintioctopunctata* Fabr. IV. *National Symp. on Growth Development and Control Technology of Insect Pests, at PG. Dept. Zoology, S.D. College, Muzaffarnagar, India,* 2–4 Oct., 1991 (Abstract): 35.

Chatra, K.C., Rami Reddy, P.K., Pillai, K.R. and Subbaraju, G.V. 1995. Insect antifeedant activity of *Justicia ligans*. *J. App. Zool. Res.* **6** (2): 138–139.

Chitra, K.C., Janardhan Rao, S., Rao, Kameswara, P. and Nagaiah, K. 1993. Field evaluation of certain plant products in the control of brinjal pest complex. *Indian J. Ent.* **55** (3): 237–240.

Chitra, K.C. and Reddy, Y.S.K. 1994. Comparative studies on the efficacy of herbal pesticides in the control of groundnut leaf-miner, *Approarema modicella* Deventer (Gelichidae: Lepidoptera). *Proc. II National Conf. of AZRA on 'Recent trends in plant, animal and human pest management: Impact on environment' at Madras Christian College, Madras* 27–29 Dec., 1994 (Abstract): 49.

Chiu, Shin-Foon. 1982. Experiments on insecticidal plants as a source of insect feeding inhibitors and growth regulation with special reference to Meliaceae. *Plant Prot. South China Agric. College, Cuangzhon, China,* p 42.

Chiu, Shin-Foon. 1989. Recent advances in research on botanical pesticides in China. In *Insecticides of plant origin, ACS Symposium Series 387 of III Chemical Congress of North America, 195th National Meeting of American Chemical Society held at Toronto, Ontario, Canada,* 5–11 June, 1988: 69–77.

Chiu, Shin-Foon. 1990. Investigation on botanical insecticides in South China — an update souvenir. *National Symposium on Problems and Prospects of Botanical Pesticides in Integrated Pest Management at C.T.R.I., Rajahmundry, India,* 21–22 Jan., 1990 (Abstract): 5–7.

Chiu, Shin-Foon., Huang, B. and Hu, M.Y. 1983. Experiments on the use of seed oils of some meliaceous plants on antifeedants in brown plant hopper control. *Acta Entomologica Sinica* **26** (1): 1–9.

Chiu, Shin-Foon., Zhang, X., Liu, S.K. and Huang, D.P. 1984. Growth disruptive effects of azardirachtin on the larvae of Asiatic corn borer, *Ostrinia furnacalis* Guen. *Acta Entomologica Sinica* **27** (3): 241–247.

Chocklingam, S. 1986. The use of extract of *Eucalyptus* in the control of *Spodroptra litura* (Noctuidae: Lepidoptera). *J. Advanced Zoology* **7** (2): 79–82.

Chopra, G. and Sood, M.L. 1980. Feeding behaviour of soft-furred field rat, *Rattus meltoda* Gray: Oils. *Bull. Grain Technol.* **18** (3): 204–207.

Chopra, R.L. 1928. *Annual Report of the Entomologist of Govt. of Punjab, Lyallpur for the Year 1925–1926, Dept. of Agric. Lahore, Punjab* **2**: 67–125.

Choudhury, B.S. 1990. Residual effect of eight vegetable oils on chickpea against pulse beetle, *Callosobruchus chinensis* Linn. *Indian J. Pl. Prot.* **18** (1): 89–92.

Cooke-Stinson, L. 1986. Tomato product repels grain weevils. *Agric. Res. USA* **34** (8): 5.

Corbett, J.R. 1974. The biochemical mode of action of pesticide. *Acad. Press, London and New York,* p 165.

Cremlyn, R. 1978. Botanical insecticides in pesticides preparation and mode of action. *John Wiley and Sons, New York,* p 39–49.

Cressman, A.W. and Brodbent, B.M. 1943. Insecticides from plant resources. *J. Econ. Entomol.* **36**: 439.

Crooker, P. 1983. Rural Development and Small Bus Trang-School Nadi, Fizi. Personal Commn. Michael et al., 1985. Plant species reportedly possessing plant control properties. *An EWC/UH DATA BASE, Univ. of Hawaii* p 249.

Crosby, D.G. 1971. Minor insecticides of plant origin. In Naturally occurring insecticides, Eds. Jacobson, M. and Crosby, D.G., 1971. *Marcel Dekker Inc., New York* p 589.

Dahlman, D.L. and Hibbs, E.T. 1967. Response of *Empoasca fabae* (Cicadelliade: Homoptera) to tomarine, solanine, leptine I, tomatidine, solanidine and demissidine. *Ann. Ent. Soc. Am.* **60** (4): 732–740.

Dahlya, R.S., Ravi Dutta and Gupta, D.C. 1986. Effect of subabool and sunnhemp leaves on cereal cyst nematode, *Heterodera avenae* (Wolt. 1924) in wheat. *Haryana Agric. Univ. J. Res.* **16**: 386–391.

Dakshinamurthy, A. 1988. Effect of some plant products on storage pests of paddy. *Tropical Science* **28** (2): 119–122.

Dale, D. 1990. Resistance factors in pest control — A case study in rice. *Presented in National Symposium on 'Problems and Prospects of Botanical Pesticides in Integrated Pest Management' at C.T.R.I., Rajahmmundary, India* 21–22 Jan., 1990 (Abstract) 30.

Dale, D. 1992. An allomone in rice plant against stem borer, *Proc. First National Symp. 'Allelopathy in Agro-ecosystems' Eds. Tauro, P. and Narwal, S.S., Indian Society of Allelopathy, H.A.U., Hissar, India*: 171–174.

Dale, D. and Nair, M.R.G.K. 1977. On the use of neem kernel powder as protectant of stored paddy. *Agric. Res. J. Kerala* **15** (1): 102–103.

Dar, D.W. 1983. Mountain State Agric. College, Benguct, Philippines. Pers. Commn. in Michael et al., 1985. Plant species reportedly possessing pest control properties. *An EWC/UH DATA BASE, Univ. of Hawaii* p 249.

Das, G.P. 1986. Effect of different concentrations of neem oil on adult mortality and oviposition of *Callosobruchus chinensis* Linn. (Bruchidae: Coleoptera). *Indian J. Agric. Sci.* **56** (10): 743–744.

Das, G.P. and Karim, M.A. 1986. Effectiveness of neem seed kernel oil as surface protectant against the pulse beetle, *Callosobruchus chinensis* Linn. *Tropical Grain Legume Bull.* No. **33**: 30–33.

Das, S.N. 1983. Dept. of Nematology, O.U.A.T., Bhubaneswar, India. Pers. Commun. in Michael et al., 1985. Plant species reportedly possessing pest control properties. *An EWC/UH DATA BASE, Univ. of Hawaii* p 249.

Dash, A., Senapati, B. and Samalo, A.P. 1995. Evaluation of neem derivatives against tissue borers of rice. *J. Appl. Zool. Res.* **6** (1): 35–37.

David, B.V., Sukumaran, D. and Kandaswamy, C. 1988. The Indian privet, *Vitex negundo* Linn. A plant possessing promising pesticidal activity. *Pesticides* **21** (2): 27–30.

David, P.M.M. 1986. Effect of slow release nitrogen fertilizers and of foliar application of neem products on rice pests. *Madras Agric. J.* **73** (5): 274–277.

Davidson, W.M. 1929. Oils and alkaloids from seeds of *Staverseacre* sp. *Delphinium staphysangria* and *Delphinium consolida* Linn. as insecticide. *J. Econ. Entomol.* **22**: 226.

Debkirtaniya, S., Ghosh, M.R., Adityachaudhary, N. and Chatterjee, A. 1980a. Extracts of garlic as possible source of insecticides. *Indian J. Agric. Sci.* **50** (6): 507–510.

Debkirtaniya, S., Ghosh, M.R., Adityachaudhary, N. and Chatterjee, A. 1980b. Note on insecticidal properties of the fruit chilli. *Indian J. Agric. Sci.* **50** (6): 510–512.

Delobel, A. and Malonga, P. 1987. Insecticidal properties of six plant materials against *Caryedon serratus* Oliv. (Coleoptera: Bruchidae). *J. Stored Prod. Res.* **23** (3): 173–176.

Del Fierro, R.S. and Morallo-Rajessus, B. 1976. Preliminary study on the insecticidal activity on makebuhai, *Tinospora rumphil* Boerl. *Youth Res. Apprenticeship Action Bogr. Rep. Soc. for Adv. Res.* p 9.

De-Ong, E.R. 1948. Chemistry and uses of insecticides. *Reinhold Publishing Corporation, New York, USA* p 345.

De-Ong, E.R. 1971. Naturally Occurring Insecticides. *Dekker, New York* p 273.

Desai, M.V., Shah, H.M. and Pillai, S.N. 1973. Nematicidal property of some plant species. *Indian J. Nematol.* **3**: 77–78.

Desai, M.V., Shah, H.M., Pillai, S.N. and Patel, A.S. 1979. Oil cakes in control of root-knot nematodes. *Trop. Res.* **5** (1): 105–108.

Deshmukh, M.G. and Prasad, S.K. 1969. Effect of water soluble extracts of oil cakes on population of *Hoplolaimus indicus* Sher. *Indian J. Ent.* **31**: 275–276.

Deshpande, S.G., Singh, B., Nanda, B. and Sharma, R.N. 1990. Antifeedant activity in the total extract and subsequent extracts of the plant, *Cassine glauca* (Leguminosae) against the castor semilooper, *Achoea janata* Linn. (Noctuidae: Lepidoptera). *Presented in National Symposium on 'Problems and Prospects of Botanical Pesticides in Integrated Pest Management' at C.T.R.I., Rajahmundry, India,* 21–22 Jan., 1990 (Abstract): 34.

Deshpande, A.D. 1967. Neem seed as a grain protectant against storage pest. *M.Sc. Thesis, IARI, New Delhi, India,* p 96.

Deshpande, R.S. Adhikary, P.S. and Tipris, N.P. 1974. Stored grain pest control agents from *Nigella sativa* and *Pogostemon heyneamus*. *Bull. Grain Tech.* **12**: 232–234.

Devakumar, C. and Mukherjee, S.K. 1985. 4-Epinimbin: a new meliacin from *Azardirachta indica* A. Juss. *Indian J. Chem.* **24B**: 1105–1106.

Devakumar, C., Goswami, B.K. and Mukherjee, S.K. 1985. Nematicidal principles from neem (*Azardirachta indica* A. Juss.) Part-I. Screening of neem kernel fractions against *Meloidogyne incognita*. *Indian J. Nematol.* **15**: 121–124.

Devakumar, C., Saxena, V.S. and Mukherjee, S.K. 1986. Evaluation of neem (*Azardirachta indica* A. Juss.) limonoids and azardiractin against safflower aphid, *Dactynotus carthami* H.R.L. *Indian J. Ent.* **48** (4): 467–470.

DevaPrasad, V., Jayaraj, S., Rabindra, R.J. and Reddy, G.P.V. 1990. Studies on the interaction of certain botanical and nuclear polyhedrosis virus against tobacco caterpillar *Spodoptera litura* Fabr. *Presented on National Symposium on 'Problems and Prospects of Botanical Pesticides in Integrated Pest Management' at C.T.R.I., Rajahmundry, India,* 21–22 Jan., 1990 (Abstract): 15–16.

Devaraj Urs, K.C. 1990. Efficacy of certain plant products in control of brinjal red spider mite, *Tetranychus urticae. In IV National Symp. on Acarology at University of Calicut, Kerala,* 12–14 Oct., 1990 (Abstract): 68.

Devaraj Urs, K.C. and Srilatha, G.M. 1990. Insecticidal, antifeedant and repellent properties of certain plant extracts against the rice moth, *Corcyra cephalonica* Staint. *Presented in National Symposium on 'Problems and Prospects of Botanical Pesticides in Integrated Pest Management' at C.T.R.I., Rajahmundry, India,* 21–22 Jan., 1990 (Abstract): 9.

Devi, D.A. and Mohandas, N. 1982. Relative efficacy of some antifeedants and deterrents against insect pests of stored paddy. *Entomon.* **7** (3): 261–264.

Dhaliwal, G.S., Pathak, M.D. and Vega, C.S. 1990a. Effect of plant extracts of rice varieties on insect pests and predator complex of rice crop. *Presented in National Symposium on 'Problems and Prospects of Botanical Pesticides in Integrated Pest Management' at C.T.R.I., Rajahmundry, India,* 21–22 Jan., 1990 (Abstract): 20.

Dhaliwal, G.S., Pathak, M.D. and Vega, C.R. 1990b. Effect of different fractions of steam distillate extracts of rice variety on brown plant hopper and striped stem borer. *Presented in National Symposium on 'Problems and Prospects of Botanical Pesticides in Integrated Pest Management' at C.T.R.I., Rajahmundry, India,* 21–22 Jan., 1990 (Abstract): 29.

Dhaliwal, G.S., Jaswant Singh and Dhaliwal, V.K. 1993. Potential of neem in insect pest management in rice. *World Neem Conf., Bangalore, India,* 24–28 Feb., 1993 (Abstract) p 16.

Dhandapani, N., Rajmohan, N.S. and Abdul Kareem, A. 1985. Plant products as antifeedants to control insects. *Pesticides* **19** (11): 53–60.

Dhara Jothi, B., Tandon, P.L. and Abraham, V. 1990a. Evaluation of different plant oils and extracts against citrus leaf miner, *Phylloenistis citrella* Staint. *Presented in National Symposium on 'Problems and Prospects of Botanical Pesticides in Integrated Pest Management' at C.T.R.I., Rajahmundry, India,* 21–22 Jan., 1990 (Abstract): 26.

Dhara Jothi, B., Abraham, V. and Tandon, P.L. 1990b. Evaluation of different plant oils and extracts against citrus aphid, *Toxoptera citricidus* Kirk. *Indian J. Pl. Prot.* **18** (2): 251–254.

Dhawan, A.K. and Simwat, G.S. 1992a. Field evaluation of neemrich 20EC for management of insect pests on cotton during reproductive phase. *Proc. First National Symposium 'Allelopathy in Agro-ecosystems'. Eds. Tauro, P. and Narwal, S.S., Indian Society of Allelopathy, H.A.U., Hissar, India*: 152–153.

Dhawan, A.K., Srimmannarayan, G., Smiwat, G.S. and Nagaiah, K. 1992. Management of cotton pests on uplant cotton, *Gossypium hirsutum* with navneem 95 EC. *Proc. First National Symposium 'Allelopathy in Agro-ecosystems'. Eds. Tauro, P. and Narwal, S.S., Indian Society of Allelopathy, H.A.U., Hissar, India*: 156–157.

Dilawari, V.K. and Dhaliwal, G.S. 1992. Allelo-chemical interactions between *Eruca sativa* Linn. and *Lipaphis erysimi* Kalt. *Proc. First National Symposium 'Allelopathy in Agro-ecosystems' Eds. Tauro, T. and Narwal, S.S., Indian Society of Allelopathy, H.A.U., Hissar, India*: 160–161.

Dimock, M.B., Kennedy, G.G. and Williams, W.G. 1982. Toxicity studies of analogues of 2-tridecanone, a naturally occurring toxicant from a wild tomato. *J. Chem. Ecol.* **8** (5): 837–842.

Doharey, K.L. and Singh, R.P. 1989. Evaluation of neem (*Azardirachta indica* A. Juss.) seed kernel extract against chafer beetles. *Indian J. Ent.* **51** (2): 217–220.

Doharey, R.B., Katiyar, R.N. and Singh, K.M. 1983. Eco-toxicological studies on pulse beetle infesting greengram. 73 VI. Effect of edible oil treatments on germination of greengram, *Vignia radiata* (Linn.) Wilczek. seeds. *Indian J. Ent.* **45** (4): 414–419.

Doharey, R.B., Katiyar, R.N. and Singh, K.M. 1984a. Eco-toxicological studies on pulse beetles infesting greengram. III. Effect of moisture content on the efficacy of edible oils for protection of greengram, *Vigna radiata* (Linn.) Wilczek. against pulse beetles, *Indian J. Ent.* **46** (3): 370–372.

Doharey, R.B., Katiyar, R.N. and Singh, K.M. 1984b. Eco-toxicological studies on pulse beetles infesting greengram. VIII. Studies on rancidity and organoleptic qualities of certain greengram, *Vignia radiata* (Linn.) Wilczek. treated with different edible oil. *Indian J. Ent.* **46** (4): 402–405.

Doharey, R.B., Katiyar, R.N. and Singh, K.M. 1987. Eco-toxicological studies on pulse beetles infesting greengram. IX. Economics of effective edible oils for the protection of greengram, *Vigna radatia* (Linn.) Wilezek. against pulse beetle. *Bull Grain Tech.* **25** (2): 152–156.

Doharey, R.B., Katiyar, R.N. and Singh, K.M. 1990. Eco-toxicological studies on pulse beetles infesting greengram. IV. Comparative efficacy of some edible oils for the protection of greengram, *Vigna radiata* (Linn.) Wilezek. against pulse beetle, *Callosobruchus chinensis* Fabr. *Bull. Grain Tech.* **28** (2): 116–119.

Don Pedro, K.N. 1985. Toxicity of some pests to *Dermestes maculatus* Sig. and *Callosobruchus maculatus* Fabr. *J. Stored Prod. Res.* **21** (1): 31–34.

Dongre, T.K. and Rahalkar, W.G. 1982. Effect of *Blumea eriantha* (Compositae) oil on reproduction of *Earias vittella* Fabr. *Experimentia* **38**: 98–99.

Dongre, T.K. and Rahalkar, W.G. 1984. Ovipositional deterrents for *Earias vitella* Fabr. in leaves of *Xanthium strumarium* (Compositae). *Indian J. Ent.* **46** (1): 111–112.

Dorn, A., Rademacher, J.M. and Sehn, E. 1986a. Ecdysteroid dependent development of the oviduct in last instar larvae of *Oncopeltus fasciatus*. *J. Insect. Physiol.* **32**: 643–647.

Dorn, A., Rademacher, J.M. and Sehn, E. 1986b. Azardirachtin on the moulting cycle, endocrine system and ovaries in last instar larvae of the milkweed bug, *Oncopeltus fasciatus*. *J. Insect. Physiol.* **32**: 231–234.

Dorn, A. and Trumm, P. 1993. Effect of Azardirachtin on neural regulation of mid gut peristalsis in *Locusta migratoria*. *World Neem Conf. Bangalore, India*, 24–28 Feb., 1993 (Abstract): 44.

Doskotch, R.W., Cheng, H.Y., Odell, T.M. and Girard, L. 1980. Nerolidol: An antifeedant sesquiterpene alcohol for gypsy moth larvae from *Melaleuca leucadendron*. *J. Chem. Ecol.* **6** (4): 845–851.

Dover, R. 1985. The responses of some lepidopterans to labiatae herb and white clover extract. *Entomologia experimentalis et Applicata* **39** (2): 177–182.

Drummond, F.A. and Casagrande, R.A. 1985. Effect of white oak extracts on feeding by Colorado potato beetle. *J. Econ. Ent.* **76** (6): 1272–1274.

Dunkel, F.V., Read, N.R. and Wittenberger, N. 1986. National storage survey of beans and sorghum in Rwanda. *Misc. Publ. Exptl. Stn. Univ. Minnesota* No. 46 p 205.

Dutta, M.K., Jat, N.R., Deshmukh, S.N. and Bhatanagar, A.K. 1978. Reetha shell powder (RSP) in pesticidal formulations. I. RSP in DDT water dispersable powder. *Indian J. Ent.* **40** (2): 142–145.

Dutta, N. 1974. Report on the performance of neem oil cake powder against *Microtermes* sp. Pers. Commn. in Ketkar 1969–1976. *Utilisation of neem and its byproducts. Directorate of Non-Edible Oils and Soap Indistries, K.V.I.C., Bombay, India*: 116–173.

Dutta, P., Bhattacharya, P.R., Rabha, L.C., Bordoloi, D.N., Barua, N.C., Choudhury, P.K., Sharma, R.B. and Barua, J.N. 1986. Feeding deterrents for *Philosamia ricini* from *Tithonia diversifolia*. *Phytoparasitica* **14**: 77–80.

Dutta, T.R., Ahmed, R., Syed, R.A. and Rao, M.K.V. 1985. Plants used by Andaman aborgines in gathering rock-bee honey. *Econ. Bot.* **39** (2): 130–134.

Edig, S.H. and Davis, G.R.F. 1980. Repellency of rapeseed extracts to adults of *Tribolium castaneum* and *T. confusum*. *Canadian Entomologist* **112** (9): 971–974.

Egumjobi, O.A. and Afolami, S.O. 1976. Effect of neem (*Azardiracta indica*) leaf extract on population of *Pratylenchus brachyurus* on the growth of maize. *Nematologica* **22**: 125–132.

El-Ghar, G.E.S.A. and El-Sheikh, A.E. 1987. Effectiveness of some plant extracts as surface protectants of cowpea seeds against the pulse beetle, *Callosobruchus chinensis*. *Phytoparasitica* **15** (2): 109–113.

Ellenby, C. 1945a. The influence of crucifers and mustard oil on the emergence of larvae of the potato root eelworm, *Heterodera rostochinensis* Wollenw. *Ann. Appl. Biol.* **32**: 67–70.

Ellenby, C. 1945b. Control of the potato root eelworm, *Heterodera rostochinensis* Wollenw. by allyl-isothiocyanate, the mustard oil of *Brassica nigra* Linn. *Ann. Appl. Bio.* **32**: 237–239.

Elliger, C.A. and Waiss, Jr. A.J. 1989. Insect growth inhibitors from *Petunia* and other solanaceous plants. *In Insecticides of plant origin. ACS Symposium Series 387 of III Chemical Congress of North America, 195th National Meeting of American Chemical Society held at Toronto, Ontario, Canada,* 5–11 June, 1988: 188–205.

Elliger, C.A. and Waiss, Jr. A.C. 1990. Insect resistance factor in *Petunia*-structure and activity. *Western Regional Research Centre USDA, 800 Buchanan Street, Albany, C A-94710 (USA) Res. Commn.* p 14.

Elliott, M., James, N.F. and Potter, C. 1978. The future pyrethroids in insect control. *Ann. Res. Ent.* **23**: 443–469.

El-Naggar, S.F., Doskotch, R.W., Odell, T.M.M. and Girard, L. 1980. Antifeedant diterpenes for the gypsy moth larvae from *Kalmia latifolia*, isolation and characterisation of ten grayanoids. *J. Natural Products* **43** (5): 617–631.

El-Sayed, E.T. 1982–1983a. Evaluation of the insecticidal properties of the common Indian neem, *Azardirachta indica* A. Juss. seed against the Egyptian cotton leaf worm, *Spodoptera littoralis* Bosd. *Bull. Entomol. Soc. Egypt Econ. Series* **13**: 39–47.

El-Sayed, E.T. 1982–1983b. Neem (*Azardirachta indica* A. Juss.) seed as antifeedant and ovipositional repellent for Egyptian cotton leaf worm, *Spodoptera littoralis* (Boisd.) *Bull. Entomol. Soc. Egypt Econ. Series* **13**: 49–58.

El-Sayed, F.M.A. and Abdel-Rahik, M. 1986. Citrus oil as protectant of cowpeas against infestation of *Callosobruchus maculatus* Fabr. (Coleoptera: Bruchidae). *Bull. Entomol. Soc. Egypt Econ. Series* **14**: 423–427.

El-Sayed, F.M.A., Etman, A.A.M. and Abdel-Razik, M. 1991. Effectiveness of natural oils in protecting some stored products from two stored product pests. *Bull. Faculty Agric. Univ. Cario* **40** (2): 409–418.

Epino, P.B. and Saxena, R.C. 1982. Neem, chinaberry and custard apple: Effect of seed oils on leaf and plant hoppers pests of rice. *Philip. Assoc. Entomol. Conv. Baguna, Philippines,* May 1982 p 23.

Fagoonee, I. 1982. Univ. Maurituis, Reduit Pers. Commn. in Michael et al., 1985. Plant species reportedly possessing pest control properties. *An EWC/UH DATA BASE, Univ. of Hawaii* p 249.

Fagoonee, I. 1983. Natural pesticides from neem tree (*Azardirachta indica* A. Juss.) and other tropical plants. *Proc. II Intn. Neem Conf. Rauischhalzhausen,* 25–28 May, 1983: 211–223.

Fagoonee, I. and Toory, V. 1984. Contribution to study of the biology and ecology of the leaf miner, *Liriomyza trifoli* and its control by neem. *Insect Science and Its Appl.* **5** (1): 23–30.

Fagoonee, I. and Umrit, G. 1981. Antigonadotropic hormones from goatweed, *Ageratum conyzoides. Insect Science and Its Application* **1** (4): 373–376.

Feinstain, L. 1952. Insecticides from Plants. *In Insects. The Year Book of Agriculture USDA, Washington, DC*: 222–229.

Feinstein, L. and Jacobson, M. 1953. In progress in the chemistry of organic natural products. *L. Zehcmeistred Vol. 10, Springer Verlag, Vienna* pp 436–447.

Feuerhake, K. and Schmutterer, H. 1982. Simple process of extraction and formulation of neem seed extract and their effect on various insect pests. *Zeitchrift fuir Pflzk. Pflzs.* **89** (12): 737–747.

Feuerhake, K. and Schmutterer, H. 1985. Development of a standardised and formulated insecticide from a crude neem kernel extract. *Z. Pflkorankn Pflesch.* **92** (6): 643–649.

Fletcher, B. 1919. Stored grain pests. *Proc. III Entomol. Meetings held at Pusa, Delhi, India*: 712–762.

Forster, H. 1983. Isolation of azardirachtins from neem (*Azardirachta indica* A. Juss.) and radio-active labelling of azardirachtin. *M.S. Thesis Univ. of Munich, West Germany* p 93.

Forster, H. 1987. Structure and biological action of Azardirachtins, a group of insect-specific growth inhibitors from neem (*Azardirachta indica* A. Juss.): Investigation of its structure, activity effects on the Mexican bean beetle, *Epilachna varivestis. Ph.D. Thesis Univ. Munich, West Germany* p 123.

Fraenkel, G.F. 1959. Investigation into chemical basis of insect host plant relationship. *Science* **129**: 1466–1470.

Frear, D.E.H. 1948. Chemistry of insecticides, fungicides and herbicides. *In Miscellaneous insecticides derived from plants. D. Van Nostrand Co. Inc. (Public) New York* p 417.

Freeborn, S.B. and Wymore, F.H. 1929. Attempt to protect sweet corn from infestation of the corn earworm, *Heliothis obsoleta* Fabr. *J. Econ. Ent.* **22**: 666–671.

Freedman, B., Reed, D.K., Powell, R.G., Madrigal, R.V. and Smith, Jr. C.R. 1982. Biological activities of *Trewia mudiflora* extracts against certain economically important insect pests. *J. Chem. Ecol.* **8** (2): 409–418.

Fry, J.S. and Sons, G. 1938. Neem leaves as an insecticide. *Gold Coast Farmer* **6** (10): 190.

Fukami, H. and Nakajima, M. 1971. Rotenone and rotenoids in naturally occurring insecticides. *Eds. Jacobson, M. and Crosby, D.G., 1971. Marcel Dekker Inc. New York* p 585.

Fukami, J., Nakastsugawa, T. and Narhashi, T. 1959. Rotenoids as insecticides. *Japan J. Entomol. Zool.* **3**: 259.

Ganiev, Sh. G. 1987. The effect of ecdysone containing extracts from *Serratulla sogdiana* Bunge and *Rhaponticum integrifolium* C. Winkl. on gypsy moth larvae. *Rastitalnye Resursy* **23** (4): 634–636.

Garciaespinosa, R. 1980. *Chenopoduim ambrosioides* Linn. a potential use in combating phyto-parasitic nematodes. *Agricultural Trop.* **2**: 92–97.

Gaskins, M.H., White, G.A., Martin, F.W., Delfel, N.E., Ruppel, E.G., and Branes, D.K. 1972. *Tepthrosia voglir. Tech. Bull. USDA (1972)* No. 1445 p 38.

Gaur, A.C. and Prasad, S.K. 1970. Effect of organic matter and inorganic fertilizer on soil and plant nematodes. *Indian J. Ent.* **32** (1): 186–189.

Gaur, H.S. and Mishra, S.D. 1989. Integrated control of nematodes in lentil with aldicarb, neem cake and treatment with thimet and its residual effect on the subsequent mung crop. *Indian J. Ent.* **51** (3): 283–287.

Gawade, M.B., Naik, D.G. and Kapadi, A.H. 1990. Anti-juvenile hormone effect of precocene-I on red cotton bug, *Dysdercus koengii. National Symp. on 'Problems and Prospects of Botanical Pesticides in Integrated Pest Management' at C.T.R.I., Rajahmundry, India,* 21–22 Jan., 1990 (Abstract): 23.

Gebreyens, T. 1980. Army worm antifeedant from *Calusena anisata* (Willd.) Hook. F. Ex. Benth. Bioassay unit, ICIPE Narobi, Kenya, in Michael et al., 1985. Plant species reportedly possessing pest control properties. *An EWC/UH DATA BASE, Univ. of Hawaii* p 249.

Geetanjali, Y., Jyotsna, D. and Srimmannarayana, G. 1987. Chemical examination and evaluation of antifeedant/insecticidal activity of *Pongamia glabra* and *Chloroxylon sweitenia. Proc. Symp. Alternative to Synthetic Insecticides in Integrated Pest Management Systems, at Madurai, India,* 1987: 27–34.

Geuskens, B.M., Luteijn, J.M. and Schoonhoven, S.M. 1983. Antifeedant activity of some ajugarin derivatives on three liptodopteran species. *Experimentia* **39** (4): 403–404.

Ghosh, S.K., Verma, G.D. and Lall, B.S. 1981. Pesticidal efficacy of some indigenous plant products against the pulse beetle, *Callosobruchus chinensis* Linn. (Larridae: Coleoptera). *Bull. Grain Tech.* **19** (2): 96–98.

Ghosh, T. and Sukul, N.C. 1992. Nematicidal principles in the leaves of *Xanthium strumarium* Linn. *Proc. First National Symp. Allelopathy in Agro-ecosystems. Eds. Tauro, P. and Narwal, S.S. Indian Society of Allelopathy, H.A.U., Hissar, India*: 178–180.

Giga, D.P. and Munetsi, M.M. 1990. The effectiveness of vegetable and citrus oils as protectants of cowpeas against infestation of *Callosobruchus rhodesianus* pri. *Plant Protection Quarterly* **5** (4): 148–151.

Giles, P.H. 1964. Plant based insecticides to protect stored grains from insects. *Tropical Agric.* **41**: 202–204.

Gill, J.S. 1972. Studies on insect feeding deterrents with special reference to the fruit extracts of the neem tree, *Azardirachta indica* A. Juss. *Ph.D. Thesis, Univ. of London, England* p 142.

Gill, J.S. and Lewis, C.T. 1971. Systemic action of an insect feeding deterrent. *Nature* **232**: 402–403.

Gillewater, H.B. and Burden, G.S. 1973. For the control of household and stored product insects. *Pyrethrum — the natural insecticide. New York, Acad*: 243–259.

Giridhar, G., Santosh and Vasudevan, P. 1988. Antifeedant properties of *Calotropis* latex. *Pesticides* **22** (1): 31–33.

Girish, G.K. and Jain, S.K. 1974. Studies on the efficacy of neem seed kernel powder to stored grain pests. *Bull. Grain Tech.* **12** (3): 226–228.

Gnanapragaram, N.C. 1981. Influence of cultivating *Eragrostis curvula* in nematode infested soils on the subsequent build-up of population in replanted tea. *Tea Quarterly* **50** (4): 160–162.

Gohokar, R.T., Thakre, S.M. and Borle, M.N. 1985. Chemical control of gram pod borer (*Heliothis armigera* Hubn.) by different synthetic pyrethroids and insecticides. *Pesticides* **19** (12): 39–40.

Gohokar, R.T., Thakre, S.M. and Borle, M.N. 1987. Chemical control of gram pod borer, *Heliothis armigera* Hubn. by different synthetic pyrethroids and insecticides. *Pesticides* **21** (1): 55–56.

Gokte, N. and Swarup, G. 1985. Effect of neem (*Azardirachta indica*) seed kernel powder treatment on penetration and development of *Anguina tritice* on wheat. *Indian J. Nematol.* **15** (1): 149.

Golob, P. and Ashman, F. 1974. The effect of oil content and insecticides on insects attacking rice. *J. Stored Prod. Res.* **10** (2): 93–103.

Golob, P. and Webley, D.J. 1980. The use of the plant and minerals traditional protectants of the stored product. *Tropical Product Institute, Publ. No.* **38**: 3, 12, 13.

Gombos, M.A. 1985. Effect of compounds isolated from the fruit of *Amorpha fruiticosa* Linn. on animal pests and pathogenic fungi. *Ph.D. Thesis, Budapest, Hungary* p 106.

Gommers, F.J. 1971. Nematicidal active principles isolated from the root of *Helenium* hybrid plants. *Phytochem.* **10**: 1945.

Gopalkumar, B., Kumaresan, D. and Varadarasan, S. 1993. Evaluation of pesticides from neem and *Lantana camara* Linn. pests of cardamom. *World Neem Conf. Bangalore, India,* 24–28 Feb., 1993 (Abstract): 11.

Gordon, H.T. 1961. Nutritional factors in insect resistance to chemicals. *Ann. Rev. Entomol.* **6**: 527–545.

Goswami, B.K. and Vijayalakshmi, K. 1983. Studies on efficacy of some indigenous plant extracts and edible oil seed cakes against root-knot nematodes on tomato. *III Nematological Symp. held at H.P.K.V.V., Solan, India,* 24–26 May 1983 (Abstract): 30.

Goswami, B.K. and Vijayalakshmi, K. 1985. In vitro and pot culture studies on the effect of dry products of *Andrographis paniculata, Calendula officinalis, Enhydra fluctuans* and *Solanum khasianum* on *M. ingonita. Indian J. Nematol.* **15** (2): 264–266.

Goswami, B.K. and Vijayalakshmi, K. 1986a. Effect of some indigenous plant materials and oil cake amended soil on the growth of tomato and root-knot nematode population. *Ann. Agric. Res.* **7** (2): 263–266.

Goswami, B.K. and Vijayalakshmi, K. 1986b. Efficacy of some indigenous plant materials and non-edible oil seed cakes against *Meloidogyne incognita* on tomato. *Indian J. Nematol.* **16**: 280–281.

Goswami, B.K. and Vijayalakshmi, K. 1987. Effect of period on decomposition of oil seed cakes in soil on *M. incognita* juveniles. *Indian J. Nematol.* **17** (1): 84–86.

Gour, A.C. and Prasad, S.K. 1970. Effect of organic matter and inorganic fertilizers on soil and plant nematodes. *Indian J. Ent.* **32**: 186–188.

Gowda, D.N. 1972. Studies on comparative efficacy of oil cakes on control of root-knot nematode, *Meloidogyne incognita* Chitwood on tomato. *Mysore J. Agric. Sci.* **6** (4): 524–525.

Gowda, D.N. and Setty, K.G.H. 1979. In vitro studies of neem cake (*Azardirachta indica* A. Juss.) extract on hatching of eggs and survival of root-knot nematode, *Meloidogyne incognita* Chitwood. *Current Res.* **8** (5): 78–79.

Govindaiah, N.S., Sharma, D.D. and Gargi. 1989. Effect of mulching of green leaves for the control of root-knot nematode in mulberry. *Indian J. Nematol.* **19** (1): 25–28.

Govindaiah, S.B., Dandin, T.P. and Datia, R.K. 1990. Effect of marigold (*Tagetes patula*) intercropping against *Meloidogyne incognita* infecting mulberry. *Indian J. Nematol.* **20** (1): 96–99.

Govindaswami, S. and Patnaik, N.K.S. 1985. Storage methods in Jeypore tract. *Rice News Teller* **6** (2): 5–7.

Goyal, R.S. 1971. Studies on better utilisation of neem (*Azardirachta indica* A. Juss.) seed cake, biological activity of alcohol extracts and its isolates. *M.Sc. Ag. Thesis, IARI, New Delhi* p 123.

Goyal, R.S., Gulati, K.C., Prakash S., Kidwai, A.M. and Singh, D.S. 1971. Biological activity of various alcohol extractive isolates of neem (*Azardirachta indica* A. Juss.) seed cake against *Rhopalosiphum nymphae* Linn. and *Achistocerca gregaria* Forsk. *Indian J. Ent.* **33** (1): 67–71.

Granich, M.S., Halpern, B.P. and Eisner, T. 1974. Gymnemic acids, secondary plant substances of dual defensive action. *J. Insect Physiol.* **20** (2): 435–439.

Grewal, P.S. and Sohi, H.S. 1988. Toxicity of some plant extracts to *Aphelenchoides composticola*. *Indian J. Nematol.* **18** (2): 354–355.

Grijpma, P. 1970. Immunity of *Toona ciliata* M. Roem var. *australis* (F.V.M) CDC. and *Khaya ivrensis* A. Chev. to the attacks of *Hypsiphyla grandella* Zell. in Turrialba, Costa. *Turrialba* **20** (1): 85–93.

Grundy, D.L. and Still, C.C. 1985. Isolation and identification of the major insecticidal compound of paleo (*Lippia stoechadifolia*). *Pesticide Biochem. and Physiol.* **23** (3): 378–382.

Gujar, G.T. 1985. Use of castor in management of root-knot nematodes. *Tob. Res.* **11** (1): 78–80.

Gujar, G.T. and Mehrotra, K.N. 1983. Juvenilizing effect of azardirachtin on noctuid moth, *Spodoptera litura* Fabr. *Indian J. Expt. Biol.* **21** (5): 292–293.

Gujar, G.T. and Mehrotra, K.N. 1985. Effect of neem seed oil on consumption, digestion and utilisation of maize (*Zea maize* Linn.) by the desert locust, *Schistocerca gregaria* Forsk. *National Acad. Sci. Letters (India)* **8** (11): 363–365.

Gujar, G.T. and Mehrotra, K.N. 1988. Biological activity of neem against the red pumpkin beetle, *Aulacophora foveicolis*. *Phytoparstica* **16** (4): 293–302.

Gujar, G.T. and Mehrotra, K.N. 1990. Toxicity and morphogenetic effects of neem oil on *Dysdercus koenigii* Fabr. *Presented in National Symp. on 'Problems and Prospects of Botanical Pesticides in Integrated Pest Management' at C.T.R.I., Rajahmundry, India,* 21–22 Jan., 1990 (Abstract): 17–78.

Gunasekaran, K. and Chelliah, S. 1985a. Juvenile hormone activity of *Tribulus terrestris* Linn. on *Spodoptera litura* Fabr. and *Heliothis armigera. The Behavioural and Physiological Approaches in Pest Management. Eds. Ragupathy, A. and Jayaraj, S. T. N. Agric. Univ. Publ.* p 149.

Gunasekaran, D. and Chelliah, S. 1985b. Antifeedant activity of *Andrographis paniculata* Nees. on *Spodoptera litura* Fabr. (Noctuidae: Lepidoptera). *In Behaviour and Physiological Approaches in Pest Management. Eds. Raghupathy, A. and Jayaraj, S. T. N. Agric. Univ. Publ.* pp 31–33.

Gunasena, H.P.M. 1983. Dept. of Crop Science, Univ. of Peradeniya, Sri Lanka. Pers. Comm. in Michael et al., 1985. Plant species reportedly possessing insect control properties. *An EWC/UH DATA BASE, Univ. of Hawaii* p 245.

Gunathilagaraj, K., Babu, P.C.S. and Jayaraj, S. 1987. Relative efficacy of neem products against early season plants of urad bean. *Neem Newsl.* **4** (2): 20–22.

Gunter, F.A. and Jepson, L.R. 1982. Modern insecticides and world food production, chapter 12, in Michael et al., 1985. Plant species reportedly possessing pest control properties. *An. EWC/UH DATA BASE, Univ. of Hawaii* p 249.

Gupta, D.C. and Kaliram. 1981. Studies on the control of *Meloidogyne javanica* infesting chickpea in different types of soil. *Indian J. Nematol.* **11** (1): 77–80.

Gupta, K.M. 1973. Neem leaves attract white grub beetles. *Indian J. Ent.* **35** (3): 276.

Gupta, P. and Bhattacharya, S. 1995. Efficacy and economics of using neem products in management of root-knot nematodes on tomato. *Presented in National symposium on nematode management with eco-friendly approaches and biocomponents. I.A.R.J., New Delhi,* 24–26 March, 1995 (Abstract): 90.

Gupta, P.R. and Dogra, G.S. 1990. Bio-activity of precocene II against the potato beetle, *Epilachna vigintioctopunctata* Fabr. *Presented in National Symp. on 'Problems and Prospects of Botanical Pesticides in Integrated Pest Management' at C.T.R.I., Rajahmundry, India,* 21–22 Jan., 1990. (Abstract): 28.

Gupta, P.R. and Dogra, G.S. 1993. Bio-activity of precocene II and *Acorus calamus* oil and its active ingredient Beta-asarone against the potato beetle, *Henosepilachna vigintioctopunctata* Fabr. *Pest Management & Economic Zool.* **1**: 50–59.

Gupta, P.S., Vimala, V., Geervani, P. and Yadagiri, B. 1988. Efficacy of vegetable oils as protectants of greengram stored in different jute bags. *J. Food Sci. Tech.* **25** (4): 194–196.

Gupta, R. and Sharma, N.K. 1985. Efficacy of garlic as nematicide against *Meloidogyne incognita. Presented in IV Nematology Symp. at Udaipur, India,* 15 May, 1985.

Gupta, R. and Sharma, N.K. 1990a. Nematicidal properties of garlic (*Allium sativum* Linn). *Presented in National Symp. on 'Problems and Prospects of Botanical Pesticides in Integrated Pest Management' at C.T.R.I., Rajahmundry, India,* 21–22 Jan., 1990 (Abstract): 29.

Gupta, R. and Sharma, N.K. 1990b. Ovicidal activity of garlic extract to *Meloidogyne incognita* (Kofoid and White) Chitwood. *J. Appl. Zool. Res.* **1** (2): 90–92.

Gupta, R. and Sharma, N.K. 1991. Nematicidal properties of garlic, *Allium sativum L. Indian J. Nematol.* **21** (1): 14–18.

Gupta, R., Sharma, N.K. and Kaur, D.J. 1985. Efficacy of garlic as nematicidal against *Meloidogyne incognita. Indian J. Nematol.* **15**: 266–267.

Gupta, S. and Rao, P.J. 1990. Effect of azardirachtin on neuro-endocrine system of *Spodoptera litura* Fabr. *Indian J. Ent.* **52** (4): 589–594.

Gupta, S.P., Prakash, A. and Rao, J. 1990. Bio-pesticidal activity of certain plant products against rice earhead bug, *Leptocorisa acuta* Thumb. *J. Appl. Zool. Res.* **1** (2): 55–58.

Hameed, S.F. 1968. Relative efficacy of different oil cakes on the incidence of root-knot nematodes. *Madras Agric. J.* **55**: 168–171.

Hameed, S.F. 1982–1983. Entomology Dept., Rajendra Agric. University, India. Pers. Commn. in Michael et al., 1985. Plant species reportedly possessing pest control properties. *An. EWC/UH DATA BASE, Univ. of Hawaii* p 249.

Hamilton, D.W. and Cleveland, M.L. 1957. Control of codling moth and other pests with Ryania. *J. Econ. Ent.* **50**: 756–759.

Hans-Jorg, F.K. 1972. Ptlanglinchez Insextizid, Nature *R. DSCH 25, Johrg/Helft*: **10** p 445.

Hariprasad, Y. and Kanaujai, K.R. 1992. Evidence for allelopathic chemicals in castor and maize against *Spilosona obliqua* Walk. *Proc. First National Symposium on 'Allelopathy in Agro-ecosystems' Eds. Tauro, P. and Narwal, S.S. Indian Society of Allopathy, H.A.U., Hissar, India*: 175–176.

Harley, K.L.S. and Thorsteinson, A.J. 1967. The influence of plant chemicals on the feeding behaviour, development and survival of the two striped grasshopper, *Melanoplus bivittatus* Say. (Acrididae: Orthoptera). *Canadian J. Zool.* **45**: 305–319.

Haroon, S.A. and Smart, Jr., G.C. 1983a. Development of *Meloidogyne incognita* inhibited by *Digitaria decumbens* c.v. *pangola. J. Nematol.* **15** (1): 102–105.

Haroon, S.A. and Smart, Jr., G.C. 1983b. Effect of root extracts of Pangola digit grass on egg hatch and larval survival of *Meloidogyne incognita. J. Nematol.* **15** (4): 646–649.

Haroon, S.A., Nation, J.N. and Smart, Jr., G.C. 1982. Isolation of natural nematicides from Pangola digit grass, *Digitaria decumbens* that affect egg hatch and larval survival of *Meloidogyne incognita* (Abstract). *J. Nematol.* **14**: 444.

Harper, S.H., Potter, C. and Gillhan, E.M. 1947. *Annona* sp. as insecticides. *Ann. Appl. Biol.* **34**: 104.

Hartley, G.S. and West, T.F. 1969. Chemicals for pest control. *Pergamon Press, Oxford* p 226.

Hasan, A. 1992. Effect of certain plant exudates and byproducts on the development of root knot nematode. *Proc. First National Symposium on 'Allelopathy in Agro-ecosystems' Eds. Tauro, P. and Narwal, S.S. Indian Society of Allopathy, H.A.U., Hissar, India*: 184–186.

Hasan, N. and Jain, R.K. 1984. Bio-toxicity of *Parthenium hysterophorus* extract against *Meloidogyne incognita* and *H. dihystare. Nematological Mediterranea* **12** 239–242.

Haseeb, A., Birpal, S., Khan, A.M. and Saxena, S.K. 1988a. Evaluation of insecticidal properties in certain alkaloid bearing plants. *Geobios* **5** (3): 116–118.

Haseeb, A., Pandey, R. and Hussain, A. 1988b. A comparison of nematicides and oil cakes for control of *Meloidogyne incognita* on *Ocimum basilium*. Nematropica **18** (1): 65–69.

Haseeb, A., Khan, A.M. and Saxena, S.K. 1982. Toxicity of leaf extracts of plants root-knot and reiniform nematode. *Indian J. Parisitol.* **6** (1): 119–120.

Haseeb, A. and Butool, F. 1990. Evaluation of nematicidal property in some members of family Lamiaceae. *Intn. Nematol. Network Newsl.* **7** (2): 24–26.

Haseeb, A. and Batool, F. 1991. Evaluation of different pesticides and oil-cake on the control of root knot nematode, *Meloidogyne incognita* infestating on davana, *Artemisia pallens. Current Entomology* **2** (2): 201–204.

Haseeb, A. and Butool, F. 1993. Use of some pesticides and oil seed cakes for the control of *Meloidogyne incognita* on *Trachyspermum ammi. Afro-Asian J. Nematol.* **3** (1): 112–114.

Haseeb, A., Pandey, R. and Hussain, A. 1988. A comparison of nematicide and oil seed cakes for the control of *Meloidogyne incognita* on *Ocimum basilicum. Nematropica* **18** (1): 65–69.

Hassanali, A. and Lwande, W. 1989. Anti-pest secondary metabolites from African plants. *In Insecticides of plant origin. ACS Symp. Series 387 of III Chem. Congr. North America 195th National Meeting of Am. Chem. Soc. held at Toronto, Ontario, Canada,* 5–11 June, 1988: 78–94.

Hassanali, A., Bentley, M.D., Slawin, A.M.Z., Williams, D.J., Sheparad, R.N. and Chapya, A.W. 1983. Pedonin, a spironotriterpenoid insect antifeedant from *Harrisonia abyssinica. Phytochem.* **26** (2): 573–575.

Head, S.W. 1967. A study of the insecticidal constituents of *Chrysanthemum cinerarae folium.* III. Their composition in different pyrethrum clones. *Pyrethrum Post* **9** (2): 3–7.

Healy, R.E., Rogers, E.F., Wallace, R.T. and Starnes, O. 1950. Insecticidal activities of certain plant extracts. *Lloydia* **13**: 89–92.

Heyde, J.V.D., Saxena, R.C. and Schmutterer, H. 1984. Effect of neem derivatives on growth and fecundity of rice pest, *Nephotettix virescens* Distt. *Bull. Soc. Entomol. Suisse* **57** (4): 423.

Heyde, J.V.D., Saxena, R.C. and Schmutterer, H. 1985. Effects of neem derivatives on growth and fecundity of rice pest, *Nephotettix virescens* (Homoptera: Cicadellidae). *Zeitschrft fur pflanzenkrankhettan und pflanzenchutz* **92** (4): 346–354.

Hiran, G.L., Knodel-Montz, J.J., Ralph, E.W. and Warthen, J.D. 1985. *Liriomyza trifoli* Burg. (Diptera: Agromyzidae) control in *Chrysanthemum* by neem seed extract applied to soil. *J. Econ. Ent.* **78** (1): 80–84.

Ho, D.T. and Kibuka, J.C. 1983. Neem (*Azardirachta indica* A. Juss.) products for control of rice stem borers. *Intn. Rice Res. Newsl.* **8** (5): 15–16.

Hoan, L. and Davide, R.G. 1979. Nematicidal properties of certain plant extracts against root-knot nematode. *Philipp. Agric.* **62** (4): 285–295.

Holman, H.J. 1940. A survey of the insecticidal material of plant origin. *Plant and Animal Products Dept., Imperial Institute, London* p 121.

Hooker, J.D. 1977. *Report on progress and conditions of the Royal Gardens at Kew* p 43.

Hosozawa, S., Kato, N. and Munakata, K. 1974a. Antifeeding active substances for insect in *Caryopteris divaricata* Masim. (Verbenaceae). *Agric. Biol. Chem.* **38**: 823.

Hosozawa, S., Kato, N., Munakata, K. and Chen, Y.N. 1974b. Antifeeding active substances for insect in plant. *Agric. Biol. Chem.* **38** (5): 1045–1048.

Hu, J., Yang, J. and Chen, L. 1983. A preliminary study on the antifeedant and toxic properties of China berry, *Melia azardirach* Linn. seed oil against major insect pests of rice. *Scientia Agric. Sinica* **5**: 63–69.

Hubert, T.D., Okunade, A.L. and Wiemer, D.F. 1987. Quadrangolide, a hergolide from *Eupatorium quadrangularae. Phytochem.* **26** (6): 1751–1753.

Hussain, M.A. 1928. *Annual report of Entomologist to Govt. of Punjab, Lyallpur for the Year 1927–1928. Report of Dept. of Agric. Punjab. Lahore,* pt. II: 57–79.

Hussain, S.I. and Masood, A. 1975. Effect of some plant extracts on larval hatching of *Meloidogyne incognita. Acta Botanica Indica* **3**: 142–146.

Hussaini, S.S., Bharata, K.S. and Prasad Rao, P.V.V. 1993. Combined efficacy of soil solarization and neem cake application against root-knot nematode, *Meloidogyne incognita* and weeds in tobacco nurseries. *World Neem Conf. Bangalore, India,* 24–28 Feb., 1993 (Abstract), p 35.

Inamori, Y., Takeuchi, K., Babu, K. and Kozawa, M. 1987. Antifeedant and insecticidal activity of daphnodorins A, B, and C. *Chem. Pharm. Bull.* **35** (9): 3931–3934.

Ines-Farias, D. 1983. INIREB-alpa Var. Mexico Pers. Commn. in Michael et al., 1985. Plant species reportedly possessing pest control properties. *An EWC/UH DATA BASE, Univ. of Hawaii* p 249.

Irvin, F.R. 1961. Woody plants of Ghana. *Oxford Univ. Press, London. Eds. Watt, J.G. and Broyer Brandwight, M.G.*: 512–527.

Islam, B.N. 1983. Dept. of Entomology, Bangladesh Agric. Univ. Pers. Commn. in Michael et al., 1985. Plant species reportedly possessing pest control properties. *An EWC/UH DATA BASE, Univ. of Hawaii* p 249.

Isman, M.B. and Prokasch, P. 1985. Deterrent and insecticidal chromens and benzofurans from *Encelia* sp. (Asteraceae). *Phytochem.* **24** (9): 1949–1951.

Isreal, P. 1965. Storage structures and their constructions. *Seminar on Post Harvest Handling of Food Grains held at Bangalore, organised by Directorate of Storage and Inspections, New Delhi,* 13–16 July, 1965 p 13.

Isreal, P. and Vedamurthy, G. 1953. Indigenous plant products as insecticides. *Rice News. Teller* **1** (4): 3–5.

Ivbijaro, M.F. 1983a. Toxicity of neem (*Azardirachta indica* A. Juss.) seed to *Sitophilus oryzae* in stored maize. *Protection Ecol.* **5** (4): 353–357.

Ivbijaro, M.F. 1983b. Preservation of cowpea, *Vigna unguiculata* (Linn.) Walp. with neem (*Azardirachta indica* A. Juss.) seed. *Protection Ecol.* **5** (2): 177–182.

Ivbijaro, M.F. 1984. Groundnut oil as a protectant of maize from damage by grain weevil, *Sitophilus zeamais* Motsch. *Protection Ecol.* **6** (4): 267–270.

Ivbijaro, M.F. 1990. The efficacy of seed oils of *Azardirachta indica* A. Juss. and *Piper guineese* Schum. & Thonn. on the control of *Callosobruchus maculatus* F. *Insect Science and Its Application* **11** (2): 149–152.

Ivbijaro, M.F. and Agbaje, M. 1986. Insecticidal activities of *Piper guineese* Schum. and Thonn. and *Capsium* sp. on the cowpea bruchid, *Callosobruchus maculatus* Fabr. *Insect Science and Its Application* **7** (4): 521–524.

Jacob, A., Bai, H.J. and Kuriyan, K.J. 1990. Plant products as an alternate to nematicide for nematode management in banana. *National Symposium on 'Problems and Prospects of Botanical Pesticides in Integrated Pest Management' at C.T.R.I., Rajahmundry, India,* 21–22 Jan., 1990 (Abstract): 45–46.

Jacob, S. and Sheila, M.K. 1990. Treatments of greengram seeds with oils against the infestation of the pulse beetle, *Callosobruchus chinensis* Linn. *Plant. Prot. Bull.* (Faridabad) **42** (3 & 4): 9–10.

Jacob, S. and Sheila, M.K. 1993. A note on the protection of stored rice from lesser grain borer, *Rhyzopertha dominica* Fabr. by indigenous plant products. *Indian J. Ent.* **55** (3): 337–339.

Jacobson, M. 1958. Insecticides from plants. A review of literature (1941–1953). *USDA Agric. Hand Book, Govt. Printing Office, Washington, DC* p 219.

Jacobson, M. 1971. The unsaturated isobutylamides. *In Naturally Occurring Insecticides. Eds. Jacobson, M. and Crosby, D.G. Marcel Dekker Inc. New York* p 585.

Jacobson, M. 1975. Insecticides from plants. A review of literature (1954–1971). *USDA Agric. Hand Book, Govt. Printing Office, Washington, DC* p 461.

Jacobson, M. 1976. Indian calamus root oil: attractiveness of constituents to oriental fruit flies, melon flies and Mediterranean fruit flies. *Lloydia* **39** (6): 412–415.

Jacobson, M. 1981. Isolation and identification of insect antifeedants and growth inhibitors from plants: An overview, in *Natural Pesticides from neem tree (Azardirachta indica A. Juss.). Eds. Schmutterer, H., Ascher, K.R.S. and Rembold, H. GTZ Press, Eschborn, West Germany*: 13.

Jacobson, M. 1986. The neem tree natural resistance par excellence, Natural resistance of plants to pests. *Am. Chem. Soc. Washington, DC*: 220.

Jacobson, M. 1988. The chemistry of neem tree. *Eds. Johns, P.S. et al., in Focus on Phyto-chemicals. Pesticides. Vol. 1. The Neem Tree*: 19–45.

Jacobson, M. 1989. Botanical pesticides: Past, present and future. In *Insecticides of plant origin ACS Symp. Series 387 of III Chem. Cong. of North America 195th National Meeting of Am. Chem. Soc. held at Toronto, Ontario, Canada,* 5–11 June, 1988: 1–10.

Jacobson, M. and Crosby, D.G. eds. 1971. Naturally occurring insecticides. *Marcel Dekker, Inc. New York* p 585.

Jacobson, M. and Pedersen, L.E.K. 1983. Synthesis and insecticidal properties of derivatives of propane-1,3-dithiol (Analogues of the insecticidal derivatives of dithiolane and trithiolane from the alga, *Chara globularis* Thuiller). *Pesticide Science* **14** (1): 90–97.

Jacobson, M., Redfern, R.E. and Mills, Jr., G.D. 1975a. Naturally occurring insect growth regulators II. Screening of insects and plant extracts as insect juvenile hormone mimics. *Lloydia* **38** (6): 455–472.

Jacobson, M., Redfern, R.E. and Mills, Jr., G.D. 1975b. Naturally occurring insect growth regulator III. Echinolone, a highly active juvenile hormone mimic from *Echinaceae angutifolia* roots. *Lloydia* **38** (6): 473–476.

Jacobson, M., Keiser, I., Miyashita, D.N. and Harris, E.J. 1976. Indian calamus root oil: attractiveness of the constituents to Oriental fruit fly, melon fly and Mediterranean fruit fly. *Lloydia* **39**: 412–413.

Jacobson, M., Crystal, M.M. and Warthen, Jr., J.D. 1981a. Boll weevil feeding deterrents from tung oil. *J. Agric. Food Chem.* **29** (3): 591–593.

Jacobson, M., Crystal, M.M. and Kleiman, R. 1981b. Effectiveness of several polyunsaturated seed oils as boll weevil feeding deterrents. *J. Am. Oil Chemists* **58** (11): 982–983.

Jacobson, M., Stokes, J.B., Warthen, Jr., J.D., Redrem, F.E., Reed, D.K., Webb, R.E. and Telek, L. 1984. Neem research in the USDA: *An update in natural pesticides from neem tree and other tropical plants. Eds. Schmutterer, H. and Asher, K.R.S., GTZ Press, Eschborn, West Germany*: 13.

Jacobson, M., Uebel, E.C., Lusby, W.R. and Waters, R.M. 1987. Optical isomers α-copaene derived from several plant resources. *J. Agric. Food Chem.* **35** (5): 798–800.

Jadhav, K.B. and Jadhav, L.D. 1984. Use of some vegetable oils, plant extracts and synthetic products as protectants for pulse beetle, *Callosobruchus maculatus* Fabr. in stored gram. *J. Food Sci. Tech.* **21** (2): 110–113.

Jagadale, G.B., Pawar, A.B. and Darekar, K.S. 1985a. Effect of organic amendments and antagonistic plants on control of the root-knot nematode infesting betel vine, *Piper betel* Linn. *Presented in IV Nematology Symposium at Udaipur, India* 17 May, 1985 (Abstract): 35.

Jagadale, G.B., Pawar, A.B. and Darekar, K.S. 1985b. Effect of organic amendments and antagonistic plants on control of root-knot nematodes infesting betel vine. *Indian J. Nematol.* **15** (2): 264.

Jain, D.C. 1987. Antifeedant active saponin from *Balanites roxburghii* stem bark. *Phytochem.* **26** (8): 2223–2225.

Jain, R.K. and Hasan, N. 1984. Toxicity of Koo-babool extracts to *M. incognita* and *H. dihystera. India J. Nematol.* **14** (2): 179–181.

Jain, S.K. and Saxena, R. 1993. Evaluation of the nematicidal potential of *Mangifera indica. Indian J. Nematol.* **23** (1): 131–132.

Jain, S. and Yadav, T.D. 1990. Effect of organic diluents on the toxicity of insecticidal dusts as seed protectant. *Bull. Grain Tech.* **28** (1): 9–12.

Jain, U., Dutta, S., Trivedi, P.C. and Tiagi, B. 1986. Allelochemic effects of some plants on hatching of *M. incognita. Indian J. Nematol.* **16** (6):275–276.

Jaipal, S., Singh, Z. and Chauhan, R. 1983. Juvenile hormone like activity of some common Indian plants. *Indian J. Agric. Sci.* **53** (8): 730–733.

Jasy, T. and Koshy, P.K. 1992. Effect of certain leaf extracts and leaves of *Glyricidia maculata* (H.B. & Kunth.) Stend. as green manure on *Radopholus similis. Indian J. Nomatol.* **22** (2): 117–121.

Javier, P.A. 1981. Isolation and bio-assay of insecticidal principle from red pepper (*Capsicum annum* Linn.) and black pepper (*Piper nigrum* Linn.) against several insect species. *M.S. Thesis, Univ. Philipp. Los Banos College, Laguna, Philippines* p 50.

Javier, P.A. and Morello-Rajessus, B. 1986. Insecticidal activity of black pepper, *Piper nigrum* Linn. extracts. *Philippine Entomologist* **6** (5): 517–525.

Jayraj, S. 1976. *Neem Products Entomological Trails Utilisation Report, Department of Entomology TNAU, Coimbatore, India* p 88.

Jayraj, S. 1991. An overview of botanical pest control research in T.N.A.U., Coimbatore, India. *Proc. Botanical Pest Control Project — Phase-II. Mid-term Project Review Meeting at Dhaka, Bangladesh; IRRI, Los Banos, Philippines,* 28–31 July, 1991: 63–78.

Jena, M. and Dani, R.C. 1994. Effectiveness of neem based formulations against rice brown plant hopper, *Nilaparvata lugens. Oryza* **31** (4): 315–316.

Jermy, T., Butt, B.A., McDonough, L., Dreyer, D.L. and Rose, A.F. 1980. Antifeedants for the Colorado potato beetle I. antifeeding constituents of some plants from sage-brush community. *Insect Sci. & Its Application* **1** (3): 237–247.

Jha, S. and Roychoudhury, N. 1988. Insecticidal and antifeedant properties of some plants. *In Integrated Pest Control Progress and Perspective, Eds. Mohandas, N. and Koshy, G. Trivendrum Assoc. Adv. Entomol.*: 453–457.

Jilani, G. 1980. Studies on natural repellents for protection of stored grains from insect attack during storage. *Ph.D. Thesis, Univ. Agric. Faisalabad, Pakistan* p 253.

Jilani, G. and Malik, M.M. 1973. Studies on neem plant as repellent against stored grain pests. *Pak. J. Sci. Indust. Res.* **16** (6): 251–254.

Jilani, G. and Saxena, R.C. 1987. Plant derivatives as protectants against insect pests of stored rice. *Annual Report of IRRI, Manila, Philippines for the Year 1987,* p 567.

Jilani, G. and Su, H.C.F. 1980. Stored product insect research and development laboratory. *SEA-USDA Savannah, GA, USA Report* 1980.

Jilani, G. and Su, H.C.F. 1983. Laboratory studies on several plant materials as repellents for protection of cereal grains. *J. Econ. Ent.* **76** (1): 154–157.

Jilani, G., Saxena, R.C. and Rueda, B.P. 1987. Plant derivatives against insect pests of stored rice. *Mid-term appraisal workshop on botanical pest control in rice based cropping system organised jointly by T.N.A.U., Coimbatore, India, IRRI, Manila, E.W. Centre, Honolulu and Asian Dev. Bank, Philippines,* 1–6 June, 1987 p 27.

Jilani, G., Saxena, R.C. and Rueda, B.P. 1988. Repellent and growth inhibiting effects of turmeric oil, sweet flag oil, neem oil and Margosan-O on red flour beetle, *Tribolum castaneum* (Tenebrionidae Coleoptera). *J. Econ. Ent.* **81** (4): 1226–1230.

Johri, R., Katiyar, A.K. and Johri, P.K. 1983. Indigenous plant as molluscicide. *J. Appl. Zool. Res.* **4** (1): 48–50.

Joshi, B.G. 1986. Use of neem products in tobacco in India. *Proc. III Intn. Neem Conf. Nairobi*: 479–494.

Joshi, B.G. and Ram Prasad, G. 1975. Neem kernel as an antifeedant against the tobacco caterpillar, *Spodoptera litura* Fabr. *Phytoparasitica* **3** (1): 59–61.

Joshi, B.G. and Rao, R.S.N. 1968. Pongamia cake can control tobacco beetle. *Indian Farming* **18** (7): 33.

Joshi, B.G., Rao, R.S.N. and Sitaramaiah, S. 1974. Relative efficacy of some synthetic insecticides and pongamia cake with and without baits for control of ground beetles. *Indian Tob. Bull.* **6** (2): 9–12.

Joshi, B.G. and Sitaramaiah, S. 1979. Neem kernel as an ovipositional repellent for *Spodoptera litura* Fabr. moths. *Phytoparositica* **7**: 199–202.

Joshi, B.G., Ram Prasad, G. and Sitramaiah, S. 1978. Neem kernel suspension protects tobacco nurseries. *Indian Farming* **28** (9): 17–18.

Joshi, B.G., Ram Prasad, G. and Nageswar Rao, S. 1984. Neem seed kernel suspension as an antifeedant for *Spodoptera litura* in planted flue-cured *Virginia* tobacco corp. *Phytoparastica* **12** (1): 3–12.

Joshi, B.G., Sitaramiah, S., Ram Prasad, G. and Rao, R.S.N. 1986. Management of *Spodoptera litura* Fabr. (Lepidoptera: Noctiudae) in tobacco nurseries by integrated *Telemomus remus* Nixon. (Hymenoptera: Scedionidae), *Chrysopa scelestes* Banles and neem seed kernel suspension. *Tob. Res.* **12** (1): 16–21.

Joshi, B.G., Sri Prakash, K. and Murthy, P.S.N. 1990. Efficacy of neem suspension, neem seed bitters, *Pongamia* cake water extract, *Bacillus thuringiensis* Berln. Nuclear Polyhedrosis Virus alone and in combination against tobacco caterpillar, *Spodoptera litura* Fabr. in nursery. *Presented in National Symposium on 'Problems and Prospects of Botanical Pesticides in Integrated Pest Management' at C.T.R.I., Rajahmudry, India* 21–22 Jan., 1990 (Abstract): 9.

Joshi, N.K., Mansukhani, H.B., Suryavanshi, D.R. and Banerji, A. 1974. Juvenile-hormone effect of a seed plant, *Psoralea coryliforlia* on the metamorphosis of red cotton bug, *Dysdercus koenigii. Symposium on Biological Approaches to Problems in Medicines, Industry and Agriculture, B.A.R.C., Bombay,* 1974 (Abstract): 36.

Joshi, N.K., Mansukhani, H.B. and Chadha, M.S. 1975. Chemosterilant activity of some juvenile-hormone mimics on red cotton bug *Dydercus koengii. All India Insect Chemosterilant Research Workers Conference, Univ. of Agric. Science, Banglore,* 1975 (Abstract): 39.

Jotwani, M.G. and Sircar, P. 1965. Neem seed as a protectant against stored grain pests infesting wheat seed. *Indian J. Ent.* **27**: 160–164.

Jotwani, M.G. and Sircar, P. 1967. Neem seed as a protectant against bruchid, *Callosobruchus maculatus* infesting some leguminous seeds. *Indian J. Ent.* **29**: 21–24.

Jotwani, M.G. and Srivastava, K.P. 1981. Neem kernel powder as grain protectant in wheat storage. *Pesticides* **15** (10): 19–23.

Jyotsna, D. and Srimmannarayana, G. 1987. Chemical examination and evaluation of antifeedant and insecticidal activities of *Vleistanthus collinus*. *Proc. Symp. Alternatives to Synthetic Insecticides in Integrated Pest Management Systems, held at Maduari, India,* 1987: 35–39.

Kaethner, 1991. Fitness reduction and mortality effects of neem base pesticides on the Colorado potato beetle, *Leptinotarsa decemlineata* Say. *J. Appl. Entomol.* **113** (5): 456–465.

Kalam, A. 1976. Head Department of Entomology Mahatma Phule Krishi Vishwavidyalaya, Rahuri, Dist. Ahmednagar (MS) referred in *Utilisation of neem and its byproducts. Report on modified neem cake manurial project (1959–1976) by C.M. Ketkar. Directorate of Non-edible Oils and Soap Industries, K.V.I.C., Bombay*: 17–176.

Kaliram, and Gupta, D.C. 1980. A note on the efficacy of fresh neem extracts in the control of *Meloidogyne javanica* infesting chickpea (*Cicer aribinum*). *Indian J. Nematol.* **10** (1): 96–98.

Kaliram, and Gupta, D.C. 1982. Efficacy of plant leaves, nematicides and fertilizers alone and in combination against *Meloidogyne javanica* infesting chickpea. *Indian J. Nematol.* **12** (2): 221–225.

Kalpana, T., Opender, K. and Saxena, P.B. 1978. The influence of *Acorus calamus* Linn. oil vapours on the histocytology of the ovaries of *Trogoderma granarium* Evert. *Bull. Grain Tech.* **16**: 3–9.

Kandalker, H.G. and Narkhede, S.S. 1989. Synthetic pyrethroids and plant extracts against sorghum shootfly: some observations. *Pesticides* **23** (4): 28.

Karasev, V.S. 1976. The role of volatile oil composition for trunk pest resistance in coniferous plants: Experiments on lumber. *Symp. Biol. Hungary* **16**: 115–119.

Kareem, A.A. 1984. Progress in the use of neem and other plant species in pest control in India. *Research Planning Works on Botanical Pest Control Project, IRRI, Los Banos, Philippines,* 6–10 Aug., 1984 p 15.

Kareem, A.A., Sadakathulla, S.S. and Subramanian, T.R. 1976. Antifeedant effect of neem (*Azardirachta indica* A. Juss.) seed against larvae of *Euproctis fraterna* Moore (Lymnantridae: Lepidoptera) in castor and *Nephantis serinopa* Meyr. (Cryptophasidae: Lepidoptera) on coconut. *Andhra Agric. J.* **21** (3 & 4): 99–102.

Kareem, A.A., Saxena, R.C. and Justo, H.D. 1987a. Evaluation of neem seed kernel (NSK) and neem bitters (NB) through seedling root-dip against rice green leaf hopper. *Presented in mid-term appraisal works on Botanical Pest Control in Rice Based Cropping Systems, organised by T.N.A.U., Coimbatore, IRRI, Manila, EW Centre, Honolulu, Hawaii and Asian Devel. Bank, Philippines, 1–6 June, 1987 at Coimbatore* p 13.

Kareem, A.A., Saxena, R.C. and Justo, H.D. 1987b. Evaluation of neem seed kernel (NSK) and neem cake (NC) as pre-sowing seed treatment with paddy to afford protection against early stage pests and their effect on seed germination and seed vigor if any. *Mid-term appraisal works on Botanical Pest Control in Rice Based Cropping Systems, organised by T.N.A.U., Coimbatore, IRRI, Manila, EW Centre, Honolulu, Hawaii and Asian Devel. Bank, Philippines, 1–6 June, 1987 at Coimbatore, India* p 11.

Kareem, A.A., Saxena, R.C. and Justo, H.D. 1987c. Cost comparison of neem oil and insecticide against rice tungro virus (RTV). *Neem Newsl.* **4** (4): 89.

Kareem, A.A., Saxena, R.C. and Boncodin, M.E.H. 1988a. Effect of neem seed treatment on rice seedling vigour and survival on brown plant hopper (BPH) and green leaf hopper (GLH). *Intn. Rice Res. Newsl.* **13** (1): 27–28.

Kareem, A.A., Boncodin, M.E.M. and Saxena, R.C. 1988b. Neem seed kernel, neem cake powder and carbofuran granules mixture for controlling green leaf hopper (GLH) and rice tungro virus (RTV). *Intn. Rice Res. Newsl.* **13** (3): 35.

Kareem, A.A., Saxena, R.C. and Palanginan, E.L. 1988c. Effect of neem seed bitters (NSB) and neem seed kernel extract (NSKE) on pest of mungbean following rice. *Intn. Rice Res. Newsl.* **13** (6): 41–42.

Kareem, A.A., Saxena, R.C., Palanginan, E.L., Boncodin, E.M. and Malayba, M.T. 1988d. Neem derivatives: effect on insect pests of rice and certain other crops in the Philippines (laboratory and field evaluation 1986–1988). *Presented at Final Works. IRRI —ADB — FWC Project on Botanical Pest Control in Rice Based Cropping Systems. IRRI, Los Banos, Philippines* 12–16 Dec., 1988 p 13.

Kareem, A.A., Saxena, R.C., Boncodin, M.E.M. and Malaba. M.T. 1989a. Effect of neem seed & leaf bitters on oviposition and development of green leaf hopper (GLH) and brown plant hopper (BPH). *Intn. Rice Res. Newsl.* **14** (6): 26–27.

Kareem, A.A., Saxena, R.C. and Boncodin, M.E.M. 1989b. Neem-carbofuran mixtures against *Nephotettix virescens* Dist. (Homoptera: Cicadelidae) and its transmission of tungo associated viruses. *J. Appl. Ent.* **108**: 68–71.

Kareem, A.A., Saxena, R.C., Boncodin, M.E.M., Krishnaswamy, C. and Sheshu, D.K. 1989c. Neem as seed treatment for rice before sowing: effects on two homopterous insects and seedling vigour. *J. Econ. Entomol.* **82** (4): 1219–1223.

Katsura, M. 1981. Nematicidal natural products. *ICIPE Conf. May 1980. Department of Agricultural Chemistry Nagoya Univ., Japan* (Abstract): 23.

Kauser, G. and Koolman, J. 1984. Ecdystenoid receptors in tissues of the blowfly, *Calliphora vicina*. In *Advances in Invertebrate Reproduction. Vol. 3, Engel, W., Ed. Elsevier Amsterdam,* p 602.

Keiser, I., Harris, J.E. and Miyassita, D.H. 1975. Attraction of ethylether extracts of 232 botanicals to oriental fruit flies, melonflies and Mediterranean fruit flies. *Lloydia* **38** (2): 141–152.

Ketkar, C.M., Kale, G.G. and Tapkiri, V.B. 1976. *Modified neem manurial project. Neem products against stored grain pests. Directorate of Non-edible Oils and Soap Industry, K.V.I.C., Bombay, India* p. 76.

Khaire, V.M., Kachare, B.V. and Patil, C.S. 1989. Studies on vegetable oils and their effect on ovipositional preference of *Callosobruchus chinensis* on pigeon pea. *Bull. Grain Tech.* **27** (2): 151–152.

Khan, A.M., Alam, M.M., Siddiqi, Z.A. and Saxena, S.J. 1967. Effect of different oil-cake on hatching of larva and on the development of root-knot caused by *Meloidogyne incognita. Intn. Symp. Pl. Path. New Delhi, India,* 1967 (Abstract): 37.

Khan, A.M., Alam, M.M. and Rais, A. 1974. Mechanism of control of plant parasitic nematodes as a result of the application of oil cakes to the soil. *Indian J. Nematol.* **4**: 93–96.

Khan, M.A.J. and Khan, R.J. 1985. Insecticidal effect of indigenous vegetable oils (*Taramira* and *Artemisia*) on some rice delphacids in Pakistan. *Pak. J. Sci. Indust. Res.* **28** (6): 428–429.

Khan, M.I. 1986. Efficacy of *Acorus calamus* Linn. rhizome powder against pulse beetle, *Callosobruchus chinensis* Linn. *P.K.V. Res. Journal* **10** (1): 72–74.

Khan, M.W. 1974. Effect of water soluble fractions of oil cake and bitter principles of neem on some fungi and nematodes. *Acta Botanica Indica* **2** (2): 120–128.

Khan, M.W. 1979. Suppression of phytophagus nematodes and certain fungi in rhizosphere of okra due to oil cake amendments. *Acta Botanica Indica* **7**: 51–56.

Khan, R.A. 1982. Pakistan CSIR, Karachi (Pakistan) Pers. Commn. in Michael et al., 1985. Plant species reportedly possessing pest control properties. *An EWC/UH DATA BASE, Univ. of Hawaii* p 249.

Khan, T.A. and Hussain, S.I. 1988a. Studies on the efficacy of seed treatments with pesticides oil cakes, neem leaf and culture filtrate of *P. lilacimus* for the control of disease caused by the presence to *R. reiniformis*, *M. incognita* and *R. solani* occurring either individually or concomitantly on cowpea. *Indian J. Nematol.* **18** (2): 192–198.

Khan, T.A. and Hussain, S.I. 1988b. Studies on efficacy of *P. lilacimus* against a disease complex caused by interaction of *R. reiniformis*, *M. incognita* and *Rhizoctionia solani* on cowpea. *Nematologia Mediterrane* **16** (2): 229–231.

Khan, W.M., Khan, A.M. and Saxena, S.K. 1973. Influence of certain oil-cake amendments on nematodes and fungi in tomato fields. *Acta Botanica Indica* **1**: 49–54.

Khan, W.M., Hug, S. and Saxena, S.K. 1976. Suppression of phytophagous nematodes and certain fungi in rhizospere in okra due to oil-cake amendments. *All India Symposum on Modern Concepts in Plant Protection,* 23–26 March, 1976 (Abstract): 88–89.

Khanvilkar, V.G. 1983. Dept. of Entomology, Konkan Krishi Vidyapeeth, Univ. of Dapoli, India. Pers. Commn. in Michael et al., 1985. Plant species reportedly possessing pest control properties. *An EWC/UH DATA BASE, Univ. of Hawaii* p 249.

Khare, B.P. 1972. Insect pests of stored grain and their control in Uttar Pradesh. *G.B. Pantnagar Agric. Tech. Univ. Pantnagar, Res. Bull.* No. 5. p 152.

Khound, J.N. 1975. Report of the Entomologist, Assam. Agric. Univ. in Ketkar C.M. 1969. *Report of the Modified Neem Cake Manurial Project. Directorate of Non-edible Oils and Soap Industry, K.V.I.C. Bombay, India*: 116–176.

Kim, M., Koh, H.S., Obota, T., Fukami, M. and Ishii, S. 1976. Isolation and identification of trans-aconitic acid as an antifeedant in barnyard grass against the brown plant hopper, *Nilaparvata lugens*. Stal. (Hompoter: Delphacidae). *Appl. Ent. Zool.* **11** (1): 563–570.

Kishor, N. and Sati, O.P. 1990. A new molluscicidal spirostanol glycocide of *Yucca aloifolia. J. Nat. Prod.* **53**: 1557–1559.

Klenk, A., Bokel, M. and Kraus, W. 1986. 3-Tigloylazardirachtol (Tiglol-2-methylcrotonyl): an insect growth regulating constituent of *Azardirachta indica. J. Chem. Soc. Chem. Commun.*: 523.

Klocke, J.A. and Kubo, I. 1982. Citrus limonoid byproducts as insect control agent. *Entomologia Expt. Appl.* **32**: 299–301.

Klocke, J.A., Baklandrin, M.F. and Yamasaki, R.B. 1989. Limonoids, phenolics and furano-coumarins as insect antifeedants, repellents and growth inhibitory components. *In insecticides of plant origin. ACS Symposium 387 of III Chem. Congr. North America 195th National Meeting of America Chemical Society held at Toronto, Ontario, Canada,* 5–11 June, 1988: 136–149.

Klocke, J.A., Hu, M.Y., Chiu, S.F. and Kubo, I. 1991. Grayanoid diterpenes insect antifeedants and insecticides from *Rhododendron molle Phytochem.* **30** (6): 1797–1800.

Koshiya, D.J. and Ghelani, A.B. 1990. Antifeedant activity of different plant derivatives against *Spodoptera litura* Fabr. on groundnut. *Presented in National Symposium on Problems and Prospects of Botanical Pesticides in Integrated Pest Management at C.T.R.I., Rajahmundry, India,* 21–22 Jan., 1990 (Abstract): 14.

Koshiya, D.J. and Chaiya, K.B. 1991. Effect of certain plant extracts on ovipositional behaviour of adults of *Spodoptera litura Fabr. IV National Symp. on Growth, Development and Control Technology of Insect Pests. S.D. College Muzaffarnager (U.P.) India,* 2–4 Oct., 1991 (Abstract): 34–35.

Koul, O. 1983. Feeding deterrence induced by plant limonoids. I. Interaction with the development of red cotton bugs. *Entomologia Expt. Appl.* **36** (1): 85–88.

Koul, O. 1984. Azardirachtin. II. Interaction with the reproduction behaviour of red cotton bugs. *Zeitschnift fur Angewandte Entomologie* **96** (2): 221–223.

Koul, O. 1985. Azardirachtin interaction with development of *Spodoptera litura* Fabr. *Indian J. Expt. Biol.* **23** (3): 160–163.

Koul, O. 1987. Antifeedant and growth inhibitory effect of calamus and neem oil on *Spodoptera litura* under laboratory conditions. *Phytoparasitica* **15** (3): 169–180.

Koul, O. 1988. Effect of neem formulation margosan-O on the development of seed cotton. *Neem Newsl.* **5** (1): 1–3.

Koul, O., Amanai, K. and Ohtaki, T. 1987. Effect of azardiarchtin on the endocrine events of *Bombax mori* JH. *Insect Physiol.* **33** (2): 1103–1108.

Kozawa, M., Baba, K., Matsuyama, Y., Kido, T., Sakai, M. and Takemoto, T. 1982. Components of the root of *Anthricus sylvestris* Hoffm. II. Insecticidal activity. *Chemical and Pharmachem. Bull.* **30** (8): 2885–2888.

Kraus, W. 1986. Constituents of neem and related species. A revised structure of Azardirachtin. *In New trends in natunal products chemistry: Studies in organic chemistry (Vol. 26). Eds. Atta-ur-Rahman and Le Quesne, P.W. Elsevier Amsterdum,* p 273.

Kraus, W. and Cramer, R. 1981. New Tetranortriterpenoid mit insect fra Bemmender, Wrikung Us Neem oil Liebigs Chem.: 181–189.

Kraus, W., Bokel, M., Klenk, A. and Pohnl, H. 1985. The structure of azardiracthin and 22,23-dihydro-23-methoxy azardirachtin. *Tetrahedron Lett.* **26**: 6435–6438.

Kraus, W., Boke, M., Bruhn, A., Cramer, R., Klaiber, I., Klenk, A., Nagl, G., Pohnl, H., Sadlo, H. and Vogler, B. 1987. Structure determination by nmr of azardirachtin and related compounds from *Azardirachta indica* A. Juss. (Meliaceae). *Tetrahedron* **43**: 2817–2819.

Krishnaiah, N.V., Kalode, M.B. and Pasalu, I.C. 1990. New approaches in utilisation of botanicals in rice insect pest control. *Presented in National Symposium on Problems and Prospects of Botanical Pesticides in Integrated Pest Management, at C.T.R.I., Rajahmmdry, India,* 21–22 Jan., 1990 (Abstract): 17.

Krishnaiah, N.V. and Kalode, M.B. 1984. Evaluation of neem oil cake and other edible oil cakes against rice pests. *Indian J. Pl. Prot.* **12** (2): 104–107.

Krishnaiah, N.V. and Kalode, M.B. 1990. Efficacy of selected botanicals against rice insect pests under green house and field conditions. *Indian J. Pl. Prot.* **18**: 197–205.

Krishnaiah, N.V. and Kalode, M.B. 1991. Efficacy of neem oil against rice insect pests under green house and field conditions. *Indian J. Pl. Prot.* **19**: 11–16.

Krishnaiah, S.R. and Ganesalingam, U.K. 1981. Laboratory evaluation and necrotic properties of steam distillation of local plant extracts to *Sitotroga cerealella* Oliv. *J. Natural Sci. Council Sri Lanka* **9** (1): 79–84.

Krishnaiah, S.R., Ganesalingam, U.K. and Senanayake, U.M. 1985. Repellancy and toxicity of some plant oils and their terpene components to *Sitrotroga cerealella* Oliv. (Lepidoptera: Gelechidae). *Tropical Science* **25** (4): 249–252.

Krishnamurthi, B. and Rao, D.S. 1950. Some important insect pests of stored grains and their control. *Entomological Series. Bull.* No. 1 p 83.

Krishnamurthy, Rao. B.H. 1982. Department of Entomology, APAU, Hyderabad (India). Pers. Commn. in Michael et al., 1985. Plant species reportedly possessing pest control properties. *An EWC/UH DATA BASE, Univ. of Hawaii* p 249.

Krishnamurthy, G.V.G. and Murthy, P.S.N. 1990. Further studies with plant extracts on root knot nematode larval *Meloidoggne javanica. Presented in National Symposium on Problems and Prospects of Botanical Pesticides in Integrated Pest Management at C.T.R.I., Rajahmundry, India,* 21–22 Jan., 1990 (Abstract): 43–44.

Kubo, I. 1983. Department of Entomology-Parasitology Univ. of California, Berkeley (USA). Pers. Commn. in Michael et al., 1985. Plant species reportedly possessing pest control properties. *An. EWC/UH DATA BASE, Univ. of Hawaii* p 249.

Kubo, I. and Nakanishi, K. 1978. Some terpenoid insect antifeedants from tropical plants. *Advances in Pesticidal Science Part-2. Pergamon Press*: 284–294.

Kubo, I. and Ganjian, I. 1982. Insect antifeedant terpenes. *Experimentia* **37** (10): 1063–1064.

Kubo, I. and Klocke, J.A. 1981. Limonoides as insect control agents. *Proc. Symp. Intn. Versailles,* 16–20 Nov., 1981: 117–129.

Kubo, I. and Klocke, J.A. 1982. Azardirachtin insect ecdysis inhibitor. *Agric. Biol. Chem.* **46** (7): 1951–1952.

Kubo, I., Uchida, M. and Klocke, J.A. 1983. An insect ecdysis inhibitor from the African medicial plant, *Plumbago copensis* (Plumaginaceae), a naturally occurring chitin synthesis inhibitor. *Agric. Biol. Chem.* **47** (4): 911–913.

Kubo, I., Matsum, T. and Matsumoto, A. 1984a. Structure of decetylazardirachtinol. Application of 2 D ^{1}H-^{1}H and ^{1}H-^{13}C shift correlation spectroscopy. *Tetrahedron Lett.* **25**: 4729–4731.

Kubo, I., Matsumoto, T., Klocke, J.A. and Kamikawa, T. 1984b. Molluscicidal and insecticidal activities of isobutylamides isolated from *Fagara macropylla. Experimentia* **40** (4): 340–341.

Kubo, I., Matsumoto, A. and Matsumoto, T. 1986. A new insect ecdysis inhibitory limonoid, deacetylazardirachtinol isolated from *Azardirachta indica* (Meliacceae) oil. *Tetrahedron* **42**: 489–496.

Kumar, A., Tewari, G.C. and Pandey, N.D. 1979. Antifeedant and insecticidal properties of bittergaurd, *Momordica charantia* Linn. against *Athalia proxina* Klug. *Indian J. Ent.* **41** (2): 103–106.

Kumar, A.R.V. and Sangappa, H.K. 1984. A note on the performance of plant products in control of gram caterpillar in Bengalgram. *Curr. Res.* **13** (4/6): 38–40.

Kumar, R. and Mehla, J.C. 1993. Comparative efficacy of some plant materials against *Sitotroga cerealella* Oliv. and *Rhizopertha dominica* Fabr. in stored milled rice. *World Neem Conf., Bangalore, India,* 24–28 Feb., 1993 (Abstract): 25.

Kumar, R., Verma, K.K., Dhindsa, K.S. and Bhatti, D.S. 1986. *Datura, Ipomea, Tagetes* and *Lawsonia* as control agents for *Tylenchulus semipenetrans* and *Anguina tritici*. *Indian J. Nematol.* **16** (2): 236–240.

Kumar, T.P. and Nair, M.R.G.K. 1976. Effect of some green leaves and inorganic wastes on root-knot nematode infestation in okra (bhindi). *Agric. Res. J. Kerala* **14** (1): 64–67.

Kumari, K., Sinha, M.M., Mehto, D.N. and Hammed, S.F. 1990. Effect of some vegetable oils as protectants against pulse beetle, *Callosobruchus chinensis* Linn. *Bull. Grain Tech.* **28** (1): 58–60.

Kundu, A.B., Ray, S., Chakrabarti, R., Nayak, L. and Chatterjee, A. 1985. Recent development in chemistry of C_{26}-trepenoids (tetra-nortriterpenoids). *J. Sci. Indian Res.* **44**: 256–263.

Kuriyan, K.J. 1988. Organic amendments for control of root-knot nematode, *Meloidogyne incognita* on bhindi. *In Integrated Pest Control Progress and Perspectives. Eds. Mohandas, N. and Koshy, G., Trivendram, Assoc. Adv. Entomol.* 1988: 133–134.

Kutsch, W. 1985. Pre-imaginal flight motor pattern in *Locusta*. *J. Insect Physiol.* **31**: 581–584.

Ladd, Jr., T.L., Jacobson, M. and Bruiff, C.R. 1978. Japanese beetle: extracts from neem tree seeds as feeding deterrents. *J. Econ. Entomol.* **71**: 810–813.

Lakhani, G.K. and Patel, N.G. 1985. Non pesticidal control of *Ephestia cautella* Walk. *Pesticides* **19** (1): 46–48.

Lal, L. 1987. Studies on natural repellents against potato tuber moth, *Phthorimaea opercullela* Zell. in country stores. *Potato Research* **30** (2): 329–334.

Lalitha, S.V., Thakur, S.S., Kishan Rao and Nagarjarao, P. 1981. Juvenile hormone mimicing activity of *Butea monosperma* flower extract on larvae of *Spodoptera litura* (Lepidoptera: Noctuidace). *Proc. Indian Acad. Parasitology* **2** (1): 39–44.

Lalitha, T., Ahmed, S.M., Dhanraj, S. and Venkatraman, L.V. 1988. Toxicity of seed extract of phulware, *Madhuca butyracea* Mach. on house fly, *Musca domestica* Linn. and pulse beetle, *Callosobruchus chinensis* Linn. *Intn. Pest Control* **30** (2): 42–45.

Lall, S.B. and Hameed, S.F. 1969. Studies on the biology of the root-knot nematodes, *Meloidogyne* sp. with special reference to host resistance and manuring. *First All India Nematol. Symp. at New Delhi* p 6.

Lange, W. and Feurhake, K. 1984. Increased effectiveness of enriched neem seed extracts by synergists piperonyl butoxide under laboratory conditions. *Zeitschrift fur Angewandte Entomologie* **98** (4): 368–378.

Lanjeswar, R.D. and Shukla, V.N. 1986. Vulnerability of larvae and eggs of *M. incognita* to some oil-cakes and fungicides. *Indian J. Nematol.* **16** (1): 69–73.

Lathrop, F.H. and Keirstead, L.G. 1946. Black pepper to control the bean weevil. *J. Econ. Entomol.* **39**: 534.

Lavie, D., Jain, M.K. and Shpan-Gabrielith, S.R. 1967. A locust phago-repellent from two *Melia* spp. *Chem. Comm.* **1**: 910–911.

Lear, B. 1959. Application of castor pomace and cropping of castor beans to reduce nematode populations. *Pl. Disease Reptr.* **43**: 459–460.

Leuschner, K. 1972. Effect of an unknown plant resistance on a shield bug. *Sonder Drunkaus. Die Naturwissenschften Springer Verlag, Berlin, Heidelberg, New York 59 Johargang Meft.* **5** (5): 217–218.

Lichtenstein, E.P., Strong, F.M. and Morgan, D.C. 1962. Identification of 2-phenyl ethyl isothiocynate as an insecticide occurring naturally in the edible parts of turnip. *J. Agric. Food Chem.* **10** (1): 30–33.

Lichtenstein, E.P., Liang, T.T., Schulz, K.R., Heinrich, K.S. and Caster, G.T. 1974. Insecticidal and synergistic components isolated from dill plants. *J. Agric. Food Chem.* **22**: 558–564.

Lidert, Z., Taylor, D.A.H. and Thirunganum, M. 1985. Insect antifeedant activity of flour prieurianin type limonoids. *J. Natural Products* **48** (5): 843–845.

Lim, G.S. and Dale, G.B. 1994. Neem pesticides in rice: potential and limitations. *IRRI Publ.* p 69.

Litsinger, J.A., Price, E.C. and Herrera, G. 1978. Filipino farmers' use of plant parts to control rice insect pest. *Intn. Rice Res. Newsl.* **3** (5): 15–16.

Liu, C.C. 1979. Use of azolla in rice production in China. In Nitrogen and Rice, *Intn. Rice Res. Inst. Los Banos, Philippines* p 378.

Loke, Wai-Hong 1983. MARDI, Malaysia. Pers. Commn. in Michael et al., 1985. Plant species reportedly possessing insect control properties. *An EWC/UH DATA BASE, Univ. of Hawaii* p 249.

Lu, R.D. 1982. A study of insect anti-juvenile hormones chemical composition of *Ageratum conyzoides* Linn. and its action against insect. *Insect Knowledge (Kunchong-Zhishi)* **19** (4): 22–25.

Lwande, W., Gebreysus, T., Chapya, A., Aacfoy, C., Hassanali, A. and Okech, M. 1983. 9-acridone insect antifeedant alkaloid from *Teclea trichocarpa* bark. *Insect Sci. Appl.* **4** (4): 393–395.

Macrae, W.D. and Towers, G.H.N. 1984. Biological activity of lignans. *Phytochem.* **23**: 1207.

Madhusudhan, V.V. and Gopalan, N. 1988. Chemical control of thrips, *Stenchaetothips biformis* in rice nursery. *Intn. Rice Res. Newsl.* **13** (3): 42.

Mahmood, I., Saxena, S.K. and Zakiruddin, R. 1979. Effect of some plant extracts on the mortality of *Meloidogyne incognita* and *Rotylenchus reiniformis. Acta Botanica Indica* **7** (2): 121–132.

Mahmood, I., Saxena, S.K. and Zakiruddin, R. 1982. Effect of certain plant extracts on the mortality of *Pratylenchus reinifermis* (Linford & Olivera, 1940) and *Meloidogyne incognita* (Kofoid and White, 1919) Chitwood, 1949. *Bangladesh J. Bot.* **11** (2): 154–157.

Majumder, S.K. 1974. Studies on insect infestation in neem powder. Pers. Commn. Ketkar, C.M. (1969–1979). Utilisation of neem and its byproducts. *Report of the modified neem cake manurial project. Directorate of Non-edible Oils and Soap Industries, K.V.I.C., Bombay*: 116–175.

Majumder, S.K. and Amla, B.C. 1977. Inhibition of the multiplication of *Callosobruchus chinensis* in vegetable oils. *J. Food Sci. Technol.* **44**: 184–185.

Mala, S. 1987. Antifeedant activity of neem oil extractive on the lepidopteran pest, *Pericallia ricini. Neem Newsl.* **4** (3): 36.

Malik, M.M. and Mujtaba, A. 1984. Screening of some indigenous plants as repellents or antifeedants to stored grain insects. *J. Stored Prod. Res.* **20** (1): 41–44.

Mallick, R.N. and Banerji, A. 1989. Effect of methanol extracts of *Ocimum sancturn* Linn. on jute semi-looper, *Anomis sabulifera* Guen. *Indian J. Ent.* **51** (1): 84–89.

Mallick, R.N., Chatterji, A., Das, P.C. and Chatterjee, S.M. 1980. Antifeedant property of jute leaf extracts against *Myllocerus discolor* Boh. *J. Ent. Res.* **4** (2): 148–152.

Mallick, S.N. and Lal, I.B. 1989. Efficacy of neem oil cake and fertilizer mixture against okra fruit borer. *Pestology* **13** (11): 6–7.

Mammen, K.V., Visalakshi, A. and Nair, M.R.G.K. 1968. Insecticidal dusts to control stored insect pests of paddy. *Agric. Res. J. Kerala* **6** (1): 59–60.

Mandal, S.N. and Bhatti, D.S. 1983. Preliminary screening of some weed shrubs for their nematicidal activity against *Meloidogyne javanica. Indian J. Nematol.* **13** (1): 123–127.

Mane, S.D. 1968. Neem seed spray as a repellent against some of the foliage feeding insects. *M.Sc. Thesis, IARI, New Delhi* p 111.

Mani, A., Kumudanathan, K. and Jagadish, C.A. 1990. Relative efficacy of neem oil and endosulfan against insect pests of mustard. *Neem Newsl.* **7** (2): 129–131.

Mansour, F.A. 1993. Effect of margosan-O, azardirachtin and repelin on spiders, predaceous and phytophagous mites. *World Neem Conf. Bangalore, India,* 24–28 Feb., 1993 (Abstract): 9.

Mansour, F.A. and Ascher, K.R.S. 1983. Effect of neem (*Azardirachta indica*) seed kernel extracts from 73 different solvents on carmine spider mite, *Tetranychus cinnabarinus* Borsd. *Phytoparistica* **11** (3/4): 177–185.

Mansour, F.A., Ascher, D.R.S. and Omari, N. 1987. Effect of neem (*Azardirachta indica*) seed kernel extracts from different solvents on the predaceous mite, *Phytoseilus persimilis* and phytophagous mite, *Tetranychus cinnabarinus. Phytoparasitica* **15** (2): 125–130.

Mariappan, V. and Saxena, R.C. 1983. Effect of custard apple oil and neem oil on survival of *Nephotettix virescens* (Homoptera Cicandellidae) and on rice tungro virus transmission. *J. Econ. Entomol.* **76** (3): 573–576.

Mariappan, V., Saxena, R.C. and Ling, K.C. 1982a. Effect of custard apple oil and neem oil on the life span of rice tungro virus transmission by *Nephotettix virescens. Intn. Rice Res. Newsl.* **7** (3): 13–14.

Mariappan, V., Saxena, R.C. and Ling, K.C. 1982b. Effect of custard apple and neem seed oils on the survival of *Nephotettix virescens* Distt. and its transmission to rice turgro virus. *Phillip. Assoc. Entomol. Conven. Baguina, Philippines* May, 1982 p 10.

Mariappan, V., Gopalan, M., Narsimhan, V. and Suresh, S. 1987. Effect of neem and other plant products on yellow mosaic virus (YMV) disease of blackgram, *Vignia mungo* Linn. *Neem Newsl.* **4** (1): 9–10.

Mariappan, V., Jayaraj, S. and Saxena, S.C. 1988. Effect of nonedible seed oils on survival of *Nephotettix virescens* (Homoptera: Cicadellidae) and on transmission of rice tungro virus. *J. Econ. Entomol.* **85** (5): 1369–1372.

Martin, H. 1964. Naturally occurring contact insecticides. *The Scientific Principle of Crop Protection. Edward Arnold (Publs.) Ltd. London* p 376.

Martin, H. 1973. Naturally occurring contact insecticides. *The Scientific Principle of Crop Protection. V. Edn. Edward Arnold (Publs.) Lt. London* p 379.

Mathew, A.S. and Chauhan, M.G. 1994. Antifeedant activity of selected medicinal plants with antihelmintic activity on tobacco caterpillar, *Spodoptera litura* Fabr. *Proc. II National Conf. of AZRA on 'Recent trends in plant, animal and human pest management: Impact on environment' at Madras Christian College, Madras, India,* 27–29 Dec., 1994 (Abstract): 49–50.

Mathews, D. 1981. The effectiveness of selected herbs and flowers in repelling garden insects. *Organic Garden and Farm Research Centre, Emmaus. Pennsylvania* p 176.

Mathur, A.C., Srivastava, J.B. and Chopra, I.C. 1961. A note on the toxicological investigation of *Pyrethrum marc. Pyrethrum Post* **6** (1): 11.

Mathur, Y.K., Shanker, K. and Ram, S. 1985. Evaluation of some grain protectants against *Callosobruchus chinensis* Linn. on blackgram. *Bull. Grain Tech.* **23** (3): 253–259.

Matsuda, K., Yamada, K., Kimura, M. and Hamada, M. 1991. Nematicidal activity of matrine and its derivatives against pine wood nematodes. *J. Agric. Food Chem.,* **39**: 189–191.

McGovern, E.R., Mayer, E.R. and Clerk, E.P. 1942. The toxicity of the natural bitter substances quassin, fernulin, helenalin and picrotoxin and some other derivatives to certain insects. *USDA Bur. Entomol. Pl. Quar. E.* 574, p 57.

McIndoo, N.E., Sievers, A.F. and Abott, W.S. 1919. Insects as insecticides. *USDA J. Agric. Res.* **17**: 177.

McIndoo, N.E. and Sievers, A.F. 1924. Plants tested for reporting to possess insecticidal properties. *USDA Publ. No.* 1201.

McIndoo, I. 1982. USDA. Bur. Entomol and Pl. Quar. Pers. Commn. in Michael et al., 1985. Plant species reportedly possessing pest control properties. *An EWC/UH DATA BASE, Univ. of Hawaii* p 249.

McLaughlin, J.L., Freedman, B., Powel, R.C. and Smith, Jr., C.R. 1980. Nerifolin and 2′acetylnerifolin insecticidal and cytotoxic agent of *Thevetia thevetoids* seeds. *J. Econ. Entomol.* **73**: 398–402.

McMillain, W.W., Bowmann, M.C., Burton, R.L., Starks, K.J. and Wiesmann, B.R. 1969. Extract of chinaberry leaf as a feeding deterrent and growth retardant for larvae of the corn ear worm and fall army worm. *J. Econ. Entomol.* **62** (3): 708–710.

Mehta, P.K. and Sandhu, G.S. 1992. Bitter gaurd leaf extract as feeding deterrent for red pumpkin beetle. *Indian J. Ent.* **54** (2): 227–230.

Meisner, J. and Mitchell, B.K. 1982. Phago-deterrent effect of neem extracts and azardirachtin on flea beetles, *Phyllotreta striolata* Fabr. *Zeitschrift fur pflanzenkarnktetien und Pflanzensachutz* **89** (8/9): 463–467.

Meisner, J. and Mitchell, B.K. 1984. Phagodeterrency induced by some secondary plant substances in adults of the flea beetle, *Phylotreta striolata. Zeitschrift fur pflanzenkarnktetien und Pflanzensachutz* **91** (3): 301–304.

Meisner, J., Wysoki, M. and Ascher, K.R.S. 1976. The residual effect of some products from neem (*Azardirachta indica* A. Juss.) seeds upon the larvae of *Boarmia (Ascotis) selenaria* Sehifg in laboratory trails. *Phytoparasitica* **4**: 185–187.

Meisner, J., Wissenberg, M., Palevitch, D. and Aharaonson, N. 1981. Phago-deterrency induced by leaves and leaf extracts of *Catharanthus roseus* in the larvae of *Spodoptera littoralis. J. Econ. Entomol.* **74** (2): 131–135.

Meisner, J., Ascher, K.R.S., Aly, R. and Warthen, Jr., J.D. 1981. Responses of *Spodoptera littoralis* (Boisd.) and *Earias insulana* (Boisd.) larvae to azardirachtin and salanin. *Phytoparasitica* **9**: 27–33.

Meisner, J., Ascher, K.R.S. and Zur, M. 1983. The residual effect of a neem seed kernel extract sprayed on fodder beet against larvae of *Spodoptera littoralis. Phytoparasitica* **11** (1): 51–54.

Meisner, J., Klein, M. and Keren, S. 1990. Effect of Margosan-O on the development of *Earies insulana. Phytoparasitica* **18** (4): 287–297.

Meisner, J., Klein, M. and Ben-moshe, E. 1992. Effect of Margosan-O on the development of the leaf hopper, *Asymmetrasca decedens. Phytoparasitica* **20** (1): 15–23.

Menasa, W. 1961. The fumigant toxicity of three pine resins to *Dendroctonus brevicornis* and *D. jeffreyi*. *J. Econ. Entomol.* **54** (2): 365–369.

Mesbach, H.A., El-Sherif, H.K. and El-Deeb, A.S. 1985. The synergistic action of senna glucosides combined with certain insecticides against the cotton leaf worm, *Spodoptera littoralis* Boisd. *Ann. Agric. Science* **23** (1): 373–379.

Metcalf, R.L., Metcalf, R.A. and Rhodes, A.M. 1980. Cucurbitacins as kairomones for diabroticite beetles. *Proc. Natl. Aca. Sci. (USA)* **77**: 3769–3772.

Mian, I.M. and Rodriguez, R. 1982. Survey of nematicidal properties of some organic materials available in Alabama as amendment to soil for control of *Meloidogyne arenaria*. *Nematropica* **12** (2): 235–246.

Michael, G., Ahmed, S., Mitchel, W.C. and Hylein, J.W. 1985. Plant species reportedly possessing pest control properties. *An EWC/UH DATA BASE, Univ. of Hawaii* p 249.

Midha, S.K. and Chabra, H.K. 1974. Effect of *Tagetes erecta* and *Sesamum indicum* on population of plant parasitic nematodes. *Indian J. Hort.* **31** (2): 196–197.

Miller, P.M. 1957. A method for quick separation of nematodes from soil samples. *Pl. Dis. Reptr.* **41**: 194.

Miller, P.M. 1979. Vegetable oils as protectants against nematode infestation. *J. Nematol.* **11** (4): 402–403.

Mirono, V.S. 1940. *Acorus calamus* used as an insecticidal and repellent preparation. *Med. Parasit. USSR* **9**: 409.

Mishra, P.K. and Singh, R.P. 1992. Antifeedant efficacy of neem, *Azardirachta indica* A. Juss. seed kernel, seed coat and fallen leaves extracts against desert locust, *Schistocerca gregaria* Forsk. *Indian J. Ent.* **54** (1): 89–96.

Mishra, R.C. and Jitender, K. 1983. Evaluation of *Mentha piperita* Linn. oil as a fumigant against the red flour beetle, *Tribolium castaneum* (Herbst.). *Indian Perfumer* **27**: 73–76.

Mishra, R.C., Masih, D.B. and Gupta, P.R. 1981. Mint oil as fumigant against *Callosobruchus chinensis* Linn. *Bull. Grain Tech.* **19** (1): 12–15.

Mishra, R.C., Masih, D.B. and Gupta, P.R. 1984. Effect of mixing mentha powder with and giving oil in water emulsion dip to the chickpea on pulse beetle *Callosobruchus chinensis* Linn. *Bull. Grain Tech.* **22** (1): 19–24.

Mishra, S.D. and Gaur, H.S. 1989. Control of nematodes infesting mung with nematicidal seed treatment and field application of aldicarb and neem cake. *Indian J. Ent.* **51** (4): 422–426.

Mishra, S.D. and Mojumder, V. 1995. Effect of the neem based pesticide viz., neemark and nimbecidine as seed treatments of cowpea against *Heterodera cajoni. Presented in national symposium on nematode problems of India — An appraisal of the nematode management with eco-friendly approaches and biocomponents. I.A.R.I., New Delhi,* 24+M26 March, 1995 (Abstract): 58.

Mishra, S.D. and Prasad, S.K. 1973. Effect of water extracts of oil seed cakes on the second stage larvae of *Meloidogyne incognita* at different concentrations and exposure times. *Indian J. Ent.* **35** (2): 101–106.

Mishra, S.D. and Mojumder, V. 1993. Nematicidal efficacy of decomposed neem products on *Meloidogyne incognita* infesting mungbean and their residual effects in chickpea. *World Neem Conf., Bangalore, India,* 24–28 Feb., 1993 (Abstract): 37.

Mitchell, E.V. and Heath, R.R. 1987. *Heliothis subflexa* Gn. (Lepidoptera: Noctuidae) L: demonstration of oviposition stimulant for ground cherry using novel bioassay. *J. Chem. Ecol.* **13** (8): 1849–1858.

Mittal, H.C. 1971. Protection of cowpeas from insect infestation with aids of fixed oils. *J. West African Sci. Assoc.* **16** (1): 45–48.

Miyakado, M., Nakayama, I., Yoshida, H. and Nakantani, N. 1979. The piperaceae amids. I. Structure of piperacide a new insecticidal amide from *Piper nigrum* Linn. *Agric. Biol. Chem.* **43** (7): 1609–1611.

Miyakado, M., Nakayame, I. and Nobuo, O. 1989. Insecticidal unsaturated isobutylamides from natural products to agro-chemical leads. *In Insecticides of Plant Origin. ACS Series 387 of III Chemical Congress of North America 195th National Meeting of Am. Chemical Society held at Toronto, Ontario, Canada,* 5–11 June, 1988: 173–187.

Mobin, M.P. and Khan, A.M. 1969. Effect of organic amendment on the population of rhyzosphere fungi and nematode around roots of some fruit trees. *All India Nematology Symposium, New Delhi,* 1969: 11–13.

Moghanoy, O. and Morallo-Rajesus, B. 1975. Insecticidal activity of extracts of *Derris Philippines Youth Res. Apprts. Act Rept. Soc. Adv. of Res. Univ. Philippines, Los Banos* p 6.

Mohan, K. and Gopalan, M. 1990. Studies on the effects of neem products and vegetable oils against major pests of rice and safety to natural enemies. *Presented in National Symposium on Problems and Prospects of Botanical Pesticides in Integrated Pest Management at C.T.R.I., Rajahmundry, India,* 22–23 Jan., 1990 (Abstract): 10–11.

Mohan, K., Gopalan, M. and Balasubramanian, G. 1991. Studies on the effects of neem products and monocrotophos against major pests of rice and their safety to natural enemies. *Indian J. Pl. Prot.* **19**: 23–30.

Mohandas, C., Rao, Y.S. and Sahu, S.C. 1982. Cultural control of root-knot nematodes, *Hirschmanniella* spp. with *Sphenochelea zeylanica. Proc. Indian Acad. Sc.* **90**: 373–376.

Mohanty, K.C. and Das, S.N. 1988a. In vitro toxicity to *Erythrina indica* extracts on *Meloidogyne incognita. Indian J. Pl. Prot.* **15**: 172–173.

Mohanty, K.C. and Das, S.N. 1988b. Nematicidal properties of *Erythrina indica* against *M. incognita* and *T. mashoodi. Indian J. Nemetol.* **18** (1): 138.

Mohanty, K.K., Chakrobarty, D.P. and Roy, S. 1988. Antifeedant activity of oil fraction of seed of some leguminous plants against *Diacrisia obliqua. Indian. J. Agric. Sci.* **58** (7): 579–580.

Mohiuddin, S., Qureshi, R.A., Khan, A., Nasir, M.K.A., Khatri, L.M. and Qureshi, S.A. 1987. Laboratory investigation of the repellency of some plant oils to red flour beetle, *Tribolium castaneum* Hrbst. *Pakistan J. Scientific Indust. Res.* **30** (10): 754–756.

Mojumder, V. and Mishra, S.D. 1995. Use of neem products as seed treatments for the management of root-knot, reniform and other plant parasitic nematodes. *Presented in national symposium on nematode problems of India — An appraisal of the nematode management with eco-friendly approaches and bio-components. I.A.R.J., New Delhi,* 24–26 March, 1995 (Abstract): 57.

Monache, F.D., Mairni, B.G.B. and Bernays, E.A. 1984. Isolation of insect antifeedant alkaloids from *Maytenus rigida* (Celastraceae). *Zeitschrift fur Angawandte Entomologie* **97** (4): 406–414.

Mookherjee, P.B., Jotwani, M.G., Yadav, T.D. and Sircar, P. 1970. Studies on incidences and extent of damage due to insect pests of stored seeds. II. Leguminous and vegetable seeds. *Indian J. Ent.* **32** (4): 350–355.

Moraes, M.V. De, 1977. A preliminary test to determine the nematicidal power of meal cakes. In *Prabalhos apresentodos a II reuniae de nematologia, Piricacaba, Brazil,* 14–16 Sept. 1976.

Morallo-Rejessus, B. 1982. Dept. of Entomology Univ. of Philippines Pers. Commn. in Michael et al., 1985. Plant species reportedly possessing pest control properties. *An EWC/UH DATA BASE, Univ. of Hawaii* p 249.

Morallo-Rajessus, B. 1984. Status and prospects of botanical pesticides in Philippines. *II SEARCA Conf. Chair Lecture 29 Aug. 1984 Univ. of Philipp. at Los Banos, College of Laguna, Philippines* p 19.

Morallo-Rajessus, B. and Decena, A. 1982. Isolation, purification and identification of the insecticidal principles from *Tagetes* spp. *Philipp. J. Crop Sci.* **7**: 31–36.

Morallo-Rajessus, B. and Eroles, L.C. 1978. Two insecticidal principles from marigold (*Tagetes*) spp. roots. *Philipp. Entomologist* **4** (1/2): 87–97.

Morallo-Rajessus, B. and Silva, D. 1979. Insecticidal activity of selected plants with emphasis on marigold (*Tagetes* spp.) and makabuhai (*Tinospara rumphii*). *NRCP Ann. Report for 1976, Univ. of Philipp. Los Banod, Mimeo, Philippines* p 25.

Mordrie, Luntz, A.J., Cottee, P.K. and Evans, K.A. 1985. Azardirachtin: Its effect on gut motility, growth and moulting in the locusts. *Physiol. Entomol.* **10** (4): 431–437.

Mordub, Luntz, A.J., Evans, K.A. and Charlett, M. 1986. Azardirachtin, ecdysteroids and ecdysis in *Locusta migratoria. Comp. Biochem. Physiol.* **85C**: 297–300.

Morgan, E.D. and Thornton, M.D. 1973. Azardirachtin in the fruits of *Melia azardirach. Phytochem.* **12**: 391–392.

Mostafa, T.S. 1991. The efficacy of neem flower and fruit powders against *Trogoderma granarium* Everts. adults infesting stored rice. *Bull. Entomol. Soc. Egypt Econ.* Series (1988–1989 Publ. 1991) **17**: 93–99.

Muckenstrum, B., Duplay, D., Mohammadi, E., Moradia, Robert, P.C., Simonis, M.T. and Kienlein, J.C. 1981. The role of natural phenyl propanoids as antifeedants for insects. *Symp. Intn. Versailles* 16–20 Nov., 1981 (recorded 1984): 131–135.

Muda, A.R. 1984. Utilisation of neem seed as a pest control agent for stored paddy. *Proc. VII. ASEAN Tech. Sem. on Grain Post Harvest Technology. ASEAN Crops Post Harvest Program and ASEAN Food Handling Bureau, Manila, Philippines*: 117–128.

Mukherjee, S.K. 1983. ICAR, New Delhi, Pers. Commn. in Michael et al., 1985. Plant species reportedly possessing pest control properties. *An EWC/UH DATA BASE, Univ. of Hawaii* p 249.

Mukherjee, S.K., Saxena, V.S. and Srivastava, K.P. 1984. Linseed oil as binding agent for pesticide granules carbofuran. *Pesticides* **18** (5): 18–19.

Mukherjee, S.N. and Sukul, N.S. 1978. Nematicidal action of three species on wild herbs. *J. Res. (India)* **2** (2): 12.

Mukherjee, S.N. and Sharma, R.N. 1990. Effect of garlic on oviposition, growth and development of *Spodoptera litura* Fabr. (Noctuidae: Lepidoptera). *Presented in National Symp. on Problems and Prospects of Botanical Pesticides in Integrated Pest Management, at C.T.R.I., Rajahmundry, India,* 21–22 Jan., 1990 (Abstract): 22.

Mukherjee, T.D. and Govinda, R. 1959. A plant insecticide *Acorus calamus* Linn. *Indian J. Ent.* **21** (3): 194–195.

Mukhopadhyaya, M.C. and Chattopadhayaya, P.R. 1979. Control of root-knot nematode, *Meloidogyne incognita* (Kofoid. & White) Chitwood with *Typhonium trilobatum* an annual herb. *Indian J. Nematol.* **9** (1): 53–59.

Mummigatti, S.G. and Raghunathan, A.N. 1977. Inhibition of multiplication of *Callosobruchus chinensis* by vegetable oils. *J. Food Sci. Tech.* **14** (4): 184–185.

Munakata, K. 1978. Nematicidal substances from plants. *Advances in Pesticide Science Part-2 Pergamon Press*: 295–302.

Murthy, K.S.R.K. 1975. Biology and control of whorl maggot, *Hyderellia* sp. on paddy. *Rice Res. Sta., Bapatla (A.P.) Pers. Comm. in Ketkar, C.M. 1976. Report of Directorate of Non-edible Oils and Soap Industries, K.V.I.C., Bombay, India*: 117–176.

Murthy, P.S.N., Ram Prasad, G., Sitaramaiah, S., Kamesware Rao, B.V., Prabhu, S.R. and Rao, N. 1989. Studies on the enrichment of neem fractions and their relative efficacy against tobacco caterpillar, *Spodoptera litura* Fabr. on Lanka and FCV tobaccos. *Tob. Res.* **15** (1): 65–70.

Murthy, P.S.N., Sitaramaiah, S., Ram Prasad, G., Nagesware Rao, S., Kameswara Rao, B.V. and Prabhu, S.R. 1990. Neem extracts for the control of *Spodoptera litura* Fabr. *Presented in National Symposium on Problems and Prospects of Botanical Pesticides in Integrated Pest Management at C.T.R.I., Rajahmundry, India* 21–22 Jan., 1990 (Abstract): 22.

Murthy, P.S.N., Ram Prasad, G., Sitaramaiah, S., Narsimharao, C.V., Kameswararao, B.V., Prabhu, S.R. and Nageswararao, S. 1993. Bio-efficacy of neem fractions against *Spodoptera litura* on tobacco. *World Neem Conf., Bangalore, India,* 24–28 Feb., 1993 (Abstract) p 8.

Mwangi, R.W. 1982. Locust antifeedant activity in fruits of *Melia volkonsii. Entomologia Expt. et. Appl.* **32** (3): 277–280.

Nadkarni, A.K. 1954. Indian Materia Medica Vol. 1. *Dhootapapeshwar Prakkar Ltd., Panvel, Bombay-7, India.*

Nagai, K. 1902. Research on poisonous principles of rohton I. *J. Tokyo Chem. Soc.* **23**: 744–777.

Nagalingam, B. and Savithri, P. 1980. Studies on the chemical control of citrus leaf miner *Phyllocnistic citrella* Staint. (Gracillaridae: Lepidoptera). *Pestology* **4** (7): 21–24.

Nagarajah, T.M. 1983. MARGA Institute, Colombo, Sri Lanka, Pers. Commn. in Michael et al., 1985. Plant species reportedly possessing pest control properties. *An EWC/UH DATA BASE, Univ. of Hawaii* p 249.

Nagasampagi, B.A., Kulkarni, M.M., Rotjatkar, S.R. and Ayyangar, N.R. 1990. Utilisation of neem in pest management. *Presented in National Symposium on Problems and Prospects of Botanical Pesticides in Integrated Pest Management at C.T.R.I., Rajahmundry, India,* 21–22 Jan., 1990 (Abstract): 23.

Naik, R.L. and Dumbre, R.B. 1984. Effect of some vegetable oils used in protecting stored cowpea on biology of pulse beetle, *Callosobruchus maculatus* Fabr. (Coleoptera: Bruchidae), *Bull. Grain Tech.* **22** (1): 25–32.

Naik, R.L. and Dumbre, R.B. 1984. Effect of some vegetable oils used in protecting against *Callosobruchus maculatus* on storability and qualities of cowpea. *Bull. Grain Tech.* **23** (1): 33–39.

Nair, M.R.G.K. 1976. Effect of neem seed kernel powder as protectant of stored grains. Pers. Commn. in Ketkar, C.M. (1969–1976). Utilisation of neem and its byproducts. *Report of the Modified Neem Cake Manurial Project. Directorate of Non-Edible Oils and Soap Industries, K.V.I.C., Bombay, India*: 116–173.

Nakajima, S. and Kawazu, K. 1977. Tridec-1-ene-3, 5,7,110-Pentayne, an ovicidal substance from *Xanthium candense. Agric. Biol.* **41** (9): 1801–1802.

Nakanishi, K. 1975. Structure of insect antifeedant azardirachtin. *Recent Adv. Phytochem.* **9**: 283–298.

Nalinasoundari, M.S., Chokalingam, S. and Durairaj, G. 1994. Juveno-mimetic activity of juvenile hormone analogues and plant extracts. *Proc. II National Conf. of AZRA on 'Recent trends in plant, animal and human pest management: Impact on environment' at Madras Christian College, Madras,* 27–29 Dec., 1994 (Abstract): 52.

Nanda, N.K., Parija, B., Nanda, B., Dash, D.D. and Pradhan, N.C. 1993. Bio-efficacy of neem derivatives against the paddy insect pest complex. *World Neem Conf., Bangalore, India,* 24–28 Feb., 1993 (Abstract): 14.

Nandal, S.N. and Bhatti, D.S. 1983. Preliminary screenings of some weeds and shrubs for their nematicidal activity against *Meloidogyne javanica. Indian J. Nematol.* **13**: 123–127.

Narasimhan, V. and Mariappan, V. 1988. Effect of plant derivatives on green leaf hopper (GLH) and rice tungo (RTV) transmissions. *Intn. Rice Res. Newsl.* **13** (1): 28–29.

Narayan, P.B.S., Sampathraj, R. and Vanithakumari, G. 1993. Rat toxicity studies with neem oil. *World Neem Conf., Bangalore, India,* 24–28 Feb., 1993 (Abstract): 55.

Narayanan, C.R., Singh, R.P. and Sawikar, D.D. 1980. Phago-deterrency of various fractions of neem oil against *Schistocerca gregaria* Fosk. *Indian J. Ent.* **42** (3): 469–472.

Nasseh, M. 1981. The effect of crude extracts of *Allium sativum* Linn. on feeding activity and metamorphosis of *Epilachna varivestis* Muls. (Coleoptora: Coccinellidae). *Zeitshrift fur Angewandte Entomologie* **92** (5): 464–471.

Nath, A. Sharma, N.K., Bhardwaj, S. and Thappa, C.N. 1982. Nematicidal property of garlic. *Nematologia* **28** (2): 253–255.

Natrajan, K. and Sundarmurthy, V.T. 1990. Effect of neem oil on cotton whitefly, *Bemisia tabaci. Indian J. Agric. Sci.* **60** (4): 290–291.

Natrajan, K. and Shanthi, P. 1994. Evaluation of neem products against *Helicoverpa armigera. Proc. II National Conf. of AZRA on 'Recent trends in plant, animal and human pest management: Impact on environment' at Madras Christian College, Madras, India,* 27–29 Dec., 1994 (Abstract): 46.

Nawrot, J., Smitalova, Z. and Holub, M. 1983. Deterrent activity of sesquiterpene lactones from the umbelliferae against storage pests. *Biochemical Systematics & Ecol.* **11** (3): 243–245.

Negherbon, W.O. 1959. *Handbook of Toxicology Vol. 3. Saunders Philadelphia* p 661.

Nelson, S.J., Sundarababu, P.C., Rajavel, D.S., Srimmannarayana, G. and Geetanjali, Y. 1993. Antifeedant and growth inhibiting effects of azardirachtin-rich neem fractions on *Sogatella furcifera* Horvath, *Spodoptera litura* Fabr. and *Helicoverpa armigera* Hubn. *World Neem Conf., Bangalore, India,* 24–28 Feb., 1993 (Abstract): 10.

Nigam, P.M. 1977. Karanja, *Pongamia glabra*: an effective cake against white grub, *Holotrichia consanguinea* Blanch. *Entomologists Newsl.* **7** (5): 20.

Nigam, P.M., Singh, S.M. and Sen, R. 1990. Effect of certain plant extracts against mustard aphid, *Lipaphis erysimi* Kalt. *Presented in National Symp. on Problems and Prospects of Botanical Pesticides in Integrated Pest Management at C.T.R.I., Rajahmundry, India,* 21–22 Jan., 1990 (Abstract): 35–36.

Nishida, R., Bowers, W.S. and Evans, P.H. 1983. Juvadence discovery of a juvenile hormone mimic in the plant *Macropiper excelsum*. *Advances of Insect Bio. Chem. Physiol.* **1** (1): 17–24.

Noor, A. and Khuswaha, K.S. 1976. Epidemic of bark eating caterpillar *Iderabela quadrinotata Walk.* and its control. *Proc. All India Symp. on Modern Concepts in Pl. Prot. at Udiapur*: 29–30.

Ntezurubanza, L. 1987. Analysis of the essential oils of some Lamiaceae used in traditional medicine in Rwanda. *Ph.D. Thesis State University of Leiden, Belgium* p 101.

Numata, A., Hokimoto, K., Takemura, T. and Fukui, M. 1983. Feeding inhibition for the larvae of yellow butterfly, *Eurena hecabe mandarina* I' Orza (Lepidoptera: Pieridae) in a flowing fern, *Osmunda japonica* Thunb. *Appl. Entomol. Zool.* **18** (1): 129–131.

Oca, G., Montes, D., Garcia, F. and Schoonhoven, A.V. 1978. Effect of four vegetable oils on *Sitophilus oryzae* and *Sitotroga cerealella* in stored maize, sorghum and wheat. *Revista colombiana de entomologia* **4** (1&2): 45–49.

Ochou, G., Hesler, L.S. and Plapp, F.W. 1986. Plant and mineral oils: effects as insecticide additives and direct toxicity to tobacco budworm larvae and house fly adults. *South-Western Entomologist* **11**: 63–68.

O'Donnell, M.J., Chambers, J.J. and McFarland, S.M. 1983. Attractancy to *Oryzaephilus surinamensis* Linn. saw toothed grain beetle to extracts of carobs, some triglycerides and related components. *J. Chem. Econ.* **I** (3): 357–379.

Ofuya, I.T. 1986. Use of wood ash, dry chilli, pepper fruits and onion scale leaves for reducing *Callosobruchus maculatas* Fabr. damage in cowpea seeds during storage. *J. Agric. Sci.* (U.K.) **107** (2): 467–468.

Oji, O. 1991. Use of *Piper guineese* in the protection of stored *zea mays* against the maize weevil. *Fitoterapia* **62** (2): 179–182.

Okabe, H., Miyahara, Y., Muchi, T.V., Miyahara, K. and Kawasaki, T. 1980. The constituents of *Momordia charentia* I. Isolation and characterisation of momordicoside-A and momordicoside-B, glucosides of pentahydroxyencurbane triterpine. *Chem. Pharma. Bull.* **28**: 2759.

Olaifa, J.I., Erhun, W.O. and Akingbohungbe, A.E. 1987. Insecticidal activity of some Nigerian plants. *Insect Science and Its Application* **8** (2): 221–224.

Olaifa, J.I. and Erhun, W.O. 1988. Laboratory evaluation of *Piper guineese* for protection of cowpea against *Callosobruchus maculatus*. *Insect Science and Its Application* **I** (1): 55–59.

Omar, O. 1983. Department of Plant Protection, Unit of Pertanian Malaysia, Pers. Commn. in Michael et al., 1985. Plant species reportedly possessing pest control properties. *An. EWC/UH DATA BASE, Univ. of Hawaii* p 249.

Omollo, J.F. 1983. Some comparative effects of plant extracts from Meliaceae family on the feeding activity, development and reproduction in *Locusta, Stomoxys, Glossina* and *Aedes*. *Neem Newsl.* (1987) **4** (2): 23.

Ormerod, E.A. 1884. Reports on observations on injurious insects. *USDA Bull. No. 8*: 43.

Oroumchi, S. and Sharifi-yazdi, M.H. 1993. Investigations on neem and china-berry in Iran. *World Neem Conf., Bangalore, India,* 24–28 Feb., 1993 (Abstract): 13.

Osmani, Z.H. and Naidu, M.B. 1956. *Pongamia glabra* (Karanjai) as insecticide. *Sci. and Culture* **22**: 235.

Osmani, Z.H., Anees, I. and Singhamony, S. 1987. Potentiation of action of juvenile hormone analogue by pepper seed extracts and mimic action of the same. *Intn. Pest. Control* **29** (6): 128–129.

Padolina, W.G. 1983. Department of Chemistry, Univ. of Philippines, Los Banos, Pers. Commn. in Michael et al., 1985. Plant species reportedly possessing pest control properties; *An EWC/UH DATA BASE, Univ. of Hawaii* p 249.

Pajini, H.R. 1964. Some observations on the insecticidal activity of the fruit of dharek, *Melia azardirach. Res. Bull. Punjab Univ. Sci.* **25**: 345–364.

Pande, Y.D. and Yadav, S.R.S. 1983. Acaricidal test of seven pesticides against *Tetranychus macfarlanei* B & P. (Acarina: Tetranychidae). *Pestology* **7**: 17–20.

Pandey, G.C. 1988. Preliminary studies on the toxicities of some plants extracts on the sugarcane top shoot borer, *Tryporhyza nivella. In Symp. on Growth, Development and Natural Resource Conservation, D.A.V. College, Muzaffarnagar (U.P.) India,* 8–10 Oct., 1988 (Abstract): 38.

Pandey, G.P. and Singh, K. 1976. Use of some plant powders, oils and extracts as protectants against pulse beetle. *Indian J. Ent.* **38** (2): 110–113.

Pandey, G.P. and Varma, B.K. 1977. *Annona reticulata* (custard apple) seed powder as protectant of mung against pulse beetle, *Callosobruchus maculatus* Fabr. *Bull. Grain Tech.* **15** (2): 100–104.

Pandey, G.P., Doharey, R.B. and Varma, B.K. 1981. Efficacy of some vegetables oils for protecting green gram against attack of *Callosobruchus maculatus* Fabr. *Indian J. Agric. Sci.* **51** (2): 910–912.

Pandey, K.C. and Farooqui, S.A. 1990. Preferential response of lucerne weevil to crude extracts of some *Medicago* sp. *Indian J. Ent.* **52** (4): 610–612.

Pandey, N.D., Krishna Pal, Pandey, S., Tripathy, R.A. and Singh, Y.P. 1985a. Use of neem, *Azardirachta indica* A. Juss. as seed protectant against moth, *Corcyra cephalonica* Staint. I. effect on the development and damage. *Bull. Grain Tech.* **23** (2): 147–153.

Pandey, N.D., Krishna Pal, Pandey, S., Tripathy, R.A. and Singh, Y.P. 1985b. Use of neem, *Azardirachta indica* A. Juss. as seed protectant against rice moth, *Corcyra cephalonica* Staint. II. effect on fecundity, fertility and longevity of adults. *Bull. Grain Tech.* **23** (3): 265–268.

Pandey, N.D., Mathur, K.K., Pandey, S. and Tripathy, R.A. 1986. Effect of some plant extracts against pulse beetle, *Callosobruchus chinensis* Linn. *Indian J. Ent.* **48** (1): 85–90.

Pandey, N.D., Singh, M. and Tewari, G.C. 1977. Antifeeding repellent and insecticidal properties of some indigenous plants materials against mustard saw fly, *Athalia proxima* Klung. *Indian J. Ent.* **39** (1): 60–63.

Pandey, N.D., Sudhakar, T.R., Tewari, G.C. and Pandey, U.K. 1979. Evaluation of some botanical antifeedants under field conditions for the control of *Athalia proxima* Klug. *Indian J. Ent.* **41** (2): 107–109.

Pandey, P.N. 1976. Effect of *Dalbergia sisso* Roxb. on development, growth and reproduction of *Utetheisa pulchella* Linn. *Zeitschrift fur Angewandte Zoologie* **63** (4): 445–449.

Pandey, P.N. 1978a. Effect of nut grass on development and growth of *Utetheisa pulchella* Linn. (Lepidoptera: Arctiidae). *Zeitschrift fur Angewandte Zoologie* **65** (4): 441–444.

Pandey, P.N. 1978b. Survival of *Utetheisa pulchella* Linn. (Lepidoptera: Arctiidae) against fractions of *Dalbergia root. Zeitschrift fur Angewandte Zoologie.* **65** (4): 445–447.

Pandey, P.N., Ansari, M.N. and Dubey, A. 1977. Influence of *Dulbergia root* on development and growth of *Utetheisa pulchella* Linn. *Zeitschrift fur Angewandte Zoologie* **64** (4): 453–457.

Pandey, R. 1995. Performance of oil seed cake, pesticides and dry leaf matter on the reproduction potential of *Meloidogyne incognita* and yield of Japanese mint HY-77. *Presented in national symposium on nematode problems in India — An appraisal of the nematode management with eco-friendly approaches and bio-components. I.A.R.I., New Delhi,* 24–26 March, 1995 (Abstract): 37.

Pandey, R. and Haseeb, A. 1988. Studies on the toxicity of certain medicinal plants to root-knot nematode, *Meloidogyne incognita. Indian J. Pl. Path.* **6** (2): 184–186.

Pandey, U.K., Srivastava, A.K., Chandel, B.S. and Lekha, C. 1982. Response of some plant origins insecticides against potato tuber moth, *Gnorimoschema operculella* Zell. (Lepidoptera: Gelechidae) infesting solanaceous plants. *Zeitschrift fur Angewandte Zoologie* **69** (3): 267–270.

Pandey, U.K., Kumar, A. and Chauhan, S.P.S. 1984. Effect of some plant extracts against sugarcane leaf hopper, *Pyrilla perpusilla* Walk. (Hemiptera: Lophopidae). *Indian. J. Ent.* **46** (4): 487–489.

Parmar, B.S. 1977. Kharanja, *Pongamia glabra* oil as a synergist for pyrethrin. *Pyrethrum Post* **14** (1): 22–23.

Parmar, B.S. and Devakumar, C. (Eds.). 1993. Botanical and bio-pesticides. *Westvill Publishing House, New Delhi* p 199.

Parmar, B.S. and Dutta, S. 1982. Evaluation of some neem oils as malathion synergists *J. Trop. Agric.* **5**: 223–226.

Parmar, B.S. and Gulati, K.C. 1969. Synergists for Pyrethrins-II: Karanjin. *Indian J. Ent.* **31** (3): 239.

Parr, J.C. and Thurston, R. 1972. Toxicity of nicotine in synthetic diets to larvae of the tobacco horn worm. *Ann. Entomol. Soc. Am.* **65**: 1185–1188.

Parr, J.C. and Thurston, R. 1968. Toxicity of *Nicotania* and *Petula* spp. to larvae of the tobacco horn worm. *J. Econ. Entomol.* **16**: 1525–1531.

Patel, C.B., Jhala, R.C., Patel, M.M. and Shah, A.H. 1990. Field bio-efficacy of neemark in comparison to chemical insecticides against sugarcane white fly, *Aleurolobus barodensis* Markel (Aleurodidae: Hompotera). *Presented at National Symposion on Problems and Prospects of Botanical Pesticides in Integrated Pest Management at C.T.R.I., Rajahmundry, India,* 21–22 Jan., 1990 (Abstract): 26.

Patel, C.B., Rai, A.B., Patel, M.B., Patel, A.J. and Shah, A.A. 1993. Acaricidal tests of botanical pesticides in comparison to conventional acaricides/pesticides against red spider mites (Acarina: Tetromychidae) on okra and Indian bean, *Indian J. Ent.* **55** (2): 184–190.

Patel, G.J., Shah, H.M., Patel, D.J. and Patel, S.K. 1988. Use of castor in management of root-knot nematodes. *Tob. Res.* **11** (1): 78–80.

Patel, G.J., Chari, M.S. and Patel, N.G. 1975. A note on toxicity of *Nicotiana gossei* Domin to the larvae *M. incognita. Tob. Res.* **44**: 784–785.

Patel, G.J., Chari, M.S. and Jaisani, B.G. 1976. Differential response of white fly, *Bemisia tabaci* Genn. to *Nicotiana sp. Gujrat Agric. Univ. Res. J.* **1** (2): 89–92.

Patel, H.K., Patel, V.C., Chari, M.S., Patel, J.J. and Patel, J.R. 1968. Neem seed pest suspension. A sure deterrent to hairy caterpiller, *Amsacta moorei* Butl. *The Madras Agric. J.* **55** (11): 509–510.

Patel, H.R. 1995a. Inhibitory effect of *Clerodendron enermi* on root-knot of okra. *The Madras Agric. J.* **72**: 470–472.

Patel, H.R. 1985b. Azolla and mustard cake against nematicides for root-knot nematode management in okra. *The Madras Agric. J.* **72**: 593–594.

Patel, H.R., Thaker, N.A. and Patel, G.C. 1987. Larval emergence and infestation of *M. incognita* as influenced by periwinkle, *Catharanthus roseus*. *Indian J. Agric. Sci.* **57**: 863–866.

Patel, H.R., Patel, D.J., Patel, C.C. and Thaker, N.A. 1990a. Effectivity of *Clerodendron inerme* Linn. and periwinkle, *Catharanthus roseus* (Linn.) G. Don. for management of root-knot nematodes on okra. *Presented in National Symposium on Problems and Prospects in Botanical Pesticides in Integrated Pest Management at C.T.R.I., Rajahmundry,* 21–22 Jan., 1990 (Abstract): 40–41.

Patel, H.R., Makwuna, M.G. and Patel, B.N. 1990b. Effect of periwinkle, *Catharanthus roseus* (Linn.) G. Don. on reiniform nematode. *Presented in National Symposium on Problems and Prospects of Botanical Pesticides in Integrated Pest Management at C.T.R.I., Rajahmundry, India,* 21–22 Jan., 1990 (Abstract): 42.

Patel, J.R., Patel, J.I., Mehata, D.M. and Shah, B.R. 1990a. Integrated management of *Amsacta moorei* Butl. with botanical insecticides. *Presented in National Symposium on Problems and Prospects of Botanical Pesticides in Integrated Pest Management at C.T.R.I., Rajahmundry, India,* 21–22 Jan., 1990 (Abstract): 23–24.

Patel, J.R., Jyani, D.B., Patel, C.C. and Borad, P.K. 1990b. Insecticidal property of some plants and their volatile formulations against *Amsacta moorei* Butl. *Presented in National Symposium on Problems and Prospects of Botanical Pesticides in Integrated Pest Management at C.T.R.I., Rajahmundry, India,* 21–22 Jan., 1990 (Abstract): 24–25.

Patel, M.B., Pastagia, J.J., Patel, H.M. and Patel, C.P. 1991. Testing insecticidal effect of *Ipomea carnea* comparison to neem products and dimethoate using bean aphids as test insects. *IV National Symp. on Growth, Development and Control Technology of Insect Pests, at S.D. College, Muzaffarnagar, India,* 2–4 Oct., 1991 (Abstract): 36–37.

Patel, S.K. 1976. Gujrat Agric. Univ., Junagarh, India, Pers. Commn. in Ketkar, C.M. 1969–1976. Utilisation of neem and its byproducts. *Modified report on neem cake manurial project. Directorate of Non-edible Oils and Soap Industries, K.V.I.C., Bombay, India*: 116–173.

Pathak, P.H. and Krishna, S.S. 1985. Neem seed oil: a capable ingredient to check reproduction of rice moth, *Corcyra cephalonica* Staint. (Galliridae: Lepidoptera). *Zeitschrift fur Angewandte Engomologie* **100** (1): 33–35.

Pathak, P.H. and Krishna, S.S. 1986. Reproductive efficacy of *Earias fabia* Stroll. (Lepidoptera: Noctuidae) affected by neem oil vapour. *Appl. Ent. Zool.* **21**: 347–348.

Pathak, P.H. and Krishna, S.S. 1987. The influence of the odours of certain botanical components or an organic solvent dichloro-methane during breeding on the reproductive efficacy of *Earias vitella* Fabr. (Lepidoptera: Noctiudae). *Uttar Pradesh J. Zool.* **7** (2): 175–179.

Patil, B.V., Nandiphalli, B.S., Parameswar, H., Thimmanagonda, B.R. and Lingappa, S. 1990. Role of plant and animal products in the control of cotton whitefly, *Bemisia tabaci* Genn. *Presented in National Symposium on Problems and Prospects of Botanical Pesticides in Integrated Pest Management at C.T.R.I., Rajahmundry, India,* 21–22 Jan., 1990 (Abstract): 14.

Patnaik, N.C., Panda, N., Patro, E.R. and Mishra, B.K. 1987a. Effect of neem, (*Azardirachta indica* A. Juss.) oil on mustard web worm, *Crocidolomia binotalis* Zell. (Lepidoptera: Pyralidae). *Neem Newsl.* **4** (2): 15–17.

Patnaik, N.C., Panda, N., Bhuyan, K. and Mishra, S.K. 1987b. Development, aberration and mortality of the mustard sawfly larvae, *Athalia lugens proxima* Klung. by neem oil. *Neem Newsl.* **4** (2): 18–19.

Paul, C.F., Agrawal, P.N. and Ausat, A. 1965. Toxicity of solvent extract of *Acorus calamus* Linn. to some grain pests and termites. *Indian J. Ent.* **27**: 114–117.

Pereira, J. 1983. The effectiveness of six vegetable oils as protectants of cowpeas and bombara groundnuts against infestation by *Callosobruchus maculatus* Fabr. *J. Stored. Prod. Res.* **19** (2): 57–62.

Petterson, J. 1976. Ethology of *Dasyneura brassicae* Linn. (Diptera: Cecidomyidae) I. Laboratory studies on olfactory section to the host plant. *Symp. Biol. Hung.* **16**: 202–208.

Pierie, R. 1981. Natural products repel the cucumber beetle. *Agric. Res.* (USA) **30** (2): 12.

Pillai, M.A.K. and Ponnia, S. 1988. Neem for control of rice thrips. *Intn. Rice Res. Newsl.* **13** (5): 33–34.

Pillai, S.N. and Desai, M.V. 1975. Antihelminthic property of marotti cake, *Hydnocarpus laurifolia. Pesticides* **9** (4): 37–39.

Pillai, S.N., Desai, M. and Saha, H.M. 1975. Nematicidal properties of tumeric. *Indian Phytopath.* **28** (1): 128–129.

Pillai, S.N. and Desai, M.V. 1976. Punnakai cake in the control of root-knot nematode. *Indian J. Mycol. Pl. Path.* **6** (1): 14–17.

Platon, N.P., Nuevo, C.R. and Birosel, D. 1970. New pesticide from the coconut. *Agric. Indust. Life.* **32**: 18–19.

Poe, S.L. 1980. Dept. of Entomology, VPI Blacksburg, VA, Pers. Commn. in Michael et al., 1985. Plant species reportedly possessing pest control properties. *An EWC/UH DATA BASE, Univ. of Hawaii,* p 249.

Poe, S.L. 1982. Dept. of Entomology, VPI Blacksburg, VA, Pers. Commn. in Michael et al., 1985. Plant species reportedly possessing pest control properties. *An EWC/UH DATA BASE, Univ. of Hawaii,* p 249.

Ponce, De Leon E.L. 1983. Further investigation of the insecticidal activity of black pepper, *Piper nigrum* Linn. and red pepper, *Capsicum anum* on major storage pests of corn and legumes. *M.S. Thesis, Univ. Philipp. Los Banos, Laguna* p 44.

Poornima, K. and Vodivelu, S. 1993. Comparative efficacy of nematicide, oil cakes and plant extracts in the management of *Meloidogyne incognita, Pratylenelus delattrei* and *Rotylenchulus reniformis* on brinjal. *Indian J. Nematol.* **23** (2): 170–173.

Potter, C. 1935. An account of constitution and use of an atomised white oil pyrethrum fluid to control *Plodia interpunctella* Hubn. and *Ephestia elutella* Hubn. in warehouses. *Ann. Appl. Biol.* **4**: 769–805.

Prabhakar, M. and Rao, P.K. 1993. Relative toxicity of botanical pesticides to *Spodoptera litura* Fabr. *J. Appl. Zool. Res.* **4** (2): 145–146.

Prabhu, S.R., Chkraborty, M.K., Sitaramaiah, S. and Joshi, B.G. 1981. Isolation and identification of toxic material from *Nicotiana gossei*. *Tob. Sci.* **25**: 114–117.

Prabhu, S.R., Chari, M.S., Ramakrishnaya, B.V., Rao, R.S.N., Kumar, D.G., Rao, K.B.V. and Murthy, P.S.N. 1990. Production and properties of nicotine sulphate in pest management. *Presented in National Symposium on Problems and Prospects of Botanical Pesticides in Integrated Pest Management at C.T.R.I., Rajahmundry, India,* 21–22 Jan., 1990 (Abstract): 23.

Pradhan, A., Das, S.N. and Dora, D.K. 1989. Effect of some organic oils on the infestivity of root-knot nematode (*Meloidogyne incognita*) affecting tomato. *Indian J. Nematol.* **19** (2): 162–165.

Pradhan, S., Jotwani, M.G. and Rai, B.K. 1963. The repellent properties of some neem products. *Bull. Reg. Lab.* **1** (2): 149–151.

Pradhan, S. and Jotwani, M.G. 1968a. Neem seed as insect deterrent. *Chemical Age. of India* **19** (9): 756–760.

Pradhan, S. and Jotwani, M.G. 1968b. Reported confirmations of our discovery of antifeedant property of neem kernel. *Entomologist Newsl.*: 75–77.

Prakash, A. 1982. Studies on insect pests of pulses and oil seeds and their management. *Ann. Tech. Report of All India Scheme of Harvest and Post-harvest Technology, Central Rice Res. Instt., Cuttack, India,* 1982 p 46.

Prakash, A. 1985. Isolation, identification and evaluation of the active components from effective plant material protection of stored paddy from insect pests. *Ph.D. Thesis by Utkal Univ., Bhubaneswar, India,* 1985 p 143.

Prakash, A. and Mathur, K.C. 1985. Active principles from plant products used in insect pest management of stored rice. *Bull. Grain Tech.* **23** (1): 77–82.

Prakash, A. and Mathur, K.C. 1986. Isolation, identification and evaluation of the active components from *Vitex negundo* leaves against storage insects. *Proc. VI Intn. Congr. Pesticide Chemistry,* 10–15 *August, 1986 at Ottawa, Canada* p 35.

Prakash, A. and Pasalu, I.C. 1979. Studies on insect pests of stored rice and methods of minimising losses caused by them. *An. Tech. Report C.R.R.I., Cuttack,* 1979: p 123.

Prakash, A., Pasalu, I.C. and Mathur, K.C. 1979a. Ovicidal activity of *Eclipta alba* Hassk. (Compositae). *Curr. Sci.* **48**: 1090.

Prakash, A., Pasalu, I.C. and Nag, D. 1979b. Improvement of storage structures and practices. *Proc. VII Ann. Works, All India Coordinated Scheme on Harvest and Post Harvest Technol. held at C.P.C.R.I., Kasargod, Kerala, India* p 29.

Prakash, A., Pasalu, I.C. and Mathur, K.C. 1980. Evaluation of some plant products as paddy grain protectants against the insect pest of stored paddy. *Bull. Grain Tech.* **18** (1): 25–28.

Prakash, A., Pasalu, I.C. and Mathur, K.C. 1981a. Plant products in insect pest management of stored grains. *Bull. Grain Tech.* **19** (3): 213–219.

Prakash, A., Pasalu, I.C. and Mathur, K.C. 1981b. Begonia leaves: a grain protectant in paddy storage. *Bull. Grain Tech.* **19** (1): 75–77.

Prakash, A., Pasalu, I.C. and Mathur, K.C. 1982. Evaluation of plant products as paddy grain protectants in storage. *Intn. J. Entomol.* **1** (1): 75–77.

Prakash, A., Pasalu, I.C. and Mathur, K.C. 1983. Use of bael (*Aegle marmelos*) leaf powder as grain protectant against storage pests of paddy. *Agric. Res. J. Kerala* **21** (2): 79–80.

Prakash, A., Pasalu, I.C. and Mathur, K.C. 1984. Allitin: a grain protectant in paddy storage. *Indian J. Ent.* **46** (3): 376–378.

Prakash, A. and Rao, J. 1984. Wild sage leaves: a paddy grain protectant in storage. *Oryza* **21**: 209–213.

Prakash, A. and Rao, J. 1986. Evaluation of plant products as antifeedants against the rice storage insects. *Proc. Symp. Resid. and Environ. Pollution*: 201–205.

Prakash, A. and Rao, J. 1987. Use of chemicals as grain protectants in storage ecosystem and its consequences. *Bull. Grain Tech.* **25** (1): 65–69.

Prakash, A. and Rao, J. 1995a. Insect pest management in stored rice ecosystems. In Stored Grain Ecosystems Eds. Jayas, D.S., White, D.D.G. and Muir, W.E. Publ. Marcel Dekker Inc. New York p. 757.

Prakash, A. and Rao J. 1995b. Minimise losses due to insects in rural rice storage in India. *Tech. Bull No. 1 (AZRA)* p 4.

Prakash, A. and Rao, J. 1994. Bio-deterioration of paddy seed quality due to insects and mites and its control using botanicals. *Annual Report of ICAR Ad-hoc Scheme (1993–1994), C.R.R.I., Cuttack, India,* p 37.

Prakash, A., Rao, J. and Pasalu, I.C. 1978–1987. Studies on stored grain pests of rice and methods of minimising losses caused by them. *Final Project Report (RPF-III) Ent-6/CRRI/ICAR (India)* p 33.

Prakash, A., Rao, J. and Kulshresthra, J.P. 1986. Evaluation of plant products for management of rice storage insects. *Proc. Research Planning Works on Botanical Pest Control of Rice Based Cropping Systems at Intn. Rice Res. Institute, Manila, Philippines,* 9–12 June, 1986 p 17.

Prakash, A., Rao, J., Pasalu, I.C. and Mathur, K.C. 1987. Rice storage and insect pest management. *B.R. Publishing Corp. New Delhi* p 337.

Prakash, A., Rao, J., Gupta, S.P. and Binh, T.C. 1989. Evaluation of certain plant products as paddy grain protectants against Angoumois grain moth, *Sitotorga cerealella* Oliv. *J. Nature Con.* **1**: 7–13.

Prakash, A., Mahapatra, P.K. and Mathur, K.C. 1990a. Isolation, identification and evaluation of an oviposition inhibitor for storage insect pests from begunia, *Vitex negundo* Linn. (Verbenaceae). *Bull. Grain Tech.* **28** (1): 33–52.

Prakash, A., Rao, J., Tewari, S.N. and Gupta, S.P. 1990b. Rice agro-ecosystem management by pesticides and its consequences. *Natcon. Publ. in Growth Develop and Natural Resource Conservation*: 131–137.

Prakash, A., Rao, J., Gupta, S.P. and Behera, J. 1990c. Evaluation of botanical pesticides as grain protectants against rice weevil, *Sitophilus oryzae* Linn. *Presented in National Symposium on Problems and Prospects of Botanical Pesticides in Integrated Pest Management at C.T.R.I, Rajahmundry, India,* 20–22 Jan., 1990 (Abstract): 8.

Prakash, A., Tewari, S.N. and Rao, J. 1990d. Exploitation of natural plant products for management of pests and diseases in rice ecosystems. *Proc. Symp. Growth, Development and Resource Conservation, at D.A.V. (PG) College, (Meerut Univ.) Muzaffarnagar, India,* 8–10 Oct., 1986: 23–36.

Prakash, A., Rao, J., Behera, J. and Gupta, S.P. 1991. Bio-pesticidal property of certain plant products against lesser grain borer, *Rhyzopertha dominica* Fabr. (Bostrichidae). *Presented in IV National Symp. on Growth, Development and Control Technology of Insect Pests, at PG. Dept. of Zoology, S.D. College, Muzaffarnagar, India,* 2–4 Oct., 1991 (Abstract): 36.

Prasad, J.S. and Rao, Y.S. 1979. Nematicidal properties of the weed, *Eclipta alba* Hassk (Compositae). *Revista di parasitologia* **XL**: 87–90.

Prasad, D., Agrawal, A.K., Buchar, V.M. and Sethi, C.L. 1987. Bio-toxicity of groundnut testas extract against *Meloidogyne incognita. Indian J. Nematol.* **17** (1): 120–124.

Prasad, D. and Khan, E. 1990. Studies on integrated control of phytonematodes infesting soyabean. *Presented in National Symposium on Problems and Prospects of Botanical Pesticides in Integrated Pest Management at C.T.R.I., Rajahmundry, India,* 21–22 Jan., 1990 (Abstract): 44–45.

Premchanda. 1989. Presence of feeding deterrent in the velvet bean, *Mucuna cochinensis* Roxb. *Indian J. Ent.* **51** (2): 217–219.

Proct, J.C. and Komprobst, J.M. 1983. Effect of *Azardirachta indica, Hannoa undulata* and *H. klaineana* seed extracts on the activity of *Meloidogyne javanica* juvenile to penetrate tomato roots. *Revide de Nematologica* **6** (2): 330–332.

Proct, J.C. and Komprobst, J.M. 1985. Effect of quassinoids extracted from *Hannoa undulata* seeds on the penetration and reproduction of *Meloidogyne javanica* on tomato. *Revide de Nematologica* **8** (4): 383–389.

Pronata, R.I. 1986. Possibility of using turmeric, *Curcuma longa* Linn. for controlling storage insects. *Biotrop. Newsl. No.* **45**: 3.

Prusty, S., Bohidar, K. and Prakash, A. 1989. Evaluation of certain botanical products as grain protectants against rice moth, *Corcyra cephalonica* Staint. (Galleridae: Lepidoptera) in stored rice. *Bull. Grain Tech.* **27** (2): 85–90.

Pruthy, H.S. 1937. Report of the Imperial Entomologist. *Sci. Rep. Agric. Res. Instt. New Delhi*: 123–127.

Pruthy, H.S. and Singh, M. 1950. Pests of stored grain and their control. *Indian J. Agric. Sci.* (Spl. issue) **18** (4): 1–52.

Puri, S.N., Butler, Jr., G.D. and Henneberry, T.J. 1991. Plant derivatives oils or soap solution as control agents for the whitefly on cotton. *J. Appl. Zool. Res.* **2** (1): 1–8.

Purohit, P., Mustafa, M. and Osmania, Z. 1983. Insecticidal property of plant extract of *Cuminum cyminum* Linn. *Science & Culture* **49** (4): 101–103.

Purohit, P., Jyostjna, D. and Srimmannaranyana, G. 1989. Antifeedant activity of indigenous plant extracts against larvae of castor semi-looper. *Pesticides* **23**: 23–26.

Purthi, I.J., Jain, R.K. and Gupta, D.C. 1987. Effect of different periods of degradation of soo-babul leaves alone or in combination with nematicides on incidences of root-knot in okra. *Indian J. Nematol.* **17** (1): 30–32.

Puttarudriah, M. and Bhatia, K.L. 1995. A preliminary note on studies of Mysore plants as source of insecticides. *Indian J. Ent.* **17** (2): 165–174.

Quadri, S.S.H. 1973. Some new indigenous plant repellents of storage pests. *Pesticides* **7** (12): 18.

Quadri, S.S.H. and Narsaiah, J. 1978. Effect of azardirachtin on the moulting process of last instar nymphs of *Periplanta americana* Linn. *Indian J. Expt. Biol.* **16**: 1141.

Quadri, S.S.H. and Rao, B. 1977. Effect of combining plant seed extracts against household insects. *Pesticides* **11** (12): 21–23.

Quadri, S.S.H. and Rao, B. 1980. The effect of oleoresin in combination with neem seed and garlic clove extracts against household and stored product pests. *Pesticides* **14** (3): 11–14.

Quisumbing, E. 1951. Medicinal plants of the Philippines. *Tech. Bull. 16, Manila Bureau of Printing* p 146.

Radhakrishnan, S., Rajamanickam, B. and Rao, S.M. 1984. Efficacy of certain insecticides against *Callosobruchus chinensis* Linn. on *Cajanus cajan* seeds in storage. *Pesticides* **17** (11): 19–20.

Rahman, S.M. and Bhattacharya, G.N. 1982. Effects of leaf extract of *Vitex negundo* on *Lathyrus sativus* Linn. *Curr. Science* **51** (8): 434–435.

Rai, A.B., Bhanderi, G.R. and Patel, C.B. 1993. Effectiveness of neem based products (Azardirachtin) against *Tetranychus macfarlanei* (Tetranychidae) on okra and their safety to predatory mites. *World Neem Conf., Bangalore, India,* 24–28 Feb., 1993 (Abstract): 26.

Rajamani, S., Pasalu, I.C., Dani, R.C. and Kulshreshtha, J.P. 1987. Evaluation of insecticides and plant products for control of insect pests of rainfed upland rice. *Indian J. Pl. Prot.* **15** (1): 43–50.

Rajamma, P. 1982. Effect of some organic materials on the control of sweet potato weevil, *Cylas formicarus* Fabr. *J. Root Crops* **8** (1–2): 64–65.

Rajanasari, P.A. and Nair, M.R.G.K. 1972. On the control of brinjal pests using deterrents. *Agric. Res. J. Kerala* **10** (2): 133–135.

Rajapakse, R.H.S. and Jayasena, K.W. 1991. Plant resistance and bio-pesticide from lemon grass oil for suppressing *Spodoptera litura* in peanuts. *Agric. Intern.* **43** (6): 166–167.

Rajasekaran, B. and Kuamarswami, T. 1985a. Control of storage insects with seed protectants of plant origin. *In Behavioural and Physiological Approaches in Pest Management, Eds. Raghupathy, A. and Jayaraj, S., T.N.A.U., Coimbatore Publication*: 15–17.

Rajasekaran, B. and Kumarswami, T. 1985b. Antifeedant properties of certain plant products against *Spodoptera litura* Fabr. *In Behavioural and Physiological Approaches in Pest Management, Eds. Raghupathy, A. and Jayaraj, S., T.N.A.U., Coimbatore Publication*: 25–28.

Rajasekaran, B. and Kumarswami, T. 1985c. Studies on increasing the efficacy of neem seed kernel extract and oil. *In Behavioural and Physiological Approaches in Pest Management, Eds. Raghupathy, A. and Jayaraj, S., T.N.A.U., Coimbatore Publication*: 29–30.

Rajasekaran, B., Jayaraj, S., Raghuraman, S. and Narayanswamy, T. 1987a. Use of neem products for the management of certain rice pests and diseases. *In Mid Term Appraisal Works on Botanical Pest Control of Rice Based Cropping System organised by T.N.A.U., Coimbatore, IRRI, Manila, E-W Centre, Honolulu, Hawaii and Asian Dev. Bank, Philippines at Coimbatore, India,* 1–6 June, 1987 p 13.

Rajasekaran, B., Jayaraj, S. and Ravindran, R. 1987b. Evaluation of neem product against the pests of black gram *Vignia mungo. In Mid Term Appraisal Works on Botanical Pest Control in Rice Based Cropping Systems organised by T.N.A.U., Coimbatore, IRRI, Manila, E-W Centre, Honolulu, Hawaii and Asian Dev. Bank, Philippines at Coimbatore, India,* 1–6 June, 1987 p 18.

Rajendran, B. and Gopalan, M. 1979. Note on juveno-mimetic activity of some plants. *Indian J. Agric. Sci.* **49** (4): 295–297.

Rajendran, B. and Gopalan, M. 1985. Juveno-mimetic effect of the plant, *Ploycias quifoyie* Bailey (Araliaceae) on ovarian development in *Dysdercus cingulatus* Fabr. *In Behavioural and Physiological Approaches in Pest Management, Eds. Raghupathy, A. and Jayaraj, S. T.N.A.U., Coimbatore Publication*: 126–130.

Ramachandra, R.G., Raghaviah, G. and Nagalingam, B. 1990. Effect of botanicals on certain behavioural responses on growth and inhibition of tobacco caterpillar, *Spodoptera litura* Fabr. *Presented in National Symposium on Problems and Prospects of Botanical Pesticides in Integrated Pest Management at C.T.R.I., Rajahmundry, India,* 21–22 Jan., 1990 (Abstract): 12–13.

Raman, K., Ganesan, S., Vyas, B.N., Godrej, N.B. and Mistry, K.B. 1993. Development of a neem formulation and its utilization for control of crop pests. *World Neem Conf., Bangalore, India,* 24–28 Feb., 1993 (Abstract): 2–3.

Ramanath, M.N., Khan, N.H., Kamalwanshi, R.S. and Trivedi, R.P. 1982. Effect of *Argemone mexicana* on *M. incognita* in okra. *Indian J. Nematol.* **12**: 205–208.

Ramesh Babu, G. 1983. Neem extract destroys snails. *Science Reporter (India)* Dec., 1983 (Issue): 23.

Ramesh Babu, T. and Hussain, S.N. 1990a. Effect of plant products on pulse beetle, lesser grain borer and seed germination —Effect of plant oils on the development of pulse beetle. *Presented in National Symposium on Problems and Prospects of Botanical Pesticides in Integrated Pest Management at C.T.R.I., Rajahmundry, India,* 21–22 Jan., 1990 (Abstract): 33.

Ram Prasad, G., Joshi, B.G. and Nageswara Rao, S. 1986. Relative efficacy of some insecticides and neem seed kernel suspension against tobacco caterpillar, *Spodoptera litura* Fabr. on tobacco. *Indian J. Pl. Prot.* **14** (2): 69–74.

Ram Prasad, G., Sitaramaiah, S., Joshi, B.G. and Nageswara Rao, S. 1987. Relative efficacy of neem seed kernel suspension, some synthetic pyrethroids and insecticides against *Spodoptera litura* Fabr. in tobacco nurseries. *Indian J. Pl. Prot.* **15**: 190–192.

Ram Prasad, G., Sitaramaiah, S. and Joshi, B.G. 1990. Effect of certain plant extracts to tobacco caterpillar, *Spodoptera litura* Fabr. *Presented in National Symposium on Problems and Prospects of Botanical Pesticides in Integrated Pest Management at C.T.R.I., Rajahmundry, India,* 21–22 Jan., 1990 (Abstract): 2.

Ramraju, K. and Sundara Babu, P.C. 1989. Effect of plant derivatives on brown plant hopper (BPH) and white-backed plant hopper (WBPH) nymphs emergence in rice. *Intn. Rice Res. Newsl.* **14** (5): 30.

Rao, A.P. and Niranjan, B. 1982. Juvenile hormone like activity of karanjin against larvae of red flour beetle, *Tribolium castaneum* Hrbst. *Comparative Physiol. Ecol.* **7** (4): 234–236.

Rao, B.H.K. and Prasad, S.P. 1969. Effect of water extracts of oil cakes on *Rotylenchus reiniformis*, Linford and Olivian. *Indian J. Ent.* **31**: 88–91.

Rao, B.T.S., Reddy, G.P.V., Murthy, M.M.K. and Prasad, V.D. 1991. Efficacy of neem products in the control of bhendi pest complex. *Indian J. Pl. Prot.* **19**: 49–52.

Rao, D.V.S., Thippeswamy, M. and Murthy, P.S.S. 1984. Studies on field evaluation of insecticides on rainfed groundnut. *Pesticides* **18** (12): 35–38.

Rao, D.V.S. and Azam, K.M. 1987. Economic control of sorghum earhead bug. *Indian J. Pl. Prot.* **15** (3): 194–197.

Rao, K.P., Mohd, A.A., Chitra, K.C. and Mani, A. 1990. Efficacy of botanical protectants against pulse beetle, *Callosobruchus chinensis* Linn. *Presented in National Symposium on Problems and Prospects of Botanical Pesticides in Integrated Pest Management at C.T.R.I., Rajahmundry, India,* 21–22 Jan., 1990 (Abstract): 18.

Rao, M.S. and Pandey, M. 1982. Comparative efficacy of karanja leaf and carbofuran on the management of *Aphelenchoides, composticola. Indian J. Nematol.* **12** (2): 158–159.

Rao, M.S., Reddy, P.P. and Tewari, R.P. 1991. Comparative efficacy of certain oil cakes against mushroom nematode, *Aphelenchoides sacchari* and their effect on yield of *Agaricus bisporus*. *Indian J. Nematol.* **21** (2): 101–106.

Rao, M.S. and Reddy, P.P. 1993. Effective utilization of welgro and RD-9 repelin for the management of root-knot nematode in tomato. *World Neem Conf., Bangalore, India,* 24–28 Feb., 1993 (Abstract): 38.

Rao, M.S., Chitra, K.C., Gunesekhar, D. and Rao, K.P. 1990. Antifeedant properties of certain plant extracts against 2nd stage larvae of *Henosepilachna vigintioctopunctata* Fabr. *Indian J. Ent.* **52** (4): 681–685.

Rao, P.J. and Rembold, H. 1983. Effect of ecdysterone on food intake of *Locusta migratoria*. *Z. Naturfosch* **38C**: 878–880.

Rao, P.J. and Subrahamaniyam, B. 1986. Azardiracthtin induced changes in the development and food utilisation and hemolym pH constituents of *Schistocerca gregaria* Forsk. *J. Appl. Ent.* **102**: 217–224.

Rao, P.R.M. and Prakasa Rao, P.S. 1979. Effect of biocides on brown plant hopper adults on rice. *Intn. Rice Res. Newsl.* **4** (3): 20.

Rao, R.S.V. and Srivastava, K.P. 1985. Relative efficacy of neem formulations against gram pod borer. *Neem Newsl.* **2** (3): 28–29.

Rao, T.K. and Rao, V.N. 1990. Repelin — a plant origin insecticide for controlling pod borer, *Helicoverpa armigera* of pigeonpea, *Cajanus cajan*. *Presented in National Symposium on Problems and Prospects of Botanical Pesticides in Integrated Pest Management at C.T.R.I., Rajahmundry, India,* 21–22 Jan., 1990 (Abstract): 21.

Rao, V.S., Thippeswamy, M. and Murty, P.S.S. 1984. Studies on field evaluation of insecticides on rainfed groundnut. *Pesticides* **18**: 35–38.

Rao, Y.S., Prasad, J.S. and Panwar, M.S. 1985. Recent researches on allelopathy of rice plant nematodes. *Interfaces* **19** (5): 28–33.

Raodev, A.K. 1973. Professor of Entomology, College of Agriculture. Parbhani (MS) India, Pers. Commun. in Ketkar C.M. 1969–1976. Utilisation of neem and its byproducts. *Modified report on neem cake manurial project. Directorate of Non-edible Oils and Soap Industries, K.V.I.C., Bombay*: 116–173.

Ravidutta, B. and Bhatti, D.S. 1986a. Effect of chemicals and phytotherapeutic substance on biological phenomenon of *M. javanica* infesting tomato. *Indian J. Nematol.* **16** (1): 19–22.

Ravidutta, B. and Bhatti, D.S. 1986b. Effective doses and time of application of nematicides and castor leaves for controlling *Meloidogyne incognita* in tomato. *Indian J. Nematol.* **16** (1): 8–11.

Reddy, A.S. and Venugopal Rao, N. 1990. Efficacy and selectivity of insecticides of natural origin against pests of cotton, *Gossypium hirsutum*. *Presented in National Symposium on Problems and Prospects of Botanical Pesticides in Integrated Pest Management at C.T.R.I., Rajahmundry, India,* 21–22 Jan., 1990 (Abstract): 10.

Reddy, G.V.P. and Urs, K.C.D. 1988. Effect of plant extracts on brown plant hopper (BPH) oviposition. *Intn. Rice Res. Newsl.* **13** (4):42.

Reddy, M.U. and Reddy, G.S. 1987. Effectiveness of selected plant materials as protectant against insect infestation and nutrient consumption during storage of green gram, *Vigna radiata*. *Bull. Grain Tech.* **25** (1): 48–57.

Reddy, P.P. and Khan, R.M. 1990. Management of root-knot nematodes infesting papaya by incorporation of some plant leaves. *Presented in National Symposium on Problems and Prospects of Botanical Pesticides in Integrated Pest Management at C.T.R.I., Rajahmundry, India,* 21–22 Jan., 1990 (Abstract): 42.

Reddy, V.R., Chitra, K.C. and Rao, K.P. 1990. Efficacy of plant extracts in the control of brinjal spotted leaf beetle, *Henosepilachna vigintiotopunctata* Fabr. *Presented in National Symposium on Problems and Prospects of Botanical Pesticides in Integrated Pest Management at C.T.R.I., Rajahmundry, India,* 21–22 Jan., 1990 (Abstract): 18–19.

Reddy, V.V.S., Ramesh Babu, T. and Narsimha Rao, B. 1993. Repelin — an effective botanical insecticide against *Carpomyia vesuviana* Costs. *World Neem Conf., Bangalore, India,* 24–28 Feb., 1993 (Abstract): 24.

Redfern, R.E., Warthen, Jr., J.D., Uebel, E.C. and Mills, Jr., G.D. 1981. The antifeedant and growth disrupting effects of azardirachtin to *Spodoptera frugiperda* and *Oncopeltus fasciatus. In Natural Pesticides from Neem Tree (Azardirachta indica* A. Juss.), *Eds. Schmutter, H., Ascher, K.R.S. and Rembold, M. GTZ Press, Eschborn, West Germany*: 129–134.

Reed, D.K., Freedman, B. and Ladd, Jr., T.L. 1982. Insecticidal and antifeedant activity of neriifolin against codling moth, striped cucumber beetle and Japanese beetle, *J. Econ. Entomol.* **75** (6): 1093–1097.

Reed, D.K., Warthen, J.D.J., Uebel, E.C. and Reed, G.L. 1982. Effect of two triterpenoids from neem on feeding by cucumber beetles (Coloptera: Chysomelidae). *J. Econ. Entomol.* **75** (6): 1109–1113.

Reed, D.K., Kwolk, W.F. and Smith, Jr., C.R. 1983. Investigation of antifeedant and other insecticidal activities of trewiasine towards the striped cucumber beetle and codling moth. *J. Econ. Entomol.* **76** (3): 641–645.

Reena, C., Afzal, J., Das, K.G. and Sukumar, K. 1983. 6-7 Dimethoxy-isochroman-3-one: antifeedant for castor semi-looper, *Achoea janata* Linn. *Indian J. Farm Chem.* **7**: 63.

Rembold, H. 1981. Modulation of JH-III titer during the gonadotrophic cycle of *Locusta migratoria* measured by gas chromatography selected on monitoring mass spectrometry. *In Juvenile Hormone Bio-chemistry, Eds. Pratt, G.E. and Brooks, G.T., Elsevier, Amsterdam, North Holland* p 11.

Rembold, H. 1984. Secondary plant products in insect control with special reference to the azardirachtins. *In Advances in Invertebrate Reproduction Vol. 3, Engels, W. Ed. Elsevier, Amsterdam, North Holland* p 481.

Rembold, H. 1987. The azardirachtins — potent insect growth inhibitor. *Mem. Inst. Oswaldo. Cruz ric de Janeiro Vol. 82 Suppl. III*: 61–66.

Rembold, H. 1989a. Isomeric azardirachtins and their mode of action. In *Neem Tree Eds. Jacobson, M. and Kubo, I., CRC Press. Boca Raton.* Vol. 1: 47–67.

Rembold, H. 1989b. The azardirachtins: their potential for insect control *Economic and Medicinal Plant Research* **3**: 57–72.

Rembold, H. 1989c. Azaridrachtins: their structure and mode of action. *In Insecticides of Plant Origin ACS Symposium Series 387 of III Chem. Congr. of North America, 195th National Meeting of American Chemical Society held at Toronto, Ontario, Canada,* 5–11 June, 1988: 150–163.

Rembold, H. and Czoppelt, Ch. 1981. Assay of plant derived inhibitors of insects growth from *Azardirachta indica* in assay test of the bee larvae. *Mitt. Dtswch. Ges. Allgem. Entomol.* **3**: 196–203.

Rembold, H. and Lackner, B. 1985. Conventional method for the determination of picomole amounts of juvenile hormone. *J. Chromatogr.* **323**: 355.

Rembold, H. and Schmutterer, H. 1981. Disruption of insect growth by neem seed components in regulation of insect development and behaviour, *Eds. Sehnal, F., Zabra, A., Menn, J.J. and Cymborowski. Wroczlaw Tech. Univ. Press Wroczlaw, Poland* p 1087.

Rembold, H. and Sieber, K.P. 1981a. Inhibition of oogenesis and ovarian ecdysteroid synthesis by azardirachtin in *Locusta migratoria. Z. Naturforsch.* **36C**: 466–469.

Rembold, H. and Sieber, K.P. 1981b. Effect of azardirachtin on oocyte development. In *Locusta migratoria migratoriodes. In Natural Pesticides from Neem Tree (Azardirachta indica A. Juss.) Eds. Schmutterer, H., Ascher, K.R.B. and Rambold, H. GTZ Press, Eschborn, West Germany:* 75–78.

Rembold, H., Eder, J. and Ulrich, G.M. 1980a. Inhibition of allatotropic activity and ovary development in *Locusta migratoria* by anti-brain antibodies. *Z. Naturforsch.* **35C**: 1117–1120.

Rembold, H., Shrama, G.K., Czoppelt, C. and Schmutterer, H. 1980b. Evidence of growth disruption in insect pests without feeding inhibition by neem seed. *Z. Pflkrankh Pflschutz* **87**: 290–297.

Rembold, H., Sharma, G.K. and Czoppelt, Ch. 1981. Growth regulating activity of azardirachtin in two homometabolous insects. *In Natural Pesticides from Neem Tree (Azardirachta indica* A. Juss.) *Eds. Schmutterer, H., Ascher, K.R.S. and Rembold, H. GTZ Press, Eschborn, West Germany*: 121–124.

Rembold, H., Sharma, G.K., Czoppelt, C. and Schmutterer, H. 1982. Azardirachtin: a potent growth regulator of plant origin. *Zeistschrift fur Angewandte Entomology* **93** (1): 12–17.

Rembold, H., Forster, H., Czoppelt, Ch., Rao, P.J. and Siber, K.D. 1984a. The azardirachtins: a group of insect regulators from neem tree. *In Natural Pesticides from Neem and Other Tropical Plants. Eds. Schmutterer, H. and Ascher, K.R.S. GTZ Press, Eschborn, West Germany*: 153–156.

Rembold, H., Subramaniyam, B. and Muller, T. 1984b. Corpurs cardiacum, a target for azardiracthin. *Experimentia* **45**: 361–363.

Rembold, H., Forster, H. and Czoppelt, Ch. 1986. Structure and biological activity of azardirachtins A & B. *Proc. III Intn. Neem Conf., Nairobi*: 149–160.

Rembold, H. Forster, H. and Sonnenbichler, J. 1987. Structure of azardirachtin-B. *Z. Naturforsch.* **42C**: 4.

Reynaud, P.A. 1986. Control of the azolla pest, *Lymnaea natalensis* with molluscicides of plant origin. *Intn. Rice Res. Newsl.* **11** (3): 27–28.

Rimpler, H. 1969. Pterosteron, polypodin B. und einnene ecdysonartigcs steroid (Viticosteron E) aus *Vitex megapotamica* (Verbenaceae). *Tetrahedron Lett.* (1969): 329–333.

Rimpler, H. 1972. Iridoides and ecdysones from *Vitex* sp. *Phytochem.* **11**: 2653–2654.

Rimpler, H. and Schulz, G. 1967. Venkommen von 20-Hydroxy ecdysone in *Vitex megapotamica. Tetrahedron Lett.* (1967): 2033–2035.

Roark, R.C. 1947. Insecticidal and fish poisoning plant. *Econ. Bot.* **1**: 437–445.

Robert, D.S., Bell, A.A., Daniel, H.O. and Likefahr, J.M. 1978. Helicocide H. A new insecticidal C_{25} terpenoid from cotton, *Gossypium hirsutum. J. Agric. Food Chem.* **26**: 115–118.

Robert, P.C., Balaisinger, P., Bouchery, Y., Simons, M.T., Kienlen, J.C., Muckersterum, B., Riss, B. and Pflieger, D. 1987. Influence of bisabolangelone, a sesquiterpenoid antifeedant on the development of the caterpiller, *Mythimna (Psudoletia) unipuncta* Haw. (Lepidoptera: Noctuidae). *Agronomie* **7** (3): 167–174.

Roberston, I.A.D. 1958. Tree planting trails for the control of red locust, *Nomadacris septemfasciata* in the Rukwa-Valley, Tanganyika. *East. Africa Agric. J.* **23** (3): 172–178.

Rohde, R.A. 1972. Expression of resistance in plants to the nematodes. *Ann. Rev. Phytopath.* **10**: 2333.

Romeo, J.T. and Simmonds, S.J.M. 1989. Non-protein amino acid feeding deterrents from *Callindra. In Insecticides of plant origin, A.C.S. Symposium Series 38 of III Chem. Congr. of North America, 195th National Meeting of Am. Chem. Soc. held at Toronto, Ontario, Canada,* 5–11 June, 1988: 59–68.

Rout, G. 1986. Comparative efficacy of neem seed powder and some common plant product mixtures against *Sitophilus oryzae* Linn. *Neem Newsl.* **3** (2): 13–14.

Routray, B.N. and Sahoo, H. 1985. Integrated control of root-knot nematode, *M. incognita* with neem cake and granular nematicides on tomato. *Indian J. Nematol.* **15** (2): 261.

Roy, A.K. 1976. Effect of de-caffenated tea waste and water hyacinth compost on the control of *Meloidogyne graminicola* (Nematoda). *Indian J. Nematol.* **6**: 73–77.

Roy, D., Sinhababu, S.P., Sakul, N.C. and Mahato, S.B. 1993. Nematicidal principle from the funicles of *Acacia auriculiformis. Indian J. Nematol.* **23** (2): 152–157.

Roy, D.C. and Pande, Y.D. 1991. Effect of *Hyptis suaveolens* (L.) Poit. leaf extract on the population of *Lipaphis erysim* Kalt. (Homoptera: Aphideae) and *Tetranychus neocaledonicus* Andre (Acarina: Tetranychidae). *IV National Symp. on Growth Development and Control Technology of Insect Pests, at P.G. Dept. Zoology, S.D. College, Muzaffarnagar (U.P.),* 2–4 Oct. 1991 (Abstract): 37.

Ruelo, J.S. 1976. Chemical, biological and integrated biochemical control of *Meloidogyne incognita. Ph.D. Thesis, Univ. Philipp. at Los Banos, Laguna* p 188.

Ruelo, J.S. and Davide, J.S. 1979. Studies on the control of *Meloidogyne incognita* II. Integration of biological and chemical control. *Philipp. Agric.* **62**: 159–165.

Rusbey, R.L. 1950. USDA Year Book of Agriculture for the Year 1950–1951: 765–772.

Ruscoe, C.N.E. 1972. Growth disruption effects of an insect antifeedant. *Nature-New. Biol.* **236**: 66.

Sachan, J.N. and Pal, S.K. 1976. Insecticides and cakes for the control of white grub, *Holotrichia insularis* Beneske, in Western Rajasthan. *Pesticides* **10** (6): 37–38.

Sachan, J.N. and Lal, S.S. 1990. Role of botanical insecticides in *Helicoverpa armigera* management in pulses. *Presented in National Symposium on Problems and Prospects of Botanical Pesticides in Integrated Pest Management at C.T.R.I., Rajahmundry, India,* 21–22 Jan., 1990 (Abstract): 15.

Sachan, J.N. and Katti, G. 1993. Neem in *Helicoverpa armigera* management: pulses ecosystem. *World Neem Conf., Bangalore, India,* 24–28 Feb. 1993 (Abstract): 20.

Saim, N. and Clifton, E.M. 1986. Compounds from leaves of sweet bay, *Laurus nobilis* Linn. as repellent for *Tribolium castaneum*, when added to wheat. *J. Stored Prod. Res.* **22** (3): 141–144.

Salem, I.E.M. and Abdel-Hafez, M.M. 1990. Natural rotenoids from *Tephrosia apollinea* and their effects against *Aphis craccivora* Koch. (Aphidae). *Mededalingen van de facultect Land bouwwetonschappen Rijksumiversiteit Gent.* **55** (26): 657–660.

Samajpathraj, R., Srimannarayan, P.B. and Vanitha Kumari, G. 1993. Effect of neem-oil: Structural and functional changes in the epididymis of rats. *World Neem Conf., Bangalore, India,* 24–28 Feb. 1993 (Abstract): 21.

Samalo, A.P., Senapati, B., Satpathy, C.R. and Jacob, T.J. 1990. Effect of neem derivatives on incidence of some major insect pests in wet rice season. *Presented in National Symposium on Problems and Prospects of Botanical Pesticides in Integrated Pest Management at C.T.R.I., Rajahmundry, India,* 21–22 Jan., 1990 (Abstract): 16.

Samuthiravelu, D.B.V. 1990. Bio-efficacy of neem oil and deltamethrin against spotted boll worm, *Earias vitella* Fabr. (Noctuidae: Lepidoptera) on cotton. *The Madras Agric. J.* **77** (7–8): 294–298.

Sandhu, G.S. and Darshan, Singh 1975a. Studies on antifeedant and insecticidal properties of neem (*Azardirachta indica* A. Juss.), dharek (*Melia azardirach* Linn.) kernel and fruit to *Pieris brassicae* larvae. *Indian J. Pl. Prot.* **3** (2): 177–180.

Sandhu, G.S. and Darshan, Singh 1975b. Studies on antifeedant effect of some indigenous plant materials against *Chrotogonus trychypterus* Blanch. *Indian J. Pl. Prot.* **3** (2): 207–208.

Sangappa, H.K. 1977. Effectiveness of oils and surface protectants against the bruchid, *Callosobruchus chinensis* Linn. infestation on red gram. *Mysore J. Agric. Sci.* **11**: 391–397.

Saradamma, K. 1988. Plant extracts for the management of insect pests of vegetables. *In Integrated Pest Control Progress and Perspectives,* Eds. Mohandas, N. and Koshi, G., *Trivendrum, Assoc. of Adv. Entomology*: 392–396.

Saramma, P.V. and Verma, A.N. 1971. Efficacy of some plant products and magnesium carbonate as grain protectants of stored wheat seed against attack of *Trogoderma granarium* Everts. *Bull. Grain Tech.* **9**: 207–210.

Sardana, H.R. and Kumar, N.K. 1989. Effectiveness of plant oils against leaf hopper and shoot and fruit borers in okra. *Indian J. Ent.* **51** (2): 167–171.

Sarra-Guzman, R. 1984. Toxicity screening of various plant extracts against *Meloidogyne incognita* Chitwood and *Radopholus similis* Cobb. and characterisation of their nematicidal components. *Ph.D. Thesis, Univ. of Philippines at Los Banos College, Laguna, Philippines* p 139.

Sasmal, S. 1991. Bionomics and control of pests of *Azolla pinnata* R.Br. *Ph.D. Thesis, Utkal University, Bhubaneswar, India* p 168.

Satpathy, J.M. 1976. Utilization of neem kernel powder for pest control in grain storage. In Ketkar, C.M. 1969–1976. Utilization of neem and its byproducts. *Report of the modified neem cake manurial project. Directorate of Non-edible Oils and Soap Industries, K.V.I.C., Bombay*: 116–173.

Satpathy, J.M. 1983. Department of Entomology, O.U.A.T., Bhubaneswar, India. Pers. Commn. in Michael et al., 1986. Plant Species Reportedly Possessing Pest Control Properties. *An EWC/UH DATA BASE, Univ. of Hawaii* p 249.

Savitir, P. 1975. Study on the efficacy of neem kernel powder and neem oil as admixture with grain and also by impregnation on paddy bags in control of certain important storage pests of paddy. *M.Sc. Thesis, College of Agric. Baptala, A.P., India,* p 129.

Savitri, P. and Sabba Rao, C. 1976. Studies on the admixture of neem kernel powder with paddy in control of important storage pests of paddy. *Andhra Agri. J.* **23** (3 & 4): 137–143.

Saxena, B.P. and Srivastava, J.B. 1972. Effect of *Acorus calamus* Linn. oil vapours on *Dysdercus koenigii* Fabr. *Indian J. Expt. Biol.* **10** (5): 359–393.

Saxena, B.P. and Srivastava, J.B. 1973. *Tagetes minuta* Linn. oil: a new source of juvenile hormone mimic substances. *Indian J. Expt. Biol.* **11**: 56–58.

Saxena, K.N. and Basit, A. 1982. Inhibition of oviposition by volatiles of certain plants and chemicals of the leaf hopper, *Amrasca devastans* Distant. *J. Chem. Ecol.* **8** (2): 329–338.

Saxena, R., Jain, S.K. and Sharma, R.K. 1993. Nematicidal effect of neem leaf extract. *World Neem Conf., Bangalore, India* 24–28 Feb., 1993 (Abstract): 39.

Saxena, R.C. 1993. Potential of neem seed extract for control of cowpea thrips. *World Neem Conf., Bangalore, India,* 24–28 Feb. 1993 (Abstract): 14.

Saxena, R.C. and Boncolin, M.E.M. 1988. Effect of neem seed bitters (NSB) on green leaf hopper (GLH) feeding. *Intn. Rice Res. Newsl.* **3** (1): 27.

Saxena, R.C. and Khan, Z.R. 1985a. Electronically recorded disturbance in feeding behaviour of *Nephotettix virescens* (Homoptera: Cicadellidae) on neem oil treated rice plants. *J. Ecol. Entomol.* **78**: 222–226.

Saxena, R.C. and Khan, Z.R. 1985b. Effect of neem oil on survival of *Nilaparvata lugens* (Homoptera: Delphacidae) and on Grassy Stunt and Ragged Stunt Virus transmission. *J. Econ. Entomol.* **78**: 647–651.

Saxena, R.C., Liquido, N.J. and Justo, Jr., H.D. 1979. Neem oil: an antifeedant to brown plant hopper. *Proc. X Ann. Conf. of the Pest Control Council of the Philippines held at Manila,* 2–5 Many, 1979 p 25.

Saxena, R.C., Liquido, N.J. and Justo, Jr., H.D. 1981a. Neem oil: a potential antifeedant for the control of the rice brown plant hopper, *Nilaparvata lugens. Proc. I. Int. Neem Conf.* (1980): 171–188.

Saxena, R.C., Waldbauer, G.P., Liquido, N.J. and Puma, B.C. 1981b. Effect of neem oil on the rice leaf-folder, *Cnaphlocrosis medinalis. Proc. I. Intn. Neem Conf. Rottach-Egern* (1980): 189–203.

Saxerna, R.C., Epicnoc, P.B., Tu, C.W. and Puma, B.C. 1984a. Neem, Chinaberry and custard apple: antifeedant and insecticidal effects of seed oils on leaf hopper and plant hopper pests of rice. *Proc. II. Intn. Neem Conf., Rauisch-Holzhousen Castle FRG,* 25–28 May, 1983: 403–412.

Saxena, R.C., Justo, H.B. and Epino, P.B. 1984b. Evaluation and utilization of neem cake against the rice brown plant hopper, *Nilaparvata lugens* (Homoptera: Diphacidae), *J. Econ. Entomol.* **77**: 502–507.

Saxena, R.C., Chiu, S.R., Mariappan, V. and Kalode, M.B. 1985. Evaluation and utilization of neem (*Azardirachta indica* A. Juss.) seed derivatives for management of rice insect pest. *Intn. Rice Res. Conf. at I.R.R.I., Manila,* 1–5 June, 1985 p 30.

Saxena, R.C., Justo, H.D. and Rueda, B.P. 1987. Neem seed bitters for management of plant hopper and leaf hopper pests of rice. *Presented in Mid-Term Appraisal Works on Botanical Pest Control in Rice Based Cropping Systems organised by T.N.A.U., Coimbatore, I.R.R.I., Manila, E-W Centre, Honolulu and Asian Dev. Bank, Philippines, Coimbatore,* 1–6 June, 1987 p 19.

Saxena, R.C., Zhang, Z.T. and Boncodin, M.E.M. 1989. Effect of neem oil on courtship signals and mating behaviour of brown planthopper (BPH) female. *Intn. Rice Res. Newsl.* **14** (6): 28–29.

Schauer, M. and Schmutterer, H. 1981. The effect of freely squeezed juices and fruit extracts of the labiate, *Ajuga remota* on the two spotted spider mite, *Tetranychus urticae* Koch. *Zatschrift fur Angewandte Emtomologie* **91** (5): 425–433.

Schaumer, M. 1984. Effect of variously formulated neem seed extracts on *Acyrthosiphon pisum* and *Aphis fabae. In Natural Pesticides from the Neem Tree and Other Tropical Plants, Eds. Schmutterer, H. and Ascher, K.R.S., G.T.Z. Press, Eschborn, West Germany*: 141–149.

Schearer, W.R. 1984. Components of oil of tansy, *Tanacetum vulgare* that repel Colorado potato beetle, *L. decemlineata. J. Natural Products* **47** (6): 964–969.

Schluter, U. 1981. Histological observations on the phenomenon of black legs and thoraxic spots: effects on pure fractions of neem kernel extracts on *Epilachna varivestis. In Natural Pesticides from Neem Tree (Azardirachta indica A. Juss.) Eds. Schmutterer, H., Ascher, K.R.S. and Rombold, H., G.T.Z. Press, Eschborn, West Germany*: 97.

Schluter, U. 1984. Disturbances of epidermal and fat body tissue after feeding azardirachtin and its consequence on larval moulting in the Mexican bean beetle, *Epilachna varivestis* (Coleoptera: Coccinellidae). *Entomol. Goner.* **10** (2): 97–110.

Schluter, U., Bidmon, H.J. and Grewe, S. 1985. Azardirachtin affects growth and endocrine events in larvae of tobacco hornworm, *Manduca sexta. J. Insect Physiol.* **31**: 773–776.

Schmeltz, I. 1971. Nicotine and other tobacco alkaloids. *In Naturally Occurring Insecticides. Eds. Jacobson, M. and Crosby, D.G. 1971, Marcel Dekker, Inc., New York* p 585.

Schmutterer, H. 1984. Neem research in the Federal Republic of Germany since the First International Neem Conference. In *Natural Pesticides from Neem Tree and Other Tropical Plants. Eds. Schmutterer, H. and Ascher, K.R.S. G.T.Z. Press, Eschborn, West Germany*: 21.

Schmutterer, H. and Rembold, H. 1980. The action of several pure fractions from seeds of *Azardirachta indica* on feeding and metamorphosis of *Epilachna varivestis* (Coleoptera: Coccinellidae). *Z. Angew. Entomol.* **89**: 179.

Schmutterer, H., Saxena, R.C. and Heyde, J. Von Der. 1983. Morphogenetic effects of some partially purified fractions and methanolic extracts of neem seeds on *Mythimna separata* Walk. and *Cnaphlocrosis medinalis* Guen. *Zeitschrift fur Agwandte Entomologie* **95** (3): 230–237.

Schoonhoven, A.V. 1978. Use of vegetable oils to protect stored beans from bruchids attack. *J. Econ. Entomol.* **71**: 254–256.

Schulz, W.D. 1981. Histo-pathological alterations in the ovaries of *Epilachna varivestis* induced by an extract from neem kernels. *Proc. I. Intn. Neem Conf. Rottach-Egern,* 1980: 81–96.

Scott, W.P. and McKibben, G.H. 1978. Toxicity of black pepper extract to boll weevils. *J. Econ. Entomol.* **71** (2): 343–344.

Seiferle, E.J., Johns, I.B. and Richardson, C.H. 1942. Poisonous alkaloids from green helebore, *Veratum viride* Ait. (Liliaceae). *J. Econ. Entomol.* **35**: 35.

Self, L.S., Guthrine, F.E. and Hodgson, E. 1964. Adaptation of tobacco hornworm to the ingestion of nicotine. *J. Insect Physiol.* **10**: 907–914.

Sen, K. and Das-Gupta, M.K. 1985. Effect of different substances on larval hatching of *M. incognita. Indian J. Nematol.* **15** (1): 100–102.

Sergent, E. 1944. Protection des cultives control les acridienspar extract de *Melia Arch. Instt. Parteur Algergie* **22** (3): 251–254.

Seyarajan, S., Sundrababu, P.C., Srimmannarayana, G. and Geetanjali. 1990. Antifeedant and morphogenetic effects of azardirachtin rich fractions on *Spodoptera litura* Fabr. *Presented in National Symposium on Problems and Prospects of Botanical Pesticides in Integrated Pest Management at C.T.R.I., Rajahmundry, India,* 21–22 Jan., 1990 (Abstract): 33–34.

Shah, A.H. and Patel, R.C. 1976. Role of tulsi plant (*Ocimum sanctum*) in control of mango fruit fly, *Dacus correctus* Bezzi. (Terphritidae). *Curr. Sci.* **45** (8): 313–314.

Shalon, U. and Pener, M.P. 1984. Sexual behaviour without adult morphogenesis in *Locusta migratoria. Experimentia* **40**: 1418–1421.

Shapan, Gabrieth. 1965. Agricultural Entomology — Section-9, *Proc. XII. Congress of Entomology, London, UK,* p 549.

Sharma, A.K. and Srivastava, R.C. 1984. Effect of groundnut oil on embryonic development of *Callosobruchus chinensis* Linn. *Bull. Grain Tech.* **22** (3): 221–224.

Sharma, A.N., Sharma, U.K. and Singh, R.S. 1971. Oil cakes and sawdust control root-knot. *Indian Farming* (March 1971): 17–20.

Sharma, G.K. 1992. Growth inhibitory activity of azandirachsin on *Coryra cophelonica, Phytoparasitica* **20** (1): 47–50.

Sharma, G.K., Czoppelt, Ch. and Rembold, H. 1980. Further evidence of insect growth disruption by neem seed fractions. *Z. Anger. Entomol.* **90**: 439–444.

Sharma, H.K. and Prasad, D. 1995. Evaluation of *Argemone mexicana* plant extract against root-knot nematode, *Meloidogyne incognita* on soybean and brinjal. *Presented in National symposium on nematode problems of India — An appraisal of the nematode management with eco-friendly approaches and bio-components. I.A.R.I., New Delhi,* 24–26 March, 1995 (Abstract): 56.

Sharma, H.L., Vimal, O.P. and Prasad, 1985. Nematicidal and manurial values of some oil cakes. *Intn. J. Crop. Pl. Diseases* **3**: 51–55.

Sharma, I.N.S., Singh, A.K. and Singh, S.P. 1992. Allelopathic potential of some plant substances as antifeedant against insect pests of jute. *Proc. I. National Symp. on Allelopathy in Agro-ecosystems. Eds. Tauro, P. and Narwal, S.S., Indian Society of Allelopathy, H.A.U., Hissar, India*: 166–168.

Sharma, M.C. and Leuschner, K. 1987. Chemical control of sorghum head bug, *Calocoris angustatus* Leth. (Hemiptera: Miridae). *Crop. Prot.* **67** (5): 334–340.

Sharma, R.K., Saxena, A. and Rard Naithani, S. 1986. Aphidicidal action of neem (*Azardirachta indica* A. Juss.) on mustard aphid, *Lipaphis erysimi* Kalt. *Neem Newsl.* **3**: 1–2.

Sharma, R.N. 1981. *Lavandula gibsonii*: a plant with insectistatic potential. *Phytoparasitica* **9** (2): 101.

Sharma, R.N., Nagasampangi, B.A., Bhosale, A.S., Kulkarni, M.M. and Tungikar, V.B. 1984. Neem-rich, the concept of enriched fractions from neem for behavioural and physiological control of insects. *Proc. II. Intn. Neem Conf. Rauiscehnholzhaussen, F.R.G.* (1983): 115–128.

Sharma, S.K., Sakhija, P.K. and Singh, I. 1981. Effect of mustard cakes on the hatching and cyst population of *Heterodera avenae* on wheat. *Indian J. Nematol.* **11**: 11–14.

Sharma, S.S. and Dahiya, B. 1986. Efficacy of some plant oils against pod borer, *Heliothis armigera* (Hubn.) on chick pea. *Neem Newsl.* **3** (2): 15–17.

Sharma, T.R. and Trivedi, P.C. 1995. Efficacy of some botanicals against *Heterodera avenae* on barley. *Presented in National symposium on nematode problems in India — An appraisal of the nematode management with eco-friendly approaches and bio-components. I.A.R.I., New Delhi,* 24–26 March, 1995 (Abstract): 4.

Sharma, V.K. and Bhatnagar, A. 1990. Studies on the effects of non-edible seed oils on *Chilo partellus* Swinh. *Presented in National Symposium on Problems and Prospects of Botanical Pesticides in Integrated Pest Management at C.T.R.I., Rajahmundry, India,* 21–22 Jan., 1990 (Abstract): 19.

Sharma, Y. 1983. Effect of aak, *Calotropis procera* on the population build up of *Rhyzopertha dominica. J. Adv. Zool.* **4** (2): 123–124.

Sharma, Y. 1985. Effect of aak flower extract on different larval stages of the lesser grain borer, *Rhyzopertha dominica* Fabr. *J. Adv. Zool.* **6** (1): 8–12.

Shayya, E. and Ikan, R. 1970. Insect control using natural products. *In Advances in Pesticides Part II.* Geissbyhar, H. (ed.), Pergamon Press, 302–312.

Shayya, E., Ravid, V., Paster, N., Juven, B., Zisman, U. and Pissarew, V. 1991. Fumigant toxicity of essential oils against four major stored product insects. *J. Chem. Ecol.* **17** (3): 499–504.

Shelke, S.S., Jakhav, L.D. and Salunkhe, G.N. 1987. Ovicidal action of some vegetable oils and extracts in storage pests of potato, *Phthorimaria operculella* Zell. *Biovigyanam* **13** (1): 40–41.

Shelke, S.S., Jakhav, L.D. and Sahinkhe, G.N. 1990. Toxicity of some vegetative oils and extracts to *Phthorimaea operculella* Zell. *Indian. J. Ent.* **52** (4): 709–110.

Shepard, H.M. 1951. Chemistry and action of insecticides. *McGraw-Hill, New York* p 171.

Sherif, A., Hall, R.G. and El Amangi, M. 1987. Drugs, insecticides and other agents from *Artemisia. Medical Hypothesis* **23** (2): 187–193.

Shi, Y.S., Wang, W.P., Liao, C. and Chiu, S.F. 1986. Effect of toosandanin on the sensory inputs of chemoreceptors of armyworm larvae *Mythimna separat. Acta Entomologica Sinica* **29** (3): 233–238.

Shin Foon, C., Xing, Z., Suiking, L. and Duanping, H. 1985. Growth-disrupting effect of azardirachtin on the larvae of Asiatic corn borer, *Ostrinia furnacalis* Guen. *Z. Angew. Entomol.* **95**: 276–279.

Shukla, B.C., Kaushik, U.K. and Gupta, R. 1991. Field evaluation of botanicals for management of rice insect pests. *Proc. Botanical Pest Control Project. Phase-II, Mid-term Project Review Meeting, Dhaka, Bangladesh,* 28–31 July, 1991: 39.

Shukla, V.N. and Banker, M.D. 1985. Studies on nematicidal properties of some plant species against *Meloidogyne incognita. P.K.V. Res. J. (India)* **9** (2): 46–52.

Shukla, V.S., Barua, N., Choudhury, P.K., Sharma, R.P. and Bordoloi, M. 1986. Absolute stereo-chemistry of the cadinenes from *Eupatorium trapezoideum. Tetrahedron* **42** (4): 1157–1167.

Siddappaji, C., Kumar, A.R.V. and Gangadhar, R. 1986. Evaluation of different insecticidal sprays against the chickpea borer, *Heliothis armigera* Hubn. *Pesticides* **20**: 13–16.

Siddiqui, M.A. and Alam, M.M. 1987. Efficacy of seed dressing with extracts of neem and persian lilac against *Meloidogyne incognita* and *Rotylenchus reiniformis. Nematologia Mediterrance* **15** (2): 399–403.

Siddiqui, M.A. and Alam, M.M. 1988a. Toxicity of different plant parts of *Tagetes lucida* to plant parasitic nematodes. *Indian J. Nematol.* **18** (2): 181–185.

Siddiqui, M.A. and Alam, M.M. 1988b. Studies on the nemato-toxicity of root exudates of certain species of *Tagetes. Indian J. Nematol.* **18** (2): 335–337.

Siddiqui, M.A. and Saxena, S.K. 1987. Effect of interculture of margosa and persian lilac with tomato and okra plant on root-knot nematodes and reiniform nematodes. *Intn. Nematol. Newsl.* **4** (2): 5–8.

Siddiqui, M.A. and Alam, M.M. 1993. Evaluation of nematicidal potential in neem allelochemicals. *World Neem Conf., Bangalore, India,* 24–28 Feb., 1993 (Abstract): 39.

Siddiqui, M.A., Haseeb, A. and Alam, M.M. 1992. Control of plant parasitic nematodes by soil amendments with latex bearing plants. *Indian J. Nematol.* **22** (1): 25–28.

Siddiqui, M.A., Haseeb, A. and Alam, M.M. 1987. Evaluation of nematicidal properties of some latex bearing plants. *Indian J. Nematol.* **17** (1): 99–102.

Siddiqui, Z.A. 1976. Nematode population and yield of certain vegetables as influenced by oil cake amendments. *Indian J. Nematol.* **6** (2): 179–182.

Sieber, K.P. 1982. Investigations of the effect of azardirachtins on the development and its regulation in *Locusta migratoria. Ph.D. Thesis, Univ. of Munich, West Germany* p 121.

Sieber, K.P. and Rembold, M. 1983. The effects of azardirachtin on the endocrine control of moulting in *Lousta migratoria. J. Insect Physiol.* **29**: 523–526.

Sighamony, S.D., Aness, I., Chandrakala, T. and Osmani, Z. 1986. Efficacy of certain indigenous plant products as grain protectants against *Sitophilus oryzae* and *Rhyzopertha dominica. J. Stored Prod. Res.* **22** (1): 21–23.

Sikdar, A.K., Govindaiah, Y.R., Madhava Rao, Jolly, M.S. and Giridhar, K. 1986. Control of root-knot nematode disease of mulberry by temik 10G-oil cakes. *Indian J. Sericulture* **25** (1): 33–35.

Simwat, G.S. and Dhawan, A.K. 1992. Use of plant insecticides against *Heliothis armigera* Hubn. infesting cotton and chickpea. *Proc. First National Symposium on Allelopathy in Agro-ecosystems, Eds. Tauro, P. and Narwal, S.S. Indian Society of Allelopathy, H.A.U., Hissar, India*: 158–159.

Singal, S.K. and Singh, Z. 1990. Studies of plant oils as surface protectants against pulse beetle, *Callosobruchus cinensis* Linn. in chickpea, *Cicer arietinum* Linn. in India. *Tropical Pest Management* **36** (3): 314–316.

Singh, B. 1985. Studies on effect of neem kernel extract and certain less persistent insecticides on the oviposition behaviour and mortality of rice desert locust, *Schistocerca gregaria* Forsk. *Ph.D. Thesis, C.S.A. Agric. Univ., Kanpur (U.P.)*: p 159.

Singh, B. and Singh, A.P. 1985. Joint action of neem kernel suspension with some insecticides against *Schistocerca gregaria* Forsk. *Neem Newsl.* **4** (1): 6–8.

Singh, B. and Singh, A.P. 1987a. Effect of neem kernel suspension on the hatchability of eggs of the desert locust *Schistocerca gregaria* Forsk. *J. Adv. Zool.* **8** (1): 52–54.

Singh, B. and Singh, A.P. 1987b. Effect of neem kernel suspension as an oviposition deterrent to the desert locust, *Schistocerca gregaria* Forsk. *Pl. Prot. Bull., India* **39** (1–2): 16.

Singh, B. and Singh, A.P. 1987c. Joint action of neem kernel suspension with some insecticides against *Schistocerca gregaria* Frosk. *Neem Newsl.* **4** (1): 6–8.

Singh, B. and Singh, A.P. 1992. Effect of neem kernel suspension on the survival of hatchlings of the desert locust, *Schistocerca gregaria* Forsk. *Indian J. Ent.* **54** (4): 384–387.

Singh, B.N., Singh, R.P., Yazdani, S.S. and Sinha, P.P. 1976. Studies on the effect of different oil cakes on incidence of root-knot nematodes. *All India Symp. on Modern Concepts of Plant Protection,* March, 1976: 69.

Singh, D. and Bhathal, S.S. 1992. Survival growth and development effects of some neem based products on *Spodoptera litura* Fabr. *Proc. I. National Symposium Allelopathy in Agro-ecosystems. Eds. Tauro, T. and Narwal, S.S., Indian Society of Allelopathy, H.A.U., Hissar, India*: 163–165.

Singh, D., Dahiya, R.S. and Dalal, M.R. 1995. Nematicidal efficacy of some plant leaves to *Heterodera avenae* in wheat. *Presented in National symposium on nematode problems of India —An appraisal of the nematode management with eco-friendly approaches and bio-components. I.A.R.I., New Delhi,* 24–26 March, 1995 (Abstract): 5.

Singh, H. and Pandey, P.N. 1979. Effect of chirata on development and growth of *Utethesia pulchella* Linn. (Lepidoptera: Arctiidae). *Zeistschrift fur Angewandte Zoologie* **66** (1): 15–20.

Singh, I. 1980. Effect of oil cake extracts on hatching of root-knot nematode, *Meloidogyne incognita. Indian J. Mycol. Pl. Path.* **12**: 115–116.

Singh, J.R., Sukhija, H.S. and Singh, P. 1990. Evaluation of neem oil rice leaf-folder and stem borer. *Presented in National Symp. on Problems and Prospects of Botanical Pesticides in Integrated Pest Management at C.T.R.I., Rajahmundry, India,* 21–22 Jan., 1990 (Abstract): 19.

Singh, K. and Sharma, P.L. 1986. Studies on antifeedant and repellent qualities of neem (*Azardirachta indica*) against aphid (*Brevicoryne brassicae* Linn.) on cauliflower and cabbage. *Res. Dev. Reptr.* **3** (1): 33–35.

Singh, K.N. and Srivastava, P.K. 1980. Neem seed kernel powder as protectant against stored grain insect pest. *Bull. Grain Tech.* **18** (2): 127–129.

Singh, M.P. and Vir, S. 1987. Effect of some natural products on oviposition and development of *Callosobruchus maculatus* Fabr. on cowpea. *Trans. Indian Soc. Desert. Technol.* **12** (2): 143–147.

Singh, O.P., Katarey, A.K. and Singh, K.J. 1988. Soybean oil as seed protectant against infestation by *Callosobruchus chinensis* Linn. on pigeonpea. *J. Insect. Sci.* **1** (1): 91–95.

Singh, R.P. 1974. *Crinum bulbispermum* (Sukhadar Shau), an antifeedant to *Schistocerca gregaria* Forsk. *Entomologists Newsl.* **4** (6): 34.

Singh, R.P. and Kataria, P.K. 1986. De-oiled neem kernel powder as protectant of wheat seeds against *Trogoderma granarium* Everts. *Indian J. Ent.* **48** (1): 119–120.

Singh, R.P. and Pant, N.C. 1980. Antifeedant properties of subfamily Amaryllidoideae against desert locust, *Schistocerca gregaria* Forsk. *Indian J. Entomol.* **42** (3): 465–468.

Singh. R.P. and Srivastava, B.G. 1983. Alcohol extract of neem (*Azardirachta indica* A. Juss.) seed oil as oviposition deterrent for *Dacus cucurbitae* Cog. *Indian J. Ent.* **45** (4): 497–498.

Singh, R.P. and Singh, S. 1985. Evaluation of de-oiled (*Azardirachta indica* A. Juss.) seed kernel against *Trogoderma granarium* Everts. *Curr. Sci.* **54** (18): 950.

Singh, R.P., Singh, Y. and Singh, A.P. 1985. Field evaluation of neem (*Azardirachta indica* A. Juss.) seed kernel extracts against the pod-borers of pigeonpea *Cajanus cajan* (Linn.) Millep. *Indian J. Ent.* **48**: 360–362.

Singh, R.P., Singh, Y. and Singh, S.P. 1988. Activity of neem (*Azardirachta indica* A. Juss.) seed kernel extracts against the mustard aphid, *Lipaphis erysimi. Phytoparasitica* **16** (3): 226–230.

Singh, R.P., Jhansi Rani, B. and Doharey, K.L. 1993. Neem dust (*Azardirachta indica* A. Juss.) for management of stored grain pests. *World Neem Conf., Bangalore, India,* 24+ 28 Feb., 1993 (Abstract): 1.

Singh, R.S. and Sitaramaiah, K. 1967. The effect of decomposition of green leaves, saw dust and urea on the incidence of root-knot of okra and tomato. *Indian Phytopath.* **20**: 349–355.

Singh, R.S. and Sitaramaiah, K. 1969. Control of root-knot nematode through organic and inorganic amendments of soil. I. Effect of oil cake and saw dusts. *First All India Symposium Nematol, New Delhi* (Abstract): 10.

Singh, R.S. and Sitaramaiah, K. 1971. Control of root-knot nematode through organic and inorganic amendments in soil. II. Effect of oil cake and saw dust. *Indian J. Mycol. Pl. Path.* **1** (1): 20–29.

Singh, R.S. and Sitaramaiah, K. 1976. Incidence of root-knot of okra and tomato in oil cake amended soil. *Plant Disease Rept.* **50** (9): 668–672.

Singh, R.V. and Kumar, V. 1995. Effect of Carbofuran and neem cake on *Meloidogyne incognita* infesting Japanese mmt. *Presented in national symposium on nematode problems of India — An appraisal of the nematode management with eco-friendly approaches and bio-components. I.A.R.I., New Delhi,* 24–26 March, 1995 (Abstract): 37.

Singh, S.P. and Singh, V. 1988. Evaluation of oil cakes and nematicides for the control of *Meloidogyne incognita* infecting eggplants. *Indian J. Nematol.* **18** (2): 338–342.

Singh, S.R., Luse, R.A., Leuschner, K. and Nanggu, D. 1978. Groundnut oil treatment for control of *Callosobruchus maculatus* Fabr. during cowpea storage. *J. Stored Prod. Res.* **14**: 77–80.

Singh, T.V.K. and Azam, K.M. 1986. Seasonal occurrence population dynamics and chemical control of citrus leaf miner, *Phyllocnistic citrella* Staint. in Andhra Pradesh. *Indian J. Ent.* **48** (1): 38–42.

Singh, V.N., Pandey, N.D. and Singh, Y.P. 1993. Efficacy of vegetable oils for the control of *Callosobruchus chinensis* Linn. infesting grain and their subsequent effect on germination. *Bull. Grain Tech.* **31** (1): 13–16.

Sinha, N.P. and Gulathi, K.C. 1963. Locust repellent action of the alcohol extract of neem cake. *Bull. Reg. Res. Lab. Jammu* **1**: 176–177.

Sinha, N.P. and Gulathi, K.C. 1964. Neem seed cakes as a source of pest control chemical aphidicidal pesticide. *Symp. 1964 (Published 1963) Ed. Majumder, S.K. Acad.: Pest Control Science, Mysore, India*: 215–220.

Sinha, S.N. 1993. Neem in the integrated management of *Helicoverpa armigera* Hubn. infesting chickpea. *World Neem Conf., Bangalore, India,* 24–28 Feb., 1993 (Abstract): 6.

Sinha, S.N. and Mehrotra, K.N. 1988. Diflubenzofuran and neem (*Azardirachta indica*) oil in control of *Heliothis armigera* infesting chickpea, *Cicer arietinum. Indian J. Agric. Sci.* **58** (3): 238–239.

Sinhababu, S.P., Ray, D., Sukul, N.C. and Mahato, S.B. 1992. Nematicidal principle from *Accaia auriculiformis Proc. I. National Symposium on Allelopathy in Agro-ecosystems Eds. Tauro, P. and Narwal, S.S. Indian Society of Allelopathy H.A.U., Hissar, India*: 181–183.

Sita, V. and Thakur, S.S. 1984. Water hyacinth *Eichhornia crassipes* Mart. as a source of insect growth regulator. *Indian J. Ent.* **46** (1): 110–111.

Sitaramaiah, K. and Singh, R.S. 1977. Response of plant parasite and soil nematodes to plant extracts in amended soil. *Pantnagar J. Res.* **2** (2): 153–157.

Sitaramaiah, K. and Singh, R.S. 1978. Effect of organic amendment on phenolic content of soil and plant in response to *Meloidogyne javanica* and its host to related compounds. *Plant and Soil* **50** (3): 671–679.

Sitaramaiah, S., RamPrasad, G. and Joshi, B.G. 1986. Efficacy of diflubenzuron and neem kernel suspension against tobacco caterpillar, *Spodoptera litura* in tobacco nurseries. *Phytoparasitica* **14** (4): 265–271.

Sivakumar, M. and Marimuthu, T. 1986. Efficacy of different organic amendments against phytonematodes associated with betelvinel (*Piper betel*). *Indian J. Nematol.* **16** (21): 278.

Siyanand, Meher, H.C., Kaushal, K.K. and Singh, G. 1995. Induction of *Tagetes patula* Linn. as mixed rotation crop — A noble technology protocol for managing root-knot nematode, *Meloidogyne incognita* in vegetables. *Presented in National Symposium on nematode problems in India. An appraisal of the nematode management with eco-friendly approaches and bio-components. I.A.A.I., New Delhi,* 24–26, March, 1995 (Abstract): 19.

Slama, K. 1978. The principles of anti-hormone action in insects. *Acta Entomol. Bohemoslova,* **75**: 65–74.

Slama, K. and Williams, C.M. 1966. Paper factor as an inhibitor of the embryonic development of the European bug, *Pyrrhocoris apterus. Nature* **210**: 329–330.

Slansky, Jr., F. 1979. Effect of the lichen chemicals atranorin and vulpinic acid upon feeding and growth of larvae of yellow striped armyworm, *Spodoptera ornighogalli. Environ. Entomol.* **8** (5): 865–868.

Smet, H., Mellaert, H. Van, Rans, M. and Loof, A. de. 1986. The effect on mortality and reproduction of beta-asarone vapours on two species of stored grain pests: *Ephestia kuchniella* (Lepidoptera) and *Triboluim confusum* Du-val (Coleoptera). *Medelelingen vande faculiets Landbouwewetanschappen Rijksuniversitat Gent.* **15** (3B): 1197–1203.

Synder, J.K. 1983. Columbia Univ. Pers. Commn. in Michael et al., 1985. Plant species reportedly possessing pest control properties. *An EWC/UH DATA BASE, Univ. of Hawaii* p 249.

Songkittisuntron, U. 1989. The efficacy of neem oil and neem extracted substances on the rice leaf hoppers *Nephotettix virescens* Dist. *J. National Res. Coun. Thailand,* **21** (2): 37–48.

Sontakke, B.K. 1993. Field efficacy of insecticide alone and in combination with neem oil against insect pests and their predators in rice. *Indian J. Ent.* **55** (3): 260–266.

Sowunmi, O.E. and Akinnusi, O.A. 1983. Studies on the use of neem kernel in control of stored cowpea beetle, *Callosobruchus maculatus* Fabr. *Trop. Grain Legume Bull.* **27**: 28–31.

Sreedevi, K., Chitra, K.C. and Kameswara Rao, P. 1993. Effect of certain plant extracts against brinjal leaf beetle, *Henosepilachna vigintioctopunctata* Fabr. under greenhouse conditions. *Indian J. Ent.* **55** (3): 329–331.

Srimmannarayana, G. and Rao, D.R. 1985. Insecticidal plant chemicals as antifeedants *Proc. National Seminar on Behavioural and Physiological Approaches in Management of Crop Pests, T.N.A.U., Coimbatore,* 1985: 18–22.

Srinivas, P., Rammurthy, M. and Rao, A.P. 1986. Effect of neem seed extract on silkworm, *Bombyx mori* Linn. *Neem Newsl.* **3** (2): 18–20.

Srinivasulu, B. and Jayarayan, R. 1988. Effect of seed oils on tungro transmission and survival of green leaf hopper of rice crop. *Indian J. Pl. Prot.* **16**: 171–175.

Srivastava, A.S. 1971. Application of organic amendments for control of root-knot nematode, *Meloidogyne javanica* Treub. *Labbev. J. Sci. and Tech.* **9B** (3/4): 203–205.

Srivastava, K.P. and Parmar, B.S. 1985. Evaluation of neem oil emulsifiable concentrate against sorghum aphid. *Neem Newsl.* **2** (1): 7.

Srivastava, K.P. and Saxena, V.S. 1986. Linseed oil as binding agent for pesticide granules endosulfan. *Indian J. Ent.* **48** (1): 43–45.

Srivastava, K.P., Agnihotri, N.P., Gajbhiye, V.T. and Jain, H.K. 1984. Relative efficacy of fenvalerate, quinalphos and neem kernel extracts for control of pod fly, *Melanagromyza obtusa* Malloch. and pod borer, *Heliothis armigera* Hubn. infesting redgram, *Cajanus cajan* (Linn.) Millsp. together with their residues. *J. Entomol. Res.* **8** (1): 1–4.

Srivastava, S., Gupta, K.C. and Agrawal, A. 1988a. Natural products as grain protectants of pulses against pulse beetles. *In review by Agrawal, A., Lal, S. and Gupta, K.C. Bull. Grain Tech.* **16** (2): 154–164.

Srivastava, S., Gupta, K.C. and Agrawal, A. 1988b. Effect of plant products on *Callosobruchus chinensis* Linn. infestation in redgram. *Seed Res.* **16** (1): 98–101.

Srivastava, S., Gupta, K.C. and Agrawal, A. 1989. Japanese mint oil as fumigant and its effect on insect infestation, nutritive value and germinability of pigeon pea seeds during storage, *Seed Res.* **17** (1): 96–98.

Srivastava, S.C., Ali, S.M. and Pandey, P.N. 1979. Survival of lepidopterans against methanol and acetone extracts of *Dalbergia* root. *Z. Angew. Zoologie* **66** (1): 5–8.

Stay, B. and Tobe, S.S. 1985. Structure and regulation of the corpus allatum. *Adv. Insect. Physiol.* **18**: 304–310.

Steets, R. 1975. The effect of crude extracts from the Meliaceae: *Azardirachta indica* and *Melia azardirach* on various species on insects. *Z. Angew. Entomol.* **77**: 306–312.

Steets, R. 1976a. The effect of components from Meliaceae and Anacardiaceae on Coleoptera and Lepidoptera. *Ph.D. Thesis, Univ. of Giessen, West Germany,* p 136.

Steets, R. 1976b. The effect of a purified extract of the fruits of *Azardirachta indica* A. Juss. on *Leptinotarsa decemilineata* Say. (Coleoptera: Chrysomelidae). *Z. Angew. Entomol.* **82** (2): 176–196.

Steets, R. and Schmutterer, H. 1975. Effect of azardirachtin on the life period and reproductive ability of *Epilachna varivestis* Mulls. *Z. Pflangenk Pflanzenchtz* **82**: 176–179.

Steffens, R.J.H. and Schmutterer, H. 1982. The effect of crude methanolic extract of neem (*Azardirachtaindica*) seed kernel extract on metamorphosis and quality of adults of the Mediterranean fruit fly, *Ceratitis capitata* Wield. (Diptera: Tephritidae). *Z. Angew. Entomol.* **94**: 98–103.

Strakhantzer, A., Bebugin, P.I., Bogolyubov, H.O. and Borozdina, K.T. 1937. Chemical-toxicological investigation of *Melia azardirach*. *Review Appl. Entomol. Ser. A (Agric)* **25**: 445–457.

Strambi, A., Strambi, C., Dayanandan, S. and Kamavar, G.K. 1993. Effects of azardirachtin on ecdysteroids level and ovarian development in adult crickets. *World Neem Conf., Bangalore, India,* 24–28 Feb., 1993 (Abstract): 45.

Su, H.C.F. 1976. Toxicity of chemical components of lemon oil to cowpea weevils. *J. Georgia Entomol Soc.* **11** (4): 297–301.

Su, H.C.F. 1977. Insecticidal properties of black pepper to rice weevils and cowpea weevils. *J. Econ. Entomol.* **70** (1): 18–21.

Su, H.C.F. 1984a. Comparative toxicity of three pepper corn extracts to four species of stored product insects. *J. Georgia Entomol. Soc.* **19** (2): 190–199.

Su, H.C.F. 1984b. Biological activity of *Zanthoxylum altum* to four species of stored product insects. *J. Georgia Entomol. Soc.* **19** (4): 462–469.

Su, H.C.F. 1985a. Laboratory evaluation of biological activity of *Cinnamomum cassia* to four species of stored products insects. *J. Entomol. Sci.* **20** (2): 247–253.

Su, H.C.F. 1985b. Laboratory study on effects of *Anethum greveolens* seeds on four species of stored product insects. *J. Econ. Entomol.* **78** (2): 451–453.

Su, H.C.F. 1986. Laboratory evaluation of the toxicity and repellency of coriander seeds to four species of stored products insects. *J. Entomol. Sci.* **21** (2): 174–196.

Su, H.C.F. 1987. Laboratory study on the long term repellency of dill seed extract to confused flour beetle. *J. Entomol. Sci.* **22** (1): 70–72.

Su, H.C.F. and Horvat, R. 1982. Isolation and characterisation of four major compounds from insecticidally active lemon peel extract. *J. Agri. Food Chem.* **35** (4): 509–511.

Su, H.C.F. and Horvat, R. 1988. Investigation of main component in insect-active dill seed extract. *J. Agric. Food Chem.* **36** (4): 752–753.

Su, H.C.F. and Robert, H. 1981. Isolation, identification and insecticidal properties of *Piper nigrum*. *J. Agric. Food Chem.* **29**: 115–118.

Su, H.C.F., Speirs, R.D. and Mahany, P.G. 1972. Citrus oil as protectant of black eyed peas against cowpea weevils: Laboratory evaluation. *J. Econ. Entomol.* **65** (5): 1433–1436.

Su, H.C.F., Robert, H. and Jilani, G. 1982. Isolation, purification and characterization of insect repellents from *Curcuma longa* Linn. *J. Agric. Food Chem.* **30**: 290–292.

Subadrabai, K. and Kundaswamy, C. 1985. Laboratory induced mortality of *Spodoptera litura* Fabr. fed on the leaf-discs of castor treated with the extracts of *Vitex negundo* Linn. and *Stachytarpheta urticifolia* (Salisb.) Smis. *Indian J. Agri. Sci.* **55**: 760–761.

Subaharan, K. and Nachiappan, R.M. 1994. Biological effect of methyl chavicol on rice brown plant hopper, *Nilaparvata lugens* Stal. *Presented in National Conf. of AZRA on Recent trends on plant, animal and human pest management: Impact on environment at Madras Christian College, Madras, India* 27–29 Dec., 1994 (Abstract): 64.

Subba Rao, B.V. 1990. Operation research trials with botanical formulation called RD 9 (*Repelin*). *Presented in National Symposium in Problems and Prospects of Botanical Pesticides in Integrated Pest Management at C.T.R.I., Rajahmundry, India,* 21–22 Jan., 1990 (Abstract): 11.

Subrahmanian, T.V. 1949. Sweet flag (*Acorus calamus*): potential source of valuable insecticide. *J. Bombay Natural Mist. Soc.* **48** (2): 333.

Subrahmaniyam, B. and Rao, P.J. 1986. Azardirachtin effects on *Schistocerca gregaria* Forsk. ovarian development. *Curr. Sci.* (India) **55**: 534–538.

Subrahmaniyam, B., Muller, T. and Rembold, H. 1989. Inhibition of turnover of neurosecretion by azardirchtin in *Locusta migratoria. J. Insect. Physiol.* **35** (6): 493–500.

Subrahmaniyam, S. 1985. Effect of *Eupatorium odoratum* extracts on *Meloidogyne incognita. Indian J. Nematol.* **15** (2): 247.

Subramaniyam, S. and Selvaraj, 1988. Effect of *Tagetes patula* Linn. leaf extract on *Radopholus similis* (Cobb, 1933) Thorne, 1949. *Indian J. Nematol.* **18** (2): 337–338.

Subrahmaniyam, T.R. 1976. Report on observations — Neem kernel powder against stored grain pests. Pers. Commn. in Ketlkar, C.M. 1969–1976. Utilization of neem and its byproducts. *Report of the modified neem cake manurial project. Directorate of Non-edible Oils and Soap Industry, K.V.I.C., Bombay, India*: 116–173.

Subrahmaniyam, T.S. 1988. Repelin for control of tobacco pest complex. *IX Intn. Tob. Sci. Congr., China,* 9–13 Oct., 1988.

Sudhakar, T.R., Pandey, N.D. and Tiwari, G.C. 1978. Antifeeding properties of some indigenous plants against mustard sawfly, *Athalia proxima* Klug. (Hymenoptera: Tentheredinidae). *Indian J. Agric. Sci.* **48** (1): 16–18.

Sujatha, A. and Punnaiah, K.C. 1984. Effect of different vegetable oils on the development of *Collosobruchus chinesis* (Linn.) in black gran. *The Andhra Agric. J.* **32** (4): 325–327.

Sujatha, A. and Punnaiah, K.C. 1985. Effect of coating stored seed of greengram with vegetable oils on the development of pulse beetles. *Indian J. Agric. Sci.,* **55** (7): 475–477.

Sukul, N.C. and Ghosh, T. 1974. Nematicidal action of some edible crops. *Nematological* **20** (2): 187–191.

Sukumaran, D., Kandaswamy, C. and Srimmannarayan, G. 1987. *Vetex negundo* Linn. — a potential plant for control of rice pest. *Proc. Symp. Alternatives to Synthetic Insecticides*: 71–74.

Sumimoto, M., Shiraga, M. and Kondo, T. 1975. Ethane in pine needles preventing the feeding of beetle, *Monochamus alternatus. J. Insect Physiol.* **21** (4) 713–722.

Sundarababu, R., Sankaranarayan, C. and Vadivelu, S. 1993. Nematode management with plant products. *Indian J. Nematol.* **23** (2): 177–178.

Sundaraju, P. and Koshy, P.K. 1985. Effect of nematicides and neem oil cakes in the control of *Radopholus similis* in yellow leaf disease affected arecanut palms. *Presented in IV Nematology Symposium at Udiapur, India,* 17 May, 1985 (Abstract): 21.

Sundramurthy, V.T. 1979. Effect of garlic extract on the development of red cotton bug. *Dysdercus cingulatus* Fabr. *Works on futurology on use of chemicals in agriculture with particular reference to future trends in pest control* p 19.

Sundararaj, R., Murugesan, S. and Ahmed, S.I. 1994. Evaluation of ethanolic extract of neem (*Azardirachta indica* A. Juss.) seed kernel powder against the rohida defoliator, *Patialus tecomella* Pajni (Coleoptera: Curculionidae). *Presented in National Conf. of AZRA on Recent trends in plant, animal and human pest management: Impact on environment at Madras Christian College, Madras, India,* 27–29 Dec., 1994 (Abstract): 50.

Sutherland, D.R.W. 1983. Dept. of Entomology, D.I.R. Auckland, New Zealand, Pers. commn. in Michael et al., 1985. Plant Species Reportedly Possessing Pest Control Properties. *An EWC/UH DATA BASE, Univ. of Hawaii* p 249.

Swarup, G. and Sharma, R.D. 1967. Effect of root extract of *Asparagus recermosus* and *Tagetes erecta* on hatching of eggs of *Meloidogyne javanica* and *M. arenaria. Indian J. Expt. Biol.* **5** (I): 59.

Swingle, M.C. and Cooker, J.F. 1935. Toxicity of fixed nicotine preparations to certain lepidopterous pests of truck crops. *J. Econ. Entomol.* **28**: 220–224.

Tada, M. and Chiba, K. 1984. Novel plant growth inhibitors and insect antifeedant from *Chrysanthemum coronarium,* Shangkul (in Japanese). *Agric. Biol. Chem.* **48** (5): 1367–1369.

Takaishi, Y., Ujita, K., Nakano, K., Murakami, K. and Tomisatsu, T. 1987. Sesquiterpene esters from *Tripterygium wilfordii* Hook. Fil. var. *regelii,* structures of tripto-fordins A-C-1. *Phytochem.* **26** (8): 2325–2329.

Talapatra, S.K., Bhattachrrya, M., Talapatra, B. and Das, B.C. 1968. The structure of Xylotenin — a new furgno — coumarin from *Chloroxylon swientenia* D.C.J. *Indian Chem.* **45**: 861.

Talati, G.M. 1976. Neem product against stored grain pest. College of Agric. Junagarh (Gujrat) India. *In Ketkar, 1976, Non-Edible Oils and Soap Industries, K.V.I.C., Bombay, Chapter 4 —Entomology*: 117–176.

Tandon, P.L. and Lal, B. 1980. Control of mango mealy bug, *Drosicha mangifera* Green. (Marygarodidae: Hompoptera) by application of insecticides in soil. *Entomol.* **5** (1): 67–69.

Tare, V. and Sharma, R.N. 1988. Antifeedant action of some oils and their constituents on some agricultural pests. *In Integrated Pest Control Progress and Perspectives. Eds. Mohan Das, N. and Koshy, G. Trivendrum Association of Advancement of Entomology*: 450–452.

Taylor, W.E. and Vickery, B. 1974. Insecticidal properties of limonin — a constituent of citrus oil. *Ghana J. Agric. Sci.* **7** (1): 61–62.

Teotia, T.P.S. and Pandey, G.P. 1978. Dharek fruit powder as a protectant of rice against the infestation of rice weevil, *Sitophilus oryzae* Linn. *Indian J. Ent.* **40** (2): 223–225.

Teotia, T.P.S. and Pandey, G.P. 1979. Insecticidal properties of sweet flat, *Acorus calamus* against rice weevil, *Sitophilus oryzae* Linn. *Indian J. Ent.* **41** (1): 91–94.

Teotia, T.P.S. and Tewari, G.C. 1972. Insecticidal properties of drups of dharek, *Melia azardirach* and rhizomes of sweet flag, *Acorus calamus* against adults of *Sitotroga cerealella* Oliv. *Indian J. Ent.* **39** (3): 222–227.

Tewari, G.C. and Moorthy, N.P.K. 1985. Plant extracts as antifeedants against *Henosepilachna vigintoctiopunctata* Fabr. and their effect on its parasite. *Indian J. Agric. Sci.* **55** (2): 120–124.

Tewari, R.K., Mathur, K. and Anand, O.P. 1986. Post coital antifertility effect of neem oil in female albino rats. *ICRS Med. Sci.* **14**: 1005–1016.

Tewari, S.N., Prakash, A. and Mishra, M. 1989. Toxicity impact of *Thevretia letifolia* on certain plant pathogenic fungi and stored grain pest. *Presented in National Symp. on Recent trends in plant disease control at School of Agric. and Rural Development, Medzephema, India,* 12–14 April, 1989 (Abstract): 13.

Tewari, S.N., Prakash, A. and Mishra, M. 1990. Biopesticidal property of *Sapindus emarginatus* Vahl. (Sapindaceae) to rice pathogens and insect pests in storage. *J. Appl. Zool. Res.* **1**: (1) 52–56.

Thakar, A.V., Ameta, O.P., Tripathi, N.N. and Vyas, A. 1992. Relative efficacy of indigenous plant extract against gram pod borers, *Heliothis armigera* Hubn. *Proc. I National Symposium Allelopathy in Agro-ecosystems, Eds. Tauro, P. and Narwal, S.S., Indian Society of Allelopathy, H.A.U., Hissar, India*: 162.

Thakar, N.A., Patel, H.R. and Patel, C.C. 1987. Azolla in management of root-knot disease in okra. *Indian J. Nematol.* **17** (1): 136–137.

Thaker, N.A. 1988. Effect of extracts of *Azolla pinnata* on egg hatching of root-knot nematodes, *Meloidogyne incognita* and *M. javanica*. *Madras Agric. J.* **75**: 297–299.

Thakre, S.M., Khan, K.M. and Bole, M. 1983. Efficacy of some insecticides in the control of pod borer complex of arhar. *Pesticides* **17** (9): 24–25.

Thangavel, P., Subramaniann, T.R. and Parameshwaran, S. 1975. Effect of certain insecticides against the incidence of cotton boll worms. *Pesticides* **9** (10): 37–38.

Thothari, K. 1973. The 'tongoge' plant of little Andaman. *Indian Forester*: 530–532.

Thurston, R. 1970. Toxicity of trichome exudates of *Nicotiana* and *Petunia* sp. to tobacco hornworm larvae. *J. Econ. Entomol.* **63**: 272–274.

Thurston, R. and Webster, J.A. 1962. Toxicity of *Nicotrinca gousie*. Domin to *Myzus persicae* Suts. *Entomol. Expt. Applicata* **5**: 233–238.

Tiitanen, K. 1964. The effect of pyrethrum preparation on certain common horticultural pests. *Ann. Agric. Farm* **3**: 272–274.

Tilak, B.D. 1977. Pest control strategy in India. *In Crop protection agents, their biological evaluation. Ed. McFarlane. NRI, London Academic Press*: 99–109.

Tiwari, R., Singh, K.N., Lal, S. and Girish, G.K. 1988. Studies on the efficacy of *Asafoetida* (Hing) against insect pests of stored spices. *Bull. Grain Tech.* **26** (2): 165–166.

Tiyagi, S.A. and Alam, M.M. 1993. Efficacy of neem cake against soil nematodes and fungi. *World Neem Conf., Bangalore, India,* 24–28 Feb., 1993 (Abstract): 39–40.

Tiyagi, S.A., Muktar, J. and Alam, M.M. 1985. Preliminary studies on the nematicidal nature of two plants of family compositae. *Intn. Nematol Network Newsl.* **2** (3): 1921.

Tripathi, A.K., Rao, S.M., Singh, D., Chakravorty, R.B., and Bhakuni, D.S. 1897. Antifeedant activity of plant extracts against *Diacrisia obliqua* Walk. *Curr. Sci. (India),* **56** (12): 607–608.

Trivedi, T.P. 1987. Use of vegetable oil cakes and cowdung ash as dust carrier for pyrethrum against *Rhysopertha dominica* and *Tribolium castaneum* Herbst. *Pl. Prot. Bull., India* **39** (1–2): 27–28.

Tsuji, G. 1966. Rice bran extract effective in terminating diapause in *Plodia interpunctella* Hubn. (Lepidoptera: Pyralidae). *Appl. Ent. Zool.* **1** (1): 51.

Turner, C.J., Temperta, M.S., Taylor, R.B., Zagorski, M.G., Termini, J.S., Schroeder, D.R. and Nakanishi, K. 1987. An *nmr* spectroscopy study of azardirachtin and its trimethyl ether. *Tetrahedron* **43**: 2788–2792.

Uhl, M. 1987. Experimental alterations of the ecdysterol and juvenile hormone titer during the egg ripening cycle of *Locusta migratoria*. *Ph.D. Thesis, University of Munich, West Germany* p 110.

Uhlenbrock, J.H. and Bijloo, J.D. 1959a. Nematicidal substances in marigold, *Tagetes erecta* Linn. *Recl. Trav. Chem. Pays Bas Belgium,* **78**: 289.

Uhlenbrock, J.H. and Bijloo, J.D. 1959b. Isolation and structure of nematicidal principle occurring in *Tagetes* roots. *Proc. IV Intn. Congr. Crop Prot. Hamburg,* 1957: 579–581.

Ulrich, G.M., Schlagintweit, B.J., Eder, T. and Rembold, H. 1985. Elimination of the allatotropic activity in locusts by micro-surgical methods: evidence of hormonal control of the corpora allata, hemolymph proteins and ovary development. *Gen. Comp. Endocrinol.* **59**: 120–126.

Usher, G. 1973. Dictionary of plants used by man. *Hasner Press, New York (USA).*

Vaishnav, M.U., Dhrug, I.U. and Lakhataria, R.P. 1993. Efficacy of cakes in the management of root-knot nematode (*Meloidogyne arenaria*) in ground nut. *World Neem Conf., Bangalore, India,* 24–28 Feb., 1993 (Abstract): 38.

Varma, B.K. and Pandey, G.P. 1978. Treatment of stored green gram seed with edible oil for the protection of *Callosobruchus maculatus* Fabr. *Indian J. Agric. Sci.* **48**: 72–75.

Varma, M.K. 1976. Organic amendments in soil and incidences of root-knot nematode (*Meloidogyne incognita*) 73 in tomato (*Lycopersican esculentum Mill.*). *Udyanika* **2**: 39–44.

Velasco-Negueruela, A., Perez-Alonso, M.J. and Buades Rodriguez, A. 1989. Continuation of a chemical study of the essential oils of Iberian nepetos, *Nepeta nepetella* Linn. and *N. amethystina* Poiret. *Anals Jordin Botanica Madrid* **47** (2): 395–400.

Velumani, S.A., Thirupathi, V., Palaniswamy, P.T. and Mohan, S. 1993. Process development for defatted neem kernel powder (DFNKP) preparation for the pest control. *Bull. Grain Tech.* **31** (1): 6–9.

Velusamy, R., Rajendran, R. and Sundarbabu, P.C. 1987. Effect of three neem products on brown plant hopper (BPH) oviposition. *Intn. Rice. Res. Newsl.* **12** (2): 36.

Venkatesan, S. 1987. Studies on the efficacy of neem products against the aphid, *Aphis gossypii* Glover on cotton. *Madras Agric. J.* **74** (45): 255–257.

Verma, A.C. and Anwar, A. 1995. Integrated nematode management by using leaves of medicinal plants in printed gourd, *Trichosanthes, dioica* field. *Presented in National Symposium on nematode problems of India — An appraisal of nematode management with eco-friendly approaches and bio-compounds. I.A.R.I., New Delhi,* 24–26 March, 1995 (Abstract): 18.

Verma, G.S. and Pandey, U.K. 1987. Insect antifeedant property of some indigenous plant products. *Z. Angeu. Zool.* **74** (1): 113–116.

Verma, G.S., Ramkrishna, U., Mulchandani, N.B. and Chadha, M.S. 1986. Insect feeding deterrents from *Tylophora asthematica* (Asclepiadaceae) *Entomol. Expt. Applicata.* **40**: 99.

Verma, R. 1986. Efficacy of organic amendments against *Meloidogyne incognita* infesting tomato. *Indian J. Nematol.* **16** (1): 105–106.

Verma, R.R. 1993. Effect of different materials on the population of root-knot nematode, *Meloidogyne incognita* affecting tomato crop. *Indian J. Nematol.* **23** (1): 135–136.

Verma, S.K. and Singh, M.P. 1985. Antifeedant effects of some plant extract on *Amsacta moorei* Walk. *Indian J. Agric. Sci.* **55** (4): 288–289.

Verma, S.P., Singh, B. and Singh, Y.P. 1983. Studies on the comparative efficacy of certain grain protectants against *Sitotroga cerealella* Oliv. *Bull. Grain Tech.* **1** (1): 37–42.

Vijayalakshmi, K. and Prasad, S.K. 1979. Nematicidal property of some indigenous plant materials against second stage larvae of *Meloidogyne incognita. Indian J. Nematol.* **9** (1): 81–82.

Vijayalakshmi, K. and Goswami, B.K. 1983. Effect of aqueous extracts of various plants and non-edible oil seed cakes on the hatching of *Meloidogyne incognita* eggs. *III. Nematol. Symp. Himachal Pradesh Krishni Vishwa Vidyalaya, Sloan,* 24–29 May, 1983 (Abstract): 32.

Vijayalakshmi, K. and Goswami, B.K. 1985. Effect of aqueous extracts of madar, *Calotropis gigantea* and amerbel, *Cuscuta reflexa* on larval mortality, hatching, penetration into tomato roots. *Indian J. Nematol.* **15** (2): 265.

Vijayalakshmi, K., Mishra, S.D. and Prasad, S.K. 1979. Nematicidal properties of some indigenous plant materials against second stage juveniles of *Meloidogyne incognita* (Kofoid and Whita) Chitwood. *Indian J. Ent.* **41** (4): 326–331.

Vir, S. 1990. Insecticidal properties of some plant extracts against insect pests under agro-climatic conditions of arid zone. *Presented in National Symposium on Problems and Prospects of Botanical Pesticides in Integrated Pest Management, C.T.R.I. at Rajahmundry, India,* 21–22 Jan., 1990 (Abstract): 30–31.

Viswanathan, R. and Kandiannan, K. 1990. Effect of urea applied with neem cake on disease intensity and insect population in rice fields. *Intn. Rice Res. Newsl.* **15** (5): 20.

Volkorky, M. 1987. Surala action acridifuge dels extraits de feuilles de *Melia azardirach. Arch. Instt. Pasteur Algier* **15** (3): 427–492.

Vyas, A.V. and Mulchandani, N.B. 1984. New chromenes from *Ageratum conyzoides. In Progress Report 1980–1983 Eds. Gonghuly, A.S.U. and Heble, M.R., Bioorganic Division, Bhabha Atomic Research Centre, Bombay* p 7.

Wada, K. and Munakata, K. 1986. Naturally occurring insect control chemicals. *J. Agric. Food Chem.* **16** (3): 471.

Waespe, J. and Hans, R. 1979. Ciba Geigy Basel, Switzerland, Pers. Commn. in Michael et al., 1985. Plant Species Reportedly Possessing Pest Control Properties. *An EWC/UH DATA BASE, Univ. of Hawaii,* p 249.

Waite, M.B. 1925. Year Book of Agriculture, *United States Department of Agriculture,* p 453.

Warthen, J.D. 1979. *Azardirchta indica*: a source of insect feeding inhibitors and growth regulators. *Agric. Reviews and Manual Science and Education Administration, USDA-ARM-NE* 4: 41.

Warthen, Jr., J.D., Redfern, R.E., Uebel, E.C. and Mills, Jr., G.D. 1978. An antifeedent for fall army worm larvae from neem seeds. *Ed. Adm. Agric. Res. Results, North East, Ser. No.* 2: 9.

Waterhouse, A.L., Holden, I. and Casida, J.E. 1985. Ryanoid insecticides: structure examination by fully coupled two dimensional ^{1}H-^{13}C shift correlation NMR spectroscopy. *J. Chem. Soc. Perkin. Trans.* II: 1011–1016.

Watt, J.M. and Breyer-Brandwijk, M.C. 1962. Medicinal and poisonous plants of Southern and Eastern Africa. *E. & S. Livingstone Ltd., Edinburgh* p 213.

Weaver, D.K., Dunkel, F.V., Ntezurubanza, L., Jackson, L. and Stock, D.T. 1991. The efficacy of linalool, major component of freshly milled *Ocimum canum* Sims. (Lamiaceae), for protection against post harvest damage by certain stored product coleoptera. *J. Stored Prod. Res.* **27** (4): 313–320.

Webb. R.C., Hinebaugh, M.A., Lindquist, R.K.S. and Jacobson, M. 1983. Evaluation of aqueous solution of neem seed against *Liriomyza sativa* and *L. trifolii* (Diptera: Agromysidae). *J. Econ. Entomol.* **76** (2): 357–362.

Wen, T.C. 1983. Department of Plant Prot. Jaingsu Acad. Sci. Nanging, China, Pers. Commn. in Michael et al., 1985. Plant Species Reportedly Possessing Pest Control Properties. *An EWC/UH DATA BASE, Univ. of Hawaii* p 249.

West, A.J., Mordue, A.J. and Luntz, A.J. 1992. The influence of azardirachtin on the feeding behaviour of cereal aphids and slugs. *Entomologia experimentalis et Applicata* **62** (1): 75–79.

Winoto, S.R. 1969. Studies on the effect of *Tagetes* sp. on plant parasitic nematode. *Bur. Publ. Weenman and Zozen., N.V. (Eds.)*: 47.

Wolfenbarger, D.A. and Guerra, A.A. 1986. Toxicity and hypoxia of three petroleum hydrocarbons and cottonseed oil to adults of boll weevil and larvae of tobacco bud worms. *South Western Entomologist* **11**: 69–74.

Worsley, R.R. and Le, G. 1934. Insecticidal properties of some South African plants. *Ann. Appl. Biol.* **21**: 649–669.

Yadav, D.N. and Patel, A.R. 1990. Effect of some potential botanical insecticides on oviposition of *Chrysopa scelestris* and their ovicidal action. *Presented in National Symposium on Problems and Prospects of Botanical Pesticides in Integrated Pest Management at C.T.R.I., Rajahmundry, India,* 21–22 Jan., 1990 (Abstract): 10.

Yadav, L.S., Yadav, P.R. and Poonia, F.S. 1979. Effectiveness of certain insecticides against the insect pests of mung bean. *Indian J. Pl. Prot.* **7** (2): 165–169.

Yadav, R.L. 1971. Use of essential oil of *Acrous calamus* Linn. as an insecticide against pulse beetle, *Bruchus chinensis* Linn. *Z. Angew. Zool.* **68** (3): 289–294.

Yadav, S.R.S. and Bhatnagar, K.N. 1987. Preliminary study on the protection of stored cowpea grains against pulse beetle by indigenous plant products. *Pesticides* **21** (8): 25–29.

Yadav, T.D. 1973. Studies on stored grain pests of pulses and their control. *Ph.D. Thesis, IARI, New Delhi,* p 153.

Yadav, T.D. 1984. Efficacy of neem (*Azardirachta indica* A. Juss.) kernel powder as seed treatment against pulse beetle. *Neem Newsl.* **1** (2): 13–15.

Yadav, T.D. 1985. Anti-ovipositional and ovicidal toxicity of neem (*Azardirachta indica* A. Juss.) oil against three species of *Calbsobruchus. Neem Newsl.* **2** (1): 5–7.

Yadav, T.D. 1993. Olfactory, ovipositional and developmental responses of five species of pulse seed beetles treated with neem oil. *World Neem Conf., Bangalore, India,* 24–28 Feb. 1993 (Abstract): 23.

Yadav, T.D. and Tanwar, R.K. 1986. Induced oviposition technique in rearing of *Lasioderma serricorne* Fabr. *Ann. Agri. Res.* **7** (2): 222–226.

Yano, K. 1987. Minor components from growing buds of *Artemisia caprillaris* that act as antifeedant. *J. Agric. Food Chem.* **35** (6): 889–891.

Yingchol, P. 1983. Department of Agronomy, Kasetsart Univ. Thailand, Pers. Commn. in Michael et al., 1985. Plant Species Reportedly Possessing Pest Control Properties. *An EWC/UN DATA BASE, Univ. of Hawaii* p 249.

Yoshio, S., Noboru, M., Kiyomitsu, I., Chukichi, K., Susuma, K. and Takahashi, N. 1982. Sterols and asparagin in the rice plant indigenous factors related to resistance against brown plant hoppers, *Nilaparvata lugens. Agric. Biol. Chem.* **46** (12): 1877–1879.

Yothers, M.A. and Carlson, F.W. 1944. Repellency of pyrethrum extract and other material to full grown codling moth larvae. *J. Econ. Entomol.* **37** (5): 617–620.

Zag, G.M. and Bhardwaj, S.C. 1976. Relative efficacy of some seed protectants of plant origin against lesser grain borer. *Proc. All India Symp. on Modern Concepts of Plant Prot. held at Udaipur, Rajasthan,* 26–28 March, 1976: 13.

Zaki, F.A. and Bhatti, D.S. 1989. Effect of castor (*Ricinus communis*) leaves in combination with different fertilizer doses on *Meloidogyne javanica* infecting tomato. *Indian J. Nematol.* **19** (2): 171–176.

Zalkow, L.H., Gordon, M.M. and Lanir, N. 1979. Antifeedants from rayless golden rod and oil of Penny-royal toxic effects for the fall army worm. *J. Econ. Entomol.* **72** (6): 812–815.

Zaman, S.M.H. 1983. Bangladesh Rice Research Institute, Pers. Commn. in Michael et al., 1985. Plant Species Reportedly Possessing Pest Control Properties. *An EWC/UH DATA BASE, Univ. of Hawaii* p 249.

Zanno, P.R., Miura, I., Nakanishi, K. and Elder, D.L. 1975. Structure of the insect phago-repellent azardirachtin: Application of PRFT/CWD Carbon-13, NMR. *J. Am. Chem. Soc.* **97**: 1975–1977.

Zehnder, G.W. and Warthen, J.W. 1990. Activity of neem extract and margosan-O for control of Colorado potato beetle in Virginia. *Proc. USDA Neem Conf.: Neem's Potential in Pest Management Programs. USDA-ARS-86, Beltsville, Maryland, USA,* 16–17 Apr., 1990: 65–75.

Zhang, X. and Zhao, S.N. 1983. Experiments on sound substance from plants for control of rice weevil. *J. Grain Storage* No. 1: 1–8.

Zhang, Y. and Chiu, S. 1985. Preliminary investigation of some meliaceous plant materials as feeding deterrents against the imported cabbage worm, *Pieris rapae* Linn. *Neem Newsl.* **2** (3): 30–32.

INDEX 1
COMMON NAMES OF PLANTS

A

B

C

D

E

F

G

H

I

J

S

Y

Z

INDEX 2
ZOOLOGICAL AND COMMON NAMES OF INSECT PESTS

A

B

C

D

E

G

H

I

L

M

N

O

P

R

S

T

U

X

Z